PHYSICS OF SOLITONS

Solitons are waves with exceptional stability properties which appear in many areas of physics, from hydrodynamic tsunamis and fibre optic communications to solid state physics and the dynamics of biological molecules. Since they were first observed in 1834 they have fascinated scientists, not only for their spectacular experimental properties and the remarkable mathematical theories that they have initiated, but also for the new insight that they provide into many physical problems.

The basic properties of solitons are introduced here using examples from macroscopic physics such as blood pressure pulses and fibre optic communications. The book then presents the main theoretical methods and discusses a wide range of applications in detail. These applications include examples from solid state and atomic physics, for example, excitations in spin chains, conducting polymers and Bose–Einstein condensates and also biological physics (e.g. energy transfer in proteins and DNA fluctuations).

In addition to knowledge on the physics of solitons, the authors aim to familiarise the reader with a new way of thinking in physics. Instead of linear approximations followed by a perturbative approach to nonlinearities it is often more efficient to treat nonlinearities intrinsically, and to base the analysis on one of the soliton equations introduced in this book. This modelling process is stressed throughout the book and also discussed in Chapter 4.

Based on the authors' graduate course, this textbook gives an instructive view of the physics of solitons for students with a basic knowledge of general physics, and classical and quantum mechanics.

THIERRY DAUXOIS is a CNRS researcher at Ecole Normale Supérieure de Lyon. He studies nonlinear waves and their consequences in thermodynamics and the physics of complex systems: condensed matter, biophysics and hydrodynamics. He has co-edited three other books on dynamics and complexity.

MICHEL PEYRARD is a physics professor at Ecole Normale Supérieure de Lyon and is a senior member of the Institut Universitaire de France. After conducting research in solid state physics, he moved on to the fundamental properties of solitons and their physical applications. He now studies the role of nonlinear excitations in statistical physics and in the physics of biological systems. He is an editor of the *Journal of Biological Physics* and has edited two other books in nonlinear science.

PHYSICS OF SOLITONS

THIERRY DAUXOIS AND MICHEL PEYRARD
Ecole Normale Supérieure de Lyon

CAMBRIDGE
UNIVERSITY PRESS

University Printing House, Cambridge CB2 8BS, United Kingdom

Cambridge University Press is part of the University of Cambridge.

It furthers the University's mission by disseminating knowledge in the pursuit of education, learning and research at the highest international levels of excellence.

www.cambridge.org
Information on this title: www.cambridge.org/9780521854214

First published in French as *Physique des Solitons* by EDP Sciences 2004
English edition first published by Cambridge University Press 2006

A catalogue record for this publication is available from the British Library

ISBN 978-0-521-85421-4 Hardback
ISBN 978-0-521-14360-8 Paperback

Contents

List of Portraits

Preface

Since the first observation of a *soliton* by John Scott Russell in 1834, these exceptionally stable solitary waves have fascinated scientists for their spectacular experimental properties and their obvious elegance. But it is perhaps the remarkable mathematical properties of integrable systems having soliton solutions which attracted most of the attention. Mathematical aspects have been put forward in most of the books dealing with solitons because they lead to beautiful theories, such as the inverse scattering transform which derives the solution of a complex *nonlinear* equation from a series of steps which are all *linear* (see Chapter 7).

However, besides mathematics, the *physics* of solitons is also very fascinating, and at the heart of modern research. For instance many experiments on Bose–Einstein condensation, for which the 2001 Nobel prize in physics was awarded, are analysed in terms of the nonlinear Schrödinger equation, introduced in Chapter 3, which is one of the basic equations of soliton theory. The role of solitons in the physics of Bose–Einstein condensates is discussed in Chapter 14. The 2000 Nobel prize in chemistry, awarded to Heeger, MacDiarmid and Shirakawa, is also closely related to solitons because the charge carriers in conducting polymers are solitons. Chapter 13, which explains these phenomena, is based on a paper by Su, Schrieffer and Heeger.

Thus the physics of solitons is a very active research topic, to which we have contributed. However this book is not a research book. Our aim is to introduce the physics of solitons in a pedagogical way, so that the book can be read by a bachelor student or a student beginning master studies. We only assume a basic knowledge of physics, analytical mechanics and quantum mechanics. This book evolved from a course given by Michel Peyrard at the University of Dijon, and then at Ecole Normale Supérieure de Lyon for students graduating in 'Statistical Physics and Nonlinear Science'. This course in now taught by Thierry Dauxois in the Master in physics and chemistry course at Ecole Normale Supérieure de Lyon.

The book does not claim to be exhaustive, however it is written is order to give a broad and fairly complete view of the topic. Part I contains the basics. It introduces the main classes of soliton equations from examples chosen in macroscopic physics. Theoretical methods are presented in Part II. Their selection has been made keeping in mind physical applications and this shows up in Parts III and IV which are devoted to selected topics in solid state or biomolecular physics. Modelling is a very important step, which starts from the physical system and leads to nonlinear equations which describe its properties. This aspect is illustrated in most of the chapters, but we also decided to devote a specific chapter to this question (Chapter 4) because this is a difficult and very important point.

Thinking in terms of solitons shines a new light on some physical systems. For instance, we show how solitons can be used to study the statistical physics of ferroelectric materials (Chapter 10) or of DNA (Chapter 16). We provide numerous bibliographical references which should allow the reader to go beyond the material presented in the book and find the elements to start research related to the physics of solitons. We also decided to include some biographical data on the scientists who founded this topic because we agree with the statement of the philosopher Whitehead who said, at the beginning of the twentieth century, 'a science which does not want to remember its pioneers is condemned'.

This book matured after years of lectures and problem-solving classes, but we are also very grateful to those who devoted time to critical reading, particularly Geneviève Peyrard for her numerous pertinent remarks. Many colleagues, Mariette Barthès, Freddy Bouchet, Hervé Courtois, Jacques Dauxois, Sébastien Dusuel, Jean-Noël Gence, Hajime Hirooka, Robin Kaiser, Ioannis Kourakis, Juan Mazo, Guy Millot, Jean-Pierre Nguenang, Sébastien Paulin, Hicham Qasmi, Florence Raynal, Stefano Ruffo, Nobuhiko Saito, Yves-Henri Sanéjouand and Nikos Theodorakopoulos, examined chapters close to their research area. We thank also Larissa Brizhik, Lincoln D. Carr, Thierry Cretegny, Bernard Deconinck, Chris Eilbeck, Ying Li, Robert I. Odom, Sylvian R. Ray, Harvey Segur, Terry Toedtemeier, Nadezhda Tsypkina, Kathleen T. Zanotti for permission to use some pictures and Martin D. Kruskal and Norman J. Zabusky who provided us precise data on the early history of 'solitons'.

A French edition of this book has been published under the title *'Physique des Solitons'* by EDP Sciences/CNRS Edition in 2004. The English edition is a translation of the French edition, which has been slightly revised and completed. Chapter 14 has been added.

Introduction

The nineteenth century and the first half of the twentieth century can be viewed as the triumph of *linear physics*, which started with Maxwell's equations and culminated with quantum mechanics, based on a linear formalism emphasising a superposition principle. The familiar mathematical tools of physics such as the Fourier transform, the linear response theory and perturbative expansions, were themselves intrinsically linear.

Of course physicists had noticed the importance of nonlinear phenomena which appeared in the Navier–Stokes equation of hydrodynamics, gravitational theory, collective effects arising from the interaction between particles in solid state physics, etc. But, in most of the cases, theoretical approaches were trying to avoid nonlinearities, or to treat them as perturbations of linear theories.

The picture dramatically changed in the last 40 years. The importance of an *intrinsic analysis* of nonlinear phenomena has been gradually understood, and led to two concepts that revolutionalised previous ideas, the strange attractor and the soliton.

Both are related to astonishing properties of nonlinear systems, and they seem to contradict each other. The strange attractor is linked to the idea of *chaos* in a system which is described by deterministic equations. It shows up in systems with a small number of degrees of freedom, which could have been viewed as 'simple', while solitons appear in systems with a very large number of degrees of freedom. Although it seems that adding degrees of freedom should make the behaviour of such systems even more complex, this is not necessarily the case. Collective effects can lead to *spatially coherent structures*, which result in a *self-organisation*. Understanding the coexistence between coherent structures and chaos in nonlinear systems is still an open question.

A soliton is a *solitary wave*, i.e. a spatially localised wave, with spectacular stability properties. Since its first observation [162] in 1834 by a hydrodynamic engineer, John Scott Russell, the soliton initiated passion and debates. John Scott

Russell himself had been so fascinated by his unexpected observation that he devoted ten years of his life to study this phenomenon while theories, based on linearised approaches, were showing . . . that solitons could not exist. The early history of solitons has been marked by long eclipses. While the first observation had been made in 1834, one had to wait until 1895 for a theory [104] that could describe solitons, thanks to an equation derived by Korteweg and de Vries. Then this phenomenon was forgotten until a numerical experiment, carried out by Fermi, Pasta and Ulam, in 1953, with one of the first computers in Los Alamos, exhibited a result that appeared to contradict thermodynamics. A one-dimensional lattice of particles coupled to each other by an anharmonic potential was not necessarily reaching thermal equilibrium. The energy, initially injected in one particular mode, was first transferred to the other modes as one would have expected, but then it was coming back, almost perfectly, to the mode that had been originally excited. It is only ten years later that an explanation could be provided by Zabusky and Kruskal [190]. As we shall see in this book, it involves solitons and it is their work that introduced the word *soliton*. This name, which sounds like the name of a particle, was chosen on purpose. A soliton is a wave, but it is also a local maximum in the energy density, which preserves its shape and velocity when it moves, exactly as a particle does. It corresponds to a solution of a *classical* field equation which simultaneously exhibits wave and quasi-particle properties. These are features that one would expect from a quantum system and not from a classical one. The quantum analogy goes so far that soliton tunnelling has been found [137].

The study of Zabusky and Kruskal is a landmark in the history of solitons. Since then solitons have stayed in the front of the scene and have been the object of a huge number of investigations in mathematics and physics.

Equations having soliton solutions, in the exact mathematical sense, provide remarkable examples of completely integrable systems with an infinite number of degrees of freedom. This is why they have interested mathematicians so much, and this also explains why most of the books on solitons are mainly devoted to their mathematical properties.

However solitons are also of major interest to physicists. They are essential to describe phenomena such as the propagation of some hydrodynamic waves, localised waves in astrophysical plasmas, the propagation of signals in optical fibres or, at the microscopic level, charge transport in conducting polymers, localised modes in magnetic crystals and the dynamics of biological molecules such as DNA and proteins, for instance. All these systems are only approximately described by the equations of the mathematical theory of solitons. One should rather speak of 'quasi-solitons'. But the remarkable feature of solitons is their exceptional stability against perturbations so that these 'quasi-solitons' exhibit most of the spectacular properties of actual solitons. Moreover they can emerge spontaneously in a physical system in which some energy is fed in, for instance as thermal energy or by an excitation with an electromagnetic wave or a mechanical stress, even if the excitation does not match exactly the soliton solution. This feature explains the interest of solitons in physics because, if a system possesses the necessary properties to allow the existence of solitons, it is highly likely that any large excitation will indeed lead to their formation. Moreover, as we shall show later, many physical systems meet the necessary criteria to sustain solitons, at least for some range of excitations.

Very often solitons provide a fruitful approach to describe the physics of a nonlinear system. Rather than making a linear approximation and then attempting to take into account nonlinearities as perturbations, it may be much more efficient to approximately describe the physics of the system by the most appropriate soliton equation and then to consider the possible perturbations of the exact soliton solution to improve the theory.

The goal of this book is to explain the *physics of solitons* by showing how this concept enters in many areas of physics. It proposes a three-step journey in the world of solitons:

- The first part introduces the main classes of soliton equations from examples chosen in macroscopic physics. For each case we start from a simple situation where a direct observation of solitons is easy, and we show how the basic laws of physics lead to nonlinear field equations having soliton solutions. This part introduces the main properties of solitons and explains the basic features that a physical system must have in order to allow their existence. The last chapter of Part I discusses in detail the process leading to modelling the physical properties of a system in terms of solitons. We consider the example of plasma physics and show in particular how *one* given system can be described by *several equations* depending on the situation of interest.
- The second part introduces some mathematical methods for the study of solitons. The mathematical aspects are not the primary aim of the book but the methods that we introduce in this part are relevant for physics because a real system is never exactly described by a soliton equation. It is therefore necessary to study and evaluate the role of the features which were neglected when a soliton equation was derived for the system. In this part we introduce some methods which are seldom discussed in the mathematical theory of solitons precisely because they are relevant for systems which are not exactly described by a soliton equation. Another topic of interest for a physicist is the time evolution of a given initial condition. An answer can be provided by a very elegant mathematical theory, the inverse scattering method, which succeeds in reducing the derivation of the solution of a nonlinear equation to a series of linear steps. We give an introduction to this method.

- The last step, presented in Parts III and IV of the book, is devoted to the physics of microscopic systems, such as atomic, solid state or biomolecular physics, where solitons have been used to study various problems. At such a scale, one cannot 'see' the solitons. They must be detected indirectly by their role in the properties of the system. Therefore, beyond the derivation of the soliton equations, as it was done in Part I, it is also necessary to discuss how solitons can be detected. Moreover, at the microscopic scale, thermal fluctuations can no longer be neglected. They can interact with the solitons, which must be studied in a background which is not at rest but coupled to a thermal bath. Besides their interactions with the thermal fluctuations, solitons themselves can play a role in the thermodynamic properties of a system, which can be significantly affected by their existence. Using the examples of ferroelectric materials and DNA we show that the concept of solitons can be a powerful tool to theoretically investigate the thermodynamic properties of some systems.

The beauty of nonlinear science is in the links that it exhibits between very different systems which share common mathematical properties. The generality of the theory is similar to that of thermodynamics which can put an extreme variety of systems in a common framework. This book shows how a few fundamental equations can be applied to a wide range of physical situations, from the macroscopic to the microscopic scale, unifying topics which are often considered as completely different such as hydrodynamics and the dynamics of biological molecules.

Of course there are still plenty of open questions. The concepts that we introduce are beginning to be applied in domains as different as biology, sociology, economy, epidemiology, ecology etc. We hope that this book will stimulate the reader to explore these open questions, keeping in mind that the astonishing ability of many systems to create extremely stable spatially coherent structures can have a profound influence on their properties.

Part I

Different classes of solitons

1

Nontopological solitons: the Korteweg–de Vries equation

This chapter introduces the concept of solitons by discussing their first experimental observation. Then it studies their main features, and particularly the conditions that are required for their existence. It also shows how the soliton solution can be derived with an elementary method which can be used to predict the possible existence of solitary waves in various systems.

Then we consider a first physical example, easy to build experimentally, which allows us to explain in detail the process that leads to the introduction of a continuous soliton equation to describe a discrete system, i.e. a system made of separate elements connected into a network. This is a very common situation, for instance in solid state physics, as discussed in Part III. A second example, the case of blood pressure waves, shows that solitons can exist even in situations where they would not be expected!

1.1 The discovery

1.1.1 John Scott Russell's observations

The very first observation of a soliton was made in 1834 by the hydrodynamic engineer John Scott Russell while he was riding his horse along a canal near Edinburgh. When a barge abruptly stopped he was struck by the sight of what he called 'the great solitary wave', that he followed for a few miles before losing it in the meanders of the canal. The description that he gave shows the enthusiasm of a scientist who then devoted about ten years of his life to study this phenomenon. As we shall see later it contains all the basic ingredients which are required to derive and solve an equation that allows the analysis of the observations:

I was observing the motion of a boat which was rapidly drawn along a narrow channel by a pair of horses, when the boat suddenly stopped – not so the mass of water in the channel which it had put in motion; it accumulated round the prow of the vessel in a state of violent

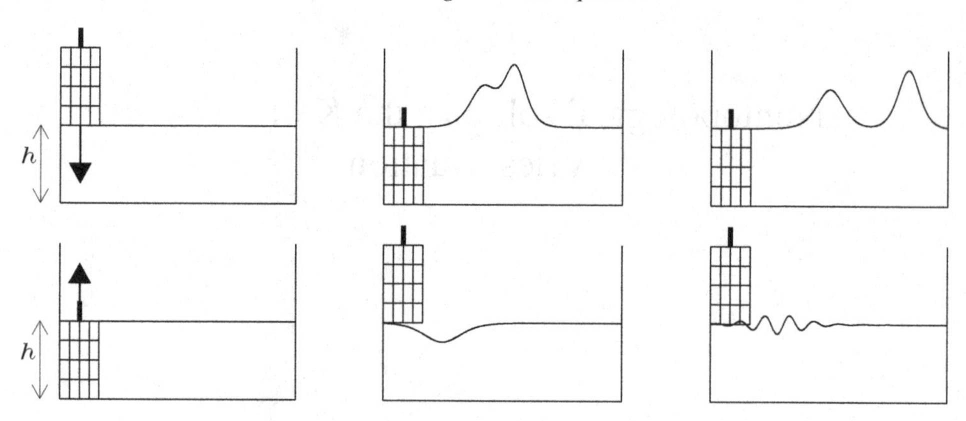

Figure 1.1. Schematic picture of the time evolution of a perturbation of the water surface in a reservoir, driven by a piston moving downward or upward.

agitation, then suddenly leaving it behind, rolled forward with great velocity, assuming the form of a large solitary elevation, a rounded, smooth and well defined heap of water, which continued its course along the channel, apparently without change of form or diminution of speed. I followed it on horseback and overtook it still rolling on at a rate of some eight or nine miles an hour, preserving its original figure some thirty feet long and a foot to a foot and a half in height. Its height gradually diminished and after a chase of one or two miles I lost it in the winding of the channel. Such in the month of August 1834 was my first chance interview with that singular and beautiful phenomenon which I have called the Wave of Translation.

A full theoretical understanding of John Scott Russell's observation had to wait until 1895 with the studies of Korteweg and de Vries who derived the equation which nowadays bears their names (abbreviated as the KdV equation). This equation was however in an implicit form [131] in the earlier studies of Joseph Valentin de Boussinesq (1842–1929) published in 1872 [33].

The KdV equation is one of the prototype equations of soliton theory because it has remarkable mathematical properties. Its study leads to the understanding of the fundamental ideas that lie behind the soliton concept, but its derivation from the basic equations of hydrodynamics is tedious. It is given in Appendix A and is only valid when the depth of the fluid and the height of the wave are small with respect to its spatial extent along the direction in which it propagates. One can notice that the second condition matches the observation of John Scott Russell who describes a wave which is thirty feet long and one and a half feet high.

Figure 1.1 shows a schematic picture of the device used by John Scott Russell to experimentally investigate 'the great solitary wave'. Waves are generated by the motion of a piston at the end of a canal. John Scott Russell observed the following features:

- Depending on its amplitude, the initial perturbation can create one, two or several solitary waves.
- Nonlinear waves have a speed higher than the speed $c_0 = \sqrt{gh}$ of long-wavelength linear waves (where g is the strength of the gravitational field and h the water depth in the canal). The deviation from c_0 is proportional to the height η of the wave, so that the speed of nonlinear waves evolves according to the law $v = c_0(1 + A\eta)$, where A is a positive constant.
- There are no solitary waves with a negative amplitude, i.e. that would move as localised pits.

The studies of John Scott Russell triggered a lot of controversy in the scientific community of his time, which assumed that nonlinear effects were of secondary importance. Many debates were raised, which highlights how surprising the properties of solitons are.

Hydrodynamic solitons are dynamic structures. They move with a constant speed and shape, but they cannot exist at rest. On both sides of the soliton the state of the medium is the same. They are called *nontopological solitons* in contrast to another class of solitons, introduced in Chapter 2, which interpolates between two different states of a medium, and can exist at rest.

Following the observation of a solitary wave on a canal near Edinburgh, the Scottish engineer JOHN SCOTT RUSSELL (1808–82) devoted many years of his life to investigate the soliton phenomenon. To him this discovery was a real revelation. Unfortunately, at the time people never shared his enthusiasm, and one had to wait more than 130 years until scientists really understood how important his discovery was. As a son of a clergyman he was expected to perpetuate this tradition [62] but his passion for sciences turned him otherwise! (1850 photograph)

John Scott Russell graduated at the age of 16 from Glasgow University, after studying in St Andrews and Edinburgh Universities. Although he had a real talent for scientific studies, he decided to work for two years in industry. After this short period he attempted to come back to Edinburgh University as a teacher. Although he had well acknowledged teaching abilities, and in spite of a very laudatory recommendation letter from John Hamilton, his application was not successful and the position was given to James David Forbes, known for contributions to the theory of heat transfer and glaciology.

In contrast his career as an engineer was very bright. He invented an improved steam-driven road carriage in 1833, which quickly gave him a good reputation as an inventor. This is why the 'Union Canal Society' of Edinburgh asked him to set up a navigation system with steam boats on the canals of Edinburgh and Glasgow, in order to replace the boats drawn by horses.

He was hired to design new barges thanks to experiments performed in a part of the canal devoted to his studies. It was during these investigations, six miles from the centre of Edinburgh, and very close to the present campus of the Heriot–Watt University, that he observed a soliton for the first time in August 1834. Following this discovery he performed several experiments on canals, rivers and lakes, but also in a specially designed, 10 m long tank, in his backyard. He gave a first report in 1838 before publishing the results of several experiments in 1844 [162].

This paper was however very badly received by two scientists who ruined all his expectations. First the well known astronomer Sir G. B. Airy (1801–92) strongly criticised his work in a paper on waves and tides which appeared in 1845. The main argument of Airy was that the formula derived by John Scott Russell from his experiments did not agree with his own theory of shallow water waves! Although he had studied the work of Russell more carefully, G. G. Stokes, one of the founding fathers of fluid mechanics, concluded that a solitary wave could not exist in a nonviscous fluid. This stopped all the research of John Scott Russell.

He turned his attention to other fields and performed one of the first experimental measurements of the frequency shift of a moving source, which the Austrian physicist Doppler (1803–53) had described earlier, and which is now called the Doppler effect. He also contributed to the design of a gigantic boat, the 'Great Eastern', 207 m long and 25 m wide, which installed the first transatlantic cable, connecting England to the United States. Nobody knew at that time that another class of solitons than the one discovered by John Scott Russell, the optical solitons, would then become the best candidates for transatlantic telecommunications in the twenty-first century!

However John Scott Russell did not die without a well deserved recognition of his achievements when the French scientist Joseph Valentine de Boussinesq (1842–1929) proposed a new theory of shallow water waves, which had solutions which agreed with his observations. These results were confirmed in 1876 by some investigations by Lord Rayleigh (1842–1919), who, ironically, was Stokes' former student. Then, in 1885, Adhémar Jean-Claude Barré de Saint Venant (1797–1886) established a correct mathematical theory for these phenomena, which explained Airy and Stokes' mistakes.

During a meeting devoted to solitons and their applications, which took place at the Heriot–Watt University of Edinburgh in 1982, scientists tried to recreate the solitary wave in the famous canal in which John Scott Russell had seen it for the first

Figure 1.2. Solitary wave created again in the Union Canal near Edinburgh in 1995, 161 years after John Scott Russell's discovery of what he called 'The great solitary wave' (Picture: Chris Eilbeck & Heriot–Watt University, 1995).

time. They failed due to technical problems! In 1995, during another conference, after a few fruitless trials, the result was more convincing as shown by the picture of a nice solitary wave (Figure 1.2). The authors of this book were among the admiring spectators on the bank, while Martin Kruskal, who is at the origin of the mathematical understanding of the properties and importance of solitons [190] (see Chapter 8), was sitting in the boat. A beautiful tribute!

1.1.2 The Korteweg–de Vries interpretation

The Euler equation for a nonviscous and incompressible fluid, the boundary conditions at the bottom and at the surface, and the assumption of an irrotational flow lead to the Korteweg–de Vries equation, which is valid in the weakly nonlinear case (see Appendix A):

$$\frac{1}{c_0}\frac{\partial \eta}{\partial t} + \frac{\partial \eta}{\partial x} + \frac{3}{2h}\eta\frac{\partial \eta}{\partial x} + \frac{h^2}{6}\frac{\partial^3 \eta}{\partial x^3} = 0, \tag{1.1}$$

where $c_0 = \sqrt{gh}$ is the propagation speed of the linear waves in the limit of long wavelengths, h the depth of the fluid and $\eta(x, t)$ the height of the surface above its equilibrium level.

Changing to a frame moving at speed c_0 by introducing the new variables $X = x - c_0 t$ and $T = t$ one gets rid of the second term and the equation simplifies into

$$\frac{1}{c_0}\frac{\partial \eta}{\partial T} + \frac{3}{2h}\eta\frac{\partial \eta}{\partial X} + \frac{h^2}{6}\frac{\partial^3 \eta}{\partial X^3} = 0. \tag{1.2}$$

Finally, if one introduces appropriate dimensionless variables ($\phi = \eta/h, \xi = X/X_0$ and $\tau = T/T_0$ where X_0 is a length and T_0 a time), one can write this equation in its standard form:

$$\frac{\partial \phi}{\partial \tau} + 6\phi\frac{\partial \phi}{\partial \xi} + \frac{\partial^3 \phi}{\partial \xi^3} = 0. \tag{1.3}$$

Among other solutions, the KdV equation has the spatially localised solution

$$\phi = A\,\mathrm{sech}^2\left[\sqrt{\frac{A}{2}}(\xi - 2A\tau)\right] \qquad (A > 0), \tag{1.4}$$

where the function $\mathrm{sech}\,x$ is defined by $\mathrm{sech}\,x = 1/\cosh x$. This solution is in quantitative agreement with the observations of John Scott Russell. Moreover the existence of the solution imposes the condition $A > 0$, in agreement with the observation that there are no dip-like solitary waves at the surface of water.[1] The soliton width, defined as $L = \sqrt{2/A}$, decreases as the amplitude increases, as shown in Figure 1.3, and tends to infinity as the amplitude A goes to zero.

Coming back to the original variables, in the laboratory frame, the solution becomes

$$\eta = \eta_0\,\mathrm{sech}^2\left[\frac{1}{2h}\sqrt{\frac{3\eta_0}{h}}\left(x - c_0\left[1 + \frac{\eta_0}{2h}\right]t\right)\right]. \tag{1.5}$$

One can notice that the propagation speed is larger than c_0; such KdV solitary waves are called *supersonic* waves and, as measured by John Scott Russell, the

[1] It has recently been shown that it is possible to create hydrodynamic solitons with a negative amplitude in experiments where the surface tension is high [63]. Such experiments have been performed at the surface of mercury, and the equation of motion is no longer exactly the KdV equation (1.3).

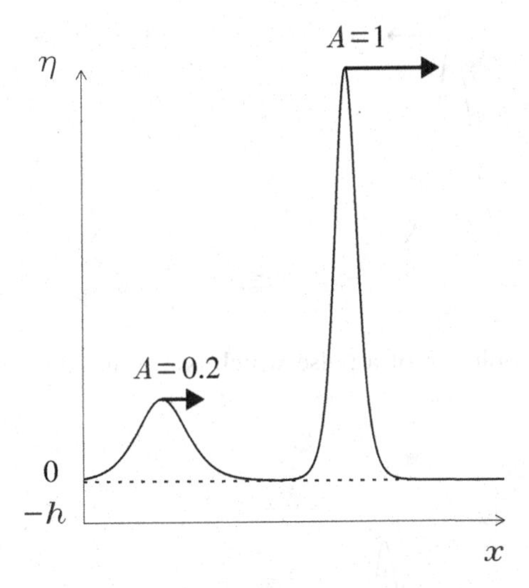

Figure 1.3. Comparison between solitons having different amplitudes A. The left pulse, having moderate amplitude and speed, is broader than the faster right pulse.

deviation of the speed from c_0 is proportional to the height η_0 of the wave above the surface.

1.1.3 Properties of the KdV equation and of its solutions

The KdV equation (1.3) shows up in many areas of physics, when waves can propagate in a *weakly nonlinear and dispersive medium* as discussed below.

Let us first try to understand the physical role of its various terms. The nonlinear term $\phi(\partial\phi/\partial\xi)$ tends to promote the formation of steep fronts or shock waves. The linear equation

$$\frac{\partial \phi}{\partial \tau} + v\frac{\partial \phi}{\partial \xi} = 0 \tag{1.6}$$

is a wave equation which has solutions of the form $\phi = f(\xi - v\tau)$, which therefore propagate at speed v. Let us now consider the equation

$$\frac{\partial \phi}{\partial \tau} + \phi\frac{\partial \phi}{\partial \xi} = 0, \tag{1.7}$$

which is called the Burgers–Hopf equation. By analogy with the linear case, one can consider that the coefficient of $\partial\phi/\partial\xi$ determines the propagation speed of the wave, so that, as a first approximation, one notices that each component of the signal moves with speed ϕ. As a result the parts of the signal which have the largest amplitude ϕ tend to move faster than the parts which have a smaller amplitude. A plot shows that this situation favours the formation of shock waves, i.e. waves

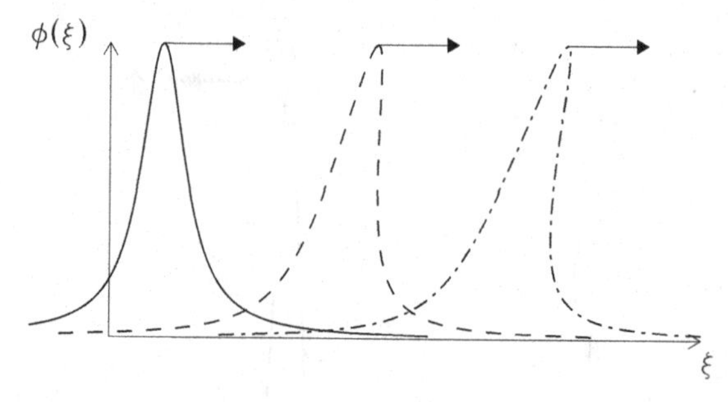

Figure 1.4. Time evolution of a pulse which is a solution of the Burgers–Hopf equation (1.7).

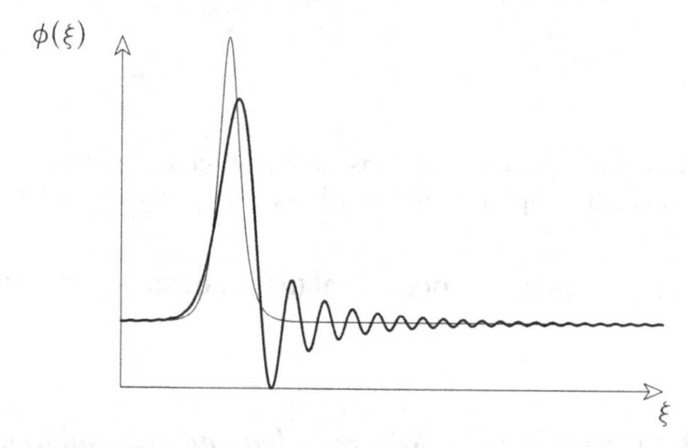

Figure 1.5. Time evolution of the same pulse as in Figure 1.4 when its dynamics are determined by the linear dispersive equation (1.8).

which exhibit discontinuities in a finite time, where the variation of the field has a vertical slope. This analysis is confirmed by the exact solution of the Burgers–Hopf equation, plotted in Figure 1.4.

Let us now consider the linearised version of Equation (1.3)

$$\frac{\partial \phi}{\partial \tau} + \frac{\partial^3 \phi}{\partial \xi^3} = 0. \tag{1.8}$$

It has plane wave solutions of the form $\phi = A\,\mathrm{e}^{\mathrm{i}(q\xi - \omega\tau)}$ provided the frequency ω and the wavenumber q are linked by the dispersion relation $\omega = -q^3$. This dispersion relation may seem surprising, but its unusual expression arises because the KdV equation is written in a frame mobile at speed c_0. From the equation in the laboratory frame (1.2), one gets the more standard dispersion relation $\omega' = c_0 q'(1 - q'^2 h^2/6)$.

Whatever the frame, one notices that the waves have a phase velocity $v_\varphi = \omega/q$ which depends on their wavenumber q. This characterises a dispersive medium. In such a medium the Fourier components of a narrow pulse propagate at different speeds, which leads to a broadening of the pulse, as shown in Figure 1.5.

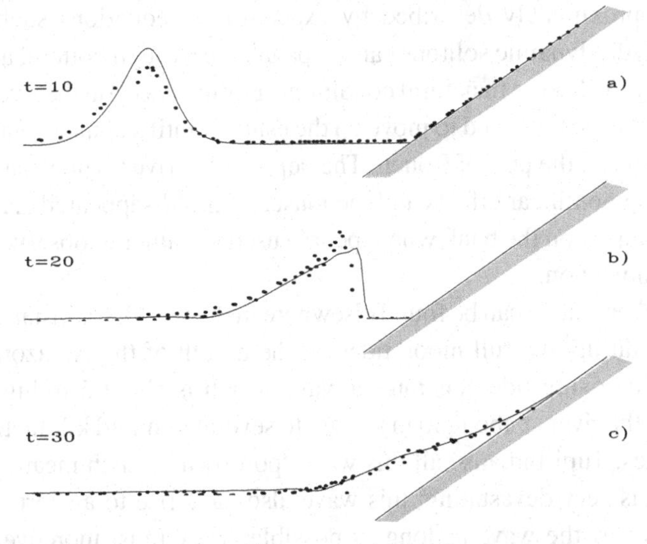

Figure 1.6. Time evolution of a solitary wave when the depth decreases next to a coast (shown as the grey area). Dots are experimental results [173] while the curves have been derived from a numerical simulation [115]. The height of the pulse above the water level is equal to 0.3 times the depth of the fluid far from the coast.

The existence of the permanent profile soliton solution of the KdV equation results from an equilibrium between these two effects, nonlinearity and dispersion: nonlinearity tends to localise the wave while dispersion spreads it out. It is important to notice that this *equilibrium is stable*. If the initial pulse is too narrow, dispersion effects dominate and it tends to broaden until equilibrium is reached. Conversely, if the initial condition is too broad, dispersion is dominated by nonlinearity which tends to localise the solution until equilibrium is reached. It is this remarkable property which makes the soliton concept so powerful and fascinating.

However there are cases where equilibrium cannot be reached due to external perturbations. This happens when a wave propagates in a medium with a depth that continuously decreases along its direction of propagation. The dispersive term, proportional to h^2 in Equation (1.1), decreases constantly whereas the nonlinear term, proportional to $1/h$ increases. As the nonlinearity dominates, the wave steepens and finally breaks (see Figure 1.6), as the waves on a beach.

The example of the beach is extreme because the perturbation systematically acts in the same way, and finally destroys the soliton. This is unusual because solitons are generally *stable* in the presence of perturbations, and particularly in the case of disorder. It is this exceptional stability which explains the breadth and the diversity of the applications of soliton theory, even though physical applications are only approximately described by exact soliton equations such as the KdV equation. Hydrodynamic solitons can propagate on a rough bottom, and this is why they are easy to observe in natural conditions. For instance, on the river Seine, a tide wave called 'mascaret' used to move up the estuary until a channel was dug to allow big boats to reach the port of Rouen. The depth of the river is nowadays too large to lead to strong nonlinear effects and the mascaret has disappeared ... together with a small restaurant on the bank where people used to gather to observe this beautiful natural phenomenon.

Similar phenomena can be found elsewhere in the world, for instance in February and March, during the full moon tides, at the mouth of the Amazon, in the north of Brazil. The rising tide generates a wave which is about 5 m high, and moves upstream in the river at about 30 m s^{-1}, up to several hundred kilometres in some of the tributaries. Tupi Indians call this wave 'pororoca', which means 'great noise'. Although it is very devastating, this wave also gave rise to a sport. The competitors have to surf the wave as long as possible. They must moreover take care to avoid the many debris carried by the Amazon as well as alligators, anacondas and piranhas!

There are other phenomena which are well described by the KdV equation. This is the case for these huge waves, called 'tsunamis' by the Japanese, which come as walls of water, devastating the coasts [54]. On 17 July 1998, one of them killed more than 2200 people on the north coast of Papua New Guinea, and, on 26 December 2004, a series of tsunamis created by a large earthquake near Sumatra killed more than 100 000 people in several countries of South-east Asia. Tsunamis are well described by the KdV equation although they move in oceans which are, on average, 4000 m deep. This does not seem to be shallow water, but this depth must be compared to the spatial extent of the wave which can reach 100 km. The propagation speed of tsunamis is larger than $c_0 = \sqrt{gh} = 200$ m s^{-1} $= 700$ km h^{-1}, which is the speed of a plane. Thus such solitary waves can cross the ocean extremely fast. They slow down when they arrive at the coast, but, as nonlinearity increases with decreasing depth, they give rise to huge waves that may break several hundred metres onto the mainland where they cause huge damage.

Most of such hydrodynamic solitons are of seismic origin [54] such as those in Izmit on 17 August 1999 and in South-east Asia on 26 December 2004. Another cause could be the sudden arrival in the sea of the enormous amount of lava emitted in some volcanic eruptions [87]. This mechanism would be, at a much larger scale, the equivalent of the experiment of John Scott Russell shown in Figure 1.1. More than 30 000 people died in 1883 due to a tsunami that followed the explosion of the volcano Krakatoa in Indonesia. The volcanic islands of the Antillas are being particularly monitored at present because their collapse might generate devastating solitons for the neighbouring islands or even the United States.

DIEDERIK JOHANNES KORTEWEG (1848–1941), son of a judge of the city of Hertogenbosch in the southern part of Netherlands started his career at the Technological University of Delft. He prepared a PhD under the supervision of J. D. Van der Waals, who later got the Nobel Prize, and was one the brightest physicists of this particularly remarkable Dutch period. His thesis work was entitled 'On the propagation of waves in elastic tubes'. He was the first Doctor of the University of Amsterdam, which had recently been granted the ability to confer this degree. Three years later he gained the first Chair in Mathematics of this University, which he occupied for 40 years. He worked on thermodynamics, fluid mechanics and classical mechanics problems, still collaborating with J. D. Van der Waals but also with J. H. van't Hoff. He also made noticeable contributions in pure mathematics on algebraic equations with real coefficients and on the properties of surfaces near singular points (Amsterdam University Drawing, 1906).

Together with his PhD student, GUSTAV DE VRIES (1866–1934), he confirmed that the statements of Airy and Stokes about Russell's observation were not correct and introduced the new equation [52, 104] which now bears their names. They derived its periodic solutions using elliptic functions, but also showed how solitons could be obtained in the long wavelength limit. Although they quoted this equation as 'very important', they did not publish anything more on this topic. G. de Vries published two papers on cyclones. Then he married a teacher of literature and French and became a teacher in a Haarlem high school (Photograph: the de Vries family).

1.2 The solutions of the KdV equation

Like all nonlinear equations, the KdV equation is very rich and has a large variety of solutions. Let us examine some of them, which are particularly important.

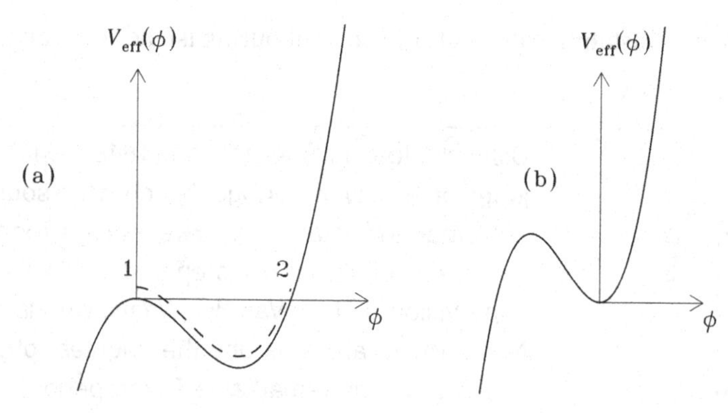

Figure 1.7. Shape of the effective potential $V_{\rm eff}(\phi)$ for $v > 0$ (a) and $v < 0$ (b).

1.2.1 Permanent profile solutions

These solutions propagate at speed v while preserving their shape. Therefore they can be written as $\phi(\xi, \tau) = \phi(\xi - v\tau)$. Introducing the new variable $z = \xi - v\tau$, the KdV equation (1.3) becomes

$$-v\phi_z + 6\phi\phi_z + \phi_{zzz} = 0, \tag{1.9}$$

(where an index denotes a partial derivative with respect to a variable, such as $\phi_z = \partial\phi/\partial z$), i.e.

$$\frac{\mathrm{d}}{\mathrm{d}z}(-v\phi + 3\phi^2 + \phi_{zz}) = 0, \tag{1.10}$$

which implies

$$\phi_{zz} + 3\phi^2 - v\phi + c_1 = 0, \tag{1.11}$$

where c_1 is an integration constant. Multiplying by ϕ_z and integrating a second time with respect to z, one gets

$$\frac{1}{2}\phi_z^2 + \phi^3 - \frac{1}{2}v\phi^2 + c_1\phi = c_2, \tag{1.12}$$

where c_2 is a second integration constant. This equation is formally identical to the equation which expresses the energy conservation for a particle with a mass equal to unity, a position denoted by ϕ, time denoted as z, subjected to the potential

$$V_{\rm eff}(\phi) = \phi^3 - \frac{1}{2}v\phi^2 + c_1\phi, \tag{1.13}$$

which is plotted in Figures 1.7. Equation (1.12) can therefore be written as

$$\frac{1}{2}\phi_z^2 + V_{\rm eff}(\phi) = c_2. \tag{1.14}$$

This approach, which introduces a 'pseudo-potential', is useful because it allows us to use our general experience of mechanics to determine the velocities v and the range of the constants c_1 and c_2 which lead to bounded ϕ solutions.

Soliton solution

The soliton solution that we introduced above, is a spatially localised solution, which means that the three fields ϕ, ϕ_z, ϕ_{zz} must decay to zero when $|z|$ tends to infinity. This implies that the integration constants must be chosen to be zero. As a result Equation (1.12) is simply

$$\frac{1}{2}\phi_z^2 + \phi^3 - \frac{1}{2}v\phi^2 = 0. \tag{1.15}$$

Let us consider a particle at rest, leaving the 'position' $\phi = 0$, denoted as point 1 in Figure 1.7(a), at an initial 'time' $z = -\infty$. According to Equation (1.12) it must evolve with a constant 'pseudo-energy'. If the particle leaves the origin on the negative side, it will fall downwards in potential while its kinetic energy will quickly increase, and ϕ will diverge. As we look for bounded solutions this would lead to a contradiction, and it confirms that the KdV equation does not have localised solutions with a negative ϕ amplitude (that would correspond to pits on the water surface) in agreement with the observations of John Scott Russell.

Conversely, if the particle leaves the origin on the positive side, it will move towards point 2, where its 'speed' ϕ_z will vanish again. Then the particle will move backward to point 1, which will be reached when z tends to infinity. This behaviour corresponds to the increase and then decrease of ϕ in the soliton solution of the KdV equation. In order to explicitly derive the solution one can integrate

$$dz = \frac{d\phi}{\sqrt{v\phi^2 - 2\phi^3}} \tag{1.16}$$

using the change of variable $\phi = (v \operatorname{sech}^2 u)/2$. This leads to the solution

$$\phi = \frac{v}{2}\operatorname{sech}^2\left(\sqrt{\frac{v}{4}}z\right). \tag{1.17}$$

Defining $A = v/2$, one recovers the soliton solution that we introduced above (Equation 1.4). The value of ϕ at point 2 determines the maximum amplitude A of the solution. It is such that $V_{\text{eff}}(A) = 0$, which agrees with the propagation speed $v = 2A$ of the solution.

When speed v is chosen to be negative, one can notice that the possibility to have a constant-energy bounded motion for a particle starting at rest from the 'position' $\phi = 0$ disappears (see Figure 1.7(b)). This different behaviour for $v > 0$ and $v < 0$ should not be considered as a symmetry breaking between the $+x$ and $-x$ directions

Figure 1.8. Solitary wave in the Maalea bay of the Maui island in Hawaï (Photograph: Robert I. Odom, 2003).

in the laboratory frame, which would be unphysical. It appears because the KdV equation is written in a frame moving at speed c_0. The difference between $v > 0$ and $v < 0$ simply means that hydrodynamic solitons are always supersonic, as observed by John Scott Russell.

General solution: cnoidal waves

If one drops the requirement that the solutions should be spatially localised, the integration constants c_1 and c_2 do not have to vanish. The equation

$$\frac{1}{2}\phi_z^2 = c_2 - c_1\phi + \frac{1}{2}v\phi^2 - \phi^3 \tag{1.18}$$

can still be solved analytically by introducing a new unknown function $u(z)$ such that $\phi - \phi_0 = -\alpha\, u^2(z)$, where α and ϕ_0 are two constants to be determined. The equation becomes

$$\left(\frac{\mathrm{d}u}{\mathrm{d}z}\right)^2 = (1 - u^2)(1 - k^2 + k^2u^2), \tag{1.19}$$

where k is another constant to be determined. The solutions of this nonlinear differential equation are the Jacobi elliptic functions which generalise the usual trigonometric functions [13].

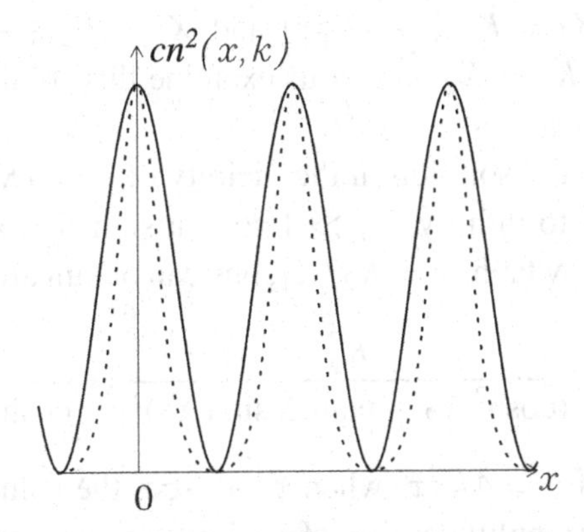

Figure 1.9. Shape of the function $\mathrm{cn}^2(x, k)$ for $k = 0$ (solid line) and $k = 0.98$ (dotted line). The horizontal axis has been scaled in each case, so that both functions have the same period.

The solution which is physically relevant for hydrodynamic waves is

$$\phi = \phi_0 - \frac{k^2 q^2}{2} \mathrm{cn}^2\left(\frac{qx}{2}, k\right) \tag{1.20}$$

which is expressed in terms of the elliptic cosine $\mathrm{cn}(x, k)$. This family of solutions, which depend on a parameter k, includes the usual cosine function for $k = 0$ and tends toward the sech function when the modulus k approaches 1. Thus one recovers the previous soliton solution. For intermediate k values, $\mathrm{cn}(x, k)$ is qualitatively similar to $\cos x$ but it is however sharper around the maxima (Figure 1.9). This shape corresponds to the 'spiky' shape that one observes on sea waves, especially near the coast where the depth decreases so that nonlinearity increases.

1.2.2 Multisoliton solutions

Besides the permanent profile solutions, the KdV equation also has an infinity of other solutions called multisoliton solutions because, when $|t|$ goes to infinity, they tend toward a superposition of several, well separated, solitons. The time evolution of such a solution describes the interaction between these solitons.

In Chapter 7 we shall introduce a systematic method to derive multisoliton solutions. Without attempting to derive it at this stage, let us consider one of them which is particularly useful for understanding the properties of solitons, the two-soliton solution:

$$\phi = \frac{2\left(K_1^2 - K_2^2\right)}{[K_1 \coth X_1 - K_2 \tanh X_2]^2}\left(\frac{K_1^2}{\sinh^2 X_1} + \frac{K_2^2}{\cosh^2 X_2}\right), \tag{1.21}$$

where we introduce $X_1 = K_1(\xi - 4K_1^2\tau)$ and $X_2 = K_2(\xi - 4K_2^2\tau)$. Let us assume for instance that $K_1 > K_2$ and let us examine this solution in the two limits $\tau \to \pm\infty$.

Let us denote by ϕ_2 the solution in the vicinity of $\xi = 4K_2^2\tau$. As $K_1 > K_2$, in the limit where τ goes to infinity, $X_1 \gg 1$ so that $\sinh X_1 \gg 1$ and $\coth X_1 \simeq 1$. As a result, defining Δ by $\tanh\Delta = K_2/K_1$, one can obtain after some calculations

$$\phi_2 \simeq 2(1 - \tanh^2\Delta)\frac{K_2^2}{(\cosh^2 X_2 - \tanh\Delta\sinh X_2)^2} = \frac{2K_2^2}{\cosh^2(X_2 - \Delta)}. \quad (1.22)$$

Thus, in the vicinity of $\xi = 4K_2^2\tau$, when $\tau \to -\infty$, the solution is identical to a single KdV soliton with amplitude $A = 2K_2^2$. Similarly, in the vicinity of $\xi = 4K_1^2\tau$, the solution takes the form

$$\phi_1 \simeq 2K_1^2\,\text{sech}^2(X_1 + \Delta'). \quad (1.23)$$

Consequently, in the limit $\tau \to -\infty$, the solution appears to be made of two separate solitons, one of them having amplitude $2K_1^2$ and speed $4K_1^2$, while the second has amplitude $2K_2^2$ and speed $4K_2^2 < 4K_1^2$.

For $\tau \to +\infty$ the same analysis shows again that the solitons are separated, but soliton 1 has passed soliton 2. In the vicinity of $\xi = 4K_2^2\tau$, the solution is now $\phi_2 \simeq 2K_2^2\,\text{sech}^2(X_2 + \Delta)$ while it is $\phi_1 \simeq 2K_1^2\,\text{sech}^2(X_1 - \Delta')$ in the vicinity of $\xi = 4K_1^2\tau$.

Each soliton has kept its velocity, however a careful analysis shows that the two solutions have indeed interacted. As the KdV equation is nonlinear, the collision *cannot be* a simple superposition. This appears clearly in Figure 1.10 because, right at the time of collision the amplitude is smaller than the amplitude of the largest soliton.

Moreover the asymptotic expressions of the solutions in the two limits $\tau \to \pm\infty$ show that the solutions experience a phase shift, which is very clearly visible on the contour plot of Figure 1.10. For $\tau \to -\infty$, replacing X_2 by its expression in Equation (1.22) one notices that soliton 2 has the position $4K_2\tau + \Delta/K_2$ while the calculation for $\tau \to +\infty$ gives its position as $4K_2\tau - \Delta/K_2$. Therefore, due to the collision, soliton 2 is displaced by $2\Delta/K_2$ with respect to the position that it would have occupied if it had not collided with soliton 1. A similar calculation shows that, on the contrary the largest soliton is moved forward by the collision.

The picture shown in Figure 1.11 taken on the sea shore shows the collision of two pulses on a beach where the water is very shallow. The exact conditions of this event are not known, but this picture clearly shows that the shape of the two solitary waves is preserved by the collision. Moreover one can also observe the shift due to the collision, although it is fairly small in this case. The qualitative agreement between Figures 1.10 and 1.11 is striking.

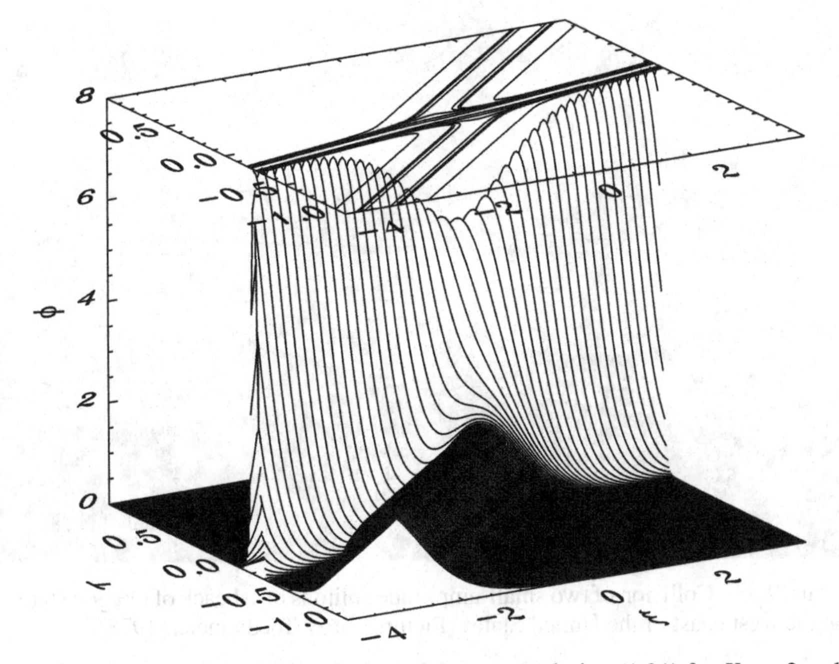

Figure 1.10. Plot of the time evolution of the exact solution (1.21) for $K_1 = 2$ and $K_2 = 1$ in the ξ, τ plane. In the range of large negative ξ and τ, one observes two pulses which are going to collide. They are still well separated. The larger has an amplitude equal to 8 and the smaller has amplitude 2. The largest pulse propagates along the whole ξ range shown (from -4 to $+4$) with only a slight change in τ because it is very fast. After the collision the smaller pulse is hidden by the larger, but it still exists, as shown by the top view which is a contour plot of the amplitude. This view clearly shows the position of the pulses versus time, and one can easily notice the phase shift due to the collision, which is particularly large for the smallest pulse.

1.3 Conservation rules

Let us introduce a Lagrangian L for a field $u(\xi, \tau)$ by

$$L = \int d\xi \; \mathcal{L} = \int d\xi \; \frac{1}{2}\left[u_\xi u_\tau + 2u_\xi^3 - u_{\xi\xi}^2\right], \tag{1.24}$$

where $\mathcal{L}$ is the Lagrangian per unit length. Writing the Lagrange equation of a continuous medium (see Appendix B for a reminder) for field u:

$$\frac{d}{d\tau}\left(\frac{\partial \mathcal{L}}{\partial u_\tau}\right) + \frac{d}{d\xi}\left(\frac{\partial \mathcal{L}}{\partial u_\xi}\right) - \frac{d^2}{d\xi^2}\left(\frac{\partial \mathcal{L}}{\partial u_{\xi\xi}}\right) - \frac{\partial \mathcal{L}}{\partial u} = 0, \tag{1.25}$$

one gets

$$u_{\xi\tau} + 6u_\xi \, u_{\xi\xi} + u_{\xi\xi\xi\xi} = 0, \tag{1.26}$$

Figure 1.11. Collision of two small-amplitude solitons on a beach of Oregon state, on the west coast of the United States (Picture: Terry Toedtemeier, 1978).

so that, defining $\phi = u_\xi$, we obtain the KdV equation (1.3). In a similar way, when the KdV equation is derived as an evolution equation for many physical systems, very often the variable which enters into the equation (here ϕ) is the spatial derivative of a physical field.

It is very useful to have a Lagrangian from which the KdV equation can be derived because it can be used in a systematic way to obtain conservation rules for the system which is described by the KdV equation [5]. However this leads to a small number of conservation rules. We shall see in Part II (Mathematical methods for the study of solitons) that the KdV system is even richer: it has an infinite number of conservation rules because it corresponds to a completely integrable field theory.

Some of these rules appear very clearly when one examines the one-soliton or two-soliton solutions. For instance, the quantity

$$M = \int_{-\infty}^{+\infty} \mathrm{d}\xi \; \phi(\xi, \tau), \tag{1.27}$$

which is called the 'mass of the wave', is a constant of the motion. This conservation rule is obvious for a one-soliton solution, which moves at constant speed and preserves its shape, but it is not so obvious for multisoliton solutions.

A field evolving according to the KdV equation also obeys the rule:

$$\frac{\mathrm{d}}{\mathrm{d}\tau} \int_{-\infty}^{+\infty} \xi \phi(\xi, \tau) \mathrm{d}\xi = 3 \int_{-\infty}^{+\infty} \phi^2(\xi, \tau) \mathrm{d}\xi = \text{constant}. \tag{1.28}$$

Using again the idea of the 'mass' of the wave, if ϕ were a mass density, $\int \xi\phi \, \mathrm{d}\xi$ would therefore be the position of the centre of mass. The conservation rule (1.28) implies that the motion of the centre of mass is uniform. This is again obvious for the one-soliton solution, but also valid for the multisoliton solutions of the KdV equation, for instance for the two-soliton solution that we studied above.

After the presentation of some of the *mathematical* properties of the KdV equation, let us now show how this equation can be applied to understand the properties of many *physical systems* which are weakly dispersive and weakly nonlinear.

1.4 Nonlinear electrical lines

1.4.1 Introduction to the physical problem

This is an interesting example because its understanding only needs basic physics and moreover it can easily be built experimentally. We have shown that, in order to sustain solitons, a system must combine dispersion and nonlinearity. Dispersion is obtained by using a device made of discrete components and nonlinearity is introduced by a nonlinear component. A passive line, i.e. a line which does not include any amplification, can be built with varicap diodes, which behave as nonlinear capacitances (Figure 1.12).

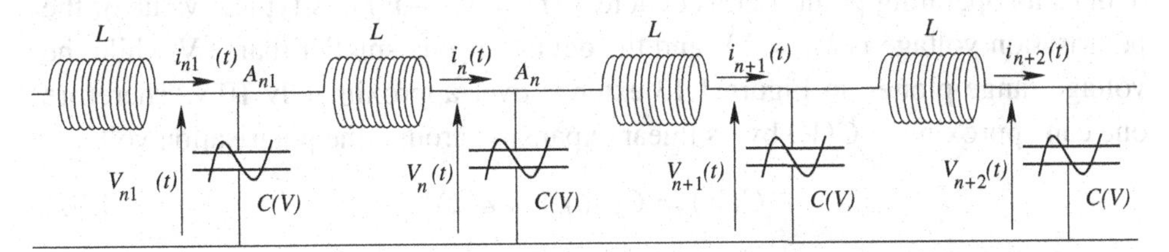

Figure 1.12. Schematic picture of a nonlinear electrical line made of inductances and varicap diodes which play the role of nonlinear capacitances $C(V)$.

The parameters of cell n are:

- The electrical current i_n in the inductance.
- The voltage V_n across the capacitance.
- The charge Q_n of the capacitance.

The conservation of the current at node A_n gives

$$i_n = \frac{\mathrm{d}Q_n}{\mathrm{d}t} + i_{n+1}. \tag{1.29}$$

Expressing the potential difference between points A_n and A_{n-1} leads to

$$V_{n-1} - V_n = L\frac{\mathrm{d}i_n}{\mathrm{d}t}. \tag{1.30}$$

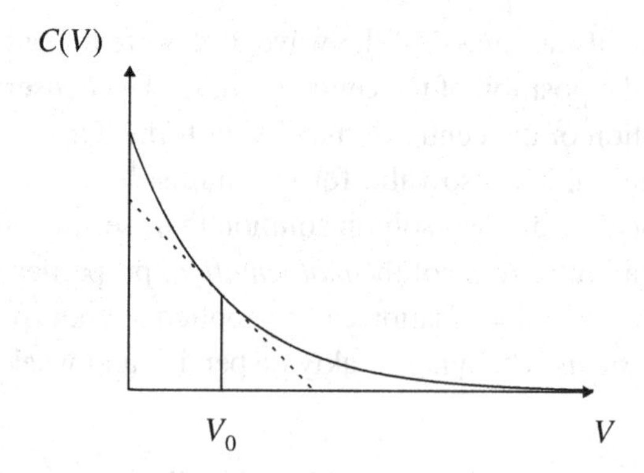

Figure 1.13. Schematic variation of the capacitance of a varicap diode versus the electrical potential difference at its connections.

A combination of these two equations gives

$$\frac{\mathrm{d}^2 Q_n}{\mathrm{d}t^2} = \frac{1}{L}\left[(V_{n-1} - V_n) - (V_n - V_{n+1})\right]. \tag{1.31}$$

The capacitance of a varicap diode depends upon the voltage V according to a law schematically shown in Figure 1.13. In practice the varicap diode is subjected to a voltage V which only slightly deviates from a polarisation voltage V_0 which defines its operating point. Let us denote $V(t) = V_0 + v(t)$. A typical value of the polarisation voltage is $V_0 = 2\,\mathrm{V}$ and the deviation v is smaller than $1\,\mathrm{V}$ while the voltage range plotted in Figure 1.13 extends over approximately $10\,\mathrm{V}$. Therefore one can approximate $C(V)$ by its linear expansion around the polarisation voltage:

$$C(V) = C(V_0)(1 - a_1 v). \tag{1.32}$$

From the definition of the dynamic capacitance $C(V) = \mathrm{d}Q/\mathrm{d}V$, the electric charge of the capacitance subjected to a voltage $V_0 + v$ is given by

$$Q = \int_0^{V_0} C(V)\mathrm{d}V + \int_{v=0}^{v} C(V_0)(1 - a_1 u)\mathrm{d}u \tag{1.33}$$

$$= Q_0 + C(V_0)\left[v - \frac{a_1}{2}v^2\right], \tag{1.34}$$

which we shall denote by

$$Q = Q_0 + C_0(v - av^2), \quad \text{with} \quad a = \frac{a_1}{2} \text{ and } C_0 = C(V_0). \tag{1.35}$$

Combining this formula with Equation (1.31) one gets an equation for the deviation v_n from the polarisation voltage at cell n:

$$\frac{\mathrm{d}^2}{\mathrm{d}t^2}\left(v_n - av_n^2\right) = \frac{1}{LC_0}(v_{n+1} + v_{n-1} - 2v_n). \tag{1.36}$$

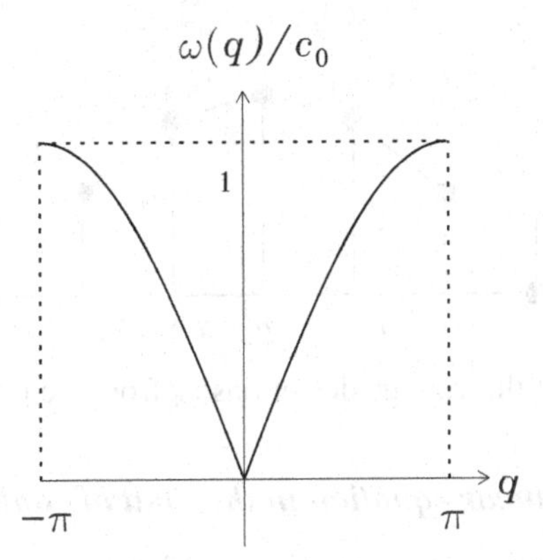

Figure 1.14. Dispersion relation of the plane waves which can propagate in an electrical line described by Equation (1.37).

For a chain of N identical cells, we get a set of N coupled nonlinear differential equations. The exact solution of this set of equations is not known so that some approximations are required. Their validity will have to be checked a posteriori.

1.4.2 Linear limit: dispersion relation

For small voltages v, a linear expansion can be used as (too?) often in physics. Such an approximation, valid if $av_n \ll 1$, leads to the equation

$$\frac{\mathrm{d}^2 v_n}{\mathrm{d}t^2} = \frac{1}{LC_0}(v_{n+1} + v_{n-1} - 2v_n) \tag{1.37}$$

which has plane wave solutions $v_n = A\mathrm{e}^{\mathrm{i}(qn-\omega t)}$ provided that ω and q are related by the dispersion relation $\omega = \pm 2c_0 \sin(q/2)$, where we define $c_0 = 1/\sqrt{LC_0}$. The frequency $\omega_0 = 2c_0$ is called the cutoff frequency since higher frequencies cannot propagate in the electrical network as shown by Figure 1.14.

The phase velocity $v_\varphi = \omega/q$ resulting from the dispersion relation depends on the wavevector q. Therefore the system is *dispersive*. However, since the phase velocity tends to the constant c_0 in the limit of long wavelengths (small q), the linear approximation shows that the electrical chain behaves as a weakly dispersive medium for long-wavelength signals. If the amplitude becomes high enough so that av can no longer be neglected with respect to 1, a small nonlinearity is also present and all conditions are met to get the KdV equation. We shall see below that it is indeed the case, but the path that leads to it has to go through several steps.

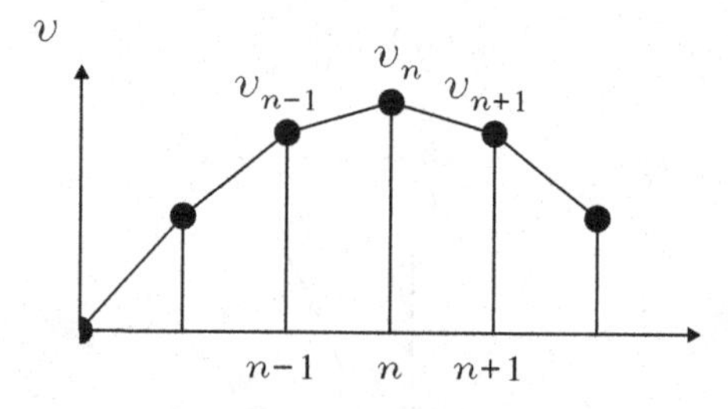

Figure 1.15. Variation of the voltage deviations v_n from the polarisation potential V_0.

1.4.3 The nonlinear equation in the limit of continuous media

The system of coupled nonlinear equations (1.36) is not exactly solvable but an approximate solution, which preserves nonlinearity, can be obtained by reducing it to a partial differential equation. Let us assume that the voltage v_n changes only slightly from one site to the next (Figure 1.15). One can then replace the set of discrete variables $v_n(t)$ by the field $v(x,t)$ such that $v_n(t) = v(x = n, t)$, where x is a dimensionless variable which measures the position along the chain, evaluated in number of cells. Assuming a slow spatial variation of $v(x,t)$ amounts to supposing that the spatial derivatives of v_n are small. Using a Taylor expansion around $x = n$ one gets

$$\begin{aligned} v_{n\pm 1}(t) = v(n \pm 1, t) = v(n,t) &\pm \frac{\partial v}{\partial x}(n,t) + \frac{1}{2}\frac{\partial^2 v}{\partial x^2}(n,t) \\ &\pm \frac{1}{6}\frac{\partial^3 v}{\partial x^3}(n,t) + \frac{1}{24}\frac{\partial^4 v}{\partial x^4}(n,t) + \cdots \end{aligned} \tag{1.38}$$

Truncating the expansion at order 4, and introducing again $c_0 = 1/\sqrt{LC_0}$, reduces the system of Equations (1.36) to a partial differential equation

$$\frac{\partial^2 v}{\partial t^2} - a\frac{\partial^2 v^2}{\partial t^2} = c_0^2 \frac{\partial^2 v}{\partial x^2} + \frac{c_0^2}{12}\frac{\partial^4 v}{\partial x^4}. \tag{1.39}$$

This approximation is called the continuum medium approximation.

The difficulty of this approach is to determine the right order to truncate the expansion. The goal is to choose the simplest possible form (i.e. the lowest possible order) while paying attention to preserve all the main physical features of the system. In this particular case, it is crucial to keep terms up to order 4 to preserve the dispersion. This can easily be checked by examining the linearised version of Equation (1.39). Looking for plane wave solutions of this equation leads to the dispersion relation

$$\omega^2 = c_0^2(q^2 - q^4/12) + \mathcal{O}(q^6), \tag{1.40}$$

which is the small amplitude expansion of the exact dispersion relation $\omega^2 = 4c_0^2 \sin^2(q/2)$ obtained previously. Therefore Equation (1.39) leads to the dispersion relation that should be expected because the continuum medium approximation only considers functions which vary slowly in space. Their spatial Fourier spectrum only includes long-wavelength contributions, i.e. it is restricted to small wavevectors q. Truncating the expansion at order 2 would have been too crude because the dispersion relation of the linearised version of Equation (1.39) would have been reduced to $\omega = c_0 q$, corresponding to a *nondispersive* medium. An essential feature of the system would have been lost.

Even a truncation of the Taylor expansion at order 4 would become insufficient if we were interested in solutions varying quickly from one cell to the next. Indeed a Fourier transform of such solutions would include significant contributions at large q, for which the fourth order expansion of $\omega^2(q)$ would lead to negative values, corresponding to instabilities.

Equation (1.39) is called the modified Boussinesq equation, or sometimes 'bad Boussinesq' due to its dispersion relation which can lead to unstable plane waves. It is very similar to the equation obtained by Boussinesq in 1895 for shallow-water hydrodynamic waves, but it differs from it because its nonlinear terms include a *time derivative* instead of a spatial derivative.

1.4.4 Quasi-soliton solutions of the electrical chain

Contrary to the KdV equation, the modified Boussinesq equation does not correspond to a completely integrable system having soliton solutions. However it has a solitary-wave solution (or quasi-soliton) which can be obtained by looking for permanent profile solutions, as we did for the KdV equation. It corresponds to a voltage pulse, propagating at constant speed c, higher than c_0:

$$v(n,t) = \frac{3}{2a}\frac{(c^2 - c_0^2)}{c^2}\,\mathrm{sech}^2\left[\frac{\sqrt{3(c^2 - c_0^2)}}{c_0}(n - ct)\right]. \tag{1.41}$$

As the study of the dispersion relation has shown that c_0 is the speed of long-wavelength linear waves in the electrical chain, we recover a feature of the solutions of the KdV equation. One can also notice that the amplitude of the voltage pulse, $A = (3/2a)(c^2 - c_0^2)/c^2$, increases with $c > c_0$. Finally, as the sign of a determines the sign of the solution, and because $a > 0$ for a varicap diode, only positive voltage pulses can propagate as solitary waves, another feature that we found for the solutions of the KdV equation.

Writing the solution as $v(n,t) = A\,\mathrm{sech}^2[(n-ct)/\ell]$ exhibits the quantity $\ell = c_0/\sqrt{3(c^2-c_0^2)}$ which determines the width of the voltage pulse. The continuum limit approximation, which requires that $v(n,t)$ varies only slightly from one site to the next, is only valid if $\ell \gg 1$. If we consider for instance a solution propagating at speed $c = 1.05\,c_0$, its width is $\ell = 1.83$, for which the continuum approximation is hardly valid. A numerical simulation of the equations of motion of the discrete chain shows that this solution, derived in the continuum limit approximation, is nevertheless a good solution for discrete chain. This is a feature that one often meets when one studies equations which have soliton or quasi-soliton solutions. These equations are astonishingly 'tolerant' to approximations!

The condition $\ell \gg 1$ is verified as long as c stays close to c_0, which corresponds to moderate amplitudes of A. Therefore, as the nonlinearity in the original equation (1.36) was $v(1-av)$, the continuum limit approximation for the electrical chain is valid for weak nonlinearities.

Although this Boussinesq equation is not completely integrable, thanks to the work of Hirota some multisoliton solutions are known. It has even been shown that the collision of two quasi-solitons given by Equation (1.41) preserves their shape and velocities, up to a phase shift, exactly as for the KdV equation. Moreover, for the electrical line, this can easily be checked *experimentally*.

1.4.5 The KdV limit for the electrical chain

The modified Boussinesq equation provides a good description of the properties of the electrical chain, but it is much less convenient than the KdV equation because much fewer mathematical tools are available for this equation than for the completely integrable KdV equation. For instance, as shown in Chapter 7, for the KdV equation there is a *systematic* scheme to study the time evolution of any initial condition, which is not available for the modified Boussinesq equation.

However, as we have shown that the Boussinesq equation describes weakly dispersive and weakly nonlinear systems, as does the KdV equation, it is natural to look for a possible transformation of the former equation to the latter. This is possible through an additional approximation that we shall now introduce.

In order to set the scale of the nonlinearity of the Boussinesq equation, let us introduce a parameter ε of the order of magnitude of the amplitude of the voltage pulse, assumed to be small. Introducing $v = \varepsilon U$, Equation (1.39) reads

$$\frac{\partial^2 U}{\partial t^2} - c_0^2 \frac{\partial^2 U}{\partial x^2} - \frac{c_0^2}{12}\frac{\partial^4 U}{\partial x^4} - \varepsilon a \frac{\partial^2 U^2}{\partial t^2} = 0. \tag{1.42}$$

The next step is to introduce appropriate changes of variables, in order to simplify this equation. One can understand where they come from by considering the dispersion relation

$$\omega = c_0(q - q^3/24) + \mathcal{O}(q^5). \tag{1.43}$$

The argument $\omega t - qx$ of the exponential which shows up in the plane wave solutions of the linearised equation can be expressed as

$$\omega t - qx = c_0(q - q^3/24)t - qx. \tag{1.44}$$

As we are looking for long-wavelength solutions, corresponding to small wavevectors, it is natural to introduce a new measure K of the wavevector through $q \equiv K\varepsilon^{\alpha}$. The value of the exponent α is not a priori known. It will be determined a posteriori by requiring that dispersion and nonlinearity appear in the equation within terms that have the same order in ε. Expression (1.44) is therefore rewritten as

$$\omega t - qx = c_0(K\varepsilon^{\alpha} - (K\varepsilon^{\alpha})^3/24)t - K\varepsilon^{\alpha}x \tag{1.45}$$

$$= \varepsilon^{\alpha}K(c_0 t - x) - \varepsilon^{3\alpha}K^3 t/24. \tag{1.46}$$

Collecting terms with different powers in ε, it seems natural to introduce the variables

$$\xi = \varepsilon^{\alpha}(x - c_0 t) \quad \text{and} \quad \tau = \varepsilon^{3\alpha}t, \tag{1.47}$$

in order to get the expression $\omega t - qx = -K\xi - K^3\tau/24$. In terms of these new variables, if one defines $\Omega(K) = -K^3/24$, the expression $\omega t - qx$ becomes $\Omega\tau - K\xi$, i.e. it recovers a form analogous to its original form. However we have made some progress because the expression of the wavevector K includes our requirement to look for an equation suitable for solutions with a slow space variation. Moreover it now seems natural to change to a frame moving at speed c_0 by introducing the variable ξ. This is reminiscent of the case of hydrodynamic waves for which the KdV equation appeared when we moved to the mobile frame at speed c_0 which was the phase velocity of long-wavelength linear waves, similar to c_0 for the electrical chain.

In the absence of any dispersion and nonlinearity, any signal would propagate at speed c_0, and therefore it would appear as *stationary* in the mobile frame. In the presence of a weak dispersion and a weak nonlinearity, it is natural to assume a slow deformation of the signal in the moving frame. This is what the new time variable τ expresses in mathematical terms.

Writing Equation (1.42) with the variables ξ and τ, we get

$$\left(\varepsilon^{6\alpha}\partial_{\tau\tau} - 2c_0\varepsilon^{4\alpha}\partial_{\tau\xi} + c_0^2\varepsilon^{2\alpha}\partial_{\xi\xi}\right)(U - a\varepsilon U^2) - c_0^2\varepsilon^{2\alpha}U_{\xi\xi} - \frac{c_0^2\varepsilon^{4\alpha}}{12}U_{\xi\xi\xi\xi} = 0 \tag{1.48}$$

or, by collecting the leading terms,

$$\varepsilon^{2\alpha}\underbrace{\left(c_0^2 U_{\xi\xi} - c_0^2 U_{\xi\xi}\right)}_{=0} - \varepsilon^{4\alpha}\left(2c_0 U_{\tau\xi} + \frac{c_0^2}{12} U_{\xi\xi\xi\xi}\right) - \varepsilon^{2\alpha+1} a c_0^2 (U^2)_{\xi\xi} = \mathcal{O}(\varepsilon^{6\alpha}, \varepsilon^{4\alpha+1}). \quad (1.49)$$

One notices that the choice $\alpha = 1/2$ brings the leading nonlinear and dispersive terms to the same order ε^2 and that, at this order, we obtain the equation

$$-c_0\left(2U_\tau + \frac{c_0}{12} U_{\xi\xi\xi} + a c_0 (U^2)_\xi\right)_\xi = 0. \quad (1.50)$$

This appears to be an appropriate choice if we look for soliton-like solutions since we expect that the effect of dispersion and nonlinearity should cancel each other.

Equation (1.50) can be readily integrated once to give

$$\frac{\partial U}{\partial \tau} + a c_0\, U \frac{\partial U}{\partial \xi} + \frac{c_0}{24}\frac{\partial^3 U}{\partial \xi^3} = 0. \quad (1.51)$$

The integration constant has been chosen to be zero because we look for spatially localised solutions. This equation can be reduced to the KdV equation by introducing scaling factors for the time and amplitude, $\tau = c_0 T/24$ and $\phi = 4aU$. Its solution

$$\phi = A\,\mathrm{sech}^2\left[\sqrt{\frac{A}{2}}\,(\xi - 2A\tau)\right] \quad (1.52)$$

confirms that, if the amplitude of the wave is of the order ε, its spatial evolution is of the order $\varepsilon^{1/2}$ while its time dependence is of the order $\varepsilon^{3/2}$ as expected. The result is therefore coherent with the spatial scale and timescale chosen in Equation (1.47).

As the path from Boussinesq to KdV may seem tortuous, it is important to check that the results which can be derived from the KdV approximation are correct. If we consider a soliton solution moving at speed c close to c_0 such that $c = c_0(1 + \alpha')$ where $\alpha' \ll 1$, the comparison of the solutions of the two equations shows that they are identical to first order in α'. When we studied the Boussinesq equation, we saw that the speed $c = 1.05\ c_0$ was the upper limit of the validity of this equation, which imposes to restrict our attention to $\alpha' \leq 0.05$. In this range the two equations, Boussinesq and KdV, give results which differ at most by a few percent. The error which is introduced by the KdV approximation is comparable to the experimental errors in the measurements on the electrical chain. But the gain is very significant because all the mathematical tools that come with the KdV equation become available. It fully justifies this approximation.

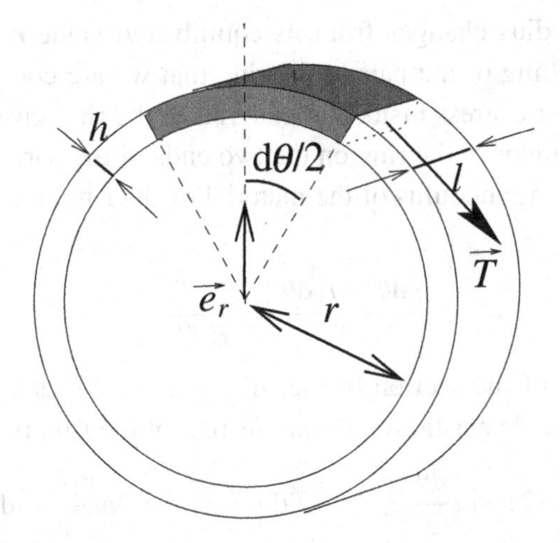

Figure 1.16. Schematic view of a small piece of artery of length ℓ along the axis of the artery, and of thickness h.

1.5 Blood pressure waves

Let us now consider a nonstandard case for which one can derive the KdV equation [143, 144]. Although it is certainly a rough approximation in this case, the KdV equation can answer a simple question: why is it possible to feel the blood pulse at the wrist or the ankle? The heart sends a pressure wave in the arteries. It moves along the arteries and induces a local expansion of the vessels. It is this deformation that we perceive when we feel the blood pulse. It is remarkable that it can propagate to the ends of the limbs without being noticeably dispersed. This can be explained by an equilibrium between the nonlinearity coming from the hydrodynamics of the blood flux and the dispersion associated with the elasticity of the arteries, as we shall see from a simple model.

A small piece of artery, of length ℓ, can be viewed as an elastic ring, having a radius r_0 at equilibrium and a thickness h (Figure 1.16). The time evolution of its radius r is determined on one hand by the pressure inside the artery and on the other hand by the elastic stress inside the material of the ring. Let us consider these two aspects successively.

The force $\mathrm{d}\vec{f_p}$ exerted by the blood pressure on a small part of the ring (shaded in Figure 1.16), defined by its angular aperture $\mathrm{d}\theta$, is directed along the radial vector $\vec{e}_r$ and is given by

$$\mathrm{d}\vec{f_p} = p\, r\mathrm{d}\theta\, \ell\, \vec{e}_r, \tag{1.53}$$

where p is the blood pressure inside the ring, or, more precisely, the difference between the maximum pressure when the heart contracts (systolic blood pressure) and the pressure in the artery when the heart is at rest (diastolic blood pressure).

When the ring radius changes from its equilibrium value r_0 to another value r, the relative stretching of the part of the ring that we are considering is $(r\mathrm{d}\theta - r_0\mathrm{d}\theta)/r_0\mathrm{d}\theta$. The elastic stress inside the material of the ring gives rise to force $\overrightarrow{T}$ applied by the remainder of the ring on the two ends of the part under study. If we denote by E the Young modulus of the material of the ring, the tensile force T is given by

$$\frac{r\mathrm{d}\theta - r_0\mathrm{d}\theta}{r_0\mathrm{d}\theta} = \frac{1}{E}\frac{T}{\ell h} \tag{1.54}$$

since ℓh is the area of the section of the ring on which T is applied. The radial component of the tensile elastic forces on this part of the ring is therefore

$$\mathrm{d}\overrightarrow{f_T} = -2T\sin\frac{\mathrm{d}\theta}{2}\,\vec{e}_r \simeq -T\mathrm{d}\theta\,\vec{e}_r = -E\ell h\frac{r-r_0}{r_0}\mathrm{d}\theta\,\vec{e}_r. \tag{1.55}$$

Writing Newton's law for the small part of the ring, and projecting it along the radial vector $\vec{e}_r$, we get

$$\mathrm{d}m\,\frac{\partial^2 r}{\partial t^2} = \mathrm{d}f_p + \mathrm{d}f_T, \tag{1.56}$$

where the mass of the ring section is $\mathrm{d}m = \rho_0\ell h r_0\mathrm{d}\theta$ if we denote by ρ_0 the mass of a unit volume of artery tissue. After some simplifications we obtain

$$\rho_0 h r_0\frac{\partial^2 r}{\partial t^2} = p\,r - Eh\frac{r-r_0}{r_0}. \tag{1.57}$$

In order to connect the deformations of the artery to the variations of the flux of the blood that it carries, it is convenient to introduce the section of the tube $A = \pi r^2$. Its time evolution obeys

$$\frac{\partial^2 A}{\partial t^2} = 2\pi r\frac{\partial^2 r}{\partial t^2} + 2\pi\left(\frac{\partial r}{\partial t}\right)^2 \simeq 2\pi r\frac{\partial^2 r}{\partial t^2} \tag{1.58}$$

if we only keep the linear part of the expression, taking into account the small values of the radial velocities. On the other hand, as the variations of r are small, we have

$$A - \pi r_0^2 = \pi(r+r_0)(r-r_0) \simeq 2\pi r_0(r-r_0), \tag{1.59}$$

which suggests rewriting Equation (1.57) as

$$2\pi r_0\frac{\partial^2 r}{\partial t^2}\rho_0 h = 2\pi p\,r_0 + 2\pi\frac{r-r_0}{r_0}\left[p r_0 - hE\right]. \tag{1.60}$$

Dividing by $\rho_0 h$, we get

$$2\pi r_0\frac{\partial^2 r}{\partial t^2} = \frac{2\pi p\,r_0}{\rho_0 h} - E2\pi\frac{r-r_0}{\rho_0 r_0} \tag{1.61}$$

if we make the approximation $pr_0 \ll hE$ which is valid for an artery because typical values [143] of p are in the range 20–40 mm Hg, r_0 in the range 2–10 mm, $h \simeq 0.1\, r_0$ and E in the range 20–130 N cm^{-2}. We finally get

$$\frac{\partial^2 A}{\partial t^2} = \frac{2\pi p\, r_0}{\rho_0 h} - \frac{E}{\rho_0 r_0^2}\left(A - \pi r_0^2\right). \tag{1.62}$$

This equation coming from the elasticity of the artery must be completed by the hydrodynamic equations of the blood flow. Viscosity, which damps the initial pulse, leads to a perturbation of the equation. We shall neglect it here to simplify the calculations, and we shall discuss below how it is actually balanced in the blood flow.

Noticing that only the z component along the axis of the artery enters into play, the Euler equation reads

$$\frac{\partial v}{\partial t} + v\frac{\partial v}{\partial z} = -\frac{1}{\rho}\frac{\partial p}{\partial z}. \tag{1.63}$$

It must be completed by the equation describing the mass conservation across a small piece of artery

$$\frac{\mathrm{d}}{\mathrm{d}t}\left[\rho A(z,t)\,\mathrm{d}z\right] = \underbrace{\rho A(z,t)v(z,t)}_{\text{incoming mass}} - \underbrace{\rho A(z+\mathrm{d}z,t)v(z+\mathrm{d}z,t)}_{\text{outgoing mass}}. \tag{1.64}$$

With the standard hypothesis of an incompressible flow (ρ constant), we finally get

$$\rho\frac{\partial A}{\partial t} + \rho\frac{\partial(Av)}{\partial z} = 0. \tag{1.65}$$

The system of Equations (1.57), (1.62) and (1.65) determines how the blood pulse propagates, but it cannot be solved in its present form. In order to control the approximations, as usual it is convenient to change to a set of dimensionless variables by defining

$$\widetilde{A} = \frac{A}{\pi r_0^2}, \quad \widetilde{p} = \frac{2r_0}{Eh}p, \quad \widetilde{v} = \frac{v}{L\Omega}, \quad \xi = \frac{z}{L} \quad \text{and} \quad \tau = \Omega t, \tag{1.66}$$

where we introduce

$$\Omega = \sqrt{\frac{E}{\rho_0 r_0^2}} \quad \text{and} \quad L = \sqrt{\frac{\rho_0}{\rho}\frac{r_0 h}{2}}. \tag{1.67}$$

The system of Equations (1.57), (1.62) and (1.65) reduces to

$$\frac{\partial^2 \widetilde{A}}{\partial \tau^2} + (\widetilde{A} - 1) = \widetilde{p} \tag{1.68}$$

$$\frac{\partial \widetilde{v}}{\partial \tau} + \widetilde{v}\frac{\partial \widetilde{v}}{\partial \xi} = -\frac{\partial \widetilde{p}}{\partial \xi} \tag{1.69}$$

$$\frac{\partial \widetilde{A}}{\partial \tau} + \frac{\partial \widetilde{A}\widetilde{v}}{\partial \xi} = 0. \tag{1.70}$$

As it is obviously nonlinear, soliton solutions can be expected provided it is also dispersive. In order to check this point, let us linearise the equations around the equilibrium values

$$\widetilde{A}_{\text{eq}} = 1, \quad \widetilde{p}_{\text{eq}} = 0 \quad \text{and} \quad \widetilde{v}_{\text{eq}} = 0. \tag{1.71}$$

The equations for the variations $\delta\widetilde{A}$, $\delta\widetilde{v}$ and $\delta\widetilde{p}$ around these equilibrium values are therefore

$$\frac{\partial^2 \delta\widetilde{A}}{\partial \tau^2} + \delta\widetilde{A} = \delta\widetilde{p} \tag{1.72}$$

$$\frac{\partial \delta\widetilde{v}}{\partial \tau} = -\frac{\partial(\delta\widetilde{p})}{\partial \xi} \tag{1.73}$$

$$\frac{\partial \delta\widetilde{A}}{\partial \tau} + \frac{\partial(\delta\widetilde{v})}{\partial \xi} = 0. \tag{1.74}$$

They have plane wave solutions

$$(\delta\widetilde{A}, \delta\widetilde{v}, \delta\widetilde{p}) = (A_0, v_0, p_0)\, e^{i(q\xi - \omega t)} \tag{1.75}$$

if the determinant

$$\begin{vmatrix} 1-\omega^2 & 0 & -1 \\ 0 & -i\omega & iq \\ -i\omega & iq & 0 \end{vmatrix} \tag{1.76}$$

vanishes. This gives the dispersion relation

$$\omega^2 = \frac{q^2}{1+q^2} \tag{1.77}$$

which yields the phase velocity $v_\varphi = \omega/q = 1/\sqrt{1+q^2}$, which indeed depends on the wavevector q. This confirms that the system is dispersive. Moreover we can notice that v_φ decreases as q increases, i.e. the dispersion relation is qualitatively similar to the dispersion relations of shallow water waves or signals in the electrical chain.

The search for soliton solutions can then be carried out with the same methods that lead from the Boussinesq to the KdV equation in the electrical chain. One first analyses the different orders of magnitude involved and then changes to a frame moving at the speed of long-wavelength linear waves, i.e. here $v_\varphi = 1$.

We still consider the case of a weak nonlinearity, i.e. we look for solutions having a magnitude of the order of ε, assumed to be small with respect to the equilibrium values. However it turns out that the expansion must be carried up to order 2 to give interesting results.

Therefore we look for solutions of the form

$$\widetilde{A} = 1 + \varepsilon A_1 + \varepsilon^2 A_2, \quad \widetilde{v} = \varepsilon v_1 + \varepsilon^2 v_2 \quad \text{and} \quad \widetilde{p} = \varepsilon p_1 + \varepsilon^2 p_2. \tag{1.78}$$

As with the study of the Boussinesq equation, changing to the moving frame amounts to looking for solutions which vary more slowly in time than in space. Thus we define the new variables

$$\chi = \varepsilon^{1/2}(\xi - \tau) \quad \text{and} \quad \eta = \varepsilon^{3/2}\tau, \tag{1.79}$$

which are then introduced into Equations (1.68), (1.69) and (1.70). The lowest order terms simply give $A_1 = v_1 = p_1$ if we look for spatially localised solutions. The next order terms lead to

$$\frac{\partial p_1}{\partial \eta} + \frac{3}{2} p_1 \frac{\partial p_1}{\partial \chi} + \frac{1}{2}\frac{\partial^3 p_1}{\partial \chi^3} = 0 \tag{1.80}$$

which is indeed the KdV equation.

Thus it turns out that the blood pressure waves appear as KdV solitons, within the approximate description that we introduced. An equilibrium between dispersion and nonlinearity explains why the blood pulse stays localised after travelling a long distance along the arteries, even down to the ends of the limbs. Introducing an appropriate value for the maximal excess of pressure generated by the heart (such as $p_{\max} = 2500$ Pa) we find a propagation speed of about 5 m s^{-1} for the blood pulse, and a spatial extent of the soliton of about 1 cm.

Besides the curiosity of considering blood pressure waves as solitons, this modelling can give useful results because it links the parameters of the soliton to the properties of the artery, such as its Young modulus. For a normal blood flow the volume of blood in the initial impulse is fixed. It is proportional to the area of the soliton solution (Figure 1.17). However, as the width and the amplitude of the soliton strongly depend on the Young modulus, it is easy to check that, if the artery becomes more rigid (owing to the effects of cigarette smoking for instance!) so that its Young modulus increases, the width of the pulse decreases while its amplitude increases, and therefore the blood pressure measured by a physician

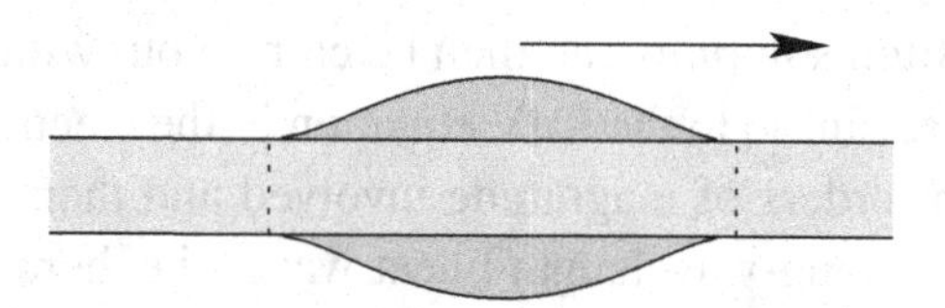

Figure 1.17. Schematic plot of the localised deformation of an artery associated with the propagation of the blood pulse. The volume of blood carried by the pulse (marked by the darker shaded area on the plot) is proportional to the area of the soliton solution.

increases. This is the effect that can lead to the breaking of small arteries in the brain.

Neglecting viscosity may seem a crude approximation, and indeed if one simply adds viscous effects to the above analysis, one finds that the blood pulse is quickly damped. But the reality of the blood flow is more subtle because there is a second effect that comes into play to balance the influence of viscosity, the conicity of the arteries. Their section gradually decreases as one moves away from the heart. A more realistic model, which includes the effect of conicity as well as viscosity shows that two perturbations of the KdV equation are generated which have opposite effects on the properties of the soliton. The balance of the two maintains a localised solution, as you can easily check by feeling your pulse.

1.6 Internal waves in oceanography

The Andaman sea, part of the Indian Ocean on the western coast of Thailand, is a place where one can observe impressive hydrodynamic solitons. The salt concentration is stratified along the vertical. Like surface waves on the sea which exist thanks to the difference between the density of water and air, *internal gravity waves* can exist within the ocean, due to the variation of the density of the water from the bottom to the surface (Figure 1.18). But, as the density difference between neighbouring layers is very small, the periods and amplitudes of these internal waves are much larger than those of surface waves. For instance amplitudes as large as 200 metres have been observed.

Sailors have known for centuries that this sea could show stripes where the surface was significantly above the average sea level. This phenomenon is for instance described in Maury's book [126] published in 1861.

> The ripplings are seen in calm weather approaching from a distance, and in the night their noise is heard a considerable time before they come near. They beat against the sides of a ship with great violence ... and a small boat could not always resist in the turbulence of these remarkable ripplings.

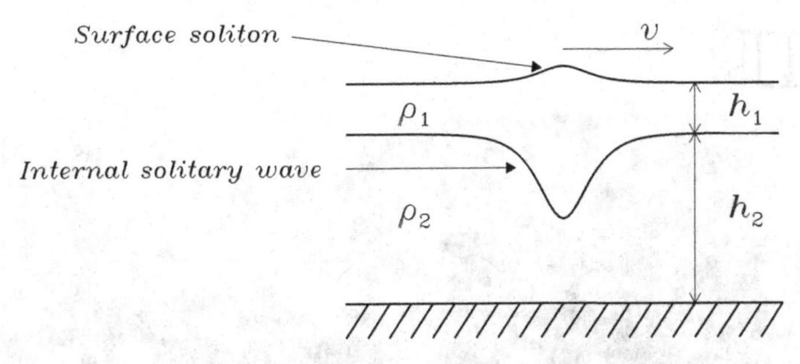

Figure 1.18. A solitary wave of amplitude η_0 in a two-layer fluid of densities $\rho_1 < \rho_2$ is a dip when the thickness of the layers are such that $h_1 < h_2$. It induces a small surface soliton which has an amplitude approximately equal to $(\rho_2 - \rho_1)\eta_0$.

As they exist in deep water far from the coast these waves cannot be attributed to standard tide phenomena. In 1965, a systematic series of measurements carried out by Perry and Schimke [148] showed that these stripes are the surface signature of some large internal waves. In 1980, Osborne and Burch analysed the data collected by the American company Exxon [141] on these waves which can damage off-shore oil-rigs. They concluded that they are indeed induced by internal solitary waves which are well described by the KdV equation. When the sea comprises layers which have different, but close, densities the internal waves also affect the surface as shown in Figure 1.18. These solitary waves are created by the tides near the coast, but then they can propagate over several hundreds of kilometres.

Measurements showed [141] that those solitons are not isolated. Groups of solitons are generated, the first solitons in a group coming with a time interval of about 40 minutes, which then decreases for the solitons which are in the tail of the group. The amplitude of the internal waves is about 60 m, and they induce surface waves which are about 2 m high. The width of one of these waves is of the order of 1 km so that they show up in a spectacular way on satellite images as shown in Figure 1.19. A measurement station installed in 1995 provides data on the creation and propagation of these remarkable water waves.

Similar internal waves exist in many oceans in the world, and for instance at the straits of Gibraltar, which connect the Mediterranean sea and the Atlantic Ocean. Such waves are presumed to be at the origin of the loss of the American submarine USS Thresher which disappeared in 1969 after a too sudden descent.

1.7 Generality of the KdV equation

There are numerous examples in physics which can be approximately described by the KdV equation. We shall meet later (see Chapter 8) the example of an atomic lattice which played a prominent role in the introduction of the soliton concept.

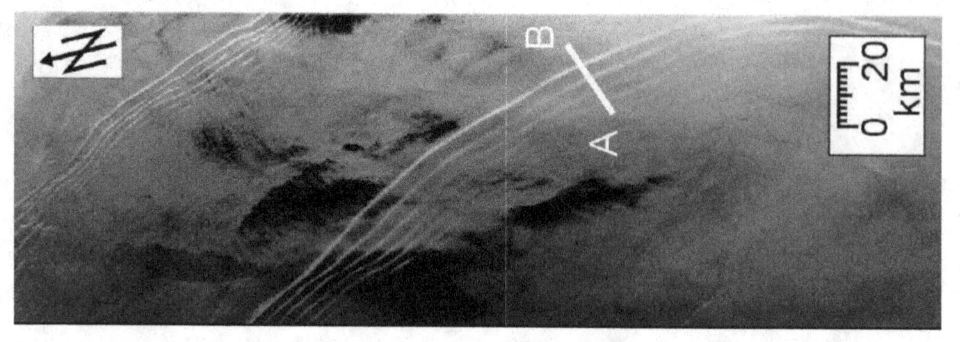

Figure 1.19. Aerial view of the surface of the Andaman sea. Two groups of internal waves are visible. They have been created by two successive tides. (Picture: Werner Alpers & ESA, 1996)

Another case which has been extensively studied is acoustic solitons in plasmas, which are longitudinal nonlinear waves in the plasma. They can be easily studied experimentally and the KdV equation provides a fairly good quantitative description of the phenomena (see Chapter 4).

The examples that we have already met are however sufficient to show the situations that lead to the KdV equation:

- It applies to systems which, at the first level of approximation, are described by a *hyperbolic linear equation* such as the wave equation $u_{tt} - c_0^2 u_{xx} = 0$.
- Moreover a *weak nonlinearity* such as $\varepsilon f(u)$, with $f(u) = Au^2 + Bu^3$ must exist.
- Finally the system must show a *weak dispersion* with a dispersion relation for small wavevectors q of the form $\omega(q) = c_0 q(1 - \lambda_0^2 q^2)$, which can arise from terms like u_{xxxx} or u_{xxtt} in the equation of motion.

In order to stay within the weakly dispersive range, let us consider signals $u(x,t)$ with a slow spatial variation. This implies that their Fourier spectrum $F(q)$ only includes components at a small wavevector q (such that $\delta = q\lambda_0 \ll 1$, i.e. $|q|$ is below some value $q_{\max}$). Therefore they can be written

$$u(x,t) = \int_{-\infty}^{+\infty} F(q) e^{i(qx-\omega t)} dq \simeq F(0) \int_{-q_{\max}}^{q_{\max}} e^{i(qx - c_0 qt + c_0 \lambda_0^2 q^3 t)} dq. \qquad (1.81)$$

Introducing dimensionless variables $X = x/\lambda_0$ and $T = c_0 t/\lambda_0$, we get

$$u(x,t) \simeq F(0) \int_{-q_{\max}}^{q_{\max}} e^{i\delta(X-T)}\, e^{i\delta^3 T}\, dq. \qquad (1.82)$$

In order to derive the KdV equation, we change to a frame moving at speed c_0 by defining $\xi = X - T$ and $\tau = T$. As we already noticed in the case of the electrical line, Equation (1.82) leads to a time variation of order δ^3 if the space variation is of order δ. This is what leads to the time variation of the order $\varepsilon^{3/2}$ once we assume a spatial variation of the order $\varepsilon^{1/2}$. The determination of δ must be done so that dispersion and nonlinearity balance each other ($\delta = \varepsilon^{1/2}$ in the cases that we investigated earlier).

Thus it appears that a weak nonlinearity and a rather general form of the dispersion relation are enough to predict that, in some range of excitation, a given physical system may show a behaviour approximately described by the KdV equation.

2

Topological solitons: the sine-Gordon equation

The KdV equation introduced a first example of solitons, but all nonlinear localised structures cannot be described by this equation or similar ones. This chapter introduces a second class of models, which are particularly useful in solid state physics, and lead to a new category of solitons, which have an exceptional stability coming from the topology of the potential-energy surface of the system.

2.1 A simple mechanical example: the chain of coupled pendula

Let us consider the chain of coupled pendula drawn in Figure 2.1. The pendula are moving around a common axis, and two neighbouring pendula are linked by a torsional spring. We denote by θ_n the rotation of pendulum n with respect to its equilibrium position. The Hamiltonian of this system is the sum of three terms:

$$H = \sum_n \frac{I}{2}\left(\frac{\mathrm{d}\theta_n}{\mathrm{d}t}\right)^2 + \frac{C}{2}(\theta_n - \theta_{n-1})^2 + mg\ell(1 - \cos\theta_n). \tag{2.1}$$

The first term is the kinetic energy associated with the rotation of the pendula, where I is the moment of inertia of one pendulum with respect to the axis. The second term describes the coupling energy between neighbouring pendula, due to the torsional spring having a torsion constant C, while the last contribution to the Hamiltonian comes from the gravitational potential energy of the pendula, ℓ being the distance of their centres of mass to the axis, m the mass of a pendulum, and g the acceleration of gravity.

Introducing the momentum $p_n = I\dot{\theta}_n$, which is the canonical conjugate of the variable θ_n, the equations of motion of the pendulum chain can be derived from this Hamiltonian, with the Hamilton equations

$$\frac{\mathrm{d}\theta_n}{\mathrm{d}t} = \frac{\partial H}{\partial p_n} \quad \text{and} \quad \frac{\mathrm{d}p_n}{\mathrm{d}t} = -\frac{\partial H}{\partial \theta_n}. \tag{2.2}$$

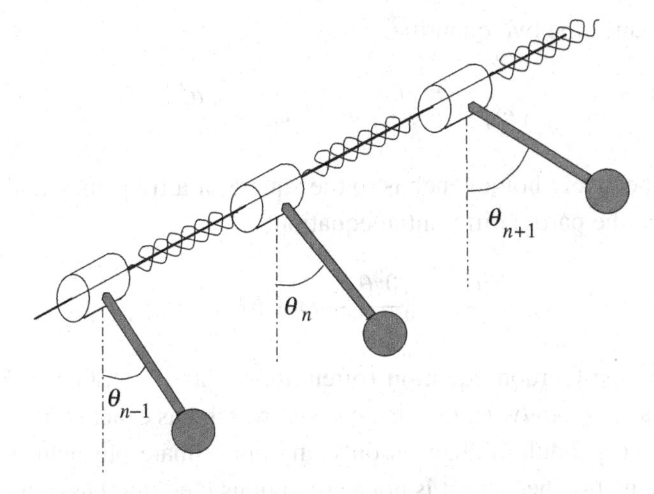

Figure 2.1. A chain of pendula sharing a common axis, coupled by torsional springs. In the continuum limit, the equations of motions of this device lead to the sine-Gordon equation.

They lead to the nonlinear coupled differential equations

$$I\frac{\mathrm{d}^2\theta_n}{\mathrm{d}t^2} - C(\theta_{n+1} + \theta_{n-1} - 2\theta_n) + mg\ell \sin\theta_n = 0. \tag{2.3}$$

Their exact solution is not known but an approximate solution can be derived from the continuum limit approximation, provided the coupling between adjacent pendula is strong enough to ensure that θ varies only slightly from one pendulum to the next.

The method is analogous to the one that we used for the electrical line in Chapter 1. Let us denote by a the distance between two pendula along the axis. We replace the discrete variables $\theta_n(t)$ by the function $\theta(x, t)$ where $\theta_n = \theta(x = na, t)$. The Taylor expansion of $\theta_{n\pm1}$ leads to

$$\theta_{n+1} + \theta_{n-1} - 2\theta_n \simeq a^2\frac{\partial^2\theta}{\partial x^2} + \mathcal{O}\left(a^4\frac{\partial^4\theta}{\partial x^4}\right), \tag{2.4}$$

if we take into account the fast decay of its successive terms which contain derivatives of increasing order of a function which is assumed to vary slowly with space. Contrary to what we did when we established the KdV equation, here we truncate the expansion to its lowest nonvanishing term. We shall show in Section 2.2.2 why this approximation is sufficient to correctly describe the physics of the pendulum chain.

Let us introduce the two quantities

$$\omega_0^2 = \frac{mg\ell}{I} \quad \text{and} \quad c_0^2 = \frac{Ca^2}{I}, \tag{2.5}$$

which are respectively homogeneous to the square of a frequency and of a speed. We obtain then the partial differential equation

$$\frac{\partial^2\theta}{\partial t^2} - c_0^2\frac{\partial^2\theta}{\partial x^2} + \omega_0^2 \sin\theta = 0, \tag{2.6}$$

known as the sine-Gordon equation (often abbreviated as 'SG'). Like the KdV equation it is a *completely integrable equation* which has exact soliton solutions.

Of course the pendulum chain is only an approximate physical model of the sine-Gordon equation because it is not a continuous medium. Discreteness effects, which arise because the chain is made of individual pendula, are not negligible. More importantly, the physical device is dissipative due to friction on the rotational axis of the pendula. However the chain is a nice device to experiment with the remarkable properties of solitons. The study of the sine-Gordon equation will show that such a simple mechanical experiment can illustrate phenomena which are typical of relativity and quantum mechanics!

2.2 Solutions of the sine-Gordon equation

2.2.1 Topology of the energy landscape

In order to analyse all the solutions of the sine-Gordon equation, it is useful to examine Figure 2.2, which shows the potential energy of the pendula as a function of θ and the spatial coordinate x along the chain. Whatever its position x, a pendulum is subjected to the same potential $V(\theta) = mg\ell(1 - \cos\theta)$ so that the energy landscape undulates regularly.

In order to completely figure out the potential energy of the system, one must also consider the harmonic coupling energy due to the torsional springs connecting pendula. In the continuum limit the pendulum chain can be viewed as an elastic string (because if θ changes in one point, this tends to induce a similar variation in the neighbouring points), which is massive and subjected to the undulations of the potential $V(\theta)$. This point of view immediately points out a feature that distinguishes the sine-Gordon model from the KdV case studied previously.

One notices that the system has *several energetically degenerate ground states*. Indeed the ground state can be achieved with $\theta = 0$ or $\theta = 2p\pi$ (p being any integer). This was not the case for the KdV model because the water in a canal only

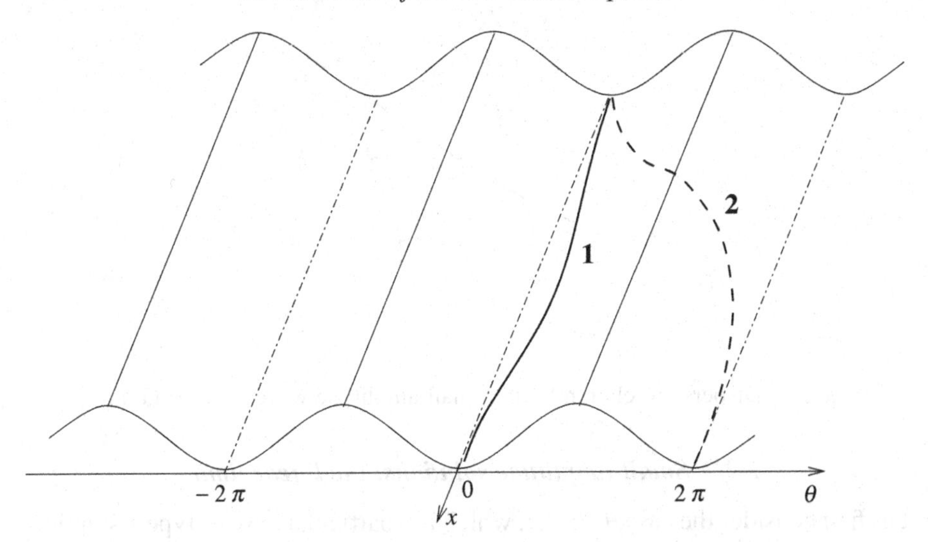

Figure 2.2. Topology of the potential energy landscape of the sine-Gordon model. Solid and dashed lines labelled 1 and 2 show the position of an imaginary massive elastic string, which would have the same motion as the pendulum chain in the continuum limit approximation.

has one possible equilibrium level. This feature of the sine-Gordon model suggests the existence of *several families of solutions:*

- Solutions in which the whole chain stays within a single potential valley (Case 1 of Figure 2.2).
- Solutions in which the chain moves from one valley to another one (Case 2 of Figure 2.2, which corresponds to a soliton solution).

More quantitatively, solutions can be distinguished by their behaviour towards the boundaries $\pm\infty$:

$$\lim_{x\to+\infty} \theta \;-\; \lim_{x\to-\infty} \theta = 0 \qquad \text{in Case 1} \tag{2.7}$$

$$\lim_{x\to+\infty} \theta \;-\; \lim_{x\to-\infty} \theta = 2p\pi \quad (p \neq 0) \ \text{in Case 2.} \tag{2.8}$$

These two solutions are said to be *topologically different* because their difference is a property of the solution as a whole. Indeed if one looks at the solutions for $|x| \to \infty$, a local view does not make any difference between the two: one sees pendula at rest in their minimal energy state. Is is only by moving along the whole pendulum chain that one can notice that there is a full turn from one end to the other in case 2.

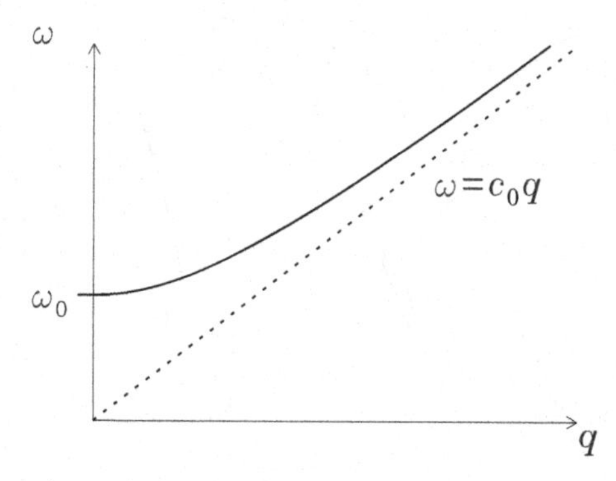

Figure 2.3. Dispersion relation for the small-amplitude waves in the SG model.

2.2.2 Small-amplitude solutions: the linear limit

Let us first consider the case $\theta \ll 2\pi$, which is a particular case of type 1 solutions. Taking the linear limit of the sinusoidal term, the SG equation (2.6) reduces to

$$\frac{\partial^2\theta}{\partial t^2} - c_0^2\frac{\partial^2\theta}{\partial x^2} + \omega_0^2\theta = 0, \tag{2.9}$$

which has plane wave solutions[1]

$$\theta = \theta_0\, e^{i(qx-\omega t)} + \text{c.c.} \tag{2.10}$$

in which the frequency ω and the one-dimensional wavevector q are related by the dispersion relation $\omega^2 = \omega_0^2 + c_0^2 q^2$, plotted in Figure 2.3. In the limit of a large wavevector q, the phase velocity of the linear waves tends towards c_0. However ω is not proportional to q, which means that the sine-Gordon equation describes *dispersive waves*. This property is due to the term $\omega_0 \sin\theta$ of the equation. As well as the KdV equation, the SG equation includes both *dispersion and nonlinearity* but, in the SG case, these two features come from the same term of the equation.

When we made the continuum approximation, it is this peculiarity which allowed us to restrict the expansion of the finite difference $(\theta_{n+1} + \theta_{n-1} - 2\theta_n)$ to the second derivative $\partial^2\theta/\partial x^2$, instead of carrying the expansion up to the fourth derivative as we did for the electrical chain in Chapter 1. For the SG equation, even if we restrict the expansion to the second order, we nevertheless keep some dispersion, so that we preserve this essential physical property of the system.

The dispersion relation of the small-amplitude waves also shows that frequencies $\omega < \omega_0$ lead to an imaginary wavevector q, i.e. to waves which are exponentially damped. This is an important property of the SG model: it exhibits a forbidden

[1] Throughout the book the notation c.c. designates the complex conjugate of the preceding expression.

region (a gap) in the spectrum of its small-amplitude excitations. This property is easy to check for a given physical system. It immediately indicates if the system may have any chance of being described by an equation of the SG category.

2.2.3 Soliton solutions

In order to derive the solutions of the SG equation, one must notice that the equation is preserved by a Lorentz transform relative to speed c_0. Therefore it is sufficient to look for static solutions, from which solutions moving at velocity v can be derived with a Lorentz transform. However, as for KdV, soliton solutions can also be obtained by looking for permanent profile solutions moving at velocity v, i.e. solutions which only depend on the single variable $z = x - vt$. Their invariance by a Lorentz transform appears naturally in the calculation.

For such permanent profile solutions, the SG equation becomes

$$v^2 \frac{d^2\theta}{dz^2} - c_0^2 \frac{d^2\theta}{dz^2} + \omega_0^2 \sin\theta = 0, \tag{2.11}$$

or

$$\frac{d^2\theta}{dz^2} = \frac{\omega_0^2}{c_0^2 - v^2} \sin\theta. \tag{2.12}$$

Multiplying by $d\theta/dz$ and integrating with respect to z we get

$$\frac{1}{2}\left(\frac{d\theta}{dz}\right)^2 = -\frac{\omega_0^2}{c_0^2 - v^2} \cos\theta + C_1. \tag{2.13}$$

The integration constant C_1 is determined by the boundary conditions that we impose on the solution. Since we are looking for a soliton, i.e. a spatially localised solution, we must have $\theta(z) \to 0 \pmod{2\pi}$ for $|z| \to \infty$ because at infinity the pendula must be in one of their ground states. For the same reason we impose $d\theta/dz \to 0$ if $|z| \to \infty$, which leads to $C_1 = \omega_0^2/(c_0^2 - v^2)$, and therefore

$$\frac{1}{2}\left(\frac{d\theta}{dz}\right)^2 - \frac{\omega_0^2}{c_0^2 - v^2}(1 - \cos\theta) = 0. \tag{2.14}$$

As we did to derive the solutions of the KdV equation, we can proceed as if this expression were the sum of the kinetic energy (with respect to the 'pseudo-time' z) and the potential energy of a fictitious particle. Thus the solution $\theta(z)$ describes the motion of this particle, having zero total energy, in the potential

$$V_{\text{eff}}(\theta) = -\frac{\omega_0^2}{c_0^2 - v^2}(1 - \cos\theta). \tag{2.15}$$

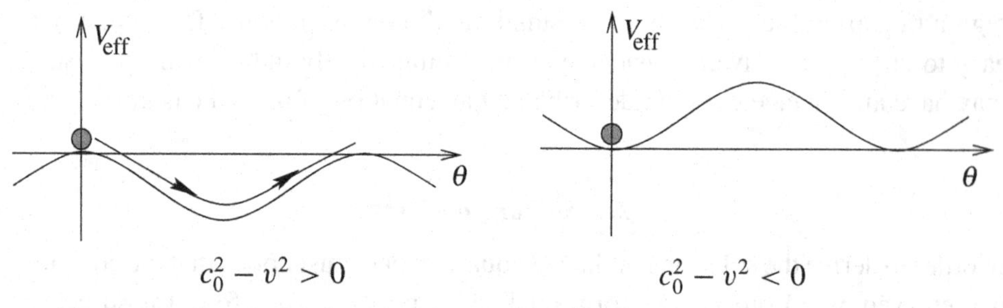

Figure 2.4. Search for the possible solutions of the SG equation by studying a fictitious particle mobile in the pseudo-potential $V_{\rm eff}(\theta)$.

Figure 2.4 shows that, for $c_0^2 - v^2 > 0$, there is a possible motion for a fictitious particle leaving $\theta = 0$ at rest. It can reach $\theta = 2\pi$ (or $\theta = -2\pi$) with a vanishing 'velocity' $\mathrm{d}\theta/\mathrm{d}z$ after an infinite fictitious 'time' z. Conversely, for $c_0^2 - v^2 < 0$, a particle initially at rest in $\theta = 0$ cannot move. This analysis shows that solitons can only travel at speeds *smaller* than c_0. Moreover it indicates that there are no permanent profile solutions which start and end in the same potential valley.[2]

For $v^2 < c_0^2$, the soliton solution can be obtained from Equation (2.14) which gives

$$\frac{\sqrt{2}\omega_0}{\sqrt{c_0^2 - v^2}}\,\mathrm{d}z = \pm\frac{\mathrm{d}\theta}{\sqrt{1-\cos\theta}}, \tag{2.16}$$

i.e.

$$\frac{\sqrt{2}\omega_0}{\sqrt{c_0^2 - v^2}}(z - z_0) = \pm\int\frac{\mathrm{d}\theta}{\sqrt{1-\cos\theta}} = \pm\int\frac{\mathrm{d}\theta}{\sqrt{2}\sin(\theta/2)} \quad (0 < \theta < 2\pi) \tag{2.17}$$

where z_0 is an integration constant. Using

$$\int\frac{\mathrm{d}\theta}{\sqrt{2}\sin(\theta/2)} = \frac{1}{\sqrt{2}}\int\frac{4\mathrm{d}t}{1+t^2}\,\frac{1+t^2}{2t} = \sqrt{2}\int\frac{\mathrm{d}t}{t} = \sqrt{2}\ln t, \tag{2.18}$$

with $t = \tan(\theta/4)$, leads to the solution

$$\frac{\omega_0}{\sqrt{c_0^2 - v^2}}(z - z_0) = \pm\ln\tan\frac{\theta}{4}, \tag{2.19}$$

[2] The analysis is performed by assuming that the equilibrium position of the pendula is $\theta = 0$, mod(2π). If one parametrically excites a pendulum by an oscillation of its suspension point, it is possible to find conditions [27, 28] for which $\theta = \pi$ is a stable position. One can then check that it is possible to generate solutions analogous to solitons on a chain of inverted pendula.

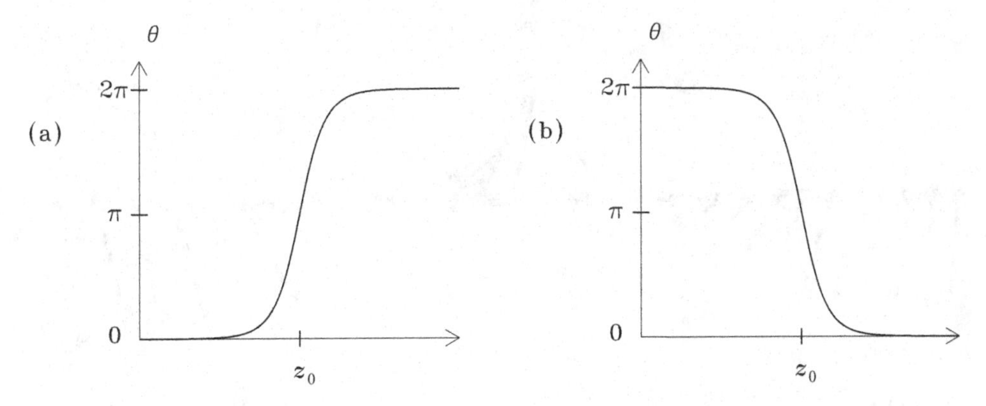

Figure 2.5. Soliton (a) and antisoliton (b) solutions of the sine-Gordon equation.

or

$$\theta = 4 \arctan \exp \left[\pm \frac{\omega_0}{c_0} \frac{z - z_0}{\sqrt{1 - v^2/c_0^2}} \right] \quad \text{with } z = x - vt. \tag{2.20}$$

The arbitrary integration constant z_0 determines the position of the soliton at time $t = 0$.

The solution exhibits the characteristic expression associated with the Lorentz invariance, as well as the validity condition $v^2 < c_0^2$ that we had found by examining the motion of a fictitious particle, and which also results from the Lorentz invariance.

The solutions 'soliton' (with a + in Expression (2.20)) and 'antisoliton' (− sign in Expression (2.20)) are plotted in Figure 2.5. As they are associated with the fast variation of some quantity, they are often called a 'kink' or an 'antikink' respectively.

Figure 2.6 shows what the pendulum chain looks like when it carries a soliton: it exhibits a local 2π torsion. Thus, as drawn in Figure 2.2, *the soliton interpolates between two different states of the system which have the same energy.*

Since the derivation of the soliton solution has shown that there are no permanent-profile solutions having their limits at $\pm\infty$ in the same potential valley, let us try to figure out this result intuitively from the potential energy landscape of the SG model. Let us suppose that we try to create such a solution. The elastic string representing the pendulum chain would climb up the side of the valley and, without going above the top, it would go down, back to the bottom. If we leave it like that, it is easy to realise that the massive string cannot stay in this position. It will fall down to the bottom, perhaps with some oscillations around the minimum. In Section 2.2.6 we shall show that such a localised oscillatory solution does indeed exist: it is the 'breather' solution.

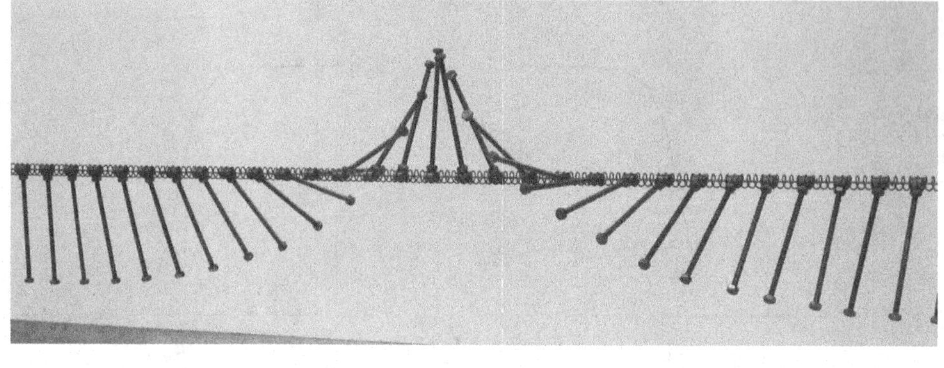

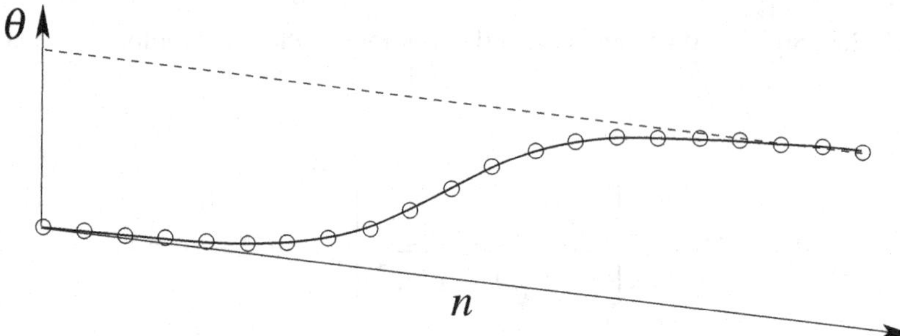

Figure 2.6. Picture of the chain of coupled pendula when it carries a soliton, and variation of θ along the chain (Picture: Thierry Dauxois & Bruno Issenman, 2003).

Solitons and antisolitons differ by their *topological charge* defined by

$$Q = \frac{1}{2\pi}\int_{-\infty}^{+\infty} \frac{\partial\theta}{\partial x}\,\mathrm{d}x = \frac{1}{2\pi}\left[\lim_{x\to+\infty}\theta(x,t) - \lim_{x\to-\infty}\theta(x,t)\right] \tag{2.21}$$

which is equal to $+1$ for a soliton and -1 for an antisoliton. This topological charge is an invariant of the system, even for the multisoliton solutions that we shall describe below, although they do not have a permanent profile. It is easy to get a qualitative understanding of this property by looking at the potential energy landscape of the sine-Gordon model shown in Figure 2.2. For the infinite chain, in order to cancel the topological charge, it would be necessary to transfer an infinite length of the chain above the potential barrier.[3] The conservation of the topological charge explains the exceptional stability of topological solitons. They are much more stable than the nontopological solitons of the KdV equation. In an infinite medium, perturbations can modify the speed of a soliton, or even bring it

[3] This picture also explains why, in a finite system, the topological charge is no longer conserved. Events occurring at the ends of the chain can change its value.

Figure 2.7. A *frozen* kink, which was naturally created by snow fallen on a horizontal bar (Picture: Thierry Cretegny, 2001).

to rest, but they cannot kill it because it would imply a change of the topological charge.

The existence of two types of solution (the soliton and the antisoliton) also distinguish the SG model from the KdV case that we studied in the previous chapter. For KdV we showed that positive pulses can be solitons, while negative pulses spread out.

As the continuum limit approximation was used to derive the sine-Gordon equation for the pendulum chain, the solutions of the SG equation are not exact solutions for the pendulum chain. Their validity for the discrete chain must be examined. Let us rewrite the solution (2.20) as

$$\theta(x,t) = 4 \arctan \exp\left(\pm \frac{x - vt}{L}\right) \quad \text{with} \quad L = \frac{c_0}{\omega_0}\sqrt{1 - \frac{v^2}{c_0^2}}. \tag{2.22}$$

In this expression L measures the spatial extent of the solution, and two properties should be noticed:

(i) The Lorentz contraction of the soliton, which is indicated by the square root in the expression of L. The width of the soliton tends to zero when its speed tends to the speed of sound c_0. This Lorentz contraction is clearly visible in the experiments performed with the pendulum chain by comparing the width of a fast soliton with the width of a static one. Therefore the pendulum chain can be used to perform an 'experiment in relativity' as we pointed out earlier.

(ii) The role of the relative weight of the coupling term c_0 with respect to the amplitude of the on-site potential measured by ω_0, which appears in the width of a soliton at rest $L_0 = c_0/\omega_0 = a\sqrt{C/(mg\ell)}$.

The continuum limit approximation is only valid if $L_0/a \gg 1$, and therefore it requires $C \gg mg\ell$. This implies that the torsional energy in the springs linking the pendula must be large with respect to their potential energy. A large value of C tends to force the angles of two neighbouring pendula, $\theta_n(t)$ and $\theta_{n+1}(t)$, to stay close to each other. The continuum limit approximation is therefore also a *strong coupling limit.*

2.2.4 Energy of the soliton

This can be calculated from the Hamiltonian (2.1). In the continuum limit approximation the contribution of a single cell, divided by the cell spacing, gives the Hamiltonian per unit length, i.e. the Hamiltonian density $\mathcal{H}$,

$$\mathcal{H}(x,t) = \frac{1}{a}\left[\frac{I}{2}\left(\frac{\partial\theta}{\partial t}\right)^2 + \frac{Ca^2}{2}\left(\frac{\partial\theta}{\partial x}\right)^2 + mg\ell(1-\cos\theta)\right], \tag{2.23}$$

if we restrict ourselves to the lowest order term in the Taylor expansion of $(\theta_{n+1}-\theta_n)^2$, as we did to derive the SG equation.

If we are only interested in the expression of $\mathcal{H}$ for a soliton, it is again convenient to introduce the variable $z = x - vt$ such that $\theta(x,t) = \theta(z)$. We get

$$\mathcal{H}(z,t) = \frac{1}{a}\left[\frac{Iv^2}{2}\left(\frac{\mathrm{d}\theta}{\mathrm{d}z}\right)^2 + \frac{Ca^2}{2}\left(\frac{\mathrm{d}\theta}{\mathrm{d}z}\right)^2 + mg\ell(1-\cos\theta)\right]. \tag{2.24}$$

Using the expression of c_0^2 and Equation (2.14), all the terms of $\mathcal{H}$ can be written as a function of $\mathrm{d}\theta/\mathrm{d}z$ which is then replaced by its value obtained during the derivation of the solution (2.20). It leads to

$$\mathcal{H} = \frac{Ic_0^2}{a}\left(\frac{\mathrm{d}\theta}{\mathrm{d}z}\right)^2 = \frac{4I\omega_0^2}{a\left(1-v^2/c_0^2\right)}\,\mathrm{sech}^2\,\frac{\omega_0(z-z_0)}{\sqrt{c_0^2-v^2}}. \tag{2.25}$$

This expression containing sech^2 does indeed describe an energy density *localised* around the centre of the soliton. This is why the soliton is called a 'quasi-particle'.

As $\int \mathrm{d}x \, \text{sech}^2 x = 2$, an integration over space gives the energy of the soliton

$$E = \int_{-\infty}^{+\infty} \mathcal{H}(x,t)\,\mathrm{d}x = \frac{8I\,\omega_0\,c_0}{a\sqrt{1 - v^2/c_0^2}} \tag{2.26}$$

which has a standard 'relativistic' expression, with respect to the speed c_0, for a particle of mass $m_0 = 8I\omega_0/c_0 a$. Therefore the SG soliton resembles a quasi-particle even more than the KdV soliton.

It is interesting to notice that the energy of the soliton can be calculated even if we don't know the analytical expression of the solution. Let us restart from the Hamiltonian density (2.25). The energy

$$E = \frac{Ic_0^2}{a} \int_{-\infty}^{+\infty} \left(\frac{\mathrm{d}\theta}{\mathrm{d}z}\right)\left(\frac{\mathrm{d}\theta}{\mathrm{d}z}\right) \mathrm{d}z, \tag{2.27}$$

can be computed by replacing one of the two factors $\mathrm{d}\theta/\mathrm{d}z$ by its analytical expression deduced from Equation (2.14). We get

$$E = \frac{Ic_0^2}{a} \frac{\sqrt{2}\omega_0}{\sqrt{c_0^2 - v^2}} \int_{-\infty}^{+\infty} \sqrt{1 - \cos\theta} \left(\frac{\mathrm{d}\theta}{\mathrm{d}z}\right) \mathrm{d}z \tag{2.28}$$

$$= \frac{Ic_0^2}{a} \frac{\sqrt{2}\omega_0}{\sqrt{c_0^2 - v^2}} \int_{\theta=0}^{\theta=2\pi} \sqrt{1 - \cos\theta}\,\mathrm{d}\theta, \tag{2.29}$$

where the integration over θ is easy to carry out. This method is also valid for other equations of the SG family which cannot be solved exactly, but which are expected to have quasi-soliton solutions owing to the shape of their potential energy landscape. Integrating the square root of the potential $V(\theta)$ over θ, it is possible to compute the energy of a soliton even though we may be unable to obtain its analytical solution.

2.2.5 Multisoliton solutions

For completely integrable equations such as the SG equation, there is a systematic method to derive multisoliton solutions. This method, called the 'inverse scattering method' will be introduced in Chapter 7. It can be used to obtain solutions with an arbitrary number of solitons and antisolitons.

At this stage, it is interesting to study one of these solutions because it will teach us something about the properties of interacting solitons. Let us consider the

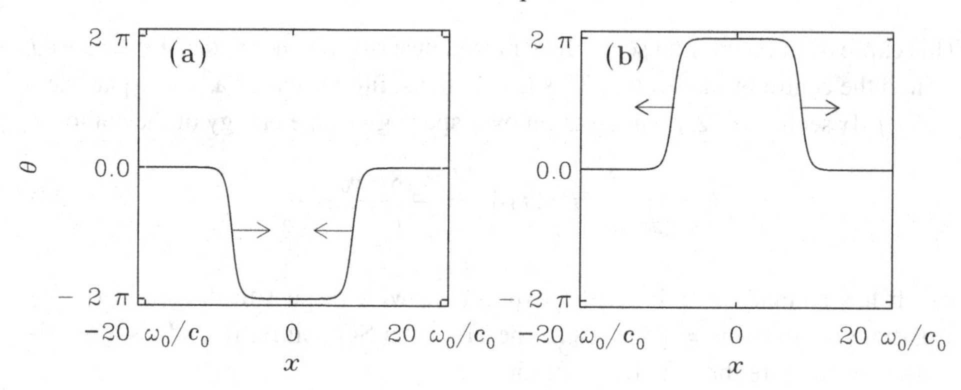

Figure 2.8. Plot of the multisoliton solution (2.30) well before the collision (a) and well after (b). One should notice that, in an infinite medium the value of the solution for $x \to \pm\infty$ is always preserved, as we discussed for the conservation of the topological charge. Due to these boundary conditions, when the two solitons pass through each other, the value of θ between them changes by 4π. Indeed the soliton and the antisoliton, which interpolate between two different states of the system, lead to a $\pm 2\pi$ transition in the value of θ. Depending whether one first meets the upward transition or the downward transition, the shape of the solution drastically changes.

particular solution of the SG equation:

$$\theta_{SA} = 4 \arctan \left[\frac{\sinh \dfrac{v\omega_0 t}{c_0} \dfrac{1}{\sqrt{1 - v^2/c_0^2}}}{\dfrac{v}{c_0} \cosh \dfrac{\omega_0 x}{c_0} \dfrac{1}{\sqrt{1 - v^2/c_0^2}}} \right]. \tag{2.30}$$

If we denote by θ_S and θ_A the soliton and antisoliton solutions that we derived previously, and using the equality $\arctan(1/z) = \pi/2 - \arctan(z)$, we get

$$\lim_{t \to +\infty} \theta_{SA} \simeq \theta_S(x + vt + \Delta) + \theta_A(x - vt - \Delta) \tag{2.31}$$

$$\lim_{t \to -\infty} \theta_{SA} \simeq \theta_S(x + vt - \Delta) + \theta_A(x - vt + \Delta). \tag{2.32}$$

where we introduce the phase shift $\Delta = \dfrac{c_0}{\omega_0} \sqrt{1 - \dfrac{v^2}{c_0^2}} \ln \left(\dfrac{v}{c_0} \right)$.

This asymptotic form shows that the solution describes the collision of a soliton and an antisoliton, which pass through each other without killing each other, as shown by the plot of the solution θ_{SA} at different instants in Figure 2.8. During the collision there is an instant when the field θ vanishes everywhere, but its derivative θ_t does not vanish. The two excitations emerge intact from the collision, but they are

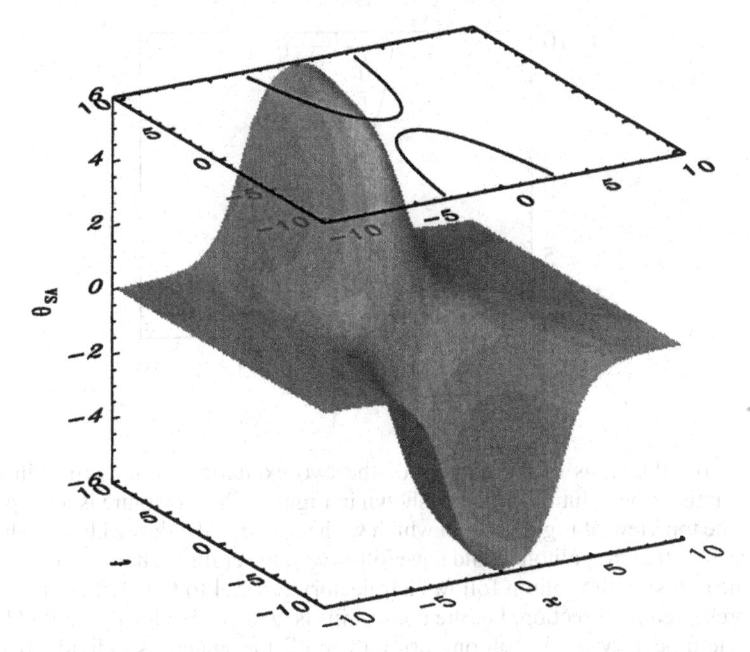

Figure 2.9. Time evolution of the multisoliton solution (2.30) (with $\omega_0 = 1$, $c_0 = 1$, $v = 0.1$). The projection on the top plane shows the time evolution of the contour $\theta = \pi$, which illustrates the motions of the centres of the two excitations. However, at the collision point, the individuality of each excitation loses its meaning, and it is no longer possible to follow their position. Their trajectories seem to be interrupted while actually the soliton and the antisoliton pass through each other, and then go on in the direction that they had initially. But their trajectories after the collision are shifted with respect to the trajectories that they would have without it.

shifted with respect to the positions that they would have occupied if the collision had not occurred, as shown in Figures 2.9 and 2.10. The collision has induced a phase shift, as did the collision of two KdV solitons shown in Figure 1.10.

The observation of the trajectories of the two excitations (as long as they can be distinguished from each other) shows that they speed up when they move toward each other and slow down when they move apart after passing through each other. This shows that the *soliton–antisoliton* interaction is attractive. A similar study with two solitons, or two antisolitons, would, on the contrary show that they repel each other. The existence of these attractive or repulsive interactions explains why there are *no permanent profile solutions with multiple solitons or antisolitons:* the interactions between the excitations that make up the multisoliton solution tend to cause a variation in the distances between its components. For instance two static solitons created on the pendulum chain tend to move apart, under the influence of their mutual repulsion. This is why, when we looked for permanent profile solutions, we only found single-soliton (or antisoliton) solutions.

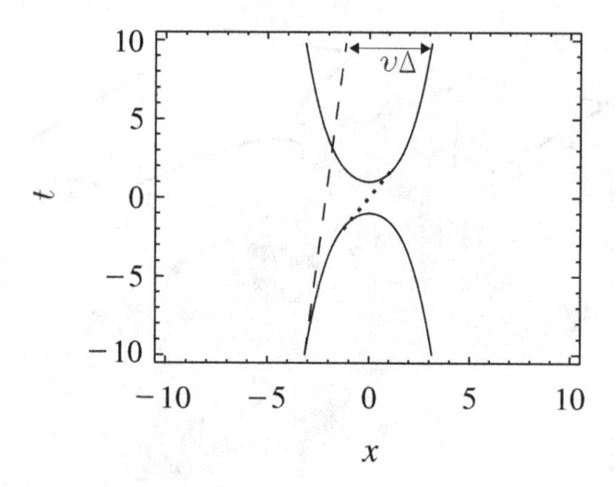

Figure 2.10. Positions of the centres of the two excitations versus time in the soliton–antisoliton solution collision shown in Figure 2.9. This figure is an expansion of the top view of Figure 2.9, in which we have added the dashed line to show the trajectory that the soliton would have followed if the collision had not occurred. After the collision the soliton follows a trajectory parallel to this dashed line, i.e. it has preserved its direction, but the phase shift is clearly visible. The dotted line shows the trajectory of the soliton during the collision, when its individuality is lost due to its interaction with the antisoliton.

2.2.6 The breather solution

The soliton–antisoliton solution has exhibited an attraction between them. It suggests that the motion of the soliton in the 'field' of the antisoliton can be viewed as the motion of a particle subjected to an effective potential (see Figure 2.11). As the total energy of the system is conserved, when the soliton approaches the bottom of the interaction potential, the decrease of its interaction energy must be balanced by an increase of its kinetic energy. The particle, coming from $-\infty$ speeds up when it approaches the position of the antisoliton, then slows down when it moves out of the attractive well, and finally recovers its initial velocity at $+\infty$. This situation, which corresponds to what we observed during the collision (Figure 2.9), corresponds to the schematic trajectory labelled (1) in Figure 2.11.

It is easy to understand that this situation is 'fragile'. Even a small dissipation can prevent the particle from moving out of the attractive well and thus prevent it from escaping to $+\infty$. This leads to the annihilation of the soliton–antisoliton pair (which is possible because its topological charge is zero). This is what is observed in the experiment performed with the pendulum chain, which is not frictionless.

Figure 2.11 also suggests another possibility. The initial energy of the soliton could be such that the soliton is trapped *inside* the attractive potential well of the antisoliton. This would lead to trajectory (2) in Figure 2.11, i.e. to an oscillatory

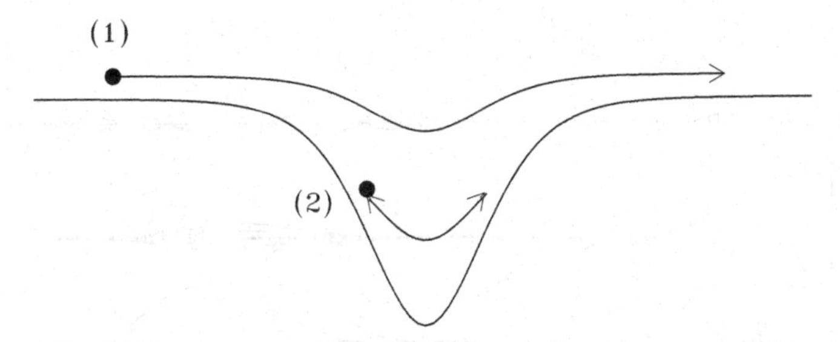

Figure 2.11. Schematic plot of a soliton–antisoliton crossing (trajectory labelled 1) and of a breather (trajectory labelled 2). The curve shows the effective potential in which the soliton moves due to its interaction with the antisoliton. The arrows suggest the motion of the soliton in a completely schematic way.

solution, which is nevertheless spatially localised because it is made of a bound soliton–antisoliton pair.

This solution does exist for the sine-Gordon model. It is the *breather*

$$\theta_B = 4 \arctan \left[\frac{\sin \dfrac{u\omega_0 t}{c_0} \dfrac{1}{\sqrt{1+u^2/c_0^2}}}{\dfrac{u}{c_0} \cosh \dfrac{\omega_0 x}{c_0} \dfrac{1}{\sqrt{1+u^2/c_0^2}}} \right], \tag{2.33}$$

where u is a parameter which does not have the meaning of a velocity. Although it is time dependent, Solution (2.33) is at rest in the sense that it corresponds to an oscillation localised around position $x = 0$, which does not propagate. It can be easily derived from the soliton–antisoliton solution (2.30) by putting $v = iu$ into Equation (2.30). It is interesting to notice that such a transformation in a soliton–soliton solution would not lead to a real solution. A bound soliton–soliton state cannot exist because their interaction is repulsive.

From Expression (2.33), one can derive the solution for a moving breather by performing the Lorentz transform

$$x \to \frac{x - vt}{\sqrt{1 - v^2/c_0^2}} \tag{2.34}$$

$$t \to \frac{t - vx/c_0^2}{\sqrt{1 - v^2/c_0^2}}. \tag{2.35}$$

This new solution will move while it oscillates.

The breather frequency Ω,

$$\Omega = \frac{\omega_0 u/c_0}{\sqrt{1 + u^2/c_0^2}}, \tag{2.36}$$

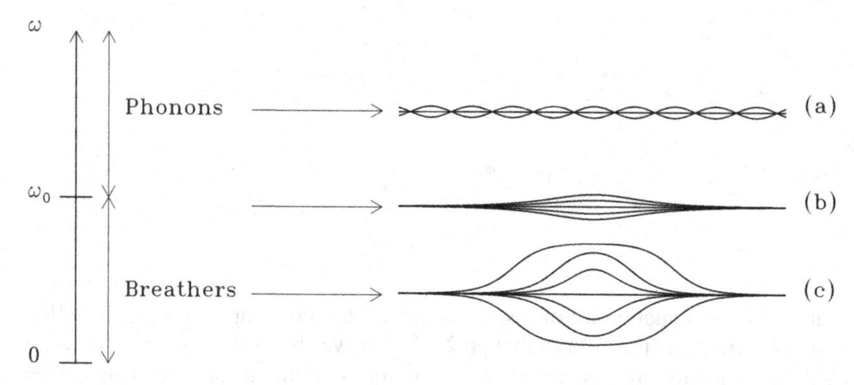

Figure 2.12. Spectrum of the periodic solutions of the SG model, and examples of such oscillatory solutions: (a) extended phonon mode, (b) breather mode with a frequency slightly below the phonon band, (c) very low frequency breather.

lies in the range $[0, \omega_0]$, i.e. it is situated below the spectrum of the phonon modes as shown in Figure 2.12. Small-amplitude breathers, which are very wide in space, have frequencies slightly below the phonon band. Conversely the frequencies of high-amplitude breathers are close to zero. A plot of such a low-frequency breather is clearly reminiscent of a soliton and an antisoliton oscillating with respect to each other.

Mobile breathers are not permanent profile solutions since they have an internal oscillation. However they are solitons because:

- They are spatially localised. A breather oscillates within an envelope, defined by

$$E_{\text{env}}(x) = 4 \arctan \left[\frac{c_0}{u} \operatorname{sech} \left(\frac{\omega_0 x}{c_0} \frac{1}{\sqrt{1 + u^2/c_0^2}} \right) \right], \tag{2.37}$$

  so that the inequality $|\theta_B(x, t)| \leq E_{\text{env}}(x)$ is valid at any time.
- They are preserved in collisions with other breathers or solitons.

Our early definition of a soliton as being 'a localised excitation which moves by preserving its shape' must therefore be extended to include breathers which have an internal oscillation. In Chapter 3 we shall meet another example of a class of solitons which have an internal time dependence when we study the nonlinear Schrödinger equation.

The solutions *phonons* (plane waves), *solitons* (as well as antisolitons) and *breathers* make up all the solutions of the sine-Gordon model. Any state of the system can be expressed in terms of these solutions. The inverse scattering method introduced in Chapter 7 shows that these solutions play the role of nonlinear 'normal modes' of the system since any state appears as a combination of these 'modes'

$$\theta(x,t) = \widetilde{\sum}\text{phonons} + \widetilde{\sum}\text{solitons} + \widetilde{\sum}\text{breathers}, \tag{2.38}$$

where we symbolically denoted by $\widetilde{\sum}$ a 'nonlinear sum' which will be defined more precisely in Chapter 7.

The origin of the name of the sine-Gordon equation

In 1926, after he established the equation for the wave function of a free particle of mass m,

$$i\hbar\frac{\partial\psi}{\partial t} = -\frac{\hbar^2}{2m}\vec{\nabla}^2\psi, \tag{2.39}$$

which now bears his name, ERWIN SCHRÖDINGER (1887–1961) looked for an equation compatible with special relativity in order to describe for instance the dynamics of relativistic electrons.

Considering the expression $\psi = \psi_0 e^{i(\omega t - \vec{p}\cdot\vec{r})}$ of a wave, it is natural to associate the momentum $\vec{p}$ with the operator $i\hbar\vec{\nabla}$, and energy E with the operator $i\hbar\partial/\partial t$. With these transforms, the Schrödinger equation (2.39) leads to the equation $E = \vec{p}^2/(2m)$ of classical dynamics. From its relativistic equivalent

$$E^2 = \vec{p}^2c^2 + m^2c^4, \tag{2.40}$$

the same transforms lead to

$$-\hbar^2\frac{\partial^2\psi}{\partial t^2} = -\hbar^2c^2\vec{\nabla}^2\psi + m^2c^4\psi \tag{2.41}$$

which can be rewritten as

$$\frac{\partial^2\psi}{\partial t^2} - c^2\vec{\nabla}^2\psi + \omega_0^2\psi = 0. \tag{2.42}$$

This equation now bears the name of the Klein–Gordon equation because it was published for the first time by OSKAR KLEIN (1894–1977) and WALTER GORDON (1893–1939). It seems that it had been derived earlier by E. Schrödinger himself, but he dropped it because it did not describe the experimental properties of an electron with a spin. More generally, all equations of the same type, with a final term which can be nonlinear, are customarily called Klein–Gordon equations.

The sine-Gordon equation (2.6) is obtained by replacing the last term of Equation (2.42) by $\omega_0^2 \sin\psi$. The similarity of the sounds 'Klein' and 'sine' explains the name of this equation, which was not discovered by any 'Mr Sine'.

2.3 Long Josephson junctions

The case of long Josephson junctions is an example where the sine-Gordon equation perfectly describes the physics of a device. Moreover the dynamics of solitons in a long Josephson junction can be directly observed and it has been used as the basis of several devices. In this section we describe the physical phenomena which underlie their operation [23, 118, 140].

A tunnel Josephson junction is made of two superconducting electrodes, often made of niobium or aluminium, separated by a thin layer of an electric insulator such as aluminium oxide, which is 10–50 Å thick (see Figure 2.13). A nondissipative tunnel current of quantum origin can flow through it. In the 1980s the theory of the Josephson junction (and particularly of junction networks) was strongly revived by the discovery of high temperature superconductivity in ceramics such as $YBa_2Cu_3O_7$ because these materials are made of superconducting microcrystals which are in imperfect electrical contact. Each contact behaves like a Josephson junction.

In order to understand how a Josephson junction operates, it is sufficient to consider the macroscopic theory proposed by London in 1935 and completed by Ginzburg and Landau in 1950. It is this theory that was used by Josephson at the beginning of the 1960s in his PhD thesis [95]. A theory within the framework of the standard microscopic theory of superconductivity was provided later. It is based on the approach developed in 1957 by Bardeen, Cooper and Schrieffer (BCS),[4] which does not describe high T_c superconductors for which a full theory is not available yet. The superconductivity arises due to the formation of pairs of carriers, called 'Cooper pairs', which are spin 0 bosons, made of two spin 1/2 electrons.

2.3.1 Dynamic equation of a junction

Let us consider a simple theory of the Josephson effect introduced by Feynman [67]. It is based on the fact that a superconductor is a *macroscopic quantum system* in which all the Cooper pairs belong to a single quantum state (or a single wave function) $|\varphi\rangle$, which is therefore a 'macro-state'. The state ket, which describes the Cooper pairs in the superconductor can be written as

$$|\psi\rangle = (\sqrt{\rho}\,\mathrm{e}^{\mathrm{i}\phi})|\varphi\rangle. \tag{2.43}$$

Inside the superconductors, only the x dependence is considered (see Figure 2.13), which amounts to saying that the state ket describes a state which has been integrated with respect to z on the thickness of each superconductor. Thus $\rho = \langle\psi|\psi\rangle$

[4] A short introduction to the BCS theory is given for instance by Kittel [100].

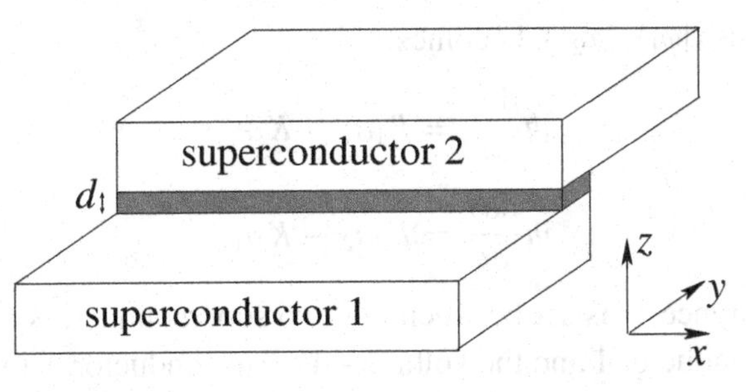

Figure 2.13. Schematic picture of a tunnel Josephson junction. The insulator is shown as the shaded part. The junction is extended along the x direction, and has negligible width in the y direction.

measures the density of Cooper pairs *per unit area*, obtained by integration of the density per unit volume with respect to z. The parameter ϕ is the phase of the state ket.

The Josephson effect

The Josephson effect appears when the coupling between the two superconductors increases because the insulating barrier between them is made thinner. When the distance between the superconductors is as low as about 10 Å, the Cooper pairs have a significant probability to tunnel through the insulator. The two states $|\psi_1\rangle$ and $|\psi_2\rangle$, describing the Cooper pairs in the two superconductors are no longer independent from each other. The system of two superconductors must be described by a new state ket $|\Psi\rangle$. Feynman's theory assumes that $|\Psi\rangle$ is fully represented by its expansion on the basis of the macro-states $|\varphi_1\rangle$ and $|\varphi_2\rangle$, characterising superconductors 1 and 2

$$|\Psi\rangle = a_1|\varphi_1\rangle + a_2|\varphi_2\rangle. \tag{2.44}$$

The junction appears therefore as a system in two entangled states.

Using the basis $(|\varphi_1\rangle, |\varphi_2\rangle)$, the Hamiltonian of the junction can be written as

$$H = \begin{pmatrix} E_1 & -K \\ -K & E_2 \end{pmatrix}, \tag{2.45}$$

where E_1 and E_2 are the energies of the pairs in superconductors S_1 and S_2, while K describes the coupling between the two superconductors due to the tunnel effect. This is a positive real parameter, which characterises the junction, and especially the thickness of the insulator.

The time evolution of the state ket $|\Psi\rangle$ is given by the Schrödinger equation

$$i\hbar \frac{\partial |\Psi\rangle}{\partial t} = H|\Psi\rangle, \tag{2.46}$$

which, in the basis $(|\varphi_1\rangle, |\varphi_2\rangle)$, becomes

$$i\hbar \frac{da_1}{dt} = E_1 a_1 - K a_2 \tag{2.47}$$

$$i\hbar \frac{da_2}{dt} = E_2 a_2 - K a_1. \tag{2.48}$$

If a voltage difference V is set between the two superconductors by applying the voltage $V/2$ to conductor 1 and the voltage $-V/2$ to conductor 2, the energy of the Cooper pairs is $E_1 = q(V/2) = -(2e)V/2 = -eV$ for conductor 1 (as the Cooper pair has a charge $(-2e)$ equal to the charge of two electrons) while the energy of the Cooper pairs in conductor 2 is $+eV$. The system of Equations (2.47) becomes

$$i\hbar \frac{da_1}{dt} = -eV a_1 - K a_2 \tag{2.49}$$

$$i\hbar \frac{da_2}{dt} = eV a_2 - K a_1. \tag{2.50}$$

Introducing $a_i = \sqrt{\rho_i}\, e^{i\phi_i}$ and multiplying Equation (2.49) by $\sqrt{\rho_1}$, we get

$$i\hbar \left[\frac{1}{2} \frac{d\rho_1}{dt} e^{i\phi_1} + i\rho_1 \frac{d\phi_1}{dt} e^{i\phi_1} \right] = -eV\rho_1 e^{i\phi_1} - K\sqrt{\rho_1 \rho_2}\, e^{i\phi_2}. \tag{2.51}$$

Separating the imaginary and the real part yields

$$\frac{\hbar}{2} \frac{d\rho_1}{dt} \cos\phi_1 - \hbar\rho_1 \frac{d\phi_1}{dt} \sin\phi_1 = -eV\rho_1 \sin\phi_1 - K\sqrt{\rho_1\rho_2}\, \sin\phi_2 \tag{2.52}$$

$$-\frac{\hbar}{2} \frac{d\rho_1}{dt} \sin\phi_1 - \hbar\rho_1 \frac{d\phi_1}{dt} \cos\phi_1 = -eV\rho_1 \cos\phi_1 - K\sqrt{\rho_1\rho_2}\, \cos\phi_2. \tag{2.53}$$

If we now multiply Equation (2.52) by $\cos\phi_1$ and Equation (2.53) by $-\sin\phi_1$ and sum them, we obtain

$$\frac{\hbar}{2} \frac{d\rho_1}{dt} (\cos^2\phi_1 + \sin^2\phi_1) = -K\sqrt{\rho_1\rho_2}\, (\sin\phi_2 \cos\phi_1 - \cos\phi_2 \sin\phi_1). \tag{2.54}$$

Introducing the variable $\theta = \phi_1 - \phi_2$, we finally get

$$\frac{d\rho_1}{dt} = \frac{2K}{\hbar} \sqrt{\rho_1\rho_2}\, \sin\theta, \tag{2.55}$$

while a similar calculation with Equation (2.50) leads to

$$\frac{d\rho_2}{dt} = \frac{2K}{\hbar} \sqrt{\rho_1\rho_2}\, \sin(\phi_2 - \phi_1) = -\frac{2K}{\hbar} \sqrt{\rho_1\rho_2}\, \sin\theta. \tag{2.56}$$

This shows that only the phase difference θ is significant, as expected.

Since ρ_1 and ρ_2 are the Cooper pair densities per unit area of the junction in each superconductor, and as each pair has the charge $-2e$, the variations $d\rho_1/dt$

and $\mathrm{d}\rho_2/\mathrm{d}t$ are associated with a current density across the insulator, due to the tunnelling of the Cooper pairs through the insulator. The current density from conductor 1 to conductor 2 is directed along the z axis and it is equal to

$$j_T = 2e\frac{\mathrm{d}\rho_1}{\mathrm{d}t} = -2e\frac{\mathrm{d}\rho_2}{\mathrm{d}t} = \frac{4eK}{\hbar}\sqrt{\rho_1\rho_2}\,\sin\theta. \tag{2.57}$$

Let us denote $j_c = 4eK\sqrt{\rho_1\rho_2}/\hbar$ the critical current density which can tunnel through the junction. It is determined by the geometrical properties of the junction and by the density of Cooper pairs. Thus we get

$$j_T = j_c\sin\theta. \tag{2.58}$$

On the other hand, if we multiply Equation (2.52) by $\sin\phi_1$ and Equation (2.53) by $\cos\phi_1$, we get

$$-\hbar\rho_1\frac{\mathrm{d}\phi_1}{\mathrm{d}t}(\sin^2\phi_1 + \cos^2\phi_1) = -eV\rho_1(\sin^2\phi_1 + \cos^2\phi_1) - K\sqrt{\rho_1\rho_2}\,(\sin\phi_1\sin\phi_2 + \cos\phi_2\cos\phi_1) \tag{2.59}$$

which simply gives

$$\frac{\mathrm{d}\phi_1}{\mathrm{d}t} = \frac{eV}{\hbar} + \frac{K}{\hbar}\sqrt{\frac{\rho_2}{\rho_1}}\cos\theta. \tag{2.60}$$

And similarly we also get

$$\frac{\mathrm{d}\phi_2}{\mathrm{d}t} = -\frac{eV}{\hbar} + \frac{K}{\hbar}\sqrt{\frac{\rho_1}{\rho_2}}\cos\theta. \tag{2.61}$$

For two similar superconductors the densities of carriers ρ_1 and ρ_2 are approximately equal ($\rho_1/\rho_2 \approx 1$), and moreover, in the experiments, they are adjusted by an external current, called the 'bias current', provided by a generator. The difference of the two Equations (2.60) and (2.61) leads to

$$\frac{\mathrm{d}\theta}{\mathrm{d}t} = \frac{2eV}{\hbar}. \tag{2.62}$$

Equations (2.58) and (2.62) are the two fundamental equations derived by Josephson. They describe effects known as continuous and oscillatory Josephson effects [175] respectively. They describe a nondissipative current of *pairs* through the insulator, with an amplitude j_c and a frequency $\nu = 2eV/h$. They have been checked in numerous experiments.

To obtain the characteristic equation of the junction, they must be completed by the Maxwell equations for the insulator,

$$\overrightarrow{\mathrm{rot}}\ \vec{E} = -\frac{\partial \vec{B}}{\partial t} \tag{2.63}$$

$$\overrightarrow{\mathrm{rot}}\ \vec{B} = \mu_0 \vec{j} + \mu_0 \varepsilon \frac{\partial \vec{E}}{\partial t} \tag{2.64}$$

if we denote by $\vec{E}$ the electric field, $\vec{B}$ the magnetic field, and ε the dielectric permittivity of the insulator. The nonmagnetic insulator has a magnetic permittivity equal to the vacuum magnetic permittivity μ_0.

'Point' junctions

If we consider a very small junction, the spatial dependence of the current in the (x, y) plane of the junction can be ignored. The total current density $\vec{j_B}$, imposed by the external voltage source, creates a magnetic field $\vec{B_0}$ such that $\overrightarrow{\mathrm{rot}}\ \vec{B_0} = \mu_0 \vec{j_B}$. With the axis defined in Figure 2.13, the current density $\vec{j_B}$ is parallel to the z axis, as well as the electric field $\vec{E} = (V/d)\vec{u_z}$.

The total current across a *real* Josephson junction is the sum of three contributions:

(i) The tunnel supercurrent of the Cooper pairs due to the Josephson effect $j_T = j_c \sin\theta$.
(ii) A standard resistive current due to normal charge carriers, i.e. electrons not involved in Cooper pairs which tunnel through the insulator, $j_R = V/(RS)$, where R is the resistance of the insulator per unit area, and S the area of the junction.
(iii) The contribution $\varepsilon(\partial E)/(\partial t)$ coming from the last term in Equation (2.64). This is a capacitive current, $(\varepsilon/d)V_t$ due to the capacitance $C = \varepsilon S/d$ of the junction.

Therefore the z component of the current leads to

$$j_B = j_T + \frac{V}{RS} + \frac{C}{S}\frac{\mathrm{d}V}{\mathrm{d}t}. \tag{2.65}$$

Thus a real junction can be viewed as an ideal junction in parallel with a resistance R and a capacitance C. This is the RCSJ model [175] ('Resistively and Capacitively Shunted Junction') shown in Figure 2.14.

Using Equations (2.58) and (2.62), we can rewrite Equation (2.65) as

$$j_B = j_c \sin\theta + \frac{1}{RS}\frac{\hbar}{2e}\frac{\mathrm{d}\theta}{\mathrm{d}t} + \frac{C}{S}\frac{\hbar}{2e}\frac{\mathrm{d}^2\theta}{\mathrm{d}t^2}, \tag{2.66}$$

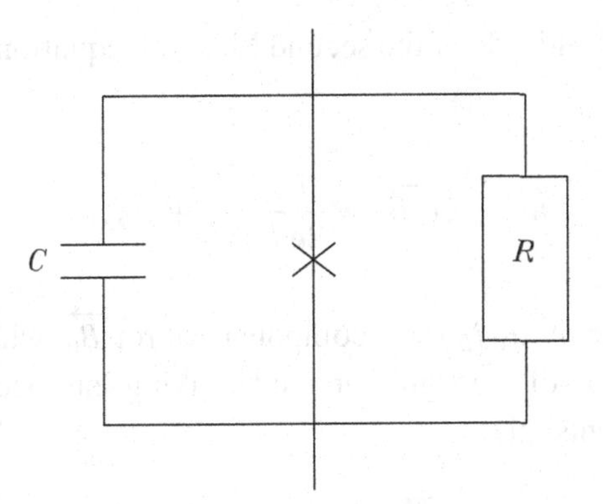

Figure 2.14. Electrical circuit equivalent to a real Josephson junction according to the RCSJ model: the ideal junction is in parallel with a resistance R and a capacitance C.

which can be simplified into

$$\frac{\mathrm{d}^2\theta}{\mathrm{d}t^2} + \frac{1}{RC}\frac{\mathrm{d}\theta}{\mathrm{d}t} + \omega_0^2 \sin\theta = \frac{2eS}{C\hbar} j_B, \tag{2.67}$$

if we introduce the frequency $\omega_0 = \sqrt{2edj_c/(\hbar\varepsilon)}$.

Hence it appears that a point Josephson junction is described by the same equation as a simple pendulum! This is the observation which has strongly revived interest in the pendulum and its nonlinear dynamics, especially because, in some of their operating regimes, junctions show a strong electrical noise. It is a consequence of the chaotic properties of the nonlinear oscillator that a junction turns out to be.

Extended junctions

Besides the studies of point junctions, many experiments have been carried out with junctions extended in one direction, which we shall denote by x. The generator maintains the average current density across the junction, but θ and V are now functions of x and t.

The y component of the first Maxwell equation (2.63) gives

$$-\frac{\partial E_z}{\partial x} = -\frac{\partial}{\partial x}\left(\frac{V(x,t)}{d}\right) = -\frac{\partial B_y}{\partial t}. \tag{2.68}$$

Using Equation (2.62), we get

$$\frac{\partial}{\partial x}\left(\frac{\hbar}{2ed}\frac{\partial\theta}{\partial t}\right) = \frac{\partial B_y}{\partial t} \tag{2.69}$$

which, after integration with respect to time, gives

$$B_y = \frac{\hbar}{2ed}\frac{\partial\theta}{\partial x} + B_{0y}(x) \tag{2.70}$$

where $B_{0y}(x)$ is an integration constant.

Projecting the left hand side of the second Maxwell equation (2.64) on the z axis leads to

$$\vec{u_z} \cdot \vec{\text{rot}}\ \vec{B} = \frac{\hbar}{2ed}\frac{\partial^2\theta}{\partial x^2} + \mu_0 j_B \tag{2.71}$$

where we again denote by $\mu_0 j_B$ the z component of $\vec{\text{rot}}\ \vec{B_0}$, which does not depend on time. The projection of the right hand side, taking into account the Josephson and the resistive currents, gives

$$\vec{u_z} \cdot \left(\mu_0 \vec{j} + \mu_0 \varepsilon \frac{\partial \vec{E}}{\partial t}\right) = \mu_0 j_c \sin\theta + \frac{\mu_0}{RS}\frac{\hbar}{2e}\frac{\partial\theta}{\partial t} + \mu_0 \frac{C}{S}\frac{\hbar}{2e}\frac{\partial^2\theta}{\partial t^2} \tag{2.72}$$

where we simply replaced the total derivatives of the point junction by partial derivatives. Combining Equations (2.71) and (2.72), we obtain

$$\frac{\partial^2\theta}{\partial t^2} - \frac{1}{\varepsilon\mu_0}\frac{\partial^2\theta}{\partial x^2} + \frac{1}{RC}\frac{\partial\theta}{\partial t} + \frac{2eS}{C\hbar} j_c \sin\theta = \frac{2eS}{C\hbar} j_B. \tag{2.73}$$

If we define the quantity $c_0^2 = 1/(\varepsilon\mu_0)$, which is homogeneous to the square of a velocity, in order to recover the notation of the sine-Gordon equation that we introduced earlier, we can notice that an extended junction is described by the equation

$$\theta_{tt} - c_0^2\theta_{xx} + \frac{1}{RC}\theta_t + \omega_0^2 \sin\theta = \frac{2eS}{C\hbar} j_B. \tag{2.74}$$

Therefore the dynamics of the Josephson junction is described by a perturbed sine-Gordon equation, which will be studied in detail in Chapter 5.

A more complete calculation, which does not use the 'two-state' approach introduced by Feynman, leads to a very similar result, but the factor c_0^2 is found to be

$$c_0^2 = \frac{1}{\varepsilon\mu_0}\frac{d}{d + 2\lambda_L} \quad \text{instead of} \quad c_0^2 = \frac{1}{\varepsilon\mu_0}, \tag{2.75}$$

where the London length λ_L measures the penetration depth of the magnetic field into the superconductor. Assuming a perfect superconductor amounts to assuming that λ_L is zero. For real superconductors λ_L should not be neglected with respect to the small thickness d of the insulator.

2.3.2 Applications to the properties of a long Josephson junction

For a large resistance R, one can expect that the junction carries solitons similar to the SG soliton,

$$\theta(x,t) = 4 \arctan \exp \left[\pm \frac{\omega_0}{c_0} \frac{x - vt}{\sqrt{1 - \frac{v^2}{c_0^2}}} \right] . \tag{2.76}$$

Such solitons can exist even in the presence of dissipation because their motion is supported by the 'bias current' j_B, as we show in Chapter 5. However, they can only be expected if their characteristic width $L = c_0/\omega_0 = \sqrt{\hbar/(\mu_0 2ed\, j_c)}$ is compatible with the dimensions of the junction. The value of j_c is difficult to compute because it depends very much on the exact geometry of the junction. In practice it is experimentally measured and it leads to L of the order of 0.1 mm for typical junctions. As the junctions generally have a length of the order of $\ell = 1–10$ mm, the soliton width is indeed much smaller than ℓ.

Let us assume that a soliton has been created in the junction. A voltage pulse

$$V(x,t) = \frac{\hbar}{2e}\theta_t = \frac{\hbar}{2e} \frac{2\omega_0 v}{c_0\sqrt{1 - \frac{v^2}{c_0^2}}} \operatorname{sech} \left[\frac{\omega_0}{c_0} \frac{x - vt}{\sqrt{1 - \frac{v^2}{c_0^2}}} \right] \tag{2.77}$$

is propagating in the junction, and is associated with the voltage across the junction

$$\overline{V} = \frac{1}{\ell} \int_{-\ell/2}^{\ell/2} V(x,t)\, \mathrm{d}x \tag{2.78}$$

which is simply given by the area of the voltage pulse $V(x,t)$. As the pulse moves like a soliton, its area is a constant that we henceforth denote by V_1.

The total current through the junction is equal to

$$I = \frac{S}{\ell} \int_{-\ell/2}^{\ell/2} j_B\, \mathrm{d}x = \frac{\hbar}{2e\ell R} \int_{-\ell/2}^{\ell/2} \theta_t\, \mathrm{d}x = \frac{\overline{V}}{R} \tag{2.79}$$

because the other terms of j_B cancel each other since θ is a solution of the SG equation. Thus we get a *dc current* proportional to $\overline{V} = V_1$, which is very small since it drops to zero for a perfect insulator.

Of course we could also create two solitons, three solitons, ... ; $\overline{V}$ and I would simply be multiplied by the number of solitons. The measured curve of the voltage versus current for a junction (see Figure 2.15) exhibits plateaux, which are called 'Fiske steps', for the voltages associated with different numbers of solitons. The current on the plateaux grows beyond the values resulting from the presence of the solitons because the external generator imposes an additional spatially uniform component, which is added to the current associated with the solitons.

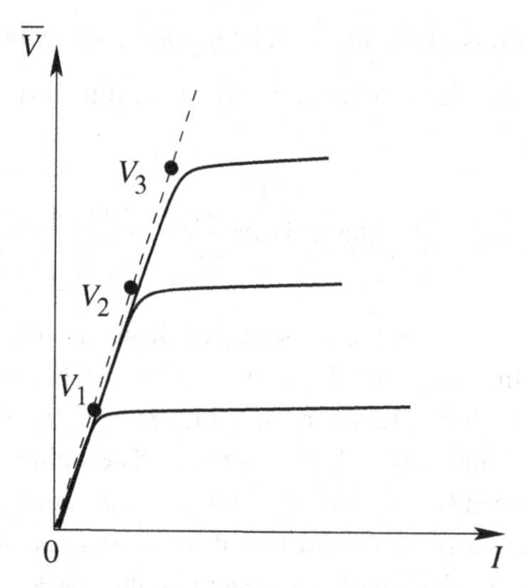

Figure 2.15. Typical measured voltage $\overline{V}$ versus current I curves for a Josephson junction. The dots correspond to the voltages and currents for 1 soliton, 2 solitons and 3 solitons (Equations (2.78) and (2.79)).

Besides their indirect experimental observations through the Fiske steps, a direct proof of the existence of the solitons is provided by the consequence of their reflections at the ends of the junction. Analysing the boundary conditions shows that the ends behave as free boundaries for the solitons. When the voltage pulse arrives at the end and is reflected, the local potential difference between the two superconductors shows a sharp peak. Therefore, although the total values of $\overline{V}$ and I across the junctions stay constant, there are fast voltage oscillations at the ends. An estimate of the propagation speed of the solitons, $c_0 = c/\sqrt{\varepsilon_r}$, assuming a relative dielectric permittivity of the insulator $\varepsilon_r = 10$ and a junction length $\ell = 10$ mm gives a frequency $\nu = c_0 \ell \simeq 10$ GHz for the reflection of the soliton at the ends. Thus a junction in which a soliton oscillates back and forth, behaves as a *microwave generator*. The microwaves emitted by a junction are easy to detect, and they provide an almost direct observation of the soliton.

We showed earlier that solitons repel each other. Therefore if we excite two solitons in the junction, we can expect them to settle as far as possible from each other, i.e. at a distance $\ell/2$, in their stationary state. As shown in Figure 2.16, this would imply that the frequency of the voltage pulses at the ends should be doubled. But experiments show that it is not the case. This prompted further studies [118] of the perturbed sine-Gordon which describes the junction in order to lift what appeared as a contradiction between theory and experiments. A complete calculation exhibits an extra term θ_{xxt}, that our simplified study did not find. It is introduced by the diffusion of the Cooper pairs at the surfaces of the superconductors. Although

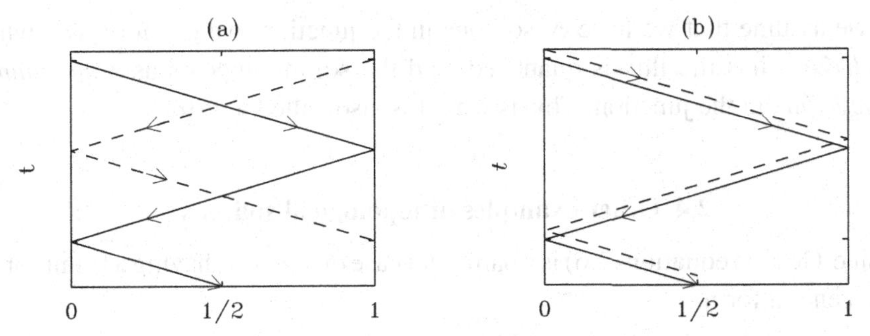

Figure 2.16. Plot of the dynamics of the solitons in a Josephson junction in which two solitons have been excited (solid line and dashed line). The horizontal axis gives their position in the junction, in units of the length ℓ of the junction, and the vertical axis corresponds to time. A voltage pulse occurs each time a trajectory hits one of the vertical boundaries which correspond to the ends of the junction. Figure (a) shows the result predicted by the sine-Gordon equation. Figure (b) shows the result predicted by the SG equation perturbed by the θ_{xxt} term: the two solitons travel together.

it is very small, this term plays a crucial role in the distribution of the solitons in the junction: they tend to get together into a 'soliton bunch'. This strong effect of a very small term can be understood if one thinks that solitons behave as nearly free particles in the junction. Their interaction potential decreases exponentially with their distance and becomes negligible as soon as they are separated by a distance larger than five or six times their width. Therefore, even a very small perturbation can overcome this weak repulsion and, as the solitons get together in a bunch and thus reflect together at the ends, the frequency of the microwave emission does not depend on the soliton number. The presence of several solitons only shows up by a more intense microwave emission, and side bands near the fundamental frequency.

2.3.3 Physical meaning of the soliton: fluxon

We have shown that the magnetic field inside the insulator has a component

$$\overrightarrow{B_1}(x,t) = \frac{\hbar}{2ed}\frac{\partial\theta}{\partial x}\,\overrightarrow{u_y}. \tag{2.80}$$

The flux of this magnetic field across the junction is therefore equal to

$$\phi = \int B_1(x,t)\, d\, \mathrm{d}x = d\int_{-\infty}^{+\infty} \frac{\hbar}{2ed}\frac{\partial\theta}{\partial x}\,\mathrm{d}x = \frac{\hbar}{2e}\,[\theta(x,t)]_{-\infty}^{+\infty} \tag{2.81}$$

where the bounds of the integral have been extended to infinity because the length ℓ of the junction is much greater than the soliton width.

If we assume that we have N solitons in the junction, we get $\phi = N\phi_0$ where $\phi_0 = h/2e$. Thus the flux is quantised, and the soliton appears as a *quantum of magnetic flux* in the junction. This is why it is also called a *fluxon*.

2.4 Other examples of topological solitons

The sine-Gordon equation (2.6) is a particular case of a system having a Hamiltonian of the general form

$$H = \int_{-\infty}^{+\infty} \left[\frac{1}{2}\phi_t^2 + \frac{1}{2}\phi_x^2 + V(\phi) \right] \mathrm{d}x, \tag{2.82}$$

which leads to the equation of motion

$$\frac{\partial^2 \phi}{\partial t^2} - \frac{\partial^2 \phi}{\partial x^2} + V'(\phi) = 0. \tag{2.83}$$

When the potential $V(\phi)$ has several degenerate minima, these nonlinear Klein–Gordon equations have permanent profile solutions which interpolate between two minima.

As for the sine-Gordon equation, it is possible to take advantage of the Lorentz invariance to look for static solutions only, and then derive the mobile ones from them by a Lorentz transform. The calculation is again equivalent to studying the dynamics of a fictitious particle in the pseudo-potential $-V(\phi)$. Many potential shapes have been investigated, but the sinusoidal potential of the SG equation is singular because it is the only one which leads to a *completely integrable system* having soliton solutions. All other potentials lead to equations having *quasi-soliton* solutions, usually called kinks because they correspond to a fast local variation of the variable ϕ. Contrary to solitons, quasi-solitons are not preserved in collisions with other kinks.

Let us consider two standard examples.

2.4.1 The ϕ^4 model

The potential $V(\phi) = 1/4(\phi^2 - 1)^2$, plotted in Figure 2.17, has two minima only, for $\phi_0 = \pm 1$. The kink connecting these minima is given by

$$\phi = \pm \tanh \frac{x - vt}{\sqrt{2}\sqrt{1 - v^2}}, \tag{2.84}$$

which is a solution of the Klein–Gordon equation

$$\frac{\partial^2 \phi}{\partial t^2} - \frac{\partial^2 \phi}{\partial x^2} - \phi + \phi^3 = 0. \tag{2.85}$$

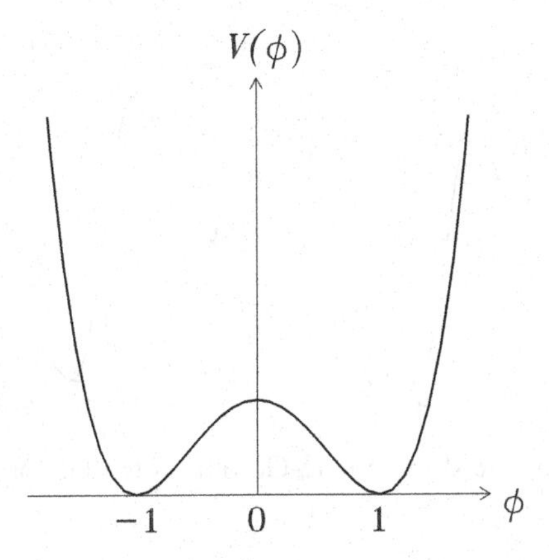

Figure 2.17. Shape of the ϕ^4 potential $V(\phi) = 1/4(\phi^2 - 1)^2$.

The existence of only two minima introduces an additional *topological constraint* with respect to the sine-Gordon equation: a multikink solution must correspond to a swing between these two minima, which means that an antikink must necessarily follow a kink. However, contrary to the sine-Gordon case, there are no exact multikink solutions. The solution plotted in Figure 2.18 is only an approximate one. It becomes better when the kinks which make it up are farther from each other.

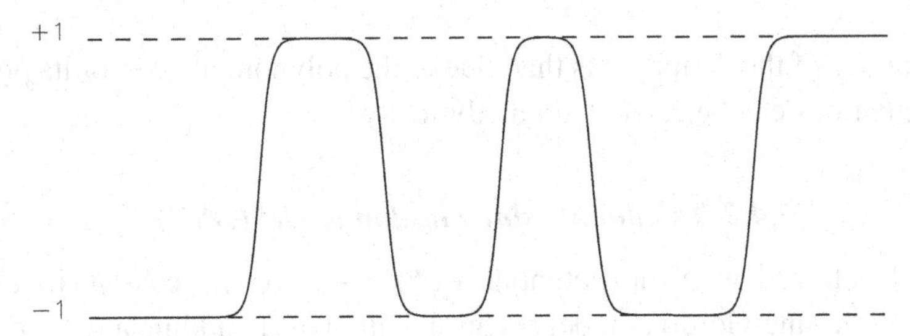

Figure 2.18. Approximate multikink solution of the ϕ^4 model. Kinks and antikinks alternate.

The ϕ^4 model is often used in physics. Its first major application is the Ginzburg–Landau theory for second order phase transitions. As discussed in Chapter 10, this model has also been successfully used in a microscopic theory of ferroelectric domain walls. The kinks are the fundamental excitations of the system and they bring an important contribution to its statistical physics. The ϕ^4 model can also describe topological defects in conducting polymers such as polyacetylene (see Chapter 13). Finally, in field theory, the same model has been proposed [46] as the simplest case of a system having stable states different from the 'vacuum' state. The

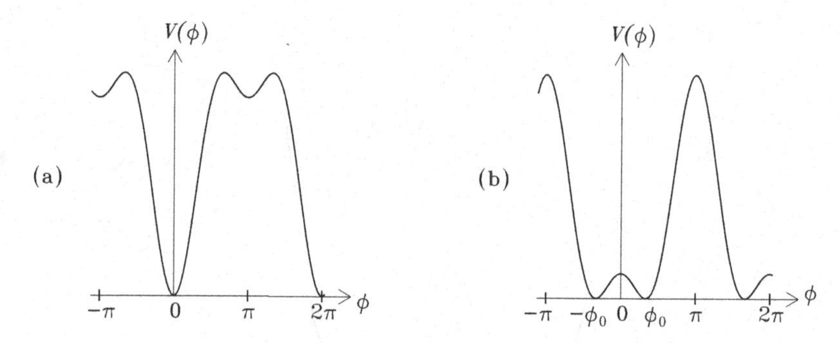

Figure 2.19. Shape of the double sine-Gordon potential for $\eta < -1/4$ (a) and $\eta > 1/4$ (b).

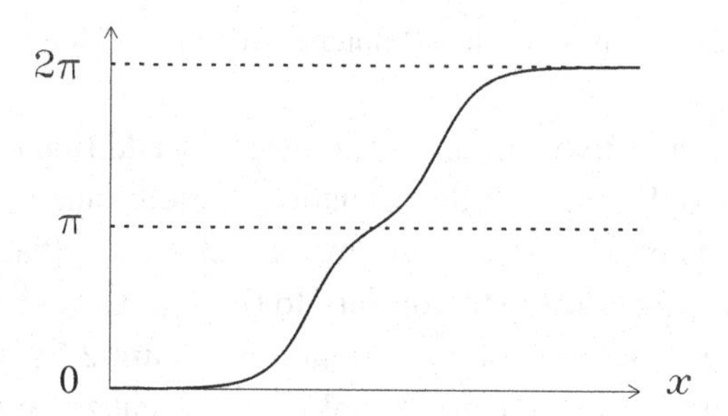

Figure 2.20. Shape of the *single soliton* solution of the double sine-Gordon for $\eta < -1/4$.

great advantage of the ϕ^4 model is that, due to the polynomial form of its potential, many calculations can be carried out analytically.

2.4.2 The double sine-Gordon model (DSG)

The model obtained with the potential $V(\phi) = -\cos\phi + \eta\cos 2\phi$, is a natural extension of the sine-Gordon model because it amounts to adding a second term in the Fourier expansion of a periodic function [42]. Moreover it is very interesting because, depending on the value of the parameter η, the topology of the potential changes qualitatively, which can lead to very peculiar properties:

- For $\eta < -1/4$, the potential has the shape plotted in Figure 2.19(a). The effective potential method shows that the soliton can only connect two absolute minima of the potential, so that it must correspond to a shift between $\phi = 0$ and $\phi = \pm 2\pi$. However the existence of the secondary minima for $\phi = \pm\pi$ has an influence on the shape of the solution. These values are energetically more favourable that the neighbouring values so that they tend to be favoured in the solution. As a result the soliton has a peculiar shape. It appears to be made of two 'sub-kinks' bound to each other, as shown in Figure 2.20.

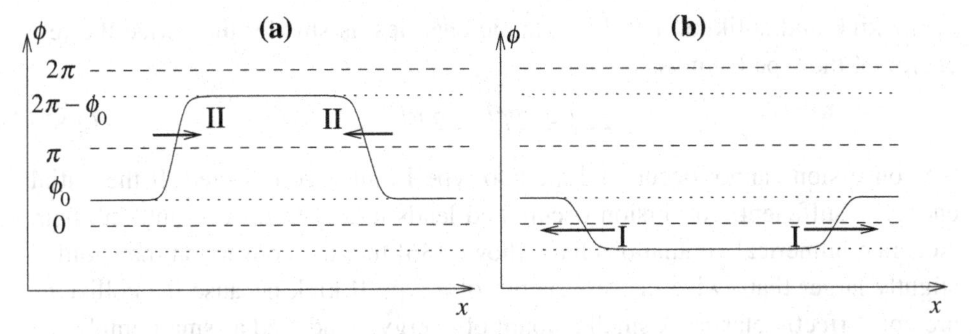

Figure 2.21. Schematic picture of the collision between a type II soliton and a type II antisoliton in the DSG model for $\eta > 1/4$. (a) Before the collisions the two excitations interpolate between the values ϕ_0 and $2\pi - \phi_0$. (b) After the collision, the value of ϕ between the two excitations has changed to $-\phi_0$ (see the corresponding Figure 2.8 for the sine-Gordon case). We now have a type I soliton and antisoliton.

- Conversely the intermediate case $-1/4 < \eta < 1/4$ is very close to the sine-Gordon case because the potential is qualitatively similar to the sinusoidal potential, the only difference being flatter maxima or minima.
- The case $\eta > 1/4$ is again very interesting because there are now two different pathways connecting the energetically degenerate absolute minima $\phi = \pm\phi_0 \pmod{2\pi}$ (see Figure 2.19(b)). They correspond to *two different kinds of kink solutions*. Since we showed that the energy of a kink is proportional to the integral of $\sqrt{V(\phi)}$ between the two minima that it connects, this implies that *the two types of kinks have different energies*. This leads to new behaviours in kink collisions. Let us denote by type II the kinks which connect the minima by passing above the global maximum of the potential, and by type I the kinks for which ϕ goes over the small secondary maximum. The rest energy of type II kinks, E_0^{II}, is therefore larger than that of type I kinks E_0^I. Looking at the shape of the potential valleys indicates that the collision between a type II kink and a type II antikink generates a type I kink and a type I antikink (see Figure 2.21). The energy balance of the collision in this conservative system gives

$$2E_0^{II} + 2T^{II} = 2E_0^I + 2T^I \tag{2.86}$$

where we denote by T^{II} and T^I the kinetic energies of the kinks before and after the collision. The condition $2E_{II} > 2E_I$ leads to $T^I > T^{II}$, which means that some potential energy has been converted into kinetic energy. As a result the two type I kinks, which have been generated in the interaction, leave the collision region with a very high speed.

The result of the collision between a type I kink and a type I antikink depends on their initial velocities. If the total initial energy, i.e. the sum of the rest energies of the kink and antikink and their kinetic energies, is smaller that twice the rest energy of the type II kinks,

$$2E_0^I + 2T^I < 2E_0^{II}, \tag{2.87}$$

the conversion cannot occur, and the two type I kinks are reflected. If the initial energy is sufficient, conversion occurs and leads to a type II kink–antikink pair. Accurate numerical simulations have shown [36] that the conversion threshold is slightly larger than twice the rest energy of a type II kink because the collisions are not perfectly elastic. A small amount of energy is radiated as small-amplitude waves. This shows the limitations of an analysis which considers solitons as quasi-particles when one wants to get precise quantitative results. Taking into account the possible emission of nonlocalised waves, we notice that Equation (2.86) is only approximate. Things get even more complicated if one realises that kinks are complex objects. They can have internal excitations (see Section 5.3). Such a mode does exist for a DSG kink, and it can store part of the energy, which travels together with the kink. But, as the energy of the internal mode is taken from the translational kinetic energy, the speed of the kink is changed.

The DSG model is still sufficiently simple to allow an analytical calculation of all the one-soliton solutions. However, as the ϕ^4 model, it is not fully integrable, and therefore it does not have exact multisoliton solutions.

The two examples that we described show the richness of the behaviour which can be observed with topological 'solitons'. In fact, except for the sine-Gordon case, it is because the solutions are quasi-solitons rather than exact solitons that their behaviour is so rich and interesting.

3

Envelope solitons and nonlinear localisation: the nonlinear Schrödinger equation

The systems that we studied up to now (surface water waves, electrical chain, pendulum chain etc.) also have small-amplitude plane wave solutions

$$\theta = A\, e^{i(qx-\omega t)} + \text{c.c.}, \tag{3.1}$$

(let us recall that we denote by c.c. the complex conjugate of the expression that precedes this symbol) which are drastically different from the soliton solutions that we investigated. Therefore one may ask what happens to these plane waves when their amplitudes grow enough to allow nonlinearity to enter into play. The answer that we shall discover in this chapter is that the plane waves may spontaneously self-modulate as shown in Figure 3.1.

This modulation, which arises due to the overtones induced by nonlinearity, can go as far as the splitting of the wave into 'wave packets' which behave like solitons (see Figure 3.1). These solitons are made of a carrier wave modulated by an envelope signal and this is why they are called *envelope solitons*.

Using the simple example of the pendulum chain introduced in Chapter 2, we shall derive the equation which describes them, which is extremely general in physics, since it appears naturally for most of the weakly dispersive and weakly nonlinear systems which are described by a wave equation in the small-amplitude limit.

3.1 Nonlinear waves in the pendulum chain: the NLS equation

Let us consider the equation which describes the dynamics of the pendulum chain in the continuum limit (SG equation):

$$\theta_{tt} - c_0^2 \theta_{xx} + \omega_0^2 \sin\theta = 0. \tag{3.2}$$

We showed in Chapter 2 that, in the linear limit, this equation has plane wave solutions. We are now interested in the medium-amplitude regime, when a weak

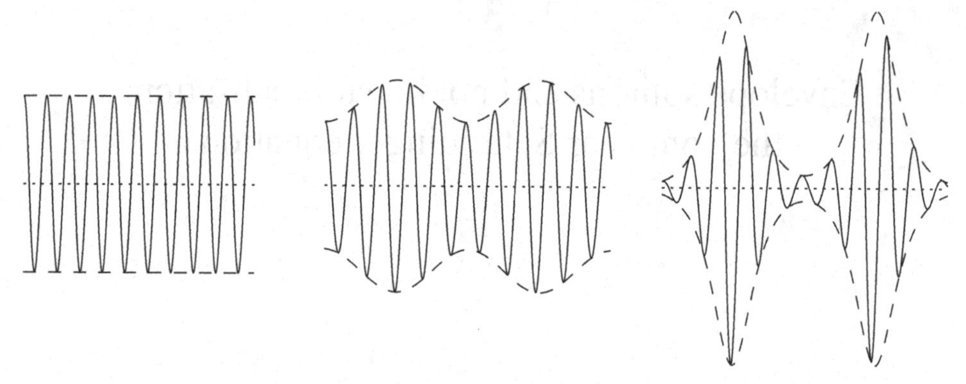

Figure 3.1. Self-modulation of a plane wave. The dashed line shows the envelope of the wave which is displayed by the solid line. The three figures show three successive stages in the evolution of the initial plane wave.

nonlinearity comes into play. Mathematically it means that we shall take into account the first nonlinear term in the expansion of $\sin\theta$ in Equation (3.2),

$$\sin\theta = \theta - \frac{\theta^3}{6} + \mathcal{O}(\theta^5). \tag{3.3}$$

Since we are considering the case of a weak nonlinearity, it may seem quite natural to look for the solution as a perturbative expansion, using the plane wave as the lowest-order term, i.e. to assume

$$\theta = \varepsilon A\, \mathrm{e}^{\mathrm{i}(qx-\omega t)} + \varepsilon^2 B(x,t) \quad (\varepsilon \ll 1). \tag{3.4}$$

However this approach fails because it leads to $B(x,t) \sim \varepsilon^2 t$, so that, in the long term, the 'small' correction would diverge [136].

Therefore it is necessary to introduce a method called a *multiple scale expansion*, suggested by the shape of the solution (see Figure 3.1), which allows us to control this divergence. A modulated wave includes two space and timescales:

(i) A fast time and space variation of the carrier wave.
(ii) A much slower variation of the envelope.

The principle of the method is to look for a solution which depends on a set of variables associated with these various scales: $T_i = \varepsilon^i t$ and $X_i = \varepsilon^i x$. Thus we look for θ as a perturbative series of functions of all these variables

$$\theta(x,t) = \varepsilon \sum_{i=0}^{\infty} \varepsilon^i\, \phi_i(X_0, X_1, X_2, \ldots, T_0, T_1, T_2, \ldots) \tag{3.5}$$

which are treated as if they were *independent variables.* This implies that the operators for the derivatives are replaced by sums of operators such as

$$\frac{\partial}{\partial t} = \frac{\partial T_0}{\partial t}\frac{\partial}{\partial T_0} + \frac{\partial T_1}{\partial t}\frac{\partial}{\partial T_1} + \frac{\partial T_2}{\partial t}\frac{\partial}{\partial T_2} + \cdots \tag{3.6}$$

$$= \frac{\partial}{\partial T_0} + \varepsilon\frac{\partial}{\partial T_1} + \varepsilon^2\frac{\partial}{\partial T_2} + \cdots \tag{3.7}$$

The crucial point is that any solution in the multidimensional space of these variables T_i, X_i will be a solution of the original problem when it is restricted to the *physical line* $T_0 = t$; $T_1 = \varepsilon t$; $T_2 = \varepsilon^2 t$; $T_3 = \varepsilon^3 t$; ... Outside of this line, the solution does not have any physical meaning. The interest of these auxiliary variables is that they allow us to impose appropriate conditions that eliminate the divergences and ensure that the asymptotic expansion converges uniformly for small values of ε.

Introducing the notation $D_i = \partial/\partial T_i$, we get

$$\frac{\partial^2}{\partial t^2} = (D_0 + \varepsilon D_1 + \varepsilon^2 D_2 + \cdots)^2 = D_0^2 + 2\varepsilon D_0 D_1 + \varepsilon^2\left(D_1^2 + 2D_0 D_2\right) + \cdots \tag{3.8}$$

Similarly we denote the spatial derivatives by $D_{Xi} = \partial/\partial X_i$. Then we put the expansion (3.5), $\theta = \varepsilon\phi_0 + \varepsilon^2\phi_1 + \varepsilon^3\phi_2 + \cdots$, into Equation (3.2) and we use expressions such as (3.6) and (3.8) to write the derivatives. The various powers of ε are identified as usual in a perturbative expansion.

At order ε, we get

$$\left(D_0^2 - c_0^2 D_{X0}^2 + \omega_0^2\right)\phi_0 = 0, \tag{3.9}$$

which is a linear equation for ϕ_0, henceforth denoted $\hat{L}\,\phi_0 = 0$. The solution of Equation (3.9) is the plane wave

$$\phi_0 = A(X_1, T_1, X_2, T_2, \ldots)\,\mathrm{e}^{\mathrm{i}(qX_0 - \omega T_0)} + \mathrm{c.c.}, \tag{3.10}$$

in which ω and q are linked by the dispersion relation $\omega^2 = \omega_0^2 + c_0^2 q^2$, identical to the dispersion relation deduced from a direct linearisation of the original equation (3.2).

There is however an important difference with the standard perturbation theory because the amplitude factor A is no longer a constant, but rather it is a function depending on the other space and timescales. This offers an additional freedom to derive the solution, which must be judiciously exploited. Instead of attempting to solve the nonlinear equation by adding an extra term to the linear solution, the method modifies the linear solution itself, before eventually adding corrective terms.

At order ε^2, we get

$$D_0^2\phi_1 + 2D_0D_1\phi_0 - c_0^2 D_{X0}^2\phi_1 - 2c_0^2 D_{X0}D_{X1}\phi_0 + \omega_0^2\phi_1 = 0 \tag{3.11}$$

which can also be written as

$$\hat{L}\,\phi_1 = -2D_0D_1\phi_0 + 2c_0^2 D_{X0}D_{X1}\phi_0 \tag{3.12}$$

$$= 2\mathrm{i}\omega\frac{\partial A}{\partial T_1}\mathrm{e}^{\mathrm{i}\sigma} + 2\mathrm{i}qc_0^2\frac{\partial A}{\partial X_1}\mathrm{e}^{\mathrm{i}\sigma} + \text{c.c.}, \tag{3.13}$$

if we introduce $\sigma = qX_0 - \omega T_0$ to simplify the expressions. Thus, to determine ϕ_1 we get a linear equation driven by terms proportional to $\mathrm{e}^{\mathrm{i}\sigma}$ which are *resonance terms* for operator $\hat{L}$. This is what occurs for the familiar example of a harmonic oscillator which is driven exactly at its resonance frequency. The response linearly grows with time. This situation is often encountered in astronomy, where such terms lead to a very slow variation of orbits: they are called secular terms. To make sure that the expansion has a meaning, one must prevent such a linear divergence. This is possible if we *impose a condition* on the amplitude A in order to cancel the right hand side terms which resonate with operator $\hat{L}$. It is called the solvability condition, and here it reads

$$\frac{\partial A}{\partial T_1} + \frac{qc_0^2}{\omega}\frac{\partial A}{\partial X_1} = 0. \tag{3.14}$$

The quantity $v_g = qc_0^2/\omega$ is the group velocity of the carrier wave $\mathrm{e}^{\mathrm{i}\sigma}$, so that Equation (3.14) means that A does not depend on X_1 and T_1 separately, but only through their combination $\xi_1 = X_1 - v_g T_1$. Therefore

$$A(X_1, T_1, X_2, T_2, \ldots) = A(X_1 - v_g T_1, X_2, T_2, \ldots). \tag{3.15}$$

After cancellation of the secular terms, Equation (3.12) reduces to $\hat{L}\,\phi_1 = 0$ which has the solution $\phi_1 = 0$. Any other solution would have the same form as ϕ_0 and could be included in the amplitude A.

At order ε^3, we get the equation

$$\hat{L}\,\phi_2 = -D_1^2\phi_0 - 2D_0D_2\phi_0 + c_0^2 D_{X1}^2\phi_0 + 2c_0^2 D_{X0}D_{X2}\phi_0 + \frac{\omega_0^2}{6}\phi_0^3$$
$$- 2D_0D_1\phi_1 + 2c_0^2 D_{X0}D_{X1}\phi_1. \tag{3.16}$$

This is again a driven linear equation, with the same operator $\hat{L}$. The terms proportional to $\mathrm{e}^{\mathrm{i}\sigma}$ in the right hand side could again lead to a resonant driving and a divergence. Therefore these secular terms must be cancelled, which leads to the equation

$$-\frac{\partial^2 A}{\partial T_1^2} + 2\mathrm{i}\omega\frac{\partial A}{\partial T_2} + c_0^2\frac{\partial^2 A}{\partial X_1^2} + 2\mathrm{i}qc_0^2\frac{\partial A}{\partial X_2} + \frac{3}{6}\omega_0^2|A|^2 A = 0, \tag{3.17}$$

if we take into account the condition $\phi_1 = 0$, obtained at order ε^2, and the resonating terms which come from the expansion of $\phi_0^3 = [A\exp(\mathrm{i}\sigma) + A^*\exp(-\mathrm{i}\sigma)]^3$.

If we had carried a standard perturbation expansion, in which A would be a constant, at order ε^2 we would not have got any resonant term. But at order ε^3 the perturbation expansion would be in trouble because the resonating nonlinear term $|A|^2 A$ would be there, unless the solution vanishes completely with $A = 0$. The particular solution of the equation

$$\hat{L}\,\phi_2 = \frac{\omega_0^2}{2}|A|^2 A\,\mathrm{e}^{\mathrm{i}\sigma} \tag{3.18}$$

would be

$$\phi_2^p = \frac{\mathrm{i}\omega_0^2}{4\omega}|A|^2 A\; T_2\,\mathrm{e}^{\mathrm{i}\sigma}\,\mathrm{e}^{\mathrm{i}(qX_2-\omega T_2)} \tag{3.19}$$

which is proportional to the time variable T_2, confirming the secular growth of the second order correction proportional to $\varepsilon^2 t$ that we mentioned earlier. The standard perturbation expansion would not have any meaning because the 'small' perturbation would finally diverge.

After the cancellation of the secular terms, i.e. after imposing that A must be a solution of Equation (3.17), we find that ϕ_2 is a function of $\mathrm{e}^{3\mathrm{i}\sigma}$. The solution at order ε^3 is therefore of the form

$$\theta = \varepsilon(A\,\mathrm{e}^{\mathrm{i}\sigma} + \text{c.c.}) + \varepsilon^3(B\,\mathrm{e}^{3\mathrm{i}\sigma} + \text{c.c.}) + \mathcal{O}(\varepsilon^4) + \cdots \tag{3.20}$$

In fact, very often only the first order solution is considered because it already includes the role of the nonlinearity through the variation of the amplitude A according to Equation (3.17).

Equation (3.17) can be simplified by moving to a frame mobile at velocity v_g. This is done by introducing the variables $\xi_i = X_i - v_g T_i$ and $\tau_i = T_i$, so that

$$\frac{\partial A}{\partial T_i} = \frac{\partial A}{\partial \tau_i} - v_g\frac{\partial A}{\partial \xi_i} \quad \text{and} \quad \frac{\partial A}{\partial X_i} = \frac{\partial A}{\partial \xi_i}. \tag{3.21}$$

We get

$$\frac{\partial A}{\partial \tau_1} = 0 \tag{3.22}$$

if we take into account the condition (3.14) which leads to the relations

$$\frac{\partial A}{\partial T_1} = -v_g \frac{\partial A}{\partial \xi_1} \quad \text{and} \quad \frac{\partial^2 A}{\partial T_1^2} = v_g^2 \frac{\partial^2 A}{\partial \xi_1^2}. \tag{3.23}$$

In the mobile frame, Equation (3.17) becomes therefore

$$\left(c_0^2 - v_g^2\right) \frac{\partial^2 A}{\partial \xi_1^2} + 2\mathrm{i}\omega \left(\frac{\partial A}{\partial \tau_2} - v_g \frac{\partial A}{\partial \xi_2} \right) + 2\mathrm{i}q c_0^2 \frac{\partial A}{\partial \xi_2} + \frac{1}{2}\omega_0^2 |A|^2 A = 0. \tag{3.24}$$

But, as the group velocity is $v_g = qc_0^2/\omega$, the terms containing $\partial A/\partial \xi_2$ cancel each other. Dividing the equation by 2ω, we arrive at

$$\mathrm{i}\frac{\partial A}{\partial \tau_2} + \frac{\left(c_0^2 - v_g^2\right)}{2\omega} \frac{\partial^2 A}{\partial \xi_1^2} + \frac{\omega_0^2}{4\omega} |A|^2 A = 0, \tag{3.25}$$

called the nonlinear Schrödinger equation (NLS), which describes the variations of the envelope A.

We carried out the calculation by starting from the sine-Gordon equation, i.e. the equation which includes the continuum limit approximation, valid for carrier waves which have a large wavelength. It is possible to perform a similar calculation from the discrete equation

$$\frac{\mathrm{d}^2 \theta_n}{\mathrm{d}t^2} - c_0^2 (\theta_{n+1} + \theta_{n-1} - 2\theta_n) + \omega_0^2 \sin \theta_n = 0. \tag{3.26}$$

The discrete carrier wave, having the dispersion relation $\omega^2 = \omega_0^2 + 4c_0^2 \sin^2(q/2)$, can be preserved, and the continuum approximation is only involved for the envelope. This leads to a nonlinear Schrödinger equation with coefficients which are functions of the wavevector q of the carrier wave. This calculation [158] using the so called *semi-discrete approximation* is presented in Section 16.2.2.

3.2 Properties of the nonlinear Schrödinger equation

Let us write Equation (3.25) with its usual notation

$$\mathrm{i}\frac{\partial \psi}{\partial t} + P \frac{\partial^2 \psi}{\partial x^2} + Q |\psi|^2 \psi = 0, \tag{3.27}$$

where P and Q are coefficients which depend on the particular problem which is being studied, as do the meanings of the variables t and x. The name of this equation is obviously linked to its structure, which appears very similar to that of the Schrödinger equation if we write it as

$$\mathrm{i}\frac{\partial \psi}{\partial t} = \left[-P \frac{\partial^2}{\partial x^2} - Q |\psi|^2 \right] \psi. \tag{3.28}$$

Equation (3.27) is formally analogous to the Schrödinger equation only if the coefficient P is positive. We can always transform the equation to make sure that this is the case because, if $P < 0$, we can change the sign of the equation and then restore the positive sign in front of the time derivative by taking the complex conjugate. This leads to an equation for ψ^* in which the coefficient of $\partial^2\psi^*/\partial x^2$ is positive. Therefore, without any restriction, we can assume $P > 0$. It should be noticed that the transformation of the equation simultaneously changes the signs of P and Q so that it does not affect the sign of the product PQ which, as shown below, is of crucial importance to determine the nature of the solutions.

The 'potential' term of the Schrödinger equation is here equal to $-Q|\psi|^2$ so that it depends on the solution ψ. This feature explains the denomination *nonlinear* for Equation (3.27).

We shall show that, when Q is positive, the ψ solution is localised, with a bell shape. It generates a 'potential well' for $-Q|\psi|^2$, which turns out to be a necessary condition for the Schrödinger equation to have a *spatially localised solution*.[1] Thus the NLS equation is such that ψ 'digs' its own potential well. This leads to a self-focusing phenomenon (also called self-trapping), which is essential for the physics of the systems described by the NLS equation. It points out the possibility to have a localisation effect due to nonlinearity, which is completely different from the well known disorder-induced localisation, also called Anderson's localisation.

3.2.1 The solution of the NLS equation

As the NLS equation is complex, let us look for a solution of the form

$$\psi = \phi(x,t)\,\mathrm{e}^{\mathrm{i}\theta(x,t)} \tag{3.29}$$

where ϕ and θ are *real* functions. If we choose a phase factor θ varying over a 2π range, we can restrict the search to positive values of the amplitude ϕ.

Putting this ansatz in the NLS equation (3.27) and separating the real and imaginary part leads to the system of equations

$$-\phi\,\theta_t + P\phi_{xx} - P\phi\,\theta_x^2 + Q\phi^3 = 0 \tag{3.30}$$

$$\phi_t + P\phi\,\theta_{xx} + 2P\phi_x\,\theta_x = 0. \tag{3.31}$$

Let us look for a particular solution in which both the carrier wave θ and the envelope ϕ are permanent profile solutions, but with different propagation velocities, u_p for θ and u_e for ϕ, i.e. we look for a solution such that

$$\phi(x,t) = \phi(x - u_e t) \quad \text{and} \quad \theta(x,t) = \theta(x - u_p t). \tag{3.32}$$

[1] For the one-dimensional Schrödinger equation this is even a necessary and sufficient condition if the potential is negative.

It is important to notice that, contrary to what we did for the KdV or SG equations, here we are not looking for a permanent profile solution because we do not assume that u_p and u_e are equal.

Thanks to these hypotheses, the system of Equations (3.30) and (3.31) simplifies into

$$u_p \phi \theta_x + P\phi_{xx} - P\phi\theta_x^2 + Q\phi^3 = 0 \tag{3.33}$$

$$-u_e \phi_x + P\phi\theta_{xx} + 2P\phi_x\theta_x = 0. \tag{3.34}$$

Equation (3.34) multiplied by ϕ is easily integrated and gives

$$-\frac{u_e}{2}\phi^2 + P\phi^2\theta_x = C, \tag{3.35}$$

where C is a constant. Up to now the method is rather general and can lead to different solutions depending on the boundary conditions, which determine the value of C. For instance this approach can lead to nonlinear wave trains, which play a role in the study of some hydrodynamic phenomena.

Here we focus our attention on spatially localised solutions, in order to derive the soliton solution of the NLS equation. This amounts to assuming that ϕ and ϕ_x should tend to zero when $|x|$ tends to infinity. As a result C must vanish and we get

$$\theta_x = \frac{u_e}{2P}, \tag{3.36}$$

excluding of course the solutions where ϕ is identically zero. An integration gives

$$\theta = \frac{u_e}{2P}(x - u_p t) + C' \tag{3.37}$$

where we can impose that the integration constant C' should be zero by an appropriate choice of the origin for time.

Putting this expression into Equation (3.33), we get

$$\frac{u_e u_p}{2P}\phi + P\phi_{xx} - \frac{u_e^2}{4P}\phi + Q\phi^3 = 0 \tag{3.38}$$

which is readily integrated after multiplication by $P\phi_x$. We get

$$\frac{P^2}{2}\phi_x^2 + V_{\text{eff}}(\phi) = 0 \tag{3.39}$$

where we introduced a 'pseudo-potential',

$$V_{\text{eff}}(\phi) = \frac{PQ}{4}\phi^4 - \frac{\left(u_e^2 - 2u_e u_p\right)}{8}\phi^2 \tag{3.40}$$

as we did for the KdV equation. The integration constant has again been set to zero because we look for a spatially localised solution.

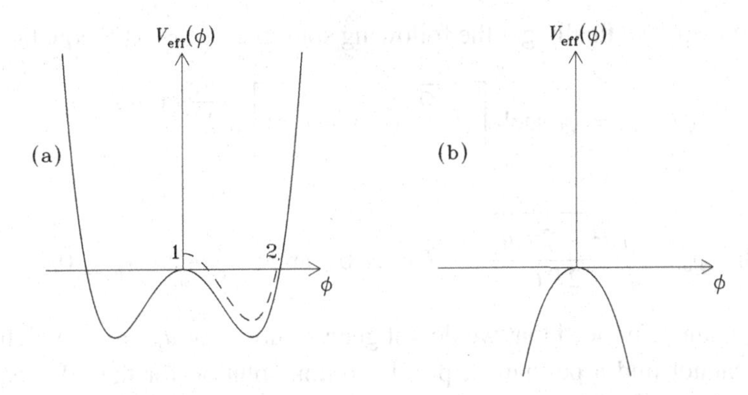

Figure 3.2. Shape of the pseudo-potential $V_{eff}(\phi)$ for (a) $PQ > 0$ and (b) $PQ < 0$. Note that, for a solution of the NLS equation in which ϕ evolves between points 1 and 2, $V_{eff}(\phi)$ is indeed negative for all the range of variation of ϕ.

As ϕ is a real number, $\phi_x^2 \geq 0$ so that Equation (3.39) implies that $V_{eff}(\phi) \leq 0$ for all ϕ values which correspond to a solution. In particular this must be true in the vicinity of $\phi = 0$ which is the limit of ϕ when $|x|$ tends to infinity. But, when ϕ tends to zero, the expression for $V_{eff}(\phi)$ is dominated by the ϕ^2 term. Therefore we must have $(u_e^2 - 2u_e u_p) \geq 0$. This does not impose a sign for u_e and u_p, which can be either positive or negative, but it shows that $u_e = u_p$ is not allowed.

Thinking again in terms of a 'pseudo-potential', we can view Equation (3.39) as the equation describing the motion of a particle with a total energy equal to zero in the potential $V_{eff}(\phi)$. It appears that a bounded motion starting from $\phi = 0$ is only possible if the product PQ is positive (see Figure 3.2). Thus this method shows that *the NLS equation can have a localised soliton-like solution only if* $PQ > 0$.

The motion starts from $\phi = 0$ (labelled 1 in Figure 3.2), reaches $\phi = \phi_0 = \sqrt{(u_e^2 - 2u_e u_p)/2PQ}$ (point 2 in the figure) and then comes back to the starting point.

The solution is obtained by integrating Equation (3.39) using the change of variable $\phi = \phi_0 \operatorname{sech} v$, which yields

$$\phi = \phi_0 \operatorname{sech}\left[\sqrt{\frac{Q}{2P}}\,\phi_0(x - u_e t) + \operatorname{arcsech}\frac{\phi(0,0)}{\phi_0}\right]. \tag{3.41}$$

The origin of the space variable x can be selected in order to eliminate the constant term, which amounts to putting the centre of the soliton $\phi = \phi_0$ at position $x = 0$ when $t = 0$.

Consequently we finally get the following solution for the NLS equation

$$\psi(x,t) = \phi_0 \operatorname{sech}\left[\sqrt{\frac{Q}{2P}}\,\phi_0(x - u_e t)\right] e^{\mathrm{i}\frac{u_e}{2P}(x - u_p t)} \tag{3.42}$$

$$\text{with} \quad \phi_0 = \sqrt{\frac{u_e^2 - 2u_e u_p}{2PQ}}, \quad PQ > 0 \quad \text{and} \quad u_e^2 - 2u_e u_p \geq 0. \tag{3.43}$$

It should again be noticed that we do not get a solution for $u_e = u_p$, which means that one cannot find a permanent profile soliton solution for the NLS equation, contrary to the case of the KdV and SG equations. The solution can also be written as

$$\psi = \phi_0 \operatorname{sech}\left(\frac{x - u_e t}{L_e}\right) e^{\mathrm{i}(\kappa x - \mu t)}, \tag{3.44}$$

with

$$L_e = \frac{1}{\phi_0}\sqrt{\frac{2P}{Q}}, \quad \kappa = \frac{u_e}{2P} \quad \text{and} \quad \mu = \frac{u_e u_p}{2P}, \tag{3.45}$$

which shows that the solution of the NLS equation is a wave packet with a width L_e which is inversely proportional to the amplitude ϕ_0. Hence if its amplitude decreases, its width increases. We recover here the result that the localisation of the solution is due to nonlinearity and, in the limit of very small amplitudes (linear limit), the solution becomes infinitely extended, as a plane wave.

Let us also point out that, if we consider a small-amplitude solution of the NLS equation ($\phi_0 \approx \varepsilon$), Equation (3.45) leads to $u_e \approx \varepsilon$ and $u_p \approx \varepsilon$. Consequently, in the solution, the variable x is multiplied by u_e, i.e. a factor of order ε while the variable t has coefficients $\phi_0 u_e$ or $u_e u_p$, which are of the order ε^2. As a result, in the solution we recover the orders of magnitude which had been obtained in the multiple scale expansion leading to the NLS equation as a function of ξ_1 and τ_2.

3.2.2 Energy localisation by modulational instability

We showed above that the NLS equation has a localised soliton solution, provided that its coefficients verify the condition $PQ > 0$. It turns out that this condition corresponds to a situation in which energy tends to spontaneously self-localise through a phenomenon of modulational instability.

Besides the soliton solution, the NLS equation also has a plane wave solution, but this solution is not always a stable one. Putting $\psi(x,t) = A_0 e^{\mathrm{i}(\kappa x - \Omega t)}$ in the

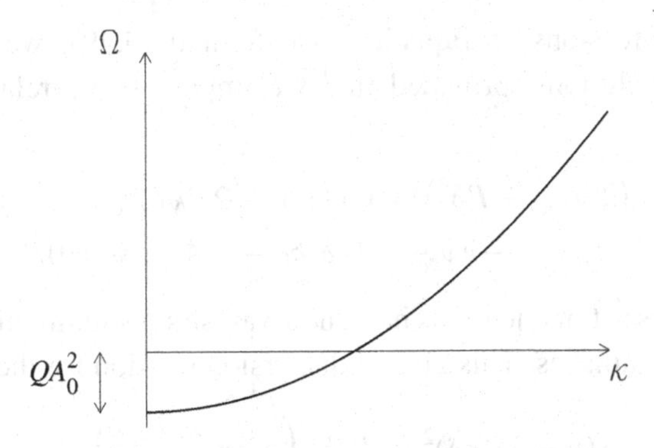

Figure 3.3. Nonlinear dispersion relation of the NLS equation.

NLS equation (3.27), shows that this is indeed an exact solution of this equation provided that Ω verifies the dispersion relation

$$\Omega = P\kappa^2 - QA_0^2. \tag{3.46}$$

This relation is the *nonlinear dispersion relation* of the plane wave. It shows that its frequency not only depends on its wavevector, but also on its amplitude A_0 (Figure 3.3). The plane wave solution exists irrespective of the sign of the product PQ, but we shall show that it is stable only if $PQ < 0$.

Let us check this statement by studying the linear stability of the plane wave. We consider a small perturbation of its phase and amplitude and look for its time evolution. We put the following expression into the NLS equation

$$\psi(x,t) = [A_0 + b(x,t)]\,\mathrm{e}^{\mathrm{i}[\kappa x - \Omega t + \theta(x,t)]}, \tag{3.47}$$

where b and θ are two real functions of the variables x and t. We keep only the linear terms in b and θ which are supposed to be small, and separate the real and imaginary parts. In the calculation some terms cancel out due to the dispersion relation (Equation 3.46) which appears as a factor of $A_0 + b$, so that we finally get the linear set of equations

$$-A_0\,\theta_t + Pb_{xx} - 2PA_0\kappa\,\theta_x + 2QA_0^2\,b = 0 \tag{3.48}$$

$$b_t + PA_0\,\theta_{xx} + 2P\kappa\,b_x = 0, \tag{3.49}$$

which has solutions of the form

$$b = b_0\mathrm{e}^{\mathrm{i}(\delta x - \nu t)} + \text{c.c.} \quad \text{and} \quad \theta = \theta_0\mathrm{e}^{\mathrm{i}(\delta x - \nu t)} + \text{c.c.} \tag{3.50}$$

Putting these expressions in Equations (3.48) and (3.49), we can check that they are indeed solutions, provided that we impose some relations linking b_0 and θ_0

$$(2QA_0^2 - P\delta^2)b_0 + \mathrm{i}A_0(\nu - 2P\kappa\delta)\theta_0 = 0 \tag{3.51}$$

$$-\mathrm{i}(\nu - 2P\kappa\delta)b_0 - PA_0\delta^2\theta_0 = 0. \tag{3.52}$$

This homogeneous set of equations has a non-vanishing solution for b_0 and θ_0 only if its determinant vanishes. This gives a dispersion relation for the perturbation

$$(\nu - 2P\kappa\delta)^2 = P^2\delta^2\left(\delta^2 - \frac{2Q}{P}A_0^2\right). \tag{3.53}$$

One immediately notices that the behaviour of ν for a given wavevector δ depends on the sign of Q/P:

- If $Q/P < 0$ (i.e. $PQ < 0$), $(\nu - 2P\kappa\delta)^2$ is positive whatever the value of δ. Solving for ν, we find a real number, which means that the values of b and θ oscillate, keeping a constant maximum amplitude. The plane wave (3.47) is thus a *stable* solution of the NLS equation if $PQ < 0$. In this case one speaks of marginal stability because the perturbations maintain a constant amplitude. They neither decay nor grow.
- For $Q/P > 0$ ($PQ > 0$), there exists a domain of the wavevector δ of the perturbation for which $(\nu - 2P\kappa\delta)^2$ is negative. In this range we have

$$\nu = 2P\kappa\delta \pm \mathrm{i}|P\delta|\sqrt{\frac{2Q}{P}A_0^2 - \delta^2}, \tag{3.54}$$

  so that ν has a nonvanishing imaginary part. Hence the term $\mathrm{e}^{-\mathrm{i}\nu t}$ leads to an exponential growth of the perturbation. The plane wave is *unstable*. It tends to self-modulate with a wavevector corresponding to the value of δ which yields the largest growth rate $\mathrm{Im}(\nu)$, i.e. $\delta = A_0\sqrt{Q/P}$, as shown in Figure 3.4.

  This phenomenon of modulational instability corresponds to an *energy localisation induced by nonlinearity* because the maximal growth rate $|Q|A_0^2$ is larger when the coefficient Q of the nonlinear term in the NLS equation and the amplitude A_0 is large. It should be noticed that the growth rate of the instability only depends on the amplitude A_0 of the plane wave but it is independent of its wavevector κ, and its frequency Ω.

In the case $PQ > 0$, which is also the condition for the existence of solitons, the study of the long-term evolution of a plane wave injected as an initial condition in the system shows that, after a stage during which the wave starts to self-modulate as predicted by the linear stability analysis of a small perturbation, this evolution goes on, until the amplitude of the plane waves reaches zero in some regions. It

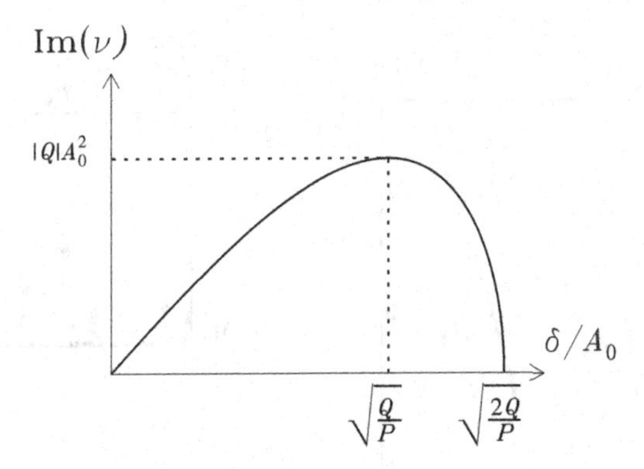

Figure 3.4. Growth rate of the perturbation Im(ν) as a function of its wavevector δ/A_0.

leads to a train of solitons, which has spontaneously emerged from the *modulational instability* of the wave. Figure 3.5(a) shows such a behaviour observed by numerical simulation. The grey scale indicates the local energy

$$E(x,t) = |\psi|^4 - |\dot{\psi}_x|^2. \tag{3.55}$$

Figures 3.5(b), 3.5(c) and 3.5(d) are three successive snapshots of the local energy along the chain. Initially the energy is uniformly distributed along the lattice (see Figure 3.5(b)). Later on, a slight modulation appears (see Figure 3.5(c)) and the homogeneous state is destabilised. Finally, as shown in Figures 3.5(a) and (d), several localised energy packets emerge. The evolution toward localisation is however complex because, as shown in the figures, the initial wave packets may split or interact with each other, leading to some intermediate stages during which the pattern is blurred.

This example shows once more that, in the conditions in which they can exist in a system, solitons are the stable excitations, and they tend to spontaneously form as soon as some energy is provided to the system. The modulational instability of a plane wave is sometimes called the Benjamin–Feir instability, carrying the names of those who first discovered it for surface water waves [26].

3.2.3 Relationship between the SG breather and the NLS soliton

We derived the NLS equation (3.25) from a multiple scale expansion of the sine-Gordon equation, and then by looking for a small-amplitude solution dominated by a term of the form

$$\theta = \varepsilon A(\xi_1, \tau_2)\, e^{i(qX_0 - \omega T_0)} + \text{c.c.} \tag{3.56}$$

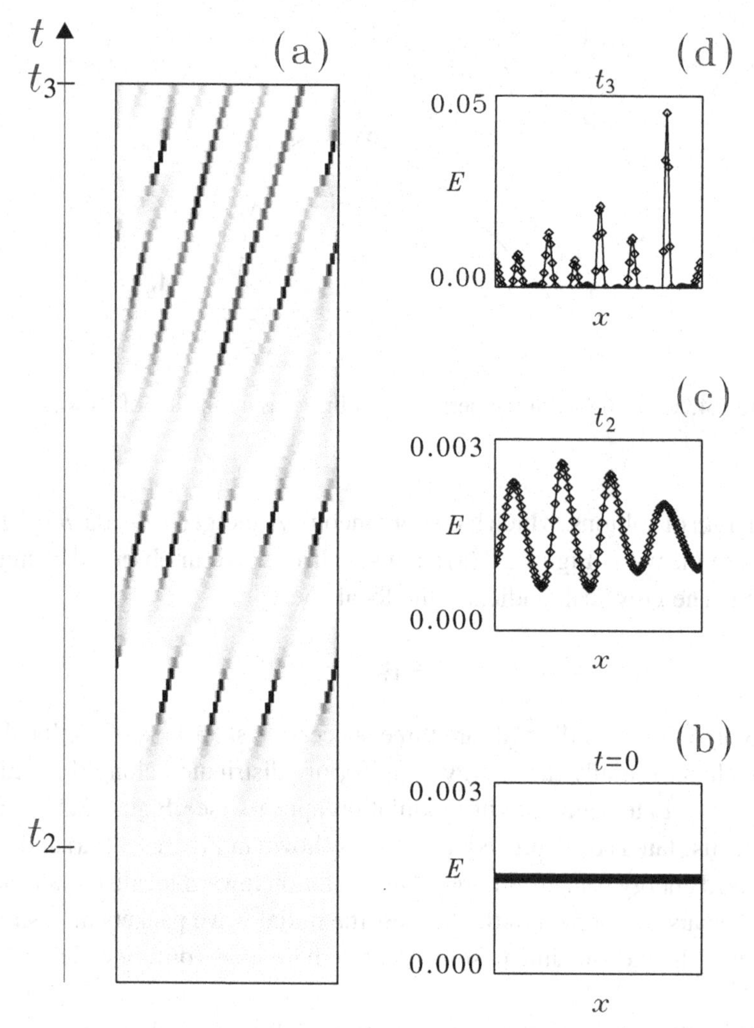

Figure 3.5. Time evolution for the field $E(x, t)$ when the initial amplitude is $A_0 = 0.25$. In panel (a), the horizontal axis indicates lattice sites and the vertical axis is time. The grey scale goes from $E = 0$ (white) to the maximum E-value (black). Panels (b), (c) and (d) show the instantaneous local energy $E(x, t)$ along the system at three different times. Notice the difference in vertical amplitude between panels (c) and (d).

Moreover we noticed that the breather solution of the SG equation can have a small amplitude, and that it corresponds to a localised oscillatory solution. It turns out to be qualitatively similar to the NLS soliton in which an envelope modulates an oscillatory carrier wave. This suggests that there must exist a relationship between these two solutions, and we shall show that the small-amplitude limit of the SG breather is identical to the soliton solution of the NLS equation. This is why the NLS soliton is sometimes called the NLS 'breather'.

First, let us notice that, if we wish to get a breather-like solution out of an NLS soliton, the spatial oscillation within the carrier wave must be eliminated, which is possible by selecting the case $q = 0$. Due to the dispersion relation $\omega^2 = \omega_0^2 + c_0^2 q^2$ of the SG equation, this means that the corresponding excitation is at the bottom of the phonon band and that its group velocity is zero. Hence variables ξ and τ in the frame moving at velocity v_g are simply the ordinary space and time variables. In this case, the NLS equation which corresponds to the SG equation is simply

$$\mathrm{i}\frac{\partial A}{\partial T_2} + \frac{c_0^2}{2\omega_0}\frac{\partial^2 A}{\partial X_1^2} + \frac{\omega_0}{4}|A|^2 A = 0. \tag{3.57}$$

In order to compare it with a breather of the SG equation which does not move in space, we must look for a solution of the NLS equation which has an envelope which is stationary in space, i.e. which is such that $u_e = 0$. At first glance, it seems impossible because the amplitude of the NLS solution is $\phi_0 = \sqrt{(u_e^2 - 2u_e u_p)/2PQ}$. If we want to preserve $\phi_0 \neq 0$ we must be careful when we take the limit $u_e \to 0$. We want to impose that $u_e u_p$ *stays constant and negative* in order to get a positive term under the root, taking into account that $PQ > 0$. The relation $u_e u_p = -PQ\phi_0^2 =$ constant, implies that the velocity of the carrier u_p should diverge in the limit. This is in agreement with the dispersion relation (3.46) that we found for the plane waves of the NLS equation. The phase of the soliton $u_e(X_1 - u_p T_2)/2P$ shows that, in the limit $u_e \to 0$, the wavevector vanishes. The dispersion relation (3.46) shows that the phase velocity indeed diverges in this case.

Taking the limit where u_e tends to zero, the solution of the NLS equation which is stationary in space becomes

$$A = \phi_0 \operatorname{sech}\left[\sqrt{\frac{Q}{2P}}\,\phi_0 X_1\right] \mathrm{e}^{\mathrm{i}\frac{Q\phi_0^2}{2}T_2}. \tag{3.58}$$

Putting it into the expression of θ which corresponds to the solution of the SG equation, and taking into account the relations $X_1 = \varepsilon x$, $T_2 = \varepsilon^2 t$, $Q = \omega_0/4$ and $P = c_0^2/2\omega_0$, we get

$$\theta = \varepsilon\phi_0 \operatorname{sech}\left[\sqrt{\frac{\omega_0^2}{4c_0^2}}\,\phi_0 \varepsilon x\right]\left(\mathrm{e}^{-\mathrm{i}\omega_0 t\left(1-\frac{\phi_0^2\varepsilon^2}{8}\right)} + \text{c.c.}\right), \tag{3.59}$$

or

$$\theta = \theta_0 \operatorname{sech}\left[\frac{\omega_0 \theta_0 x}{4c_0}\right] \cos\left[\omega_0\left(1 - \frac{\theta_0^2}{32}\right)t\right] \tag{3.60}$$

if we introduce $\theta_0 = 2\varepsilon\phi_0$.

Thus we find a spatially localised excitation, which oscillates with a frequency *smaller* than the frequency ω_0, so that it lies in the phonon gap, as does the SG breather. Moreover, as for the SG breather, the frequency decreases when the amplitude increases.

It is possible to make the comparison more quantitative by considering the small-amplitude expansion of the SG breather (2.33),

$$\theta_B = 4\tan^{-1}\left[\frac{c_0}{u}\operatorname{sech}\frac{\omega_0 x}{c_0\sqrt{1+u^2/c_0^2}}\sin\frac{\omega_0 t u}{c_0\sqrt{1+u^2/c_0^2}}\right], \tag{3.61}$$

in the domain $c_0/u \ll 1$. Expanding up to order 2 with respect to the variable c_0/u, we get

$$\theta_B \simeq \theta_0 \operatorname{sech}\frac{\omega_0\theta_0 x}{4c_0}\sin\left[\omega_0 t\left(1 - \frac{\theta_0^2}{32}\right)\right], \tag{3.62}$$

by denoting the quantity $4c_0/u$ by θ_0. Therefore we exactly get the solution (3.60) of the NLS equation, up to a shift of the time origin which turns the sine into a cosine. This shows the complete coherence of the analysis: whether we take the small-amplitude limit on the equation itself (which leads to the NLS equation and then to its breather solution), or on the exact SG solution, we get the same result.

3.3 Conservation laws

3.3.1 The NLS Lagrangian

The NLS equation can be deduced from a Lagrangian, from which a Legendre transform can derive a Hamiltonian. It order to show this, it is convenient to first simplify the NLS equation by an appropriate change of the spatial and amplitude scales. We restrict our attention to the case where P and Q are positive, and we define

$$X = \frac{x}{\sqrt{2P}} \quad \text{and} \quad \psi = \varphi\sqrt{Q}. \tag{3.63}$$

The NLS equation (3.27) reduces to

$$\mathrm{i}\varphi_t + \frac{1}{2}\varphi_{XX} + |\varphi|^2\varphi = 0. \tag{3.64}$$

Let us check that this equation can be derived from a variational principle applied to an action that we shall specify. Unfortunately there is no systematic process to derive a Lagrangian from a dynamic equation. A second difficulty is the possible existence of several Lagrangians leading to the same dynamic equations because they are defined up to the gradient of a function.

We denote by $\mathcal{L} = \mathcal{L}(\varphi, \varphi_t, \varphi_X, \varphi^*, \varphi_t^*, \varphi_X^*, X, t)$ the Lagrangian density, from which we can define

$$\text{the Lagrangian} \quad L = \int \mathrm{d}X\, \mathcal{L} \tag{3.65}$$

$$\text{and the action} \quad S = \int \mathrm{d}t\, L = \int \mathrm{d}t \int \mathrm{d}X\, \mathcal{L}. \tag{3.66}$$

The NLS equation is complex, and therefore it is an equation with two unknown fields, the real part of φ and its imaginary part. It is convenient to consider instead the fields φ and φ^* which are treated as *independent from each other*. This may seem surprising, but one must think that choosing φ amounts to selecting a point in the complex plane. If the real axis is given a priori, then one can deduce the quantities $\mathrm{Re}\,\varphi$ and $\mathrm{Im}\,\varphi$. But, on the contrary if we are given a priori φ and φ^*, the real axis is *then* determined as the bisector of the vectors defined by φ and φ^* in the complex plane. In order to simplify some formulae, in this section we denote by φ^ℓ ($\ell = 1$ or 2) the variables φ and φ^*.

If we choose the Lagrangian density

$$\mathcal{L} = \mathrm{i}(\varphi\varphi_t^* - \varphi_t\varphi^*) + |\varphi_X|^2 - |\varphi|^4 \tag{3.67}$$

the Lagrange equations (see Appendix B)

$$\frac{\mathrm{d}}{\mathrm{d}t}\left(\frac{\partial \mathcal{L}}{\partial \varphi_t}\right) + \frac{\mathrm{d}}{\mathrm{d}X}\left(\frac{\partial \mathcal{L}}{\partial \varphi_X}\right) = \frac{\partial \mathcal{L}}{\partial \varphi} \tag{3.68}$$

$$\frac{\mathrm{d}}{\mathrm{d}t}\left(\frac{\partial \mathcal{L}}{\partial \varphi_t^*}\right) + \frac{\mathrm{d}}{\mathrm{d}X}\left(\frac{\partial \mathcal{L}}{\partial \varphi_X^*}\right) = \frac{\partial \mathcal{L}}{\partial \varphi^*}, \tag{3.69}$$

give

$$\frac{\mathrm{d}(-\mathrm{i}\varphi^*)}{\mathrm{d}t} + \frac{\mathrm{d}(\varphi_X^*)}{\mathrm{d}X} = \mathrm{i}\varphi_t^* - 2\varphi\varphi^{*2} \Rightarrow -\mathrm{i}\varphi_t^* + \frac{1}{2}\varphi_{XX}^* + \varphi\varphi^{*2} = 0, \tag{3.70}$$

$$\frac{\mathrm{d}(\mathrm{i}\varphi)}{\mathrm{d}t} + \frac{\mathrm{d}(\varphi_X)}{\mathrm{d}X} = -\mathrm{i}\varphi_t - 2\varphi^*\varphi^2 \Rightarrow \mathrm{i}\varphi_t + \frac{1}{2}\varphi_{XX} + \varphi^*\varphi^2 = 0, \tag{3.71}$$

i.e. Equation (3.64) and its complex conjugate.

It is convenient to chose a *real* Lagrangian density, but other choices are possible, such as

$$\mathcal{L}' = 2\mathrm{i}\varphi\varphi_t^* + |\varphi_X|^2 - |\varphi|^4 \quad \Rightarrow \quad \mathrm{i}\varphi_t + \frac{1}{2}\varphi_{XX} + \varphi^*\varphi^2 = 0. \tag{3.72}$$

The equation which results from $\mathcal{L}'$ is identical to Equation (3.64) because the difference between the two Lagrangians $\mathcal{L}' - \mathcal{L} = 2\mathrm{i}\varphi\varphi_t^* - \mathrm{i}(\varphi\varphi_t^* - \varphi_t\varphi^*) = \mathrm{i}\partial_t|\varphi|^2$, can be expressed as the gradient of some function. On another hand, the Lagrangian $\mathcal{L}'' = \mathrm{i}\varphi\varphi_t^* + |\varphi_X|^2 + |\varphi|^4$ leads to $\mathrm{i}\varphi_t + \varphi_{XX} - \varphi^*\varphi^2 = 0$ which is the NLS equation with a negative PQ product, which must be studied separately.

3.3.2 The NLS Hamiltonian

From the Lagrangian, we can establish a Hamiltonian H, or a Hamiltonian density $\mathcal{H}$ as usual. One first introduces the canonical momentum densities which are conjugated to φ and φ^*,

$$p_\varphi = \frac{\partial\mathcal{L}}{\partial\varphi_t} = \frac{\partial}{\partial\varphi_t}[\mathrm{i}(\varphi\varphi_t^* - \varphi_t\varphi^*) + |\varphi_X|^2 - |\varphi|^4] = -\mathrm{i}\varphi^* \tag{3.73}$$

$$p_{\varphi^*} = \frac{\partial\mathcal{L}}{\partial\varphi_t^*} = \mathrm{i}\varphi. \tag{3.74}$$

Then the Hamiltonian density is derived from the canonical transform

$$\mathcal{H} = \sum_\ell p_{\varphi^\ell}\frac{\mathrm{d}\varphi^\ell}{\mathrm{d}t} - \mathcal{L} \tag{3.75}$$

$$= p_\varphi\varphi_t + p_{\varphi^*}\varphi_t^* - \mathcal{L} = -\mathrm{i}\varphi^*\varphi_t + \mathrm{i}\varphi\varphi_t^* - \mathcal{L}$$

$$= |\varphi|^4 - |\varphi_X|^2. \tag{3.76}$$

Trying to derive equations of motions from Hamiltonian (3.76), can lead to difficulties, with wrong pre-factors that differ by a factor of two from the correct ones, unless one proceeds with caution. This is a standard problem for Hamiltonians which depend on the first derivatives of the field.

This problem arises because, contrary to an essential assumption made to derive the usual Hamilton equations (see Appendix B), the variables φ and φ^* are here directly linked to their conjugate variables, through Equations (3.73) and (3.74), instead of being related to them by derivative operators as usual. The phase space does not have satisfactory properties and new Poisson brackets should be derived following a method introduced by Dirac [55, 172] to deal with such cases. However, here a direct procedure can be proposed.

Let us attempt to derive Hamilton's equation when the variables are not independent from each other. The definition of the Hamiltonian density (Equation 3.75) directly leads to

$$\frac{\partial \mathcal{H}}{\partial p_\varphi} = \frac{\partial}{\partial p_\varphi}\left[\sum_\ell p_{\varphi^\ell}\varphi_t^\ell - \mathcal{L}\right] \tag{3.77}$$

$$= \varphi_t + \sum_\ell p_{\varphi^\ell}\frac{\partial \varphi_t^\ell}{\partial p_\varphi} - \frac{\partial \mathcal{L}}{\partial p_\varphi} \tag{3.78}$$

$$= \varphi_t + \sum_\ell p_{\varphi^\ell}\frac{\partial \varphi_t^\ell}{\partial p_\varphi} - \sum_\ell \frac{\partial \mathcal{L}}{\partial \varphi_t^\ell}\frac{\partial \varphi_t^\ell}{\partial p_\varphi} \tag{3.79}$$

$$= \varphi_t + \sum_\ell p_{\varphi^\ell}\frac{\partial \varphi_t^\ell}{\partial p_\varphi} - \sum_\ell p_{\varphi^\ell}\frac{\partial \varphi_t^\ell}{\partial p_\varphi} \tag{3.80}$$

$$\frac{\partial \mathcal{H}}{\partial p_\varphi} = \varphi_t. \tag{3.81}$$

Thus this Hamilton equation is not modified by the constraints discussed above.

The second equation is more complex to establish. Using Lagrange equation (3.68), we get

$$\frac{\partial \mathcal{H}}{\partial \varphi} = \frac{\partial}{\partial \varphi}\left[\sum_\ell p_{\varphi^\ell}\varphi_t^\ell - \mathcal{L}\right] \tag{3.82}$$

$$= \sum_\ell \frac{\partial p_{\varphi^\ell}}{\partial \varphi}\varphi_t^\ell + 0 - \frac{\partial \mathcal{L}}{\partial \varphi} \tag{3.83}$$

$$\frac{\partial \mathcal{H}}{\partial \varphi} = \sum_\ell \frac{\partial p_{\varphi^\ell}}{\partial \varphi}\varphi_t^\ell - \frac{\mathrm{d}}{\mathrm{d}t}\left(\frac{\partial \mathcal{L}}{\partial \varphi_t}\right) - \frac{\mathrm{d}}{\mathrm{d}X}\left(\frac{\partial \mathcal{L}}{\partial \varphi_X}\right). \tag{3.84}$$

We also have

$$\frac{\partial \mathcal{H}}{\partial \varphi_X} = \sum_\ell \frac{\partial p_{\varphi^\ell}}{\partial \varphi_X}\varphi_t^\ell + \sum_\ell p_{\varphi^\ell}\frac{\partial \varphi_t^\ell}{\partial \varphi_X} - \frac{\partial \mathcal{L}}{\partial \varphi_X} = -\frac{\partial \mathcal{L}}{\partial \varphi_X}. \tag{3.85}$$

From Equations (3.84) and (3.85) we get the second Hamilton equation

$$\frac{\partial \mathcal{H}}{\partial \varphi} = \sum_\ell \frac{\partial p_{\varphi^\ell}}{\partial \varphi}\varphi_t^\ell - \frac{\mathrm{d}}{\mathrm{d}t}\left(\frac{\partial \mathcal{L}}{\partial \varphi_t}\right) + \frac{\mathrm{d}}{\mathrm{d}X}\left(\frac{\partial \mathcal{H}}{\partial \varphi_X}\right) \tag{3.86}$$

which can also be expressed as

$$\frac{\mathrm{d}p_\varphi}{\mathrm{d}t} = \frac{\mathrm{d}}{\mathrm{d}X}\left(\frac{\partial \mathcal{H}}{\partial \varphi_X}\right) - \frac{\partial \mathcal{H}}{\partial \varphi} + \sum_\ell \frac{\partial p_{\varphi^\ell}}{\partial \varphi}\varphi_t^\ell. \tag{3.87}$$

Taking into account the constraints set by Equations (3.73) and (3.74) the last term does not vanish and it corrects the equation with respect to the equations that would have been derived by a hasty application of the standard equations. It leads to

$$2\varphi\varphi^{*2} = \mathrm{i}\varphi_t^* - \frac{\mathrm{d}(-\mathrm{i}\varphi^*)}{\mathrm{d}t} + \frac{\mathrm{d}(-\varphi_X^*)}{\mathrm{d}X} \tag{3.88}$$

$$2\varphi\varphi^{*2} = \mathrm{i}\varphi_t^* + \mathrm{i}\varphi_t^* - \varphi_{XX}^* \tag{3.89}$$

which is the expected Equation (3.70).

It is surprising, but instructive, to note that this difficulty would have been completely hidden if we had chosen the imaginary Lagrangian (3.72). In this case the canonical variables are $p_\varphi = \partial\mathcal{L}/\partial\varphi_t = 0$ and $p_{\varphi^*} = \partial\mathcal{L}/\partial\varphi_t^* = 2\mathrm{i}\varphi$. The Hamiltonian (3.76) is not affected by the change of Lagrangian, and using the standard Hamilton equation for p_φ^* leads to

$$\frac{\mathrm{d}p_{\varphi^*}}{\mathrm{d}t} = \frac{\mathrm{d}}{\mathrm{d}X}\left(\frac{\partial\mathcal{H}}{\partial\varphi_X^*}\right) - \frac{\partial\mathcal{H}}{\partial\varphi^*} \quad \Rightarrow \quad 2\mathrm{i}\varphi_t = -\varphi_{XX} - 2\varphi\varphi^{*2} \tag{3.90}$$

which is the NLS equation that we are looking for.

On the other hand the equation for p_φ would give

$$\frac{\mathrm{d}p_{\varphi}}{\mathrm{d}t} = \frac{\mathrm{d}}{\mathrm{d}X}\left(\frac{\partial\mathcal{H}}{\partial\varphi_X}\right) - \frac{\partial\mathcal{H}}{\partial\varphi} \quad \Rightarrow \quad 0 = -\varphi_{XX}^* - 2\varphi^*\varphi^2. \tag{3.91}$$

The two equations are not complex conjugates of each other, and this reveals the difficulty that we pointed out above.

3.4 Nœther's theorem

The NLS equation belongs to the class of completely integrable systems, for which an infinity of conservation laws can be obtained by the inverse scattering method (Chapter 7). However some of these conservation laws, which are important for physical applications, can be derived more easily from the invariance properties (symmetries) of the equation, thanks to Nœther's theorem for a Lagrangian system.

3.4.1 A reminder of the theorem

Let us study a system which depends on two fields, φ and φ^*, and two coordinates X and t. We shall consider an infinitesimal transform of the fields and coordinates of the form

$$\begin{aligned} X &\rightsquigarrow X' = X + \varepsilon\chi \\ t &\rightsquigarrow t' = t + \varepsilon\tau \\ \varphi &\rightsquigarrow \varphi' = \varphi + \varepsilon\psi \\ \varphi^* &\rightsquigarrow \varphi^{*\prime} = \varphi^* + \varepsilon\psi^*. \end{aligned}$$

If the functional expression of the Lagrangian density and the action are invariant through the transform, the following conservation law holds [78]:

$$\frac{\mathrm{d}}{\mathrm{d}t}\left\{\frac{\partial\mathcal{L}}{\partial\varphi_t}[\psi-\chi\varphi_X-\tau\varphi_t]+\frac{\partial\mathcal{L}}{\partial\varphi_t^*}[\psi^*-\chi\varphi_X^*-\tau\varphi_t^*]+\mathcal{L}\tau\right\}$$
$$+\frac{\mathrm{d}}{\mathrm{d}X}\left\{\frac{\partial\mathcal{L}}{\partial\varphi_X}[\psi-\chi\varphi_X-\tau\varphi_t]+\frac{\partial\mathcal{L}}{\partial\varphi_X^*}[\psi^*-\chi\varphi_X^*-\tau\varphi_t^*]+\mathcal{L}\chi\right\}=0. \tag{3.92}$$

Let us integrate this relation with respect to space. If we choose boundary conditions such that $|\varphi| \to 0$, $|\varphi_X| \to 0$, $|\varphi_t| \to 0$ at infinity, or with periodic boundary conditions, the contribution of the second term (in $\mathrm{d}/\mathrm{d}X$) vanishes and we get the integral expression of the conservation law

$$\int \mathrm{d}X\left\{\frac{\partial\mathcal{L}}{\partial\varphi_t}[\psi-\chi\varphi_X-\tau\varphi_t]+\frac{\partial\mathcal{L}}{\partial\varphi_t^*}[\psi^*-\chi\varphi_X^*-\tau\varphi_t^*]+\mathcal{L}\tau\right\}=C, \tag{3.93}$$

where C is a constant.

3.4.2 Application to the NLS equation

Let us consider a few interesting transforms. First we shall choose *an infinitesimal change in the phase of the fields* which does not change the coordinates ($\chi = 0$, $\tau = 0$) while the fields are multiplied by a phase factor $\mathrm{e}^{\mathrm{i}\varepsilon}$ with $\varepsilon \ll 1$:

$$\varphi \rightsquigarrow \varphi' = \varphi\mathrm{e}^{\mathrm{i}\varepsilon} \quad \text{i.e.} \quad \varphi' = \varphi(1+\mathrm{i}\varepsilon)+\mathcal{O}(\varepsilon^2) \quad \text{or} \quad \psi = \mathrm{i}\varphi \tag{3.94}$$

$$\varphi^* \rightsquigarrow \varphi^{*\prime} = \varphi^*\mathrm{e}^{-\mathrm{i}\varepsilon} \text{ i.e. } \varphi^{*\prime} = \varphi^*(1-\mathrm{i}\varepsilon)+\mathcal{O}(\varepsilon^2) \quad \text{or} \quad \psi^* = -\mathrm{i}\varphi^*. \tag{3.95}$$

The Lagrangian density and action are not modified in this transform. The conservation law (3.93) gives

$$\int_{-\infty}^{+\infty} \mathrm{d}X\{(-\mathrm{i}\varphi^*)(\mathrm{i}\varphi)+(+\mathrm{i}\varphi)(-\mathrm{i}\varphi^*)\} = C, \tag{3.96}$$

which amounts to the conservation of the norm, usual for the Schrödinger equation in quantum mechanics,

$$\int_{-\infty}^{+\infty} \mathrm{d}X |\varphi|^2 = C'. \tag{3.97}$$

Let us now consider *an infinitesimal time translation* which does not change X and the fields φ and φ^* but turns t into $t' = t + \varepsilon$. It corresponds to $\tau = 1$, $\chi = 0$ and $\psi = \psi^* = 0$. Since $\partial/\partial t = \partial/\partial t'$, the Lagrangian density and the action are preserved in this transform. The conservation law gives

$$\int \mathrm{d}X \left\{ \frac{\partial \mathcal{L}}{\partial \varphi_t} [-\varphi_t] + \frac{\partial \mathcal{L}}{\partial \varphi_t^*} [-\varphi_t^*] + \mathcal{L} \right\} = C''. \tag{3.98}$$

Expanding this expression we obtain

$$\int \mathrm{d}X \left\{ |\varphi_X|^2 - \frac{1}{2} |\varphi|^4 \right\} = C'', \tag{3.99}$$

which, except for the sign, is the energy conservation because the left hand side is the integral of the Hamiltonian density (3.76). This is a particular case of the general result which connects energy conservation and invariance by time translation.

It should be noticed that the Hamiltonian density is not bounded from below. This might appear as a problem for the statistical physics of a system described by the NLS equation, however this is not the case due to the conservation of the norm, Equation (3.97), which imposes an extra condition which bounds the energy.

3.5 Nonlinear electrical lines

As mentioned in the introduction to this chapter, the NLS equation is very general and can apply to numerous weakly nonlinear physical systems which are described by a weakly dispersive wave equation in the linear limit. The list of its applications is therefore very long. We shall discuss some of them here. They have been chosen either because they allow simple experimental investigations (electrical lines), because they have a large technological importance (optical fibres), or to illustrate the generality of the NLS equation in more than one spatial dimension (self-focusing in optics).

The case of nonlinear electrical lines is particularly interesting. On one hand it is easy to build such a line in a laboratory, and on the other hand, it shows that two types of solitons can coexist in the same physical system, 'pulse' solitons, and 'envelope' solitons.

Let us again consider the electrical line that we studied in Chapter 1, described in the continuum limit approximation by Equation (1.39) which is

$$v_{tt} - c_0^2 v_{xx} - \frac{c_0^2}{12} v_{xxxx} - a(v^2)_{tt} = 0. \tag{3.100}$$

In the small-amplitude limit, where the voltage v is of order ε ($\varepsilon \ll 1$) but nevertheless large enough to excite nonlinearities, this equation can be reduced to the NLS equation by looking for a solution with a multiple scale expansion

$$v = \varepsilon V_1(X_0, T_0, X_1, T_1, X_2, T_2, \ldots) + \varepsilon^2 V_2(T_0, T_1, T_2, X_0, X_1, X_2, \ldots) + \cdots \tag{3.101}$$

The calculation is more tedious than for the sine-Gordon equation due to the even power of the nonlinear term in Equation (3.100), but it can be carried out in a similar way. It yields

$$V_1 = A_1(X_1, T_1, X_2, T_2, \ldots)\, \mathrm{e}^{\mathrm{i}(qX_0 - \omega T_0)} + \text{c.c.} \tag{3.102}$$

where q and ω are related by the linear dispersion relation $\omega^2 = c_0^2 q^2 - c_0^2 q^4/12$ which gives the group velocity $v_g = c_0^2(q - q^3/6)/\omega$.

The calculation shows that, in order to satisfy the equation at order ε^2, one *must* introduce the term

$$V_2 = A_2 \mathrm{e}^{2\mathrm{i}(qX_0 - \omega T_0)} + \text{c.c.} \quad \text{with} \quad A_2 = \frac{4a\omega^2}{c_0^2 q^4} A_1^2. \tag{3.103}$$

Changing to the frame moving at the group velocity, as for the SG case, i.e. introducing the variables $\xi_i = X_i - v_g T_i$ and $\tau_i = T_i$, one gets the following equation for A_1:

$$\mathrm{i}\frac{\partial A_1}{\partial \tau_2} + \frac{1}{2\omega}\left[c_0^2\left(1 - \frac{q^2}{2}\right) - v_g^2\right]\frac{\partial^2 A_1}{\partial \xi_1^2} - \frac{4a^2\omega^3}{c_0^2 q^4}|A_1|^2 A_1 = 0. \tag{3.104}$$

Replacing the group velocity by its value, we arrive to an NLS equation with the coefficients

$$P = -\frac{c_0^2 q^2}{72 - 6q^2}(18 - q^2) \quad \text{and} \quad Q = -\frac{4a^2\omega^3}{c_0^2 q^4}. \tag{3.105}$$

As Q is always negative, as well as P for long-wavelength carrier waves (which was initially assumed in order to apply the continuum limit approximation), the condition that $PQ > 0$ in order to have solitons is verified, in agreement with experimental observations [9].

It may seem surprising to find that *a given physical system* can be described by *two equations* as different as the modified Boussinesq equation which has pulse soliton solutions and the NLS equation which has envelope solitons. However, experiments confirm the validity of these two approximations, which are relevant for different initial excitations of the electrical chain. Moreover they show that the two types of solitons can be created *simultaneously* in the chain. We shall come back to this point in the discussion of Chapter 4, devoted to modelling.

3.6 Solitons in optical fibres

This is certainly one of the main applications of solitons, and an example where the idea of a theoretician of nonlinear science opened a multi-million-euro market! The next transatlantic optical fibre is likely to be based on the notions that we introduce in this section.

3.6.1 *Origin of the nonlinearity: nonlinear polarisation*

In order to study the propagation of an electromagnetic wave in a dielectric material, one must take into account the polarisation $\vec{P}(\vec{r}, t)$ of the material under the effect of the field $\vec{E}(\vec{r}, t)$. In most cases, only the linear contribution is considered, which amounts to keeping only the first term in the expansion of $\vec{P}$ versus $\vec{E}$

$$P_j(t) = \epsilon_0 \int_{-\infty}^{+\infty} \chi_{jk}^{(1)}(t-\tau) E_k(\tau) \mathrm{d}\tau = \epsilon_0\, \chi_{jk}^{(1)}(t) \star E_k(t) \tag{3.106}$$

where P_j and E_k are the components of $\vec{P}$ and $\vec{E}$, and the integral takes into account the delayed response of the polarisation when the field varies. The $\star$ sign denotes a convolution product, and summation over repeated indices have been omitted.

Denoting $\widetilde{F}(\omega) = \int F(t) e^{i\omega t} \mathrm{d}t$ as the Fourier transform of the function $F(t)$, this equation can be rewritten in Fourier space as

$$\widetilde{P}_j(\omega) = \epsilon_0\, \widetilde{\chi}_{jk}^{(1)}(\omega) \widetilde{E}_k(\omega) \tag{3.107}$$

where $\widetilde{\chi}_{jk}(\omega)$ is the susceptibility tensor of the material.

When a signal propagates in an optical fibre [14], nonlinear effects can become important for two reasons:

(i) Due to the very small cross-section of a fibre (10^{-6} cm^2), an incoming power of a few watts leads to power density of the order of a MW cm^{-2} in the fibre.
(ii) Applications often involve propagation over very long distances, from a kilometre to thousands of kilometres, so that even very small nonlinear terms may have a huge cumulative effect.

In this case, the next terms in the expansion of the polarisation cannot be ignored, and the expansion becomes

$$\begin{aligned} P_j(t) &= \epsilon_0 \int_{-\infty}^{+\infty} \chi_{jk}^{(1)}(t-\tau)E_k(\tau)\mathrm{d}\tau \\ &+ \epsilon_0 \int_{-\infty}^{+\infty} \chi_{jkl}^{(2)}(t-\tau_1, t-\tau_2)E_k(\tau_1)E_l(\tau_2)\mathrm{d}\tau_1\,\mathrm{d}\tau_2 \\ &+ \epsilon_0 \int_{-\infty}^{+\infty} \chi_{jklm}^{(3)}(t-\tau_1, t-\tau_2, t-\tau_3)E_k(\tau_1)E_l(\tau_2)E_m(\tau_3)\mathrm{d}\tau_1\,\mathrm{d}\tau_2\,\mathrm{d}\tau_3 \end{aligned} \tag{3.108}$$

which suggests that we should introduce the dielectric susceptibility tensors at order 2, $\chi_{jkl}^{(2)}$, and 3, $\chi_{jklm}^{(3)}$ in our calculations.

Let us consider for instance the Fourier transform of the second term in the expansion, denoted by $\widetilde{T}_2$. Expressing $\chi^{(2)}(t-\tau_1, t-\tau_2)$ in terms of its Fourier transform, we get

$$\widetilde{T}_2(\omega) = \int \mathrm{d}t \int \mathrm{d}\tau_1 \int \mathrm{d}\tau_2 \frac{1}{4\pi^2} \int \mathrm{d}\omega_1 \int \mathrm{d}\omega_2\, \widetilde{\chi}^{(2)}(\omega_1,\omega_2)\, \mathrm{e}^{-\mathrm{i}\omega_1(t-\tau_1)-\mathrm{i}\omega_2(t-\tau_2)} \\ E_1(\tau_1)E_2(\tau_2)\mathrm{e}^{\mathrm{i}\omega t} \tag{3.109}$$

$$= \frac{1}{4\pi^2} \int \mathrm{d}\omega_1 \int \mathrm{d}\omega_2\, \widetilde{\chi}^{(2)}(\omega_1,\omega_2) \int \mathrm{d}\tau_1 E_1(\tau_1)\mathrm{e}^{\mathrm{i}\omega_1\tau_1} \int \mathrm{d}\tau_2 E_2(\tau_2)\mathrm{e}^{\mathrm{i}\omega_2\tau_2} \\ \underbrace{\int \mathrm{d}t\, \mathrm{e}^{\mathrm{i}(\omega-\omega_1-\omega_2)t}}_{2\pi\,\delta(\omega-\omega_1-\omega_2)} \tag{3.110}$$

$$= \frac{1}{2\pi} \int \mathrm{d}\omega_1 \int \mathrm{d}\omega_2\, \widetilde{\chi}^{(2)}(\omega_1,\omega_2)\, \widetilde{E}_1(\omega_1)\widetilde{E}_2(\omega_2)\, \delta(\omega-\omega_1-\omega_2), \tag{3.111}$$

using the result that $2\pi\,\delta(\omega)$ is the Fourier transform of unity. Similarly

$$\widetilde{T}_3(\omega) = \frac{1}{4\pi^2} \int \mathrm{d}\omega_1 \int \mathrm{d}\omega_2 \int \mathrm{d}\omega_3\, \widetilde{\chi}^{(3)}(\omega_1,\omega_2,\omega_3)\, \widetilde{E}_1(\omega_1)\widetilde{E}_2(\omega_2)\widetilde{E}_3(\omega_3) \\ \delta(\omega-\omega_1-\omega_2-\omega_3). \tag{3.112}$$

These tensors must have the symmetry of the material, which imposes some constraints on their components. An optical fibre is made of an *isotropic* material, so that a rotation of the electric field must induce the same rotation of the polarisation. This implies that the tensors are diagonal, with all components equal: they are simply *scalars*. Moreover, as a change in the sign of the electric field must reverse the polarisation, the tensor $\chi^{(2)}$ must vanish, and the first non-linear term of the Expansion (3.108) is the third order term $\chi^{(3)}$. This is why second harmonic experiments, which require the presence of a $\widetilde{T}_2$ term, are made with anisotropic crystals. This technology, which we shall not discuss here, is called the '$\chi^{(2)}$ effect'. It is a research area in nonlinear optics which is currently blooming.

In an isotropic material, if we apply a field $\vec{E}$ of frequency ω, and which therefore includes components proportional to $\exp(-i\omega t)$ and $\exp(i\omega t)$, the Dirac δ function of Equation (3.112) implies the presence of two contributions in the third order term of the polarisation

$$P^{(3)}(t) = \widetilde{\chi}^{(3)}(-\omega, -\omega, -\omega)E_0^3 e^{-3i\omega t} + \widetilde{\chi}^{(3)}(\omega, -\omega, -\omega)|E_0|^2 E_0 e^{-i\omega t} + \text{c.c.}, \tag{3.113}$$

where the factor $4\pi^2$ coming from the Fourier transforms has been included in the definition of the dielectric susceptibilities in order to simplify the notation. Therefore the polarisation includes both a third harmonic contribution, and a nonlinear contribution which has the same frequency as the fundamental excitation.

The growth of the third harmonic is generally very weak due to the absence of so-called 'phase matching': third harmonics generated in two different places in the dielectric do not arrive in phase at a given point so that, on average, they cancel each other out because their interferences are either constructive or destructive. Moreover the dispersion relation $\omega(k)$ of the medium is seldom satisfied by the pair $\{3\omega, 3k\}$. On the contrary, nonlinear effects at the frequency of the fundamental have a cumulative effect. This is why, in optical fibres, the second term of Expression (3.113) is the dominant term.

Hence, if an optical fibre is excited at one end by an electric field $\vec{E} = \vec{E}_0(r) e^{i(\vec{k}.\vec{r} - \omega t)} + \text{c.c.}$, the leading term of the polarisation is

$$\vec{P}(t, \vec{r}) = \epsilon_0 \widetilde{\chi}^{(1)}(\omega)\vec{E}_0(r) e^{i(\vec{k}.\vec{r} - \omega t)} + \epsilon_0 \widetilde{\chi}^{(3)}(\omega)\vec{E}_0|\vec{E}_0|^2 e^{i(\vec{k}.\vec{r} - \omega t)}. \tag{3.114}$$

This expression of the polarisation can be analysed by saying that the optical index of the medium contains a contribution proportional to $|\vec{E}_0|^2$; this is the optical Kerr effect.

3.6.2 Structure of the electric field in the fibre

The expression of $\vec{P}$ already suggests that the NLS equation might apply to the fibre, but a complete calculation to establish the equation for the propagation of an optical signal in a fibre is tedious because the electromagnetic field is a *vector* field while, up to now, we have restricted our attention to scalar fields. However, for a mono-mode optical fibre and a polarised electromagnetic wave, the equation can be reduced to the scalar case because the structure of the wave in the fibre, i.e. the variation of the field within a section, is determined by transverse boundary conditions, and along the z axis of the fibre only the variation of its amplitude has to be analysed. Moreover the structure of the wave can be derived to an excellent approximation by *only considering the linear part of the polarisation* because it involves propagation in the transverse direction, i.e. over very small distances, which do not allow nonlinear effects to build up enough to play any significant role.

Consequently the study of the propagation of the electromagnetic wave in a fibre can be split in two steps:

(i) Determining the transverse structure of the wave by solving *linearised equations*.
(ii) Studying the propagation along the axis of the fibre taking into account *nonlinear effects*.

The calculation, which is long and rather technical, is presented in detail in references [103] and [138]. Here, as the transverse structure is of secondary importance, we shall only list the basic ideas for the first step and study the second step more completely.

In order to find out the structure of the wave in the fibre, let us start from the equation

$$\nabla^2 \vec{E} - \vec{\nabla}\left(\vec{\nabla} \cdot \vec{E}\right) - \frac{1}{\epsilon_0 c^2} \frac{\partial^2 \vec{D}}{\partial t^2} = \vec{0}, \tag{3.115}$$

which results from the two Maxwell equations

$$\vec{\nabla} \wedge \vec{E} = -\frac{\partial \vec{B}}{\partial t} \quad \text{and} \quad \vec{\nabla} \wedge \vec{H} = \frac{\partial \vec{D}}{\partial t}, \tag{3.116}$$

since we are interested in a nonconducting ($\vec{j} = \vec{0}$) and nonmagnetic ($\vec{B} = \mu_0 \vec{H}$) material. The expression $\vec{D} = \epsilon_0 \vec{E} + \vec{P}$, which only includes the linear part of the polarisation, leads to

$$\vec{D} = \epsilon_0 \vec{E} + \epsilon_0 \chi(\omega) \vec{E} = \epsilon(\omega) \vec{E}(\vec{r}, t). \tag{3.117}$$

The dielectric permittivity ϵ may depend on the distance to the axis of the fibre, but these variations are negligible on the scale of the wavelength of the light. For instance in fibres using an index gradient to confine the field, the variation of the index occurs over distances which are of the order of magnitude of the radius of the fibre, which is much larger than the wavelength of light. The relation $\vec{\nabla} \cdot \vec{D} = 0$ gives

$$\epsilon \vec{\nabla} \cdot \vec{E} + \vec{E} \cdot \vec{\nabla} \epsilon = \left(\epsilon \vec{k} + \vec{\nabla} \epsilon\right) \vec{E} = \vec{0}, \tag{3.118}$$

and, using the hypothesis $|\vec{\nabla}\epsilon| \ll \epsilon/\lambda$, we neglect the second term with respect to the first one because $k = 2\pi/\lambda$. Therefore we get $\vec{\nabla}\cdot\vec{E} = 0$. This approximation simplifies the calculations but it is not essential for the validity of the results that we shall derive for nonlinear propagation in a fibre.

Taking the time Fourier transform of Equation (3.115) gives

$$\nabla^2\vec{E}(r,\omega) + \frac{\omega^2}{\epsilon_0 c^2}\,\epsilon(\vec{r},\omega)\vec{E}(\vec{r},\omega) = \vec{0}\,. \tag{3.119}$$

We can look for solutions of this equation under the form of a *guided wave*

$$\vec{E}(\vec{r},\omega) = \vec{U}(\vec{r_\perp},\omega)\,\mathrm{e}^{\mathrm{i}(kz-\omega t)}, \tag{3.120}$$

where we denote by $\vec{r_\perp}$ the vector that specifies the position in a section orthogonal to the z axis of the fibre ($\vec{r} = \vec{r_\perp} + z\vec{e_z}$). These waves are guided by the fibre if $|\vec{U}(\vec{r_\perp},\omega)|$ exponentially decreases when we move away from the axis. This boundary condition in the plane orthogonal to the fibre imposes a dispersion relation for each possible propagation mode $\vec{U_n}$, which can be written

$$k_n = \frac{\omega\beta_n(\omega)}{c}. \tag{3.121}$$

For each mode, the direction of $\vec{E}$ and its variation over a transverse section are well defined, but the amplitude of the field is not because the mode has been derived by solving *a linear equation* (3.119). Therefore, if we are interested in one particular mode, it is sufficient to study a *scalar* which measures the amplitude of the mode.

Let us chose $\vec{U}$ so that $\vec{U}(\vec{r_\perp} = 0,\omega) = 1$ in order to set the amplitude scale, and let us denote by $\tilde{\phi}(z,\omega)$ the multiplicative factor which gives the amplitude in a section so that

$$\vec{E}(\vec{r},\omega) = \tilde{\phi}(z,\omega)\,\vec{U}(\vec{r},\omega)\,\mathrm{e}^{\mathrm{i}(kz-\omega t)}. \tag{3.122}$$

The amplitude of a single mode does not depend on time. Thus it cannot be used to transfer information along a fibre. The amplitude of the signal has to be modulated, and this is why soliton fibres carry light pulses as *wave packets* which combine several modes having frequencies centred around a reference frequency ω_0 associated with the wavevector k_0 (Figure 3.6). Such a superposition of modes can be written

$$\vec{E}(\vec{r_\perp},z,t) = \int_{\Delta\Omega} \mathrm{d}\Omega\,\tilde{\phi}(z,\Omega)\,\vec{U}(\vec{r_\perp},\omega_0+\Omega)\,\mathrm{e}^{\mathrm{i}[(k_0+Q(\Omega))z-(\omega_0+\Omega)t]}. \tag{3.123}$$

Hence we sum different components having frequencies $\omega = \omega_0 + \Omega$, each of them corresponding to an amplitude factor $\tilde{\phi}(\Omega)$ and a wavevector

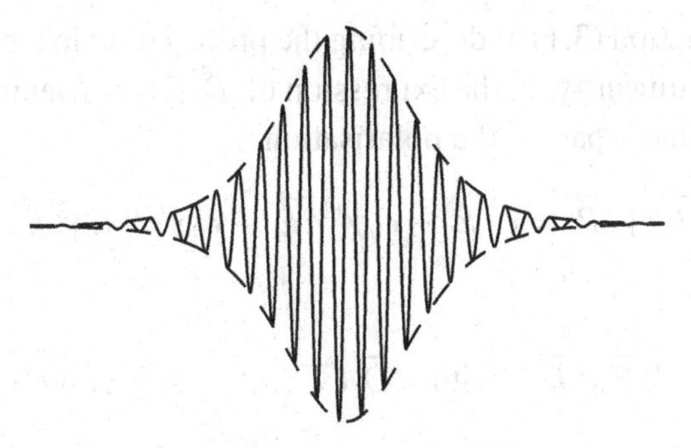

Figure 3.6. Plot of a typical light pulse, which is a wave packet with an envelope $\phi(z,t)$ plotted as a dashed line, while the solid line shows $\widetilde{\phi} U e^{i(k_0 z - \omega_0 t)}$.

$k(\omega_0 + \Omega) = k_0 + Q(\Omega)$. The modulation is selected to make sure that the variation of the frequency around the frequency of the carrier wave stays small, i.e. $\Delta\Omega \ll \omega_0$.

Thus we can assume that the spatial structure of each component of the wave packet is the same as the structure of the mode (ω_0, k_0), which amounts to assuming that $\vec{U}(\vec{r_\perp}, \omega_0 + \Omega) \simeq \vec{U}(\vec{r_\perp}, \omega_0)$. Therefore

$$\vec{E}(\vec{r_\perp}, z, t) = \vec{U}(\vec{r_\perp}, \omega_0)\, e^{i(k_0 z - \omega_0 t)} \int_{\Delta\Omega} d\Omega\, \widetilde{\phi}(z, \Omega)\, e^{i[Q(\Omega) z - \Omega t]}. \quad (3.124)$$

The integral gives the amplitude factor $\phi(z,t)$ of the wave packet, expressed as the Fourier transform of its envelope $\widetilde{\phi}(z, \Omega)$, which is a slow varying function, and therefore only includes low frequencies ($\Delta\Omega \ll \omega_0$). As a result the integral is a function *which varies slowly in space and time with respect to the carrier wave* which is in front of it. In order to express this result more formally, we can consider that ϕ depends on slow variables, $\phi(\varepsilon z, \varepsilon t)$ with $\varepsilon \ll 1$, or, in other words, that its space and time derivative are of order ε.

These approximations are well verified for a wave packet carried by an optical fibre. In order to transfer data at a rate of 1 Gbit s^{-1}, i.e. 10^9 bits s^{-1}, one must use light pulses as short as, for instance, $\delta t = 2 \cdot 10^{-10}$ s which still leaves enough time between the pulses to identify them. The spatial extent of such a pulse is $c\,\delta t = 3 \cdot 10^8 \times 2 \cdot 10^{-10} = 60$ mm. Such a distance contains about 40 000 wavelengths $\lambda = 1.5\,\mu m$ of a carrier wave in the near infrared, which is a wavelength heavily used in optical communications. Thus the space–time variation of $\phi(z,t)$ would be about 40 000 times slower than that of the carrier wave.

3.6.3 Nonlinear propagation along the fibre

Knowing the field distribution in a plane orthogonal to the fibre, the goal is now to determine how it evolves along the z axis. Thus we sit on the axis $\vec{r_\perp} = \vec{0}$ and again consider Equation (3.115) describing the propagation in the fibre, now taking into account its nonlinearity. In the expression of $\overrightarrow{D}$ it is convenient to separate the linear and the nonlinear part of the polarisation.

$$\overrightarrow{D} = \epsilon_0 \overrightarrow{E} + \overrightarrow{P} = \epsilon_0 \overrightarrow{E} + \epsilon_0 \chi^{(1)} \overrightarrow{E} + \epsilon_0 \chi^{(3)} | \overrightarrow{E} |^2 \overrightarrow{E}, \tag{3.125}$$

which we shall write

$$\overrightarrow{D} = \overrightarrow{D_\ell} + \epsilon_0 \chi^{(3)} | \overrightarrow{E} |^2 \overrightarrow{E} \quad \text{with} \quad \overrightarrow{D_\ell}(\overrightarrow{r}, \omega) = \epsilon(\overrightarrow{r}, \omega) \overrightarrow{E}(\overrightarrow{r}, \omega). \tag{3.126}$$

The wave equation becomes

$$\nabla^2 \overrightarrow{E} - \frac{1}{\epsilon_0 c^2} \frac{\partial^2 \overrightarrow{D_\ell}}{\partial t^2} = \frac{\chi^{(3)}}{c^2} \frac{\partial^2 | \overrightarrow{E} |^2 \overrightarrow{E}}{\partial t^2} \tag{3.127}$$

in which we want to consider a solution of the form

$$\overrightarrow{E}(\overrightarrow{r}, t) = \phi(z, t) \overrightarrow{U}(\overrightarrow{r_\perp}, \omega_0) \, \mathrm{e}^{\mathrm{i}(k_0 z - \omega_0 t)} + \text{c.c.} \tag{3.128}$$

Remembering that $\chi^{(3)}$ is of order ε^2 and that the derivatives of ϕ are of order ε, we can carry the calculations up to order ε^2. We shall henceforth omit the signs $\rightarrow$ above the electric field and U because we are only interested in one component of the field.

The right hand side of Equation (3.127) easily gives

$$\frac{\chi^{(3)}}{c^2} \frac{\partial^2 |E|^2 E}{\partial t^2} \simeq -\frac{\omega_0^2}{c^2} \chi^{(3)} |\phi|^2 \phi \, U(\overrightarrow{r_\perp} = 0, \omega_0) \, \mathrm{e}^{\mathrm{i}(\mathrm{k}_0 \mathrm{z} - \omega_0 \mathrm{t})}, \tag{3.129}$$

where $|U|^2 = 1$ on the axis of the fibre. Let us now examine the terms of the left hand side of the wave equation. We have

$$\nabla^2 \overrightarrow{E} = \nabla^2 \phi U \mathrm{e}^{\mathrm{i}(\mathrm{k}_0 \mathrm{z} - \omega_0 \mathrm{t})} + 2 \overrightarrow{\nabla \phi} . \overrightarrow{\nabla} U \mathrm{e}^{\mathrm{i}(\mathrm{k}_0 \mathrm{z} - \omega_0 \mathrm{t})} + \phi \nabla^2 \left(U \mathrm{e}^{\mathrm{i}(\mathrm{k}_0 \mathrm{z} - \omega_0 \mathrm{t})} \right) \tag{3.130}$$

$$= \frac{\partial^2 \phi}{\partial z^2} U \mathrm{e}^{\mathrm{i}(\mathrm{k}_0 \mathrm{z} - \omega_0 \mathrm{t})} + 2\mathrm{i} \frac{\partial \phi}{\partial z} k_0 U \mathrm{e}^{\mathrm{i}(\mathrm{k}_0 \mathrm{z} - \omega_0 \mathrm{t})} + \phi \nabla^2 \left(U \mathrm{e}^{\mathrm{i}(\mathrm{k}_0 \mathrm{z} - \omega_0 \mathrm{t})} \right) \tag{3.131}$$

because the gradient of U is orthogonal to the z axis since U only depends on $\overrightarrow{r_\perp}$.

The calculation of $\partial^2 \overrightarrow{D_\ell} / \partial t^2$ must be carried out with care because, if we introduce *nonlinearity* in to the equation, we must not neglect *dispersion* otherwise we shall get shock-wave solutions. It is convenient to perform the calculation in Fourier space by introducing the inverse Fourier transform $\mathcal{F}^{-1}$ so that we have

$$D(z, t) = \mathcal{F}^{-1} \left[\epsilon(\omega) E(z, \omega) \right]. \tag{3.132}$$

Owing to its wave packet shape, the electric field $E(z, \omega)$ has a spectrum which is sharply peaked at the frequency ω_0. This suggests expanding $\epsilon(\omega)$ around its value for $\omega = \omega_0$. The expansion must be carried up to the second order to stick to the order of the previous calculations. It yields

$$D(z,t) = \mathcal{F}^{-1}\left[\left(\epsilon(\omega_0) + (\omega-\omega_0)\frac{\partial\epsilon}{\partial\omega}_{|\omega_0} + \frac{(\omega-\omega_0)^2}{2}\frac{\partial^2\epsilon}{\partial\omega^2}_{|\omega_0} + \cdots\right)E(z,\omega)\right]$$
$$= \epsilon(\omega_0)\mathcal{F}^{-1}[E(z,\omega)] + \frac{\partial\epsilon}{\partial\omega}_{|\omega_0}\mathcal{F}^{-1}[(\omega-\omega_0)E(z,\omega)]$$
$$+ \frac{1}{2}\frac{\partial^2\epsilon}{\partial\omega^2}_{|\omega_0}\mathcal{F}^{-1}[(\omega-\omega_0)^2E(z,\omega)]. \tag{3.133}$$

Equation (3.124) for $\Omega = \omega - \omega_0$, gives

$$\mathcal{F}^{-1}[(\omega-\omega_0)E(z,\omega)] = U(\omega_0)\,\mathrm{e}^{\mathrm{i}(k_0 z-\omega_0 t)}\,\mathcal{F}^{-1}\left[\int_{\Delta\Omega}\mathrm{d}\Omega\;\Omega\phi(\Omega)\,\mathrm{e}^{\mathrm{i}[Q(\Omega)z-\Omega t]}\right]$$
$$= U(\omega_0)\,\mathrm{e}^{\mathrm{i}(k_0 z-\omega_0 t)}\frac{\partial\phi}{\partial t}. \tag{3.134}$$

A similar procedure for the factor $(\omega-\omega_0)^2$ finally gives

$$D(z,t) = \left[\epsilon(\omega_0)\phi(z,t) + \mathrm{i}\frac{\partial\epsilon}{\partial\omega}_{|\omega_0}\frac{\partial\phi}{\partial t} - \frac{1}{2}\frac{\partial^2\epsilon}{\partial\omega^2}_{|\omega_0}\frac{\partial^2\phi}{\partial t^2}\right]U(\omega_0)\,\mathrm{e}^{\mathrm{i}(k_0 z-\omega_0 t)}. \tag{3.135}$$

The expression that we have derived is indeed of order ε^2 because it contains the second derivative of ϕ. There are still a few steps to go. We compute $\partial^2 D(z,t)/\partial t^2$ by neglecting all the derivatives of ϕ of order higher than 2 since we are carrying out the calculation at order ε^2. This leads to

$$\frac{\partial^2 D(z,t)}{\partial t^2} = \left[-\omega_0^2\epsilon(\omega_0)\phi - 2\mathrm{i}\omega_0\left(\epsilon + \frac{\omega_0}{2}\frac{\partial\epsilon}{\partial\omega}\right)\frac{\partial\phi}{\partial t}\right.$$
$$\left. + \left(\epsilon + 2\omega_0\frac{\partial\epsilon}{\partial\omega} + \frac{\omega_0^2}{2}\frac{\partial^2\epsilon}{\partial\omega^2}\right)\frac{\partial^2\phi}{\partial t^2}\right]U(\omega_0)\,\mathrm{e}^{\mathrm{i}(k_0 z-\omega_0 t)}. \tag{3.136}$$

Putting the three Expressions (3.129), (3.131) and (3.136) in Equation (3.127) we reach (at last!) the final stage. First we get a term proportional to ϕ, but its pre-factor vanishes thanks to Equation (3.119) which gives the spatial structure of the wave. We can then simplify by the factor $U\,\mathrm{e}^{\mathrm{i}(k_0 z-\omega_0 t)}$, which appears in all remaining terms, to get the equation determining the time evolution of the envelope

$$\frac{\partial^2\phi}{\partial z^2} + 2\mathrm{i}k_0\frac{\partial\phi}{\partial z} + \frac{2\mathrm{i}\omega_0}{\epsilon_0 c^2}\left(\epsilon + \frac{\omega_0}{2}\frac{\partial\epsilon}{\partial\omega}\right)\frac{\partial\phi}{\partial t} - \frac{1}{\epsilon_0 c^2}\left(\epsilon + 2\omega_0\frac{\partial\epsilon}{\partial\omega} + \frac{\omega_0^2}{2}\frac{\partial^2\epsilon}{\partial\omega^2}\right)\frac{\partial^2\phi}{\partial t^2}$$
$$+ \frac{\omega_0^2}{c^2}\chi^{(3)}|\phi|^2\phi = 0. \tag{3.137}$$

Thus we get an equation which looks very complicated at a first glance. However examining the physical meaning of its coefficients will allow us to simplify it.

Let us consider the dispersion relation of electromagnetic waves propagating freely in the dielectric medium of the fibre in which the wave speed is c_{medium}

$$k^2 = \frac{\omega^2}{c^2_{\text{medium}}} = \frac{\omega^2 \epsilon(\omega)}{c^2 \epsilon_0}. \tag{3.138}$$

It yields

$$\frac{\partial k}{\partial \omega} = \frac{1}{v_g} = \frac{\omega \epsilon(\omega)}{k c^2 \epsilon_0} + \frac{1}{2k} \frac{\omega^2}{c^2 \epsilon_0} \frac{\partial \epsilon}{\partial \omega} \tag{3.139}$$

where v_g is the group velocity of a wave packet centred around frequency ω. The term proportional to $\partial \phi / \partial t$ in Equation (3.137) has therefore a simple expression in terms of $1/v_g$.

Moreover we have

$$\frac{\partial}{\partial \omega}\left(\frac{k}{v_g}\right) = \frac{1}{v_g} \underbrace{\frac{\partial k}{\partial \omega}}_{1/v_g} + k \frac{\partial}{\partial \omega}\left(\frac{1}{v_g}\right) = \frac{\epsilon(\omega)}{c^2 \epsilon_0} + \frac{2\omega}{c^2 \epsilon_0} \frac{\partial \epsilon}{\partial \omega} + \frac{\omega^2}{2c^2 \epsilon_0} \frac{\partial^2 \epsilon}{\partial \omega^2}. \tag{3.140}$$

For $\omega = \omega_0$, we recover the coefficient of $\partial^2 \phi / \partial t^2$, so that Equation (3.137) can finally be written as

$$\frac{\partial^2 \phi}{\partial z^2} + 2\mathrm{i} k_0 \left[\frac{\partial \phi}{\partial z} + \frac{1}{v_g} \frac{\partial \phi}{\partial t}\right] - k_0 \frac{\partial}{\partial \omega}\left(\frac{1}{v_g}\right) \frac{\partial^2 \phi}{\partial t^2} - \frac{1}{v_g^2} \frac{\partial^2 \phi}{\partial t^2}$$
$$+ \frac{\omega_0^2}{c^2} \chi^{(3)} |\phi|^2 \phi = 0, \tag{3.141}$$

which reminds us of the NLS equation, except for a change of frame.

Without nonlinearity and group velocity dispersion, the solution would simply be

$$\phi(z, t) = \phi\left(t - \frac{z}{v_g}\right) = \phi(\tau) \quad \text{setting} \quad \tau = t - \frac{z}{v_g}. \tag{3.142}$$

The envelope would keep a permanent profile and move at the group velocity v_g, in agreement with its definition. In such a case, observing the envelope passing through any section of the fibre, we would always observe the same function, simply shifted by the amount z/v_g, depending on the observation point. This suggests making a frame change to switch to variables

$$\tau = t - \frac{z}{v_g} \quad \text{and} \quad \xi = z. \tag{3.143}$$

Without the dispersion and the nonlinearity, ϕ would only depend on τ and not on ξ. In their presence the envelope will, in general, depend on the observation point, but we can predict that its variation with ξ will be very slow. In the new frame it is reasonable to assume that it is one order of magnitude slower than its variation with z which was assumed to be of order ε.

With the new variables we have

$$\frac{\partial^2 \phi}{\partial z^2} = \frac{\partial^2 \phi}{\partial \xi^2} - \frac{2}{v_g}\frac{\partial^2 \phi}{\partial \xi \partial \tau} + \frac{1}{v_g^2}\frac{\partial^2 \phi}{\partial \tau^2}. \tag{3.144}$$

If we take into account that the variation of ϕ with respect to τ is of order ε (as was its variation with respect to t) and that its variation with respect to ξ is of order ε^2, and if we stay at the order ε^2 at which we carry out all the calculations, the derivative $\partial^2\phi/\partial z^2$ becomes

$$\frac{\partial^2 \phi}{\partial z^2} = \frac{1}{v_g^2}\frac{\partial^2 \phi}{\partial \tau^2}, \tag{3.145}$$

which simplifies Equation (3.141) into

$$\mathrm{i}\frac{\partial \phi}{\partial \xi} - \frac{1}{2}\frac{\partial}{\partial \omega}\left(\frac{1}{v_g}\right)\frac{\partial^2 \phi}{\partial \tau^2} + \frac{\omega_0^2}{2k_0c^2}\chi^{(3)}|\phi|^2\phi = 0. \tag{3.146}$$

Hence it is a nonlinear Schrödinger equation, as a function of variables $\xi = z$ and $\tau = t - z/v_g$, in which the role of time and space have been switched with respect to the usual NLS equation. As shown earlier, this equation has the following solution

$$\phi = \phi_0 \operatorname{sech}\left[\sqrt{\frac{Q}{2P}}\,\phi_0\xi\right] \mathrm{e}^{\mathrm{i}\frac{Q\phi_0^2}{2}z}. \tag{3.147}$$

Notes:

- The calculation, as we carried it out above, is reminiscent of a multiple scale expansion. A. Newell and J. Moloney [138] have done it in this way. The procedure is even more cumbersome but there is no need to assume a priori that $\partial\phi/\partial\xi$ is of order ε^2.
- The NLS equation in the optical fibre could be derived much faster, starting from the dispersion relation expanded around $\omega = \omega_0$ and taking into account nonlinear effects which introduce a dependence of k upon the electric field. In our study of the polarisation in the medium, we showed that the optical index depends on the square of the modulus of the field. This indicates that k should depend on $|\phi|^2$ in addtion to its usual dependence on ω. Its expansion leads to

$$k(\omega, |\phi|^2) = k_0(\omega_0) + \frac{\partial k}{\partial \omega}(\omega - \omega_0) + \frac{1}{2}\frac{\partial^2 k}{\partial \omega^2}(\omega - \omega_0)^2 + \frac{\partial k}{\partial |\phi|^2}|\phi|^2, \tag{3.148}$$

which can be written as

$$(k - k_0) - (\omega - \omega_0)\frac{1}{v_g} - \frac{1}{2}\frac{\partial}{\partial \omega}\left(\frac{1}{v_g}\right)(\omega - \omega_0)^2 - \frac{\partial k}{\partial |\phi|^2}|\phi|^2 = 0. \quad (3.149)$$

This relation can be viewed as the dispersion relation coming from a wave equation for a wave centred around the frequency ω_0 and the wavevector k_0, if we associate

$$(k - k_0) \text{ with } \frac{1}{\mathrm{i}}\frac{\partial}{\partial z}, \quad (\omega - \omega_0) \text{ with } -\frac{1}{\mathrm{i}}\frac{\partial}{\partial t} \text{ and } (\omega - \omega_0)^2 \text{ with } -\frac{\partial^2}{\partial t^2} \quad (3.150)$$

applied to ϕ. This dispersion relation would result from

$$\mathrm{i}\left(\frac{\partial \phi}{\partial z} + \frac{1}{v_g}\frac{\partial \phi}{\partial t}\right) + \frac{1}{2}\frac{\partial}{\partial \omega}\left(\frac{1}{v_g}\right)\frac{\partial^2 \phi}{\partial t^2} + \frac{\partial k}{\partial |\phi|^2}|\phi|^2\phi = 0, \quad (3.151)$$

which is identical to Equation (3.141) except for the two terms $(\partial^2\phi/\partial z^2)$ and $(1/v_g^2)(\partial^2\phi/\partial t^2)$ which could not be obtained by this expansion of k limited to order 2 in ω because they are of higher order.

Starting from the dispersion relation thus appears an ideal method to derive the NLS equation ... if the result is already known! But it cannot explicitly give all the coefficients of the equation, such as the expression of the nonlinear term versus $\chi^{(3)}$. In spite of its limitation, this method is nevertheless interesting because it can easily tell if a given physical system can be described by the NLS equation.

- The NLS equation that we derived by direct calculation includes the group velocity v_g defined from the dispersion relation in the medium in free space, while the derivation from the dispersion relation includes the group velocity in the fibre, which is the correct result. The discrepancy arises from the approximation $\vec{\nabla}.\vec{E} = 0$ used to separate the different components of $\vec{E}$. If one does not make this assumption, the three components of $\vec{E}$ must be treated together [138]. The calculation is longer, although it is carried out along the same lines, and it introduces the group velocity in the fibre. It shows that the group velocity dispersion has two origins: the dispersion coming from the material itself, and a geometrical dispersion resulting from the transverse boundary conditions.

3.6.4 Confrontation with experiments

The NLS equation for an optical fibre was proposed [85] in 1973 by two theoreticians, A. Hasegawa and F. Tappert, but the first experimental checks were only made in 1980 by L. Mollenauer [133] because two technical problems had to be overcome:

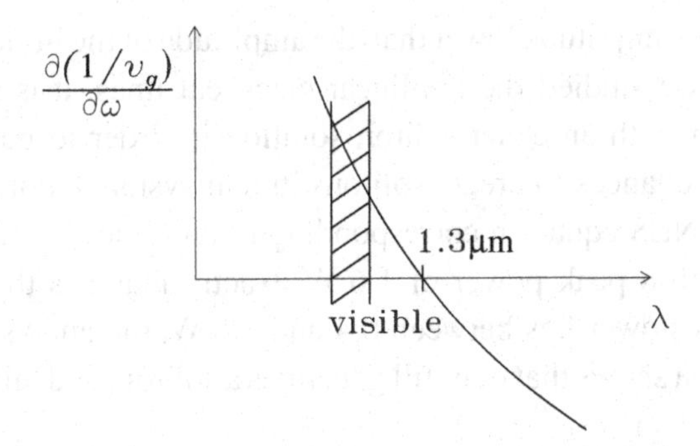

Figure 3.7. Schematic plot of the group velocity dispersion versus the wavelength λ for a silica fibre.

(i) Mono-mode fibres, very thin and with a very small loss, were not available until fairly recently.
(ii) Working in the conditions where the NLS equation (3.146) has soliton solutions, i.e. when $PQ > 0$, was difficult. Dielectric materials are such that $\chi^{(3)} > 0$ so that the coefficient Q is necessarily positive. In order to have P positive too, $\partial(1/v_g)/\partial\omega$ must be negative. This means that the experiment had to be carried out in a region of *anomalous dispersion* because, for ordinary materials, $\partial(1/v_g)/\partial\omega$ is positive. As shown in Figure 3.7, to have a positive P one has to work with a wavelength λ greater than 1.3 μm, i.e. in the near infrared region. Another possibility would be to take advantage of the geometrical dispersion. The first method was chosen for experiments carried out with neodyme Yag lasers. Their strongest emission is at $\lambda = 1.06$ μm but they also have a weaker emission at $\lambda = 1.32$ μm which was introduced in a fibre having the zero of its dispersion curve for $\lambda = 1.27$ μm. Nowadays experiments are performed with micro-structured fibres, also called photonic crystal fibres, which have their zero dispersion point λ_{ZD} in the visible domain.

When these conditions are met, experiments confirm that the NLS equation provides a very good description of the behaviour of a pulse of light sent into the fibre.

Satsuma and Yajima [166] studied the evolution of an initial condition of the form

$$\phi = a\,\phi_0\,\mathrm{sech}\left(\sqrt{\frac{Q}{2P}}\phi_0\tau\right) \tag{3.152}$$

injected at one end of a fibre ($\xi = 0$), which is, except for the factor a, the soliton solution of the NLS equation. They showed that, in the range $a \in [0.5, 1.5]$, such a pulse gives rise to a single soliton together with dispersive waves. For $a < 1$, a soliton which has an amplitude lower than the amplitude of the initial pulse emerges. As noticed when we studied the nonlinear electrical lines, it is not necessary to excite the medium with an exact soliton solution in order to generate a soliton. This enhances the chances to create solitons in real systems. For instance, for the parameters of the NLS equation corresponding to an optical fibre, a pulse lasting 1 ps (10^{-12} s), with a peak power of 1.6 W exactly matches the properties of a soliton. If the peak power lies between 0.4 and 3.6 W, the analysis carried out by Satsuma and Yajima shows that one still generates a soliton, and this is confirmed by experiments.

However the propagation of a single soliton is not a highly sensitive test of the validity of the NLS description because a small distortion of the pulse as it propagates would be difficult to detect owing to experimental errors. Obtaining *multisoliton* solutions is, on the contrary, a much more stringent test. The calculations carried out by Satsuma and Yajima [166] show that the number of solitons which emerge from a given initial condition is the largest integer N such that $N \leq a + 1/2$. The cases $N = 2$ and $N = 3$ are particularly interesting because the analytical solution can be viewed as a set of two or three solitons which move together while they oscillate with respect to each other, similarly to the kinks making up the SG breather.

Due to this internal oscillation, when the solitons which compose the solution are exactly on top of each other, the amplitude of the solution becomes very high while its width is very small because nonlinear localisation is very strong. This phenomenon shows up regularly, with a spatial recurrence period which is a characteristic of the initial condition. The observation of this phenomenon is a *very sensitive test of the validity of the NLS description* [169]. The sharpening of the pulse is even more remarkable when the number of solitons in the multisoliton solution grows. It has been investigated experimentally [170] up to $N = 13$!

This situation provides a mechanism to compress a light pulse. It is used in *soliton lasers* which include an optical fibre within the laser cavity. This is the basis of the femtosecond lasers which are now made.

3.6.5 Application to optical fibre communications

In spite of its remarkable success in analysing experimental data, the NLS equation that we derived for an optical fibre does not describe all the properties which are observed in long distance communication because very small phenomena may have a significant impact when propagation over thousands of kilometres is involved. The equation must be completed by a few additional terms, which can be inferred by carrying further the Expansion (3.148) of the dispersion

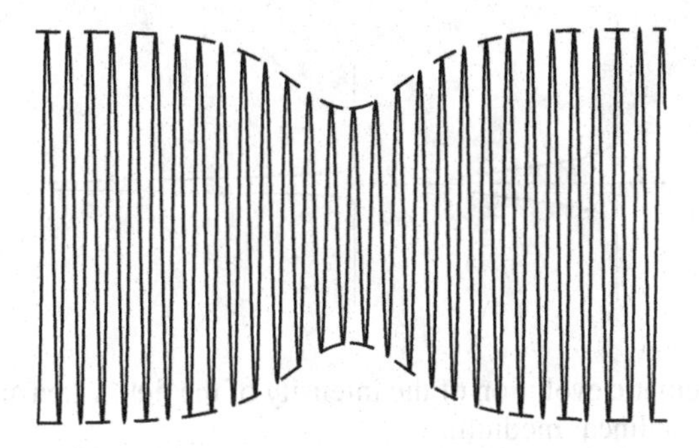

Figure 3.8. Schematic plot of a 'grey soliton'.

relation

$$k(\omega, |\phi|^2) - k_0(\omega_0) = \frac{\partial k}{\partial \omega}(\omega - \omega_0) + \frac{1}{2}\frac{\partial^2 k}{\partial \omega^2}(\omega - \omega_0)^2 + \frac{\partial k}{\partial |\phi|^2}|\phi|^2$$
$$+ \frac{1}{6}\frac{\partial^3 k}{\partial \omega^3}(\omega - \omega_0)^3 + \frac{\partial^2 k}{\partial \omega \partial |\phi|^2}(\omega - \omega_0)|\phi|^2. \quad (3.153)$$

The two extra terms add two contributions to the NLS equation,

$$-\frac{\mathrm{i}}{6}\frac{\partial^3 k}{\partial \omega^3}\frac{\partial^3 \phi}{\partial t^3} + \mathrm{i}\left(\frac{\partial^2 k}{\partial \omega \partial |\phi|^2}\right)\frac{\partial(|\phi|^2 \phi)}{\partial t}. \quad (3.154)$$

Moreover losses in the fibre cannot be neglected. They are associated with the imaginary part of $\epsilon(\omega)$ which leads to an $\mathrm{i}\Gamma\phi$ term in the NLS equation.

The role of all these terms can be studied in a perturbative calculation of the NLS equation [84]. As expected the dissipative term leads to an exponential decay of the amplitude of the soliton, while the extra dispersive term introduces a contribution to the propagation speed of the soliton which depends on its amplitude. The last term in (3.154) has little effect on a single soliton, but it breaks the stability of multisoliton solutions leading to a splitting of the solitons which make them up [103].

Finally, it is also necessary to consider the possible interactions between the different components of the electromagnetic field if the fibre has a small birefringence. All these effects have been studied very accurately both theoretically and experimentally, due to their technological importance. The example of optical fibres illustrates the interest of a close collaboration between theory and experiments, and between fundamental and applied science.

We have carried out analysis of soliton solutions in optical fibres in the anomalous dispersion regime. The normal dispersion regime ($P < 0$, $Q > 0$) also allows a soliton-like solution [85], of a different nature. It is the so-called 'grey soliton' plotted in Figure 3.8, which is a local drop in the amplitude of a plane wave. Its energy density is not localised so that, strictly speaking, it is not a soliton. Such excitations

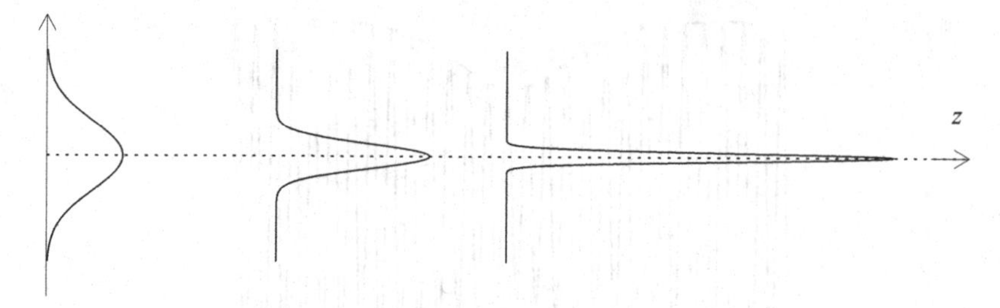

Figure 3.9. Schematic evolution of the intensity of the field when an intense beam propagates in a nonlinear medium.

have been exhibited in experiments using very long wave trains, which are obviously finite although the theoretical solution involves infinite plane waves [101]. They have also been observed in Bose–Einstein condensates, as discussed in Section 14.5.2, and in some experiments in fluids [63].

3.7 Self-focusing in optics: the NLS equation in two space dimensions

The intensity of the pulses which create solitons in fibres is moderate, and propagation over long distances is necessary to allow nonlinear effects to play a role while, in directions transverse to the fibre, they can be neglected. But, if one excites a medium with much more intense fields, or in highly nonlinear materials such as liquid crystals, nonlinear effects can strongly influence the distribution of the electric field in the transverse direction, and lead to self-focusing phenomena (Figure 3.9).

Let us consider a light beam propagating in a nonlinear medium which is assumed to be infinite because its size orthogonal to the beam is much larger than the size of the beam. Moreover we assume that the beam is *monochromatic*, that its diameter is much larger than the wavelength of the light, and that the light is polarised so that we only have to consider one component of the electric field

$$E = \phi(x, y, z)\, e^{i(k_0 z - \omega_0 t)} + \text{c.c.}, \tag{3.155}$$

where the function ϕ describes the structure of the field in the beam.

The wave equation in the nonlinear medium is Equation (3.127) written for a single component,

$$\nabla^2 E - \frac{1}{\epsilon_0 c^2}\frac{\partial^2 D_\ell}{\partial t^2} = \frac{\chi^{(3)}}{c^2}\frac{\partial^2}{\partial t^2}|E|^2 E. \tag{3.156}$$

Since we are studying a monochromatic wave, the calculation is much simpler than for a wave packet because we can use the simple relation $D_\ell = \epsilon(\omega_0)E$. Simple rules of vector calculus give

$$\nabla^2 E = \nabla^2\phi\, \mathrm{e}^{\mathrm{i}(k_0 z-\omega_0 t)} + \left(2\mathrm{i}k_0\frac{\partial\phi}{\partial z} - k_0^2\phi\right)\mathrm{e}^{\mathrm{i}(k_0 z-\omega_0 t)}. \tag{3.157}$$

Putting this expression into the wave equation (3.156), and using the dispersion relation $k_0^2 = \omega_0^2\epsilon(\omega)/c^2\epsilon_0$, we get

$$2\mathrm{i}k_0\frac{\partial\phi}{\partial z} + \nabla^2\phi + \frac{\omega_0^2\chi^{(3)}}{c^2}|\phi|^2\phi = 0, \tag{3.158}$$

which belongs to the family of equations of the form

$$\mathrm{i}\frac{\partial\phi}{\partial z} + P\nabla^2\phi + Q|\phi|^2\phi = 0 \tag{3.159}$$

where the operator ∇^2 acts in a D dimensional space.

For $D = 1$, we recover a NLS equation where the space variable z replaces time. We have shown that it has soliton solutions if the product PQ is positive. This case is however very peculiar because for $D > 1$ a phenomenon much more dramatic than the modulational instability that we found for $D = 1$ appears: the 'collapse' of the wave. The case $D = 2$ corresponds to self-focusing in optics and $D = 3$ is relevant for self-focusing in plasmas. In this last case, the first order derivative in the equation is a derivative with respect to time, as in the standard NLS equation (3.27).

For a light beam in a nonlinear medium [194], the variation of ϕ with space in the transverse direction is much slower than the space variation of the exponential factor of Expression (3.155). While the exponential factor varies over a length of the order of the wavelength of light (a micron or below), the variation of ϕ occurs over a length of the order of the diameter of the beam, such as a millimetre. In order to express this difference in the equation, we can assume that ϕ depends on the variables εx and εy, with $\varepsilon \ll 1$. For a parallel beam, the energy conservation implies that ϕ does not depend on z. In a nonlinear medium, we shall show that the section of the beam changes as it propagates, but this variation takes place over distances which are much larger than the transverse dimension of the beam. Therefore we assume that ϕ depends on z through the variable $\varepsilon^2 z$. This analysis of the relevant scales for the beam shows therefore that its amplitude is of the form $\phi(\varepsilon x, \varepsilon y, \varepsilon^2 z)$, which expresses quantitatively that the transverse and longitudinal directions play a very different role in the variation of the field. Consequently the derivatives with respect to variables x and y arising from $\nabla^2\phi$ in the wave equations are of order ε^2 while the derivative with respect to z is of order ε^4 and can be neglected with respect to the others. This is why we must consider ∇^2 *in two spatial dimensions* in the wave equation (3.158) describing an optical beam in

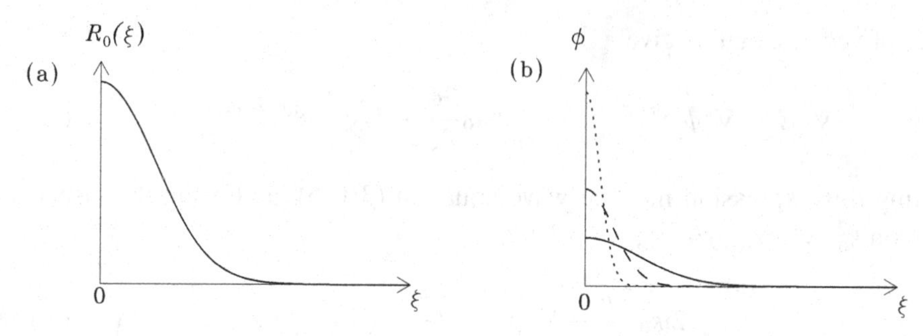

Figure 3.10. a) Plot of the solution $R_0(\xi)$ of Equation (3.165). b) Plot of ϕ for three values of z: the dashed and dotted lines correspond to increasing z.

a nonlinear medium,

$$2\mathrm{i}k_0\frac{\partial\phi}{\partial z}+\frac{\partial^2\phi}{\partial x^2}+\frac{\partial^2\phi}{\partial y^2}+\frac{\omega_0^2\chi^{(3)}}{c^2}|\phi|^2\phi=0, \tag{3.160}$$

which is of the NLS type.

Taking into account the cylindrical symmetry of the beam, and after an appropriate change of scales, this 2D NLS equation can be written

$$2\mathrm{i}\frac{\partial\phi}{\partial z}+\frac{1}{r}\frac{\partial}{\partial r}\left(r\frac{\partial\phi}{\partial r}\right)+|\phi|^2\phi=0 \tag{3.161}$$

which, similarly to its one-dimensional counterpart, has the following conservation laws [194]

$$N=\int_0^{+\infty} r|\phi|^2\mathrm{d}r \tag{3.162}$$

and

$$H=\int_0^{+\infty}\left(|\nabla\phi|^2-\frac{1}{2}|\phi|^4\right)r\mathrm{d}r=\int_0^{+\infty}\left(\left|\frac{\partial\phi}{\partial r}\right|^2-\frac{1}{2}|\phi|^4\right)r\mathrm{d}r. \tag{3.163}$$

Equation (3.161) has a set of exact solutions of the form

$$\phi_n(r,z,\lambda)=\lambda R_n(\lambda r)\,\mathrm{e}^{\mathrm{i}\lambda^2 z/2} \tag{3.164}$$

provided that R_n is a solution of the equation

$$\frac{\mathrm{d}^2R_n}{\mathrm{d}\xi^2}+\frac{1}{\xi}\frac{\mathrm{d}R_n}{\mathrm{d}\xi}+R_n^3(\xi)-R_n(\xi)=0 \quad \text{with} \quad \xi=\lambda r. \tag{3.165}$$

The functions $R_n(\xi)$ have n roots for positive values of ξ.

The solution $R_0(\xi)$, plotted in Figure 3.10(a) shows the spatial distribution of the amplitude in the beam. But, if one starts from an initial condition for which the quantity H defined by (3.163) is negative, this solution is *unstable*. This situation occurs for high intensity beams, and it leads to a *singularity* after the beam has travelled a *finite distance* z_0 in the nonlinear medium.

The calculation [194] shows that, in the vicinity of the singularity, ϕ behaves as

$$\phi(r = 0, z) \sim \frac{A}{(z - z_0)^{2/3}}. \tag{3.166}$$

The amplitude of the electric field diverges in the centre of the beam.

But, as the norm N defined by (3.162) is a constant, the divergence must go together with a drop in the beam diameter, as schematised in Figure 3.10(b).

The condition $H < 0$ is obtained when the incident power exceeds the power corresponding to the solution ϕ_0, which appears as a *critical power:*

$$P_{cr} = 2\pi \int_0^{+\infty} r|\phi_0(r, z)|^2 \, \mathrm{d}r. \tag{3.167}$$

The collapse of the beam is the extreme limit of *nonlinear energy localisation* that we first met with the example of modulational instability. In the case of optics it leads to the breaking of the beam into one or several filaments, depending on the shape of the initial beam.

Of course, when we find such a divergence in the solution of an equation corresponding to a physical system, it indicates that the equation is beyond its limit of validity. In the case of an intense laser beam in a fibre, this limit can even show up practically, by a breaking of the fibre! In most of the cases there are saturation effects in the nonlinearity, which prevent a true divergence. For instance, for an optical medium which polarises by the orientation of some polar molecules along the field, there is a maximum polarisation when all the molecules are parallel to the field.

As an example of such a case, Zakharov and Synaleh [194] have studied the equation

$$2\mathrm{i}\frac{\partial \phi}{\partial z} + \frac{1}{r}\frac{\partial}{\partial r}\left(r\frac{\partial \phi}{\partial r}\right) + [|\phi|^2 - \gamma|\phi|^4]\phi = 0, \tag{3.168}$$

in which the nonlinearity saturates. In this case the collapse is followed by a very sharp drop in the amplitude, and then by another collapse, and so on. The phenomenon is similar to the 'bursts' observed in some mechanisms of the transition to chaos [124], and this behaviour has been observed in plasmas.

A more general form of Equation (3.159) has been studied [156] in order to specify the roles of the dimension D of the operator ∇^2, and of the nonlinearity controlled by a parameter σ. It is the equation

$$\mathrm{i}\frac{\partial \phi}{\partial z} + P\nabla^2\phi + Q|\phi|^{2\sigma}\phi = 0. \tag{3.169}$$

The results show that, for $PQ < 0$ an initial excitation never diverges (this was true in one dimension since there is no modulational instability in this case). For $PQ > 0$ the excitation may diverge if $\sigma D \geq 2$:

- The case $\sigma D = 2$, which is the case of self-focusing in optics, is the marginal case because the divergence occurs only if $\int |\phi|^2 \mathrm{d}\mathcal{V}$ exceeds some threshold (where $\mathrm{d}\mathcal{V}$ denotes the volume element for the variables entering into the ∇^2 operator).
- Conversely, for $\sigma D > 2$ the divergence always occurs. This can be understood in the following way:
  - For large σ, the nonlinearity is high and, as we know, this generally favours energy localisation.
  - The dimension of the space enters through the volume in which the energy is collected when it becomes concentrated in the centre. Assuming that we collect energy up to distance ℓ around the centre, the volume in which energy is collected grows as ℓ^D. For a high dimension, a larger amount of energy can be brought to the centre.

3.8 Conclusion

This study of envelope solitons and of the NLS equation has shown the broad generality of the nonlinear Schrödinger equation because it can be derived for numerous physical systems using a multiple scale expansion.

It exhibits the peculiar role of dimension $D = 1$, but one must not conclude too hastily that nonlinear localised excitations do not exist in dimensions $D > 1$ because our results are based on the continuum limit approximation. In a discrete lattice, the existence of a minimal length scale introduces an additional dispersion, which allows the existence of localised excitations in dimensions 2 or 3. But the question of their mobility as solitons in the lattice is then open because the continuous translational invariance is lost.

As the NLS equation is very easy to derive for many physical systems, before drawing conclusions in a particular case, one should not forget the maxim:

> 'Live by Non Linear Schrödinger, die by Non Linear Schrödinger'.
> D. K. Campbell, Los Alamos

4

The modelling process: ion acoustic waves in a plasma

4.1 Introduction

In the previous chapters we introduced three main classes of soliton equations and, using examples from macroscopic physics, we showed how they could be applied to physical problems. The presentation was therefore oriented from the equation toward the physical system which could illustrate it.

Of course the physicist's approach is different. It starts from a given real system and attempts to determine what is the appropriate theoretical description. The goal is to go from the system toward the equation, which may not be unique because it may depend on the type of excitation which is applied to the system or on the external conditions which are imposed on it. The case of electrical lines has shown us that a particular device can be described by both the KdV equation and the NLS equation. However this example was a bit artificial because the device had been specially designed to exhibit both dispersion, thanks to the discrete character due to the individual units, and nonlinearity, provided by the choice of nonlinear capacitances. Hence it was not surprising to find that solitons could exist in such a system!

Our goal in this chapter is to show how the modelling process is carried out, i.e. how one goes from the physical system to the equations which describe it. We will study a physical system which is much more complex than electrical lines, and the first step will be to identify the dominant physical phenomena before we can attempt to describe them quantitatively by equations. Ion acoustic waves in a plasma provide an interesting example which shows how solitons can play a role in a natural medium which is important for the physics of the high atmosphere as well as astrophysics. At the end of this chapter we shall come back to some examples that we introduced in the previous chapters and discuss their modelling.

4.2 The plasma

4.2.1 Physics of a plasma

Plasma physics investigates the properties of electromagnetic waves in a conducting fluid. A plasma is a medium made of ionised atoms and electrons, on which an electromagnetic field applies forces which can lead to large-scale displacements inside the medium. These collective motions react on the field according to Maxwell's equations and therefore plasma physics must study a complex system of coupled matter and fields.

When we write Ohm's law $\overrightarrow{j} = \sigma \overrightarrow{E}$ in a medium, we implicitly assume that the motion of the charges under the effect of the field $\overrightarrow{E}$ is damped by an effective frictional force which can be written $-\gamma m \overrightarrow{v}$, where $\overrightarrow{v}$ is the velocity of a charge carrier and γ a constant which has the dimension of a frequency. This frictional force arises from collisions between the charge carriers, and possibly from diffusion of the carriers by fixed charges, as for instance with positive ions in a solid metal. Under the effect of the field and this frictional force, the carriers reach a limiting velocity which is proportional to the field, hence Ohm's law. This picture is only valid if the frequency ν of the applied field is much smaller than γ. For high frequency electromagnetic fields, inertial effects play an important role in the dynamics of the carriers, and the notion of an effective frictional force loses its meaning. For $\nu \gg \gamma$, electrons and ions which are accelerated in opposite directions by the field, tend to move apart from each other, which generates large electrostatic forces acting against the separation of the charges. This competition can lead to oscillations, called plasma oscillations. This frequency range is the domain of 'plasma physics' while magnetohydrodynamics considers fluids where collective motions occur at much lower frequencies, without a separation of the charges [94].

The field of plasma physics is easily reached in rarefied gases, such as in the high layers of the atmosphere or in astrophysics, because the density of electrons and ions is low enough to reduce the probability of a charge carrier being diffused by another charge, which leads to low values of γ.

Let us study a fully ionised hydrogen plasma which has a low enough density to allow us to neglect all collision processes, so that the interactions between the particles are only determined by the Coulomb forces. The behaviour of the system is then described by the equations of motion of the particles, the conservation of the particle numbers in each species (ions and electrons) and the Poisson equation for the electrostatic potential [112, 113].

The first physical phenomenon which must be studied is the *mechanical motion* of the charge carriers. Projecting the momentum balance equation for species α (which will henceforth be denoted by i for the ions and e for the electrons), i.e. the equation of motion for a small volume element in the fluid, we get

$$\rho_\alpha \frac{dv_{\alpha,k}}{dt} = \rho_\alpha \left[\frac{\partial v_{\alpha,k}}{\partial t} + \sum_j v_{\alpha,j} \frac{\partial v_{\alpha,k}}{\partial x_j} \right] = F_{\alpha,k} - \sum_j \frac{\partial P_{\alpha,kj}}{\partial x_j} \tag{4.1}$$

where:

- we denote by ρ_α the specific mass, n_α the density of particles, m_α the mass and q_α the charge of species α which is equal to $+e$ for the protons of the hydrogen plasma and $-e$ for the electrons.
- $F_{\alpha,k}$ is the component of the force per unit volume, which comes from the electric field, i.e. $\overrightarrow{F} = n_\alpha q_\alpha \overrightarrow{E}$ since there are n_α particles of charge q_α in the unit volume.
- $P_{\alpha,kj}$ is the pressure tensor in the fluid of particles α. As we neglect the viscosity of the plasma, it reduces to the electrostatic pressure P_α so that it is diagonal. Thus only the contribution $\partial P_\alpha / \partial x_k$ remains in the equation.
- The plasma is globally neutral, i.e. $\int n_i(\mathbf{x})d\mathbf{x} = \int n_e(\mathbf{x})d\mathbf{x}$ but, at the high frequencies $\nu \gg \gamma$ that we want to study, the local neutrality is not observed: the densities $n_i(\mathbf{x})$ and $n_e(\mathbf{x})$ may be different (where we designate by $\mathbf{x}$ the set of position coordinates for a charge).
- The density is assumed to be very low so that collisions can be neglected. The particles only interact through the electromagnetic field, which is in fact a self-coherent force because this field depends on the positions of the particles.

It should be noticed that the velocity $\overrightarrow{v_\alpha}(\mathbf{x}, t)$ varies because it explicitly depends on time, but also due to the motion of the fluid, so that there are two contributions in its total derivative with respect to time, as is usual in hydrodynamics.

As a result the momentum balance equation becomes

$$\frac{\partial \overrightarrow{v_\alpha}}{\partial t} + \left(\overrightarrow{v_\alpha} . \overrightarrow{\nabla}\right) \overrightarrow{v_\alpha} = -\frac{q_\alpha}{m_\alpha} \overrightarrow{\nabla} \varphi - \frac{1}{n_\alpha m_\alpha} \overrightarrow{\nabla} P_\alpha, \tag{4.2}$$

where we introduce the electric potential φ, from which the electric field derives through $\overrightarrow{E} = -\overrightarrow{\nabla} \varphi$.

Let us now write the *conservation of the particle number*, i.e. the mass conservation for each species

$$\frac{\partial \rho_\alpha}{\partial t} + \overrightarrow{\nabla} . \left(\rho_\alpha \overrightarrow{v_\alpha}\right) = 0 \quad \text{or} \quad \frac{\partial n_\alpha}{\partial t} + \overrightarrow{\nabla} . \left(n_\alpha \overrightarrow{v_\alpha}\right) = 0. \tag{4.3}$$

The second physical aspect which is essential in a plasma is *the electromagnetism of the medium.* The *Poisson equation* completes the set of equations for the plasma.

It gives

$$\Delta\varphi = \nabla^2\varphi = -\frac{\rho}{\varepsilon_0} = -\frac{n_i e - n_e e}{\varepsilon_0}. \tag{4.4}$$

4.2.2 Temperatures and equations of state

The ratio $\mu = m_e/m_i$ of the electron and ion masses allows us to consider two regimes in the dynamics of a nonmagnetic plasma, depending on the timescale of interest which is defined by one of the two plasma frequencies $\omega_{p,\alpha}^2 = n_\alpha q_\alpha^2/m_\alpha\varepsilon_0$:

(i) The dynamics associated with the fast electrons occur around the electronic frequency ω_{pe} which is so high that the ions look almost frozen so that they only bring a second order contribution in the equations of motion.
(ii) The other dynamics occur typically at the ion frequency ω_{pi}. In this frequency range, at the first level of approximation, the electrons can be viewed as infinitely mobile.

Here we shall study the second case, i.e. we are interested in the frequency range of the *ion* acoustic waves in a plasma. This amounts to neglecting all inertial effects for the electrons, as if their mass was strictly zero. Their momentum balance equation (4.1) is simply reduced to the equilibrium between the electronic pressure gradient and the electrostatic force,

$$e\vec{\nabla}\varphi - \frac{1}{n_e}\vec{\nabla}P_e = \vec{0}. \tag{4.5}$$

As the electrons are extremely mobile, their thermal conductivity is very efficient to ensure a uniform temperature in the electron 'gas', which can be treated as isothermal to a good approximation. In order to study the plasma, we need to know its equation of state. Since we assumed a low density plasma, the electron gas is very well described by an ideal gas equation of state, i.e.

$$P_e = n_e T_e, \tag{4.6}$$

expressing temperature in energy units, so that the Boltzmann constant k_B does not appear in the equation.

Combining Equations (4.5) and (4.6), we get

$$e\vec{\nabla}\varphi - \frac{T_e}{n_e}\vec{\nabla}n_e = \vec{0} \tag{4.7}$$

which leads to the following expression for the density of electrons

$$n_e = n_0 \exp\left(\frac{e\varphi}{T_e}\right), \tag{4.8}$$

where n_0 is an integration constant.

Ions have a mass which is much larger than electrons so that, for them, inertial effects are, on the contrary, very important. In the fairly low frequency range of the ion acoustic waves that we chose to investigate, thermal fluctuations in the motion of the ions are negligible with respect to the coherent displacement imposed by the acoustic wave. This is because the thermal velocity of the ions, $v_i = \sqrt{T_i/m_i}$ is small, even when the ionic temperature is as high as the electronic temperature T_e. Hence, in the momentum balance for the ions, the term due to the pressure gradient is negligible with respect to the electromagnetic forces. Neglecting this term, i.e. assuming that $P_i \simeq 0$, amounts to assuming that $T_i \simeq 0$ because P_i and T_i are linked by the equation of state of the ions. In a low density plasma this is the ideal gas equation of state, $P_i = n_i T_i$, as for the electrons.

At this stage we have identified the physical phenomena which enter into the dynamics of the plasma and, focusing our study on the frequency range of ion acoustic modes, we have simplified the equations which derive from the laws of mechanics, electromagnetism and thermodynamics. Furthermore, we shall henceforth assume that the excitation of the plasma is uniform in the (y, z) plane, so that only the x spatial coordinate has to be considered. In this case, if we simply denote by v the velocity of the ions, the equations of motions of the ions in the approximation $P_i \simeq 0$ are given by

$$\frac{\partial v}{\partial t} + v\frac{\partial v}{\partial x} = -\frac{e}{m_i}\frac{\partial \varphi}{\partial x}, \tag{4.9}$$

$$\frac{\partial^2 \varphi}{\partial x^2} = -\frac{e}{\varepsilon_0}(n_i - n_e), \tag{4.10}$$

$$\frac{\partial n_i}{\partial t} + \frac{\partial (n_i v)}{\partial x} = 0. \tag{4.11}$$

They must be completed by the expression for the electronic density (4.8) in order to make a complete set of equations for the problem that we are investigating.

4.2.3 The dimensionless set of equations

Studying the physics of the system and determining the dominant phenomena is of course the first stage for modelling, but it is seldom sufficient because, in most of the cases, the equations which are derived from this analysis cannot be solved in their initial form. Approximations are often necessary, but, in order to control them, it is essential to correctly evaluate the weight of the different terms. For this purpose, it is best to switch to a dimensionless set of equations by introducing appropriate scales for the problem.

In the absence of a magnetic field, a plasma is characterised by a single length scale, the Debye length $\lambda_D = (\varepsilon_0 T_e / n_0 e^2)^{1/2}$ which is a screening length: in equilibrium an ion is surrounded by an electronic cloud with a density which decreases as $\exp(-r/\lambda_D)$ as a function of the distance r to the ion. The plasma can be characterised by two timescales $\tau_i = 1/\omega_{p,i} = (m_i\epsilon_0/n_0e^2)^{1/2}$, which is the inverse of the ionic plasma frequency, and the inverse of the electronic plasma frequency. For the study of ion acoustic waves, it is the ionic timescale τ_i which is relevant. The space and timescales define a natural speed scale $v_s = \lambda_D/\tau_i = (T_e/m_i)^{1/2}$, which is the speed of sound in the plasma. Finally the energy scale can be defined by the electronic temperature T_e, which we expressed in energy units. This determines a natural scale for the electrical potential, T_e/e.

A numerical calculation gives an idea of the order of magnitude of these various scales. Let us choose a typical ionic density for a low density plasma, $n_0 = 10^{20}\ \mathrm{m}^{-3}$, and an electronic temperature $T_e = 2 \cdot 10^3$ K, which is a moderate value for a plasma. We get $\lambda_D = 0.3\ \mu\mathrm{m}$, $\tau_i = 7.6 \cdot 10^{-11}$ s and $v_s = 4000\ \mathrm{m\ s}^{-1}$, which is a good order of magnitude for the speed of sound in a plasma, although it sounds high for standard acoustic waves in a gas.

The scales that we have defined allow us to introduce the dimensionless variables

$$X = \frac{x}{\lambda_D}, \quad T = \frac{t}{\tau_i}, \quad V = \frac{v}{v_s}, \quad \phi = \frac{e}{T_e}\varphi, \quad N = \frac{n_i}{n_0}, \quad \text{and} \quad N_e = \frac{n_e}{n_0}. \tag{4.12}$$

Writing the dynamic equations of the plasma as a function of these new variables, we get a set of dimensionless equations

$$\text{Eq. (4.8)} \quad \Rightarrow \quad \phi = \ln N_e \tag{4.13}$$

$$\text{Eq. (4.10)} \quad \Rightarrow \quad \phi_{XX} = N_e - N \tag{4.14}$$

$$\text{Eq. (4.11)} \quad \Rightarrow \quad N_T + (NV)_X = 0 \tag{4.15}$$

$$\text{Eq. (4.9)} \quad \Rightarrow \quad V_T + VV_X = -\phi_X. \tag{4.16}$$

4.3 Study of the linear dynamics

Although we stressed earlier the importance of considering nonlinearity intrinsically, in the modelling process one should not neglect the information which can be learned from a study of the small-amplitude perturbations around an equilibrium state. Besides the results on the response of the weakly perturbed system, they can give useful hints to carry out the study of the nonlinearities.

The simplest equilibrium state for ion acoustic waves in a plasma is $V = 0$, $\phi = 0$ and $N = N_e = 1$. Let us study small perturbations around this state

$$V = \delta V, \quad N = 1 + \delta N, \quad N_e = 1 + \delta N_e, \quad \text{and} \quad \phi = \delta\phi. \tag{4.17}$$

Such perturbations can be analysed with a *linearised* version of the dynamic equations of the plasma written for δV, δN, δN_e and $\delta\phi$,

$$\delta\phi = \delta N_e \tag{4.18}$$

$$\delta V_T = -\delta\phi_X \tag{4.19}$$

$$\delta N_T + \delta V_X = 0 \tag{4.20}$$

$$\delta\phi_{XX} = \delta\phi - \delta N. \tag{4.21}$$

Note that Equation (4.18) has been used to write Equation (4.21).

If we take the derivative of Equation (4.19) with respect to variable X, we obtain $\delta V_{TX} = -\delta\phi_{XX}$, which can be combined with Equation (4.21), to get

$$-\delta V_{TX} = \delta\phi - \delta N. \tag{4.22}$$

Taking again the derivative of this equation with respect to X and T yields

$$-\delta V_{TTXX} = \delta\phi_{TX} - \delta N_{TX} \tag{4.23}$$

which finally gives

$$-\delta V_{TTXX} = -\delta V_{TT} + \delta V_{XX} \tag{4.24}$$

if we take into account the derivatives of Equations (4.20) and (4.21), respectively, with respect to T and X.

This equation has the wave solution $\delta V = a e^{\mathrm{i}(\omega \mathrm{T} - \mathrm{kX})}$ provided that the dispersion relation

$$\omega^2 = \frac{k^2}{1+k^2} \tag{4.25}$$

holds. As shown in Figure 4.1, $\omega(k)$ has the usual shape for the dispersion relation of *acoustic waves*, which passes through the origin and tends to $\omega \propto k$ in the limit of small k. Moreover the dispersion relation shows that ω^2 is always positive, which means that any small perturbation of the equilibrium state will oscillate without growing over time. Thus the linear analysis concludes that the equilibrium state is *stable*.

The dispersion relation yields the phase velocity

$$v_\varphi = \frac{\omega}{k} = \frac{1}{\sqrt{1+k^2}} \tag{4.26}$$

which depends on k. Therefore the linear waves in a plasma are *dispersive waves*. Since we know that the full equations of motion for the medium are nonlinear, we can expect to find some situations in which dispersion and nonlinearity balance

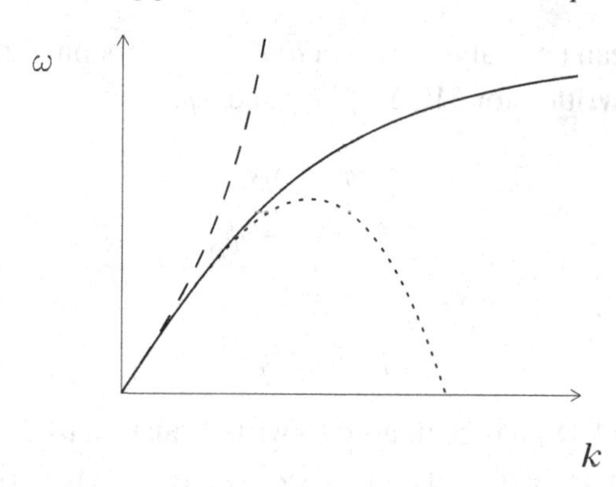

Figure 4.1. Dispersion relation for ion acoustic waves in a plasma. The solid line shows the dispersion relation $\omega(k)$ given by (4.25). The dotted line shows the dispersion relation (4.47) while the dashed line is a plot of Equation (4.65).

each other. This shows that it is necessary to carry on the study in the presence of the nonlinear terms in the dynamic equations.

4.4 Nonlinear study

4.4.1 The plasma can be described by a KdV equation

Let us now investigate *nonlinear* acoustic waves in order to look for a regime where nonlinearity and dispersion compete, which might lead to solitons.

The linear study showed us the way because, if we expand the dispersion relation for low k we get

$$\omega(k) = k\left(1 - \frac{1}{2}k^2\right) + \mathcal{O}(k^5). \tag{4.27}$$

Consequently we obtain the type of dispersion relation that we met earlier when we discussed the generality of the KdV equation in Section 1.7. This suggests that we should apply to Equations (4.13)–(4.16) the method that led to the KdV equations in the examples that we treated earlier.

The idea is to focus on the weakly dispersive case. For a dispersion relation such as (4.27), it means that we consider small values of k, i.e. we look for solutions which vary slowly in space. Moreover we saw in Section 1.7 that it is convenient to change to a reference frame moving at the speed of sound, which, in our reduced units, is equal to unity. Therefore we introduce the variable $\xi = \varepsilon^{\alpha}(X - T)$, where the factor ε^{α} ensures a slow spatial variation. The appropriate value of α will have to be determined later. In the discussion of Section 1.7 we showed that, for a dispersion relation such as (4.27), the appropriate time variable in the moving frame is $\tau = \varepsilon^{3\alpha} T$. The procedure is to write the equations in terms of these new variables and, if we look for a possible soliton solution, determine α so that dispersive and nonlinear terms appear at the same order in the resulting equations. This shows that the proper value of α is $\alpha = 1/2$, as in the nonlinear electrical lines. The complete calculation is straightforward but tedious and thus, here we shall only check that the choice $\alpha = 1/2$ is correct.

Let us introduce the slow space and time variables $\xi = \varepsilon^{1/2}(X - T)$ and $\tau = \varepsilon^{3/2}T$. As a result the former space and time derivative operators become $\partial_X = \varepsilon^{1/2}\partial_\xi$ and $\partial_T = -\varepsilon^{1/2}\partial_\xi + \varepsilon^{3/2}\partial_\tau$, which leads to new expressions of the equations of motion (4.14), (4.15) and (4.16)

$$\varepsilon\phi_{\xi\xi} = e^{\phi} - N \tag{4.28}$$

$$\varepsilon N_\tau - N_\xi + (NV)_\xi = 0 \tag{4.29}$$

$$\varepsilon V_\tau - V_\xi + V V_\tau = -\phi_\xi . \tag{4.30}$$

These equations include terms at different orders. Let us now perform a perturbative expansion around the equilibrium state

$$N = 1 + \varepsilon N_1 + \varepsilon^2 N_2 + \mathcal{O}(\varepsilon^3) \tag{4.31}$$

$$V = 0 + \varepsilon V_1 + \varepsilon^2 V_2 + \mathcal{O}(\varepsilon^3) \tag{4.32}$$

$$\phi = 0 + \varepsilon\phi_1 + \varepsilon^2\phi_2 + \mathcal{O}(\varepsilon^3). \tag{4.33}$$

Introducing these expressions into Equations (4.28), (4.29) and (4.30), at order ε we get

$$\phi_1 = N_1, \quad N_{1\xi} = V_{1\xi} \quad \text{and} \quad V_{1\xi} = \phi_{1\xi}, \tag{4.34}$$

which leads to

$$N_1 = \phi_1 = V_1 + A(\tau), \tag{4.35}$$

where A is an arbitrary function of variable τ, which must be set to zero if we are looking for solutions with an amplitude which does not depend on time, such as solitons.

At order ε^2, the three equations lead to

$$\phi_{1\xi\xi} = \phi_2 + \frac{1}{2}\phi_1^2 - N_2, \tag{4.36}$$

$$N_{1\tau} - N_{2\xi} + (N_1 V_1)_\xi + V_{2\xi} = 0, \tag{4.37}$$

$$V_{1\tau} - V_{2\xi} + V_1 V_{1\xi} = -\phi_{2\xi} . \tag{4.38}$$

Combining Equations (4.36) and (4.38) with Equation (4.35), we get

$$V_{1\tau} + V_{1\xi\xi\xi} - V_{2\xi} + N_{2\xi} = 0. \tag{4.39}$$

If we sum Equations (4.37) and (4.39), we obtain the Korteweg–de Vries equation

$$V_{1\tau} + V_1 V_{1\xi} + \frac{1}{2} V_{1\xi\xi\xi} = 0, \tag{4.40}$$

that was suggested by the expression of the dispersion relation at low k and the presence of a nonlinearity in the system. Both nonlinearity and dispersion appear at the same order ε^2, which confirms a posteriori that the choice $\alpha = 1/2$ is appropriate.

Let us now show that the calculation can also be carried out without making the simplifying assumption $A(\tau) = 0$ and that its leads nevertheless to the KdV equation. The equation at order ε^2 which results from Equations (4.36), (4.37) and (4.38), *without assuming* $N_1 = V_1$, is

$$N_{2\xi} - N_{1\xi} N_1 + N_{1\xi\xi\xi} + V_{1\tau} + V_1 V_{1\xi} = V_{2\xi}. \tag{4.41}$$

Taking into account Equation (4.38) and the relation $N_1 = V_1 + A(\tau)$, we get

$$V_{1\tau} + \frac{A_\tau}{2} + \frac{1}{2} V_{1\xi\xi\xi} + V_1 V_{1\xi} = 0. \tag{4.42}$$

We can introduce new variables,

$$\eta = \tau, \qquad \chi = \xi + \frac{1}{2}\int_0^\tau A(\alpha)\mathrm{d}\alpha \quad \text{and} \quad V = V_1 + \frac{A}{2}, \tag{4.43}$$

and, after some simplifications we obtain

$$V_\eta + V V_\chi + \frac{1}{2} V_{\chi\chi\chi} = 0, \tag{4.44}$$

which is again the KdV equation.

4.4.2 Dispersion relation

If we linearise Equation (4.40) and look for a plane wave solution $V = V_0\, \mathrm{e}^{\mathrm{i}(\Omega\tau - \mathrm{K}\xi)}$ we get the dispersion relation

$$\Omega = -\frac{K^3}{2}. \tag{4.45}$$

Coming back to the original X and T variables the solution is

$$V = V_0\, \mathrm{e}^{\mathrm{i}(\Omega\varepsilon^{3/2}\mathrm{T} - \mathrm{K}\varepsilon^{1/2}(\mathrm{X}-\mathrm{T}))} \tag{4.46}$$

which can be identified with the expression $a\ e^{i(\omega T - kX)}$, used to derive the dispersion relation (4.25), giving

$$k = K\varepsilon^{1/2} \quad \text{and} \quad \omega = k - \frac{1}{2}k^3, \tag{4.47}$$

which agrees with the expansion (4.27) of the dispersion relation (4.25). Figure 4.1 confirms that the dispersion relation derived from the linearisation of the physical equations of the plasma, and the dispersion relation deduced from the linearisation of the KdV equation agree for small values of k. This is consistent with our choice of slow variables. The KdV equation indeed describes excitations with $k \ll 1$ in the plasma, i.e. with a wavelength which is large with respect to the Debye length λ_D.

Let us stress that the result that we obtained for the plasma is a particular case of the general property that we discussed in Section 1.7: when the linear approximation shows that a nonlinear physical system has dispersive waves with a dispersion relation $\omega = ak + bk^3$, where a and b are constants, then *a regime for which dispersion and nonlinearity balance each other may exist* and such a regime is described by the KdV equation. It can be derived by an expansion similar to the one that we made for the plasma.

4.5 Derivation of the nonlinear Schrödinger equation

The KdV description combines dispersion and nonlinearity in the plasma and it exhibits solutions which explicitly take nonlinear effects into account. However, it does not properly describe all the excitations which can be produced by an electromagnetic wave in a plasma because they look like wave packets rather than pulses similar to the KdV solutions.

But we know one equation which describes 'nonlinear wave packets'. It is the NLS equation. The physical properties of the plasma suggest that, if the KdV equation correctly describes its nonlinearity, it must be possible to go from this equation to another one, more appropriate for wave packets, the NLS equation.

We shall check that it is indeed the case, and moreover it will show us that the main classes of soliton equations are not disconnected from each other; there are links between them. We already noticed such a link between the SG and the NLS equations since we derived the latter from an expansion of the former and we showed that a small amplitude expansion of the SG breather gives an NLS soliton. Here we shall establish another of those links by examining how the analysis of the successive orders of a multiple scale expansion of the KdV equation finally leads to an NLS equation for the amplitude.

Let us look for a solution of Equation (4.40) as a perturbative expansion

$$V = \varepsilon\theta_1 + \varepsilon^2\theta_2 + \varepsilon^3\theta_3 + \cdots, \tag{4.48}$$

which does not include any constant term because $V = 0$ in the equilibrium state of the plasma. The multiple scale method is a natural choice to look for wave packet solutions because a wave packet indeed includes several scales, a fast changing carrier wave modulated by an envelope which varies more slowly. Here we shall give a more general view of the multiple scale expansion than in Chapter 3 because it is a powerful method, which has many other applications, such as the study of perturbed KdV equations.

Let us assume that the solution depends on several timescales $\tau_i = \varepsilon^i\,\tau$ and several spatial scales $\xi_i = \varepsilon^i\,\xi$. Then the derivations with respect to time and space take the form

$$\partial_\xi = \partial_{\xi_0} + \varepsilon\partial_{\xi_1} + \varepsilon^2\partial_{\xi_2} + \cdots \tag{4.49}$$

$$\partial_{\xi\xi\xi} = \partial^3_{\xi_0\xi_0\xi_0} + 3\varepsilon\partial^3_{\xi_0\xi_0\xi_1} + 3\varepsilon^2\partial^3_{\xi_0\xi_1\xi_1} + 3\varepsilon^2\partial^3_{\xi_0\xi_0\xi_2} + \cdots \tag{4.50}$$

We introduce Expansion (4.48) into Equation (4.40), taking into account the different scales, and then we identify the terms order by order.

Order ε gives

$$\mathcal{L}\,\theta_1 = \theta_{1\tau_0} + \frac{1}{2}\theta_{1\xi_0\xi_0\xi_0} = 0 \tag{4.51}$$

which introduces an operator $\mathcal{L}$ which will be important in the next steps of the calculation, and leads to

$$\theta_1 = A(\tau_1, \tau_2, \ldots, \xi_1, \xi_2, \ldots)\,\mathrm{e}^{\mathrm{i}(\Omega\tau_0 - \mathrm{K}\xi_0)} + \text{c.c.} \tag{4.52}$$

with the dispersion relation $\Omega = -K^3/2$ that we already met earlier. It is not surprising that the lowest order yields the linear dispersion relation because, at this order, the analysis is restricted to linear terms.

Order ε^2 gives the equation

$$\theta_{2\tau_0} + \frac{1}{2}\theta_{2\xi_0\xi_0\xi_0} = -\theta_{1\tau_1} - \theta_1\theta_{1\xi_0} - \frac{3}{2}\theta_{1\xi_0\xi_0\xi_1} \tag{4.53}$$

$$\mathcal{L}\,\theta_2 = \mathcal{NL}\,(\theta_1) \tag{4.54}$$

in which the linear operator $\mathcal{L}$ appears again and $\mathcal{NL}$ is a nonlinear operator. In Chapter 3 we pointed out that, in a perturbative expansion, it is crucial to avoid secular terms, i.e. terms which are resonant for the operator $\mathcal{L}$, because they lead to the divergence of some parts of the expansion. If we cancel the terms of the nonlinear operator $\mathcal{NL}$ which can resonate with $\mathcal{L}$, we get

$$-A_{\tau_1} + \frac{3K^2}{2} A_{\xi_1} = 0 \qquad \text{i.e.} \qquad A_{\tau_1} + v_g A_{\xi_1} = 0, \tag{4.55}$$

where v_g is the group velocity of the waves which obey the dispersion relation obtained at order ε. This equation shows that, for variables at order 1, the function A only depends on $\xi_1' = \xi_1 - v_g \tau_1$, which means that, at this order, it keeps a permanent profile in the frame moving at the group velocity of the wave packet. Therefore we shall write the amplitude A as

$$A = A(\xi_1 - v_g \tau_1, \xi_2, \ldots, \tau_2, \ldots) \tag{4.56}$$

After the secular terms have been eliminated, Equation (4.53) is reduced to

$$\theta_{2\tau_0} + \frac{1}{2}\theta_{2\xi_0\xi_0\xi_0} = \mathrm{i}KA^2\, \mathrm{e}^{2\mathrm{i}(\Omega\tau_0 - K\xi_0)} + \text{c.c.} \tag{4.57}$$

which has a particular solution of the form $\theta_2 = B \exp[2\mathrm{i}(\Omega\tau_0 - K\xi_0)] + \text{c.c.}$ Putting this expression in Equation (4.57), we easily get $B = A^2/3K^2$.

It is useful to notice that, when we applied the multiple scale expansion to the sine-Gordon equation, at this order, after cancellation of the secular terms, we simply had $\mathcal{L}\,\theta_2 = 0$ so that we could chose $\theta_2 = 0$ because any other solution could be included into the solution θ_1. The SG and the KdV cases are different because the KdV nonlinearity is *quadratic* instead of being cubic as in the SG equation.

Order ε^3 leads to the equation

$$\theta_{3\tau_0} + \frac{1}{2}\theta_{3\xi_0\xi_0\xi_0} = \mathcal{L}\,\theta_3 \tag{4.58}$$

$$\begin{aligned} &= -\theta_{2\tau_1} - \theta_{1\tau_2} - \theta_2\theta_{1\xi_0} - \theta_1\theta_{2\xi_0} - \theta_1\theta_{1\xi_1} \\ &\quad - \frac{3}{2}[\theta_{1\xi_0\xi_1\xi_1} + \theta_{1\xi_0\xi_0\xi_2} + \theta_{2\xi_0\xi_0\xi_1}]. \end{aligned} \tag{4.59}$$

The secular terms of the right hand side, proportional to $\exp[\pm\mathrm{i}(\Omega\tau_0 - K\xi_0)]$, must again be cancelled, which leads to

$$-2A_{\tau_2} + \frac{2\mathrm{i}}{3K}|A|^2 A + 3\mathrm{i}KA_{\xi_1\xi_1} + 3K^2 A_{\xi_2} = 0. \tag{4.60}$$

Changing again to the frame moving at the group velocity v_g by defining the variables $\xi_2' = \xi_2 - v_g\tau_2$ and $\tau_2' = \tau_2$, we can simplify this equation into

$$-A_{\tau_2'} + \frac{3\mathrm{i}K}{2} A_{\xi_1\xi_1} + \frac{\mathrm{i}}{3K}|A|^2 A = 0. \tag{4.61}$$

If we divide it by i, we recover the usual NLS equation

$$\mathrm{i}A_{\tau_2'} + \frac{3K}{2}A_{\xi_1\xi_1} + \frac{1}{3K}|A|^2 A = 0. \tag{4.62}$$

The product of the coefficients of the second order derivative and of the nonlinear term (PQ product in the standard notation) is *positive*.

This result indicates that envelope solitons can propagate in plasma. Moreover, as shown in Chapter 3 on the NLS equation, it indicates that an acoustic plane wave is *unstable* with respect to a small-wavevector modulation, and tends to split into solitons.

For plasma, moving from a KdV description to the NLS equation is not only useful to model the response of the medium when it is excited by an electromagnetic wave. It has a deeper physical significance because it shows that plane waves are unstable in the plasma, and that *a calculation in the linear approximation gives a result which is qualitatively wrong* since it concludes, on the contrary, in stability. Of course the linear stability analysis is not fully meaningless. In plasma, if we create a plane wave with a very tiny amplitude, the nonlinear instability, which is a second order effect, will be extremely slow to grow, and may even stay unnoticed over the limited timescale of an experiment. But one should however, expect a long term instability, even for very weak signals, which points out that nonlinear effects should never be completely neglected.

Moreover, in practice they are likely to play a higher role than the one found with the hypothesis that the plasma is excited by a plane wave, invariant by translation along y and z, which reduces the study to a one-dimensional problem. In an experiment, the exciting wave does not interact with the full transversal extent of the plasma (in particular for atmospheric plasmas of course) so that a two- or three-dimensional study should be performed. In such a case we can expect much more violent events because, as shown in Chapter 3, self-modulation results in a 'collapse', i.e. an infinite concentration of the energy. In practice the energy density stays finite due to effects which have been neglected in our calculations, such as the ionic temperature which is not strictly zero. An arbitrary perturbation of the equilibrium state is generally a superposition of plane waves with various wavevectors, and each component tends to show self-focusing, contributing to the instability of the equilibrium state. This is a well known situation, which can lead to a chaotic response inducing an intense 'noise' in the radio waves which propagate in a plasma.

4.6 Experimental observations

Exciting a plasma by modulated waves is common, in particular in order to transmit electromagnetic signals. To observe the behaviour described by the KdV equation one must first use excitations much stronger than the excitations that lead to the NLS regime (let us recall that the NLS equation was derived from the KdV equation by looking for a solution at order ε) and second, one must use pulses, which are qualitatively similar to the solitons of the KdV equation instead of sinusoidal waves. Some experiments have been performed [91] by applying a strong voltage pulse at one end of a plasma confined in a container in which the ratio between the electronic and ionic temperatures was about 30, in agreement with our hypothesis $T_e \gg T_i$; otherwise the damping of the ion acoustic wave would have been too strong. The results confirm the predictions of the KdV equation because they show that:

- A strong positive pulse, associated with a compressive wave in the plasma, tends to steepen under the effect of nonlinearity until it breaks into a train of solitons which propagate at supersonic speed. Their amplitude is proportional to the excess of their velocity beyond the speed of sound, in agreement with the KdV solution.
- Conversely a negative voltage pulse, which is associated with a local drop in the ionic density, disperses into a set of nonlocalised waves as it propagates.

Experiments have also been carried out by applying a pulse at both ends of the container. This leads to the head-on collision of two solitons which move in opposite directions. Such a configuration cannot be described within the formalism that we have presented because, when we derive the KdV equation, we change the frame to a frame moving at the speed of sound. At this stage we need to choose one direction. This indicates that *modelling must not only be appropriate for the physical system of interest, the plasma, but it must also take into account the conditions of its excitation.* This is a remark that we have already made about the choice between a KdV description, adapted to pulse excitations, and an NLS description, valid for a sinusoidal excitation. Boundary conditions must be considered from the start of the modelling process and not only at the end, when an equation has been obtained.

An equation which preserves the invariance by changing x into $-x$ in the laboratory frame, contrary to the KdV equation, has been obtained by Makhankov [123] who started from the linear dispersion relation. Rewriting the dispersion relation (4.25) under the form $\omega^2(1+k^2) = k^2$, and then looking for the differential operators which could lead to such a dispersion relation, as we did for optical fibres (Section 3.6.3), we get

$$\left[\frac{\partial^2}{\partial t^2}\left(1-\frac{\partial^2}{\partial x^2}\right)-\frac{\partial^2}{\partial x^2}\right]\phi = 0. \tag{4.63}$$

A nonlinear equation for the plasma is then obtained by adding to Equation (4.63) the nonlinear term $\partial\phi^2/\partial x$ that we found in the KdV equation. It leads to a Boussinesq-type equation which has a correct linear spectrum (because it has been built for that!).

Its solutions are, however, not exactly preserved in collisions. A small energy loss occurs, but the quasi-solitons are nevertheless well preserved, in agreement with the experimental observations [91].

A quantitative comparison between the experimental results and the predictions of the KdV equation, shows however, that the modelling that we introduced is not perfect. For instance, if one measures the difference between the soliton speed and the speed of sound, and plot it versus the soliton amplitude, one gets a straight line as predicted by the KdV equation, but its slope is 10 to 20% higher than KdV would predict [178]. Many improvements to the KdV equation for a plasma have been proposed [178]. They lead to modified KdV equations, which have a qualitative behaviour very similar to the KdV equation that we derived, but they give a better quantitative agreement with the experiments. The main correction comes from a finite ionic temperature which introduces a term $-3T_i/(T_e n_i) \times (\partial n_i/\partial x)$ in the dynamic equation (4.9). Its role is to renormalise the nonlinear term of the KdV equation. Moreover dissipation cannot be ignored. It tends to reduce the amplitude of the soliton but it also creates a 'tail' behind it, which grows as it propagates. When it is included in the calculation, it leads to an additional diffusive term in the KdV equation, which correctly describes the experiments. Finally, we restricted our study to a one-component plasma, but solitons also exist in multicomponent plasmas, such as the Ar–He plasma. It is remarkable to notice that, in this case too, the nonlinear acoustic perturbations are described by a KdV equation.

It may seem appropriate to try to set up the most complete equation, taking into account all the contributions of the plasma physics, such as the nonzero ionic temperature, dissipation, possible inhomogeneities in the electronic temperature because the mobility of the electrons is not infinite, etc. However the 'improvement' which is obtained is often an illusion, first because it introduces many model parameters, which usually cannot be determined a priori, and second because the equation which is derived can seldom be solved analytically to make general predictions on the behaviour of the system. *The best model is not necessarily the most complete,* but rather the model that addresses the right questions on the system and properly takes into account the experimental possibilities. When we discussed the electrical lines we noticed that it can be better to use a KdV equation which comes with a full mathematical arsenal, rather than a modified-Boussinesq description, which is in principle better, but makes predictions which differ from those of the KdV equations by margins which are smaller than the experimental error. On the other hand, if we want to describe the collision of two solitons moving in opposite directions, it is the Boussinesq description which is appropriate.

4.7 Discussion

Since modelling a given physical system can lead to equations as different from each other as the KdV, the NLS or the Boussinesq equations, it is legitimate to ask if the description which is obtained has a real meaning or whether it is only the fruit of some ad hoc hypotheses. As shown in the previous section, there is some validity in this remark if, by 'ad hoc hypotheses', we mean hypotheses adapted to what we intend to describe. We emphasised that it can be a legitimate and correct approach. Of course, although the study of the different classes of nonlinear equations, and their families, provide a full set of tools among which one can choose, the possibility to adapt the model to the system is restricted by precise validity criteria. For instance, it should be noticed that equations of the sine-Gordon family did not enter into our discussion of plasma because such a medium only has a single equilibrium state. In another respect, the choice between a KdV or a NLS description is not only determined by the shape of the excitation, i.e. whether it is a pulse or a wave, because its amplitude must also be considered. The NLS description has been obtained from the KdV equation with a perturbative expansion which assumed a leading term at order ε. Therefore it should be applied to smaller excitations than the ones that are described by KdV. It is not by chance that its solutions look closer to the plane waves of the linearised equation than the KdV solitons. The NLS equation appears as a bridge between the linear limit and the nonlinearity represented by the KdV equation.

In order to obtain information on the dominant behaviours of a given physical system, irrespective of its external excitation, a standard approach in physics is to study simplified asymptotic models, valid in some specific ranges. Let us examine two examples, in hydrodynamics and for electrical lines.

4.7.1 Hydrodynamic waves

Two regimes can be considered for hydrodynamic waves:

(i) *For a shallow layer of fluid*, as shown in Appendix A, the dynamics of the height $u(x, t)$ of the water above the equilibrium surface is well described by the Korteweg–de Vries equation studied in Chapter 1,

$$u_t + u_x + uu_x + u_{xxx} = 0. \quad (4.64)$$

It turns out that, to the same order of approximation, one can derive [25] the Benjamin–Bona–Mahony (BBM) equation

$$u_t + u_x + uu_x - u_{xxt} = 0. \quad (4.65)$$

This equation is less known, and therefore less used, than its cousin (4.64) and it does not have its remarkable mathematical properties. However it has a much better behaviour for highly localised excitations which contain short wavelength components, i.e. large wavevectors k in their spectrum, as clearly shown in Figure 4.1. Indeed for a sufficiently large value of k, according to the dispersion relation deduced from the KdV equation, ω can vanish for a nonzero value of k, which is associated with an instability, while the BBM equation does not exhibit this pathological behaviour.

In order to determine the limit of validity of the KdV equation, it is also necessary to study the stability of its solitary waves with respect to transverse perturbations. Let us introduce a characteristic length μ in the transverse direction y, while staying within the validity of the hypotheses made in Appendix A in order to derive the KdV equation. Assuming that μ is large with respect to the spatial extent of the KdV soliton, one gets the Kadomtsev–Petviashvili (KP) equation [96],

$$(u_t + u_x + uu_x + u_{xxx})_x = \pm u_{yy}. \tag{4.66}$$

This KP equation has a large variety of quasi-periodic exact solutions, each having several independent phase factors. Recent comparisons with experiments [82, 83] have shown that the two-phase solutions give a very accurate description of the experimental observations for shallow water waves when they develop structures which are no longer strictly one-dimensional. It is likely that the sophisticated multiphase solutions could describe even more complex phenomena.

(ii) On the contrary, *for deep water waves*, a nonlinear Schrödinger equation can be derived if we are interested in the envelope of the free surface. One must assume that the envelope varies slowly in space, and set a maximum δk for the wavenumber of the envelope of the oscillations with respect to the wavenumber k_0 of the carrier wave. More precisely the hypotheses are:

(1) $\varepsilon = A/L \ll 1$ where A is an amplitude and L a characteristic spatial extent as shown in the figure of Appendix A.

(2) $\delta k/k_0 = \mathcal{O}(\varepsilon)$, which defines the range where a balance between the nonlinear effects and the dispersion of the slowly varying envelope is possible.

In the limit where the depth h is such that h/L tends to infinity, one gets a nonlinear Schrödinger equation [51]. The associated modulational instability is known in hydrodynamics under the name of 'Benjamin–Feir instability'. Although it occurs in deep water, its effects are noticeable even on the coast. Observing the waves on the beach, one often notices a regular modulation of their strength, with a period of 6 or 7 waves.

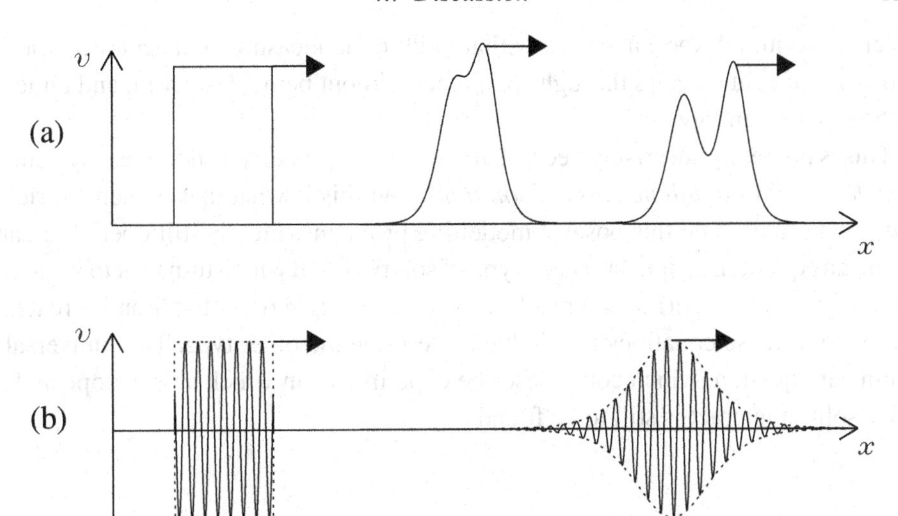

Figure 4.2. Schematic plot of the signals observed in a nonlinear electrical line when the initial excitation is of (a) 'pulse'-type or (b) 'envelope'-type. In each case the signal on the left is the signal which is injected with a standard source and the signal(s) on the right show the voltage in the line after solitons have been formed and separated from the small linear components which were also created by the excitation.

4.7.2 Electrical lines

Electrical lines provide an interesting example where one can experimentally observe a coexistence of two types of solitons in the same physical system, 'pulse' solitons and 'envelope' solitons. We have shown that the electrical lines introduced in Chapter 1 can be described in the continuum limit by the equation

$$v_{tt} - c_0^2 v_{xx} - \frac{c_0^2}{12} v_{xxxx} - a(v^2)_{tt} = 0. \tag{4.67}$$

- In Chapter 1, we showed that this Boussinesq equation appeared as an intermediate step in an analysis leading to the KdV equation, by appropriately choosing time, space and amplitude scales.
- In the small-amplitude limit, when the voltage v is of order ε but nevertheless large enough to excite nonlinearities, this equation can be reduced to an NLS equation with $PQ > 0$ (see Chapter 3). This condition ensures the existence of envelope solitons, in agreement with the experimental observations [9].

Therefore, in this system we predict the possible existence of two types of solitons, which can have similar amplitudes. The creation of one type or the other is only determined by the shape of the electrical signal applied at one end of the line, as schematised in Figure 4.2. Using two different generators, one at each end of the line, we can therefore generate a head-on collision between a pulse soliton and an envelope soliton. Experiments show that, within the measurement accuracy, these two types of solitons pass through each other without being destroyed, and almost without losing energy.

This is not really surprising because it is a general property of nonlinear systems, *they do not have a unique class of solutions*, and this is what makes them so rich. But, at the same time this poses a modelling problem which is still open. We can obtain an equation, adapted for each type of solution, and which turns out to be good for analysing the experimental results as long as *a single type* of soliton is created. But none of these equations is suitable to describe the other type. The 'universal' nonlinear equation, which could describe experiments in which an envelope and a pulse soliton interact is still to be found.

Part II

Mathematical methods for the study of solitons

Introduction

Soliton solutions are interesting for their mathematical properties because they describe systems with an infinite number of degrees of freedom which are fully integrable, and which can be investigated with elegant mathematical methods, such as the inverse scattering transform that we shall introduce in this second part.

For a physicist however, solitons could seem 'boring' because, once they have been created, they survive permanently and only experience phase shifts in collisions. But physical applications show that things are not so simple. Systems obeying nonlinear field equations are only approximately described by the integrable soliton equations and soliton solutions only represent the physical properties locally and for a finite time. They are only a first approximation of the actual system.

Phenomena which have been neglected when deriving the soliton equations such as:

- Higher order terms in the equation.
- Dissipation.
- A weak space or time variation of the parameters.
- Discreteness effects arising from an underlying lattice, for instance in solid state physics applications, etc.

may play an important role and modify the properties of the solitons in the long term or over long distances.

These effects can be treated as perturbations around the soliton solution and investigated by perturbation theory, or one can introduce collective coordinate methods which study the dynamics of the soliton considered as a well defined 'object' which keeps its individuality. These methods will be presented in the first two chapters of this mathematical part.

Another important problem for a physicist is the time evolution of an arbitrary initial condition. We know that it will generally create solitons, but one may ask how many, what are their properties? A method which can answer these questions is the 'inverse scattering transform' which is described in Chapter 7. We shall also show that this method, which is only valid for fully integrable systems, may be a good starting point for a perturbative analysis.

5

Linearisation around the soliton solution

The first idea that comes to mind when we want to study the effect of a perturbation on a soliton is to use linear response theory, i.e. to linearise around the soliton solution. This method is therefore equivalent to a linear stability analysis of the soliton. However it does not mean that we neglect nonlinear effects because we *linearise* around a *nonlinear* solution.

5.1 Spectrum of the excitations around a sine-Gordon soliton

Let us consider, for instance, the sine-Gordon equation

$$\theta_{tt} - c_0^2\,\theta_{xx} + \omega_0^2 \sin\theta = 0. \tag{5.1}$$

We know that it has a soliton solution

$$\theta^{(S)} = 4 \arctan \exp \left[\pm \frac{\omega_0}{c_0} \frac{(x - vt)}{\sqrt{1 - v^2/c_0^2}} \right], \tag{5.2}$$

as well as small-amplitude plane wave solutions $\theta = A \exp \mathrm{i}(kx - \omega t)$ which have the dispersion relation $\omega^2 = \omega_0^2 + c_0^2 k^2$.

Let us now investigate small excitations *around the soliton solution* by looking for a solution of the form

$$\theta(x,t) = \theta^{(S)}(x,t) + \psi(x,t) \quad \text{with} \quad |\psi(x,t)| \ll 1. \tag{5.3}$$

As the sine-Gordon equation is invariant by a Lorentz transform, we can choose to work in the rest frame of the soliton, i.e. set the velocity v to zero, so that the solution $\theta^{(S)}$ can be assumed to be independent of time. Putting this solution into Equation (5.1), we get

$$\psi_{tt} - c_0^2 \left(\theta_{xx}^{(S)} + \psi_{xx}\right) + \omega_0^2 \left(\sin\theta^{(S)} + \psi \cos\theta^{(S)}\right) = 0, \tag{5.4}$$

if we only keep the linear terms for the perturbation ψ. Taking into account that $\theta^{(S)}$ is a solution of Equation (5.1), the equation simplifies into

$$\psi_{tt} - c_0^2 \psi_{xx} + \omega_0^2 \cos\theta^{(S)} \ \psi = 0. \tag{5.5}$$

Using

$$1 - \cos\theta^{(S)} = 2\sin^2\left(\theta^{(S)}/2\right) = 2\left[2\tan\left(\theta^{(S)}/4\right)/\left(1 + \tan^2\left(\theta^{(S)}/4\right)\right)\right]^2 \tag{5.6}$$

$$= 2\,\mathrm{sech}^2(\omega_0 x/c_0), \tag{5.7}$$

and looking for solutions under the form

$$\psi(x,t) = f(x)\mathrm{e}^{-\mathrm{i}\omega t}, \tag{5.8}$$

we obtain an equation for the function f

$$-c_0^2 \frac{\mathrm{d}^2 f}{\mathrm{d}x^2} + \omega_0^2 \left(1 - 2\,\mathrm{sech}^2 \frac{\omega_0 x}{c_0}\right) f(x) = \omega^2 f(x), \tag{5.9}$$

which is formally analogous to the stationary Schrödinger equation which determines the quantum states of a particle in the potential

$$V(x) = \omega_0^2 \left(1 - 2\,\mathrm{sech}^2(\omega_0 x/c_0)\right). \tag{5.10}$$

Thus the soliton acts as a potential well for the linear waves, and it should be noticed that the equation (and therefore the potential too) is the same for a soliton or an antisoliton.

In Chapter 2 we showed that the sine-Gordon equation describes the dynamics of a chain of pendula coupled by a torsional spring, so that it appears that this very simple device can be used to illustrate 'quantum mechanics' by showing the diffusion of a wave packet by a potential well, or 'special relativity' by demonstrating the Lorentz contraction of a moving soliton, all in a table-top mechanical experiment!

The potential $V(x)$ is one of the few potentials which can be fully solved analytically. This is a standard problem of quantum mechanics [108, 134] which yields the following results:

- There is only one bound state, corresponding to the eigenvalue $\omega_b^2 = 0$ and to the localised eigenstate

$$f_b(x) = (2\omega_0/c_0)\,\mathrm{sech}(\omega_0 x/c_0). \tag{5.11}$$

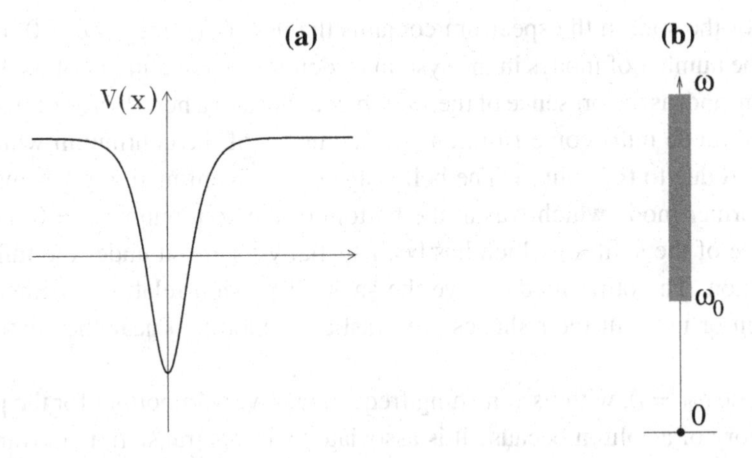

Figure 5.1. (a) Shape of the potential $V(x) = \omega_0^2(1 - 2\,\mathrm{sech}^2 \frac{\omega_0 x}{c_0})$. (b) Schematic plot of the spectrum of small excitations around a sine-Gordon soliton. It includes the frequencies higher than ω_0 as well as the frequency $\omega = 0$.

It is the lowest bound state of a one-dimensional Schrödinger equation, therefore we know that its eigenfunction should not have any root, and indeed this is true.

- The eigenfunctions of the continuum part of the spectrum, associated with the eigenvalues $\omega_k^2 = \omega_0^2 + c_0^2 k^2$, are given by

$$f_k(x) = \frac{1}{\sqrt{2\pi}} \frac{c_0}{\omega_k} e^{ikx} \left(k + i\frac{\omega_0}{c_0} \tanh \frac{\omega_0 x}{c_0} \right) \tag{5.12}$$

and they are such that $\lim_{x \to \pm\infty} f_k(x) = (1/\sqrt{2\pi}) \exp[i(kx \pm \phi_0)]$ where $\phi_0 = \arctan(\omega_0/c_0 k)$.

Consequently, far from the soliton, the eigenfunctions of the continuum are similar to plane waves. This result could have been predicted because, in these regions, the soliton solution is almost a constant corresponding to a minimum of the sinusoidal potential of the sine-Gordon model. This is why, far from the centre of the soliton, the solutions of Equation (5.5) are identical to the plane waves which are solutions of the SG equation linearised around an equilibrium state. For these waves the soliton only causes a phase shift $2\phi_0$ but it does not have any effect on the amplitude of the wave. The potential created by the soliton is said to be *reflectionless* and this is a peculiarity of integrable systems. However this is not a sufficient condition for a system to be fully integrable because some solitary wave solutions of nonintegrable equations, such as the ϕ^4 equation, also create a reflectionless potential for the linear waves.

The spectrum plotted in Figure 5.1(b) is qualitatively different from the spectrum in the absence of a soliton, which only includes frequencies $\omega \geq \omega_0$. Due to the presence of the soliton the spectrum contains the new frequency, $\omega_b = 0$. In fact, because the number of modes in the system is identical to the number of its degrees of freedom, and, as the presence of the soliton does not add a new degree of freedom, this 'extra' mode must come from one of the modes of the continuum which has been shifted due to the soliton. The bell shape of its eigenfunction gives the clue. It is the former mode which was at the bottom of the continuum ($k = 0$ mode in the absence of the soliton) which has been spatially localised under the influence of the soliton. The other modes have the same dispersion relation whether there is a soliton or not, but their shapes are drastically modified near the centre of a soliton.

The mode $\omega_b = 0$, with its vanishing frequency, is very important for the perturbation theory of a soliton because it is associated with its translational invariance. As the SG equation is invariant by space translation, its solution $\theta^{(S)}(x - a)$ is a solution whatever the value of a, with the same energy as the solution $\theta^{(S)}(x)$. Therefore, although the soliton solution is an extremum of the Hamiltonian, it is not associated with a minimum. Instead it shows *marginal stability* with respect to translation. If we perform a small translation of the soliton, the variation of the solution is given by

$$\Delta\theta^{(S)}(x) = \theta^{(S)}(x - \delta a) - \theta^{(S)}(x) \simeq -\delta a \frac{\partial \theta^{(S)}}{\partial x} = -\delta a\, 2\frac{\omega_0}{c_0} \operatorname{sech}\left(\frac{\omega_0 x}{c_0}\right) \tag{5.13}$$

which is proportional to the eigenfunction $f_b(x)$ given by Equation (5.11). The proportionality constant measures the amplitude of the translation. The zero frequency mode is thus the translational mode of the soliton. It is the so-called *Goldstone mode* which is found in all field theories which are translationally invariant.

5.2 Application: perturbation of a soliton

5.2.1 Method

The functions $f_b(x)$ and $f_k(x)$, solutions of the Schrödinger equation (5.9), are the eigenfunctions of the Hermitian operator

$$D = -c_0^2 \frac{\mathrm{d}^2}{\mathrm{d}x^2} + \omega_0^2 \left(1 - 2\operatorname{sech}^2 \frac{\omega_0 x}{c_0}\right). \tag{5.14}$$

As shown in quantum mechanics courses they make up an orthogonal basis for the space of functions which describe small deviations around a soliton solution.

As the function f_b is real while the others are complex, the orthogonality conditions are

$$\int_{-\infty}^{+\infty} \mathrm{d}x \; f_b(x) f_b(x) = \frac{8\omega_0}{c_0} \tag{5.15}$$

$$\int_{-\infty}^{+\infty} \mathrm{d}x \; f_k^*(x) f_{k'}(x) = \delta(k - k') \tag{5.16}$$

$$\int_{-\infty}^{+\infty} \mathrm{d}x \; f_k(x) f_b(x) = 0, \tag{5.17}$$

and the completeness property of the basis gives

$$\int_{-\infty}^{+\infty} \mathrm{d}k \; f_k^*(x) f_k(x') + \frac{c_0}{8\omega_0} f_b(x) f_b(x') = \delta(x - x'). \tag{5.18}$$

It should be noticed that the function $f_b(x)$ is not normalised. Its coefficient is chosen so that $f_b(x)$ is exactly equal to the derivative of the soliton solution (Goldstone mode).

This basis is very convenient for studying the effect of perturbations on sine-Gordon solitons [68] because the results have a direct meaning:

- The coefficient of the eigenfunction $f_b(x)$ is the opposite of the translation of the soliton under the influence of the perturbation, according to Relation (5.13).
- The contributions of the continuum spectrum, proportional to $f_k(x)$, give the deformations of the soliton.

5.2.2 Example: Response of a soliton to an external force in the presence of dissipation

Let us consider the perturbed sine-Gordon equation

$$\theta_{tt} - c_0^2 \theta_{xx} + \omega_0^2 \sin\theta + \Lambda \theta_t = E, \tag{5.19}$$

which is similar to the equation that we obtained for a long Josephson junction (see Chapter 2), i.e. an SG equation driven by a constant term, and including damping.

It is convenient to eliminate some factors by introducing dimensionless variables

$$t' = \omega_0 t, \quad x' = \frac{\omega_0}{c_0} x \quad \text{and defining} \quad \Gamma = \frac{\Lambda}{\omega_0}, \quad \chi = \frac{E}{\omega_0^2}. \tag{5.20}$$

The equation is reduced to

$$\theta_{t't'} - \theta_{x'x'} + \sin\theta + \Gamma \theta_{t'} = \chi. \tag{5.21}$$

Our goal is to determine what happens to a soliton mobile at the velocity β in the presence of driving and damping. If we assume that the driving is small, we can look for a perturbed soliton solution,

$$\theta(x', t') = \theta^{(S)}(x', t') + \psi(x', t'). \tag{5.22}$$

In order to benefit from our knowledge of a basis appropriate for the expansion of the small deviations around a static soliton, it is convenient to move to the rest frame of the soliton by performing a Lorentz transform. Let us introduce the variables

$$\xi = \gamma(x' - \beta t'), \quad \tau = \gamma(t' - \beta x') \quad \text{where} \quad \gamma = 1/\sqrt{1 - \beta^2}. \tag{5.23}$$

Equation (5.21) leads to

$$\theta_{\tau\tau} - \theta_{\xi\xi} + \sin\theta + \Gamma\gamma\left(\theta_\tau - \beta\theta_\xi\right) = \chi. \tag{5.24}$$

Putting the ansatz (5.22) in this equation, and because $\theta^{(S)}$ is a solution of the sine-Gordon equation, we get

$$\psi_{\tau\tau} - \psi_{\xi\xi} + (1 - 2\,\mathrm{sech}^2\,\xi)\psi + \gamma\Gamma\psi_\tau - \beta\gamma\Gamma\psi_\xi = \chi + 2\beta\gamma\Gamma\,\mathrm{sech}\,\xi, \tag{5.25}$$

if we keep only linear terms in ψ and use the identity $\theta_\xi = \psi_\xi + 2\,\mathrm{sech}\,\xi$.

In order to solve this equation we can expand ψ on the basis (f_b, f_k) that we defined above. The expression of the eigenfunctions in dimensionless variables is simply obtained by setting $\omega_0 = c_0 = 1$. We get

$$\psi(\xi, \tau) = \frac{1}{8}\phi_b(\tau) f_b(\xi) + \int_{-\infty}^{+\infty} \mathrm{d}k\ \phi_k(\tau) f_k(\xi). \tag{5.26}$$

As we expand the function $\psi(\xi, \tau)$ for a given τ on the basis of space dependent functions $f(\xi)$, the coefficients depend on τ. The factor 1/8 is introduced to make sure that the $\phi_b(\tau)$ coefficient of the expansion keeps its usual expression $\phi_b(\tau) = \int \mathrm{d}\xi\ f_b(\xi)\psi(\xi, \tau)$ although the function f_b is not normalised.

We put Expansion (5.26) into Equation (5.25) and project on the different basis functions. The projection on f_b, i.e. the product by f_b followed by an integration over ξ, determines the translation of the soliton. The $\psi_{\tau\tau}$ term simply gives

$$\begin{aligned}\int \mathrm{d}\xi\ \psi_{\tau\tau}(\xi, \tau) f_b(\xi) &= \frac{1}{8}\phi_{b,\tau\tau}(\tau) \int \mathrm{d}\xi\, f_b(\xi) f_b(\xi) + \int \mathrm{d}k\ \phi_{k,\tau\tau}(\tau) \int \mathrm{d}\xi\ f_b(\xi) f_k(\xi) \\ &= \frac{1}{8}\phi_{b,\tau\tau}\end{aligned} \tag{5.27}$$

if we take into account the orthogonality relations of the basis.

The $\psi_{\xi\xi}$ term leads to

$$\int \mathrm{d}\xi\ \psi_{\xi\xi}(\xi,\tau) f_b(\xi) = \frac{1}{8}\phi_b(\tau) \int \mathrm{d}\xi f_b(\xi) f_{b,\xi\xi}(\xi) + \int \mathrm{d}k\ \phi_k(\tau) \int \mathrm{d}\xi\ f_b(\xi) f_{k,\xi\xi}(\xi). \tag{5.28}$$

However, taking into account Equation (5.9) verified by the functions f, the combination of this expression with the term $(1-2\,\mathrm{sech}^2\,\xi)\psi$ reduces to

$$\frac{1}{8}\phi_b(\tau)\omega_b^2 \int \mathrm{d}\xi\ f_b(\xi) f_b(\xi) + \int \mathrm{d}k\ \phi_k(\tau)\ \omega_k^2 \int \mathrm{d}\xi\ f_b(\xi) f_k(\xi) \tag{5.29}$$

which vanishes because $\omega_b = 0$ and the orthogonality condition cancels the second term.

The contribution of the term containing ψ_τ is simple because, owing to the orthogonality relations, only $\gamma\Gamma\phi_{b,\tau}$ subsists.

The next term, containing ψ_ξ, requires an explicit calculation of the projection of the derivatives f_ξ on f_b. On the one hand we get

$$\int \mathrm{d}\xi\ f_b(\xi)\frac{\mathrm{d}f_b}{\mathrm{d}\xi} = 4\int_{-\infty}^{+\infty} \mathrm{d}\xi\ \mathrm{sech}\,\xi \times (-\mathrm{sech}\,\xi\,\tanh\xi) = 0 \tag{5.30}$$

and, on the other hand, using

$$\frac{\mathrm{d}f_k}{\mathrm{d}\xi} = \frac{\mathrm{i}k}{\sqrt{2\pi}\,\omega_k}\mathrm{e}^{\mathrm{i}k\xi}(k+\mathrm{i}\tanh\xi) + \frac{\mathrm{e}^{\mathrm{i}k\xi}}{\sqrt{2\pi}\,\omega_k}(\mathrm{i}\,\mathrm{sech}^2\,\xi) \tag{5.31}$$

$$= \mathrm{i}k f_k(\xi) + \frac{\mathrm{i}\mathrm{e}^{\mathrm{i}k\xi}}{\sqrt{2\pi}\,\omega_k}\,\mathrm{sech}^2\,\xi, \tag{5.32}$$

the first term of the projection $\int \mathrm{d}\xi\ f_b(\mathrm{d}f_k/\mathrm{d}\xi)$ vanishes due to the orthogonality condition because it is proportional to f_k. The second term gives

$$\int \mathrm{d}\xi\ f_b\frac{\mathrm{d}f_k}{\mathrm{d}\xi} = \mathrm{i}k\underbrace{\int \mathrm{d}\xi\ f_b f_k}_{=0} + \frac{2\mathrm{i}}{\sqrt{2\pi}\,\omega_k}\underbrace{\int \mathrm{d}\xi\ \mathrm{e}^{\mathrm{i}k\xi}\,\mathrm{sech}^3\,\xi}_{\widetilde{F}(k)} \tag{5.33}$$

if we introduce $\widetilde{F}(k)$, the Fourier transform of the function $\mathrm{sech}^3\,\xi$.

Finally the right hand side of Equation (5.25) leads to

$$\chi\int \mathrm{d}\xi\ f_b(\xi) = 2\chi\int_{-\infty}^{+\infty} \mathrm{d}\xi\ \mathrm{sech}\,\xi = 2\pi\,\chi \tag{5.34}$$

and

$$\int \mathrm{d}\xi\ f_b(\xi)\,\mathrm{sech}\,\xi = 2\int_{-\infty}^{+\infty} \mathrm{d}\xi\ \mathrm{sech}^2\,\xi = 4. \tag{5.35}$$

As a result the projection of Equation (5.25) on $f_b(\xi)$ gives

$$\frac{\mathrm{d}^2\phi_b(\tau)}{\mathrm{d}\tau^2} + \gamma\Gamma\frac{\mathrm{d}\phi_b(\tau)}{\mathrm{d}\tau} = 2\pi\chi + 8\beta\gamma\Gamma + \frac{2\mathrm{i}\beta\gamma\Gamma}{\sqrt{2\pi}}\int \mathrm{d}k\,\frac{\widetilde{F}(k)}{\omega_k}\phi_k(\tau). \tag{5.36}$$

The projection on the eigenfunctions belonging to the continuum proceeds in the same way. It gives an equation for $\phi_k(\tau)$ which also includes $\phi_b(\tau)$. Therefore the calculation leads to a set of coupled equations for the coefficients $\phi_b(\tau)$ and $\phi_k(\tau)$ which cannot be solved exactly. However it is possible to compute the perturbation of the soliton using an approximate method which proceeds in two steps:

(i) In the first step we neglect the contribution of the continuum in the equation for $\phi_b(\tau)$, which amounts to studying the response of the soliton *without taking into account its deformation*.
(ii) In the second step, as $\phi_b(\tau)$ is known, it is possible to compute $\phi_k(\tau)$ which gives the deformation of the soliton. Then it is even possible to obtain a correction to $\phi_b(\tau)$ by examining again Equation (5.36) with the values of $\phi_k(\tau)$ which have been computed at the previous stage. Hence the solution is obtained in an iterative process.

Let us consider step (i). We have to solve the differential equation

$$\frac{\mathrm{d}^2\phi_b(\tau)}{\mathrm{d}\tau^2} + \gamma\Gamma\frac{\mathrm{d}\phi_b(\tau)}{\mathrm{d}\tau} = 2\pi\chi + 8\beta\gamma\Gamma \tag{5.37}$$

which gives

$$\phi_b(\tau) = \left(8\beta + \frac{2\pi\chi}{\gamma\Gamma}\right)\left[\tau + \frac{\mathrm{e}^{-\gamma\Gamma\tau} - 1}{\gamma\Gamma}\right]. \tag{5.38}$$

But we have seen that the coefficient of the translational mode $f_b(\xi)$ in Equation (5.26), $\phi_b(\tau)/8$, is equal to the *opposite of the translation of the soliton under the effect of the perturbation.* Thus the displacement due to the driving χ in the presence of damping is

$$d = -\beta\left(1 + \frac{\pi\chi}{4\beta\gamma\Gamma}\right)\left[\tau + \frac{\mathrm{e}^{-\gamma\Gamma\tau} - 1}{\gamma\Gamma}\right]. \tag{5.39}$$

This expression shows that, after a transient, for times $\tau \gg 1/\gamma\Gamma$ the soliton reaches a limiting velocity β^e equal to

$$\beta^e = -\beta\left(1 + \frac{\pi\chi}{4\beta\gamma\Gamma}\right) \tag{5.40}$$

in the frame mobile at velocity β. We can check that this result agrees with some simple predictions which can be made from the physics of the system. If we had

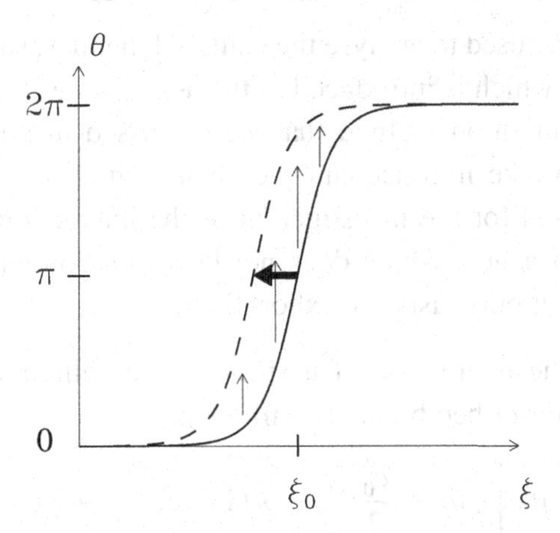

Figure 5.2. Negative displacement of a sine-Gordon soliton subjected to a positive uniform driving $\chi > 0$.

some damping but no driving ($\chi = 0$) we would obtain $\beta^e = -\beta$, which corresponds to a soliton at rest in the laboratory frame, as expected. If we consider the limit of a vanishing damping ($\Gamma \to 0$), the solution leads to a divergence of β^e because a stable equilibrium is no longer possible.

Coming back to the dissipative driven case ($\Gamma \neq 0$, $\chi > 0$), we notice that the soliton slows down under the effect of the perturbation, as we can expect from an examination of Figure 5.2: a positive driving χ, which tends to increase θ, moves the soliton backward.

The calculations of step (ii) are very tedious, and we shall not give them here but they can be found in reference [68]. The main result is that the driven soliton loses its symmetry. Moreover, far from the centre, θ is slightly shifted, which is not surprising because the driving shifts the equilibrium values of θ.

This change occurring on the soliton wings gave rise to a debate on the properties of the soliton as a Newtonian quasi-particle. Indeed when it is subjected to an external force, the soliton first changes its shape without actually moving as a whole, and motion only starts subsequently. The first stage seems to be in contradiction with a quasi-particle behaviour. It points out one important aspect of soliton dynamics: although their stability and immunity to collisions immediately suggests that solitons behave like particles, one should not forget that they are *deformable particles*. We shall show that, in some cases such as the ϕ^4 model, this possibility for a soliton to have some internal shape change may deeply influence its dynamics.

The result (5.38) points out the *limits of the perturbative approach* because the function $\phi_b(\tau)$ grows linearly with τ. Over a long time scale, $\phi_b(\tau)$ can no longer be considered as a perturbation. Therefore the result is only useful to determine how the soliton begins to move, i.e. it is restricted to short timescales. Beyond this stage, other methods must be used to analyse the soliton dynamics, such as the collective coordinate approach which is introduced in the next chapter.

However the perturbation method that we discussed in this chapter is useful because it is able to take into account the change in shape of the soliton. It is particularly well suited for the investigation of the interaction between a soliton and a localised disturbance, which does not lead to a continuous growth of the perturbation because it only lasts for a short time.

▷ *Exercise*: Study the interaction of a sine-Gordon soliton with a perturbative potential $g(x)$, described by the Hamiltonian

$$H = \int \mathrm{d}x \left[\frac{1}{2}\theta_t^2 + \frac{c_0^2}{2}\theta_x^2 + \omega_0^2(1 - \cos\theta) - \lambda g(x)\,\theta_x \right], \tag{5.41}$$

with $g(x) = Y(x - x_0) - Y(x + x_0)$, where $Y(x)$ is the Heaviside step function.

1. Derive the dynamic equation for θ.
2. Expand the perturbation of the soliton on the basis $(f_b(x),\ f_k(x))$, and derive the equation verified by the coefficient of $f_b(x)$.
3. Integrate this equation and show that the soliton only experiences a phase shift when it passes through the perturbative potential. Study qualitatively how the velocity varies in the vicinity of the defect for a soliton or an antisoliton. It should be noticed that it is not necessary to get an explicit expression of all integrals which enter into the calculation to answer this question.

The solution can be found in reference [68].

5.3 Spectrum of the excitations around a ϕ^4 soliton

The study of the ϕ^4 model, described by the Hamiltonian

$$H = \int \mathrm{d}x \left[\frac{1}{2}\phi_t^2 + \frac{1}{2}\phi_x^2 + \frac{1}{4}(\phi^2 - 1)^2 \right] \tag{5.42}$$

can be carried out along the same lines as for the sine-Gordon case. However, the results exhibit a qualitative difference, which has far-reaching consequences.

Putting the ansatz $\phi = \phi_S(x) + \psi(x, t)$ into the ϕ^4 equation of motion leads to the equation linearised in ψ

$$\psi_{tt} - \psi_{xx} + \left(2 - 3\,\mathrm{sech}^2 \frac{x}{\sqrt{2}}\right) \psi = 0, \tag{5.43}$$

using of course the expression of the soliton solution $\phi_S = \tanh(x/\sqrt{2})$.

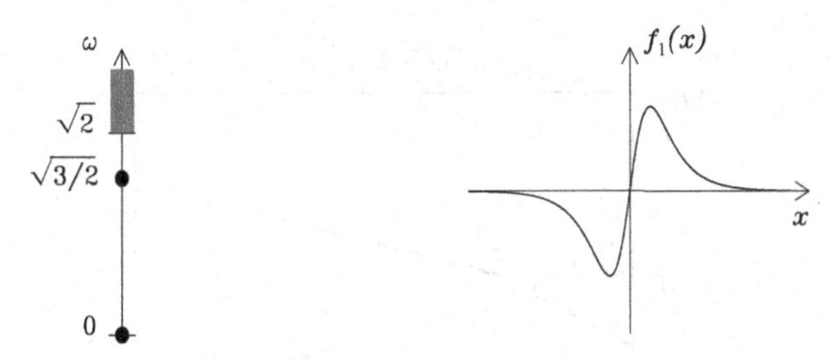

Figure 5.3. Spectrum of the excitations around a ϕ^4 soliton, and shape of the first excited state $f_1(x)$.

If we look for a solution of the form $\psi(x,t) = \mathrm{e}^{-\mathrm{i}\omega t} f(x)$ as we did for the sine-Gordon case, we get the Schrödinger-like equation

$$-f_{xx} + \left(2 - 3\,\mathrm{sech}^2 \frac{x}{\sqrt{2}}\right) f(x) = \omega^2 f(x). \tag{5.44}$$

It is remarkable to notice that the shape of the potential that we obtained for the sine-Gordon model was the same, but here the ratio between the depth of the well and its width is larger, which allows the existence of two bound states [108, 134].

Besides the delocalised states of the continuum, which lie above $\omega_c^2 = 2$, we find again a bound state for $\omega = 0$ which is the translational mode of the soliton

$$f_b(x) = \frac{\mathrm{d}}{\mathrm{d}x}\left(\tanh \frac{x}{\sqrt{2}}\right) = \frac{1}{\sqrt{2}}\,\mathrm{sech}^2 \frac{x}{\sqrt{2}}. \tag{5.45}$$

But here we also find a second bound state for $\omega_1^2 = 3/2$, associated with the eigenfunction

$$f_1(x) = \frac{1}{2^{3/4}} \tanh \frac{x}{\sqrt{2}}\,\mathrm{sech}\,\frac{x}{\sqrt{2}}, \tag{5.46}$$

which is a *localised shape change of the soliton.* The position of the centre of the soliton is not modified by this odd mode which verifies $\int \mathrm{d}x f_1(x) = 0$. This mode, represented in Figure 5.3, leads to an oscillation of the slope of the soliton at frequency ω_1.

Consequently, contrary to the sine-Gordon case where the only possible excitations of the soliton are a translation or a delocalised mode of the continuum, the ϕ^4 soliton can react to a perturbation by an excitation of its *internal mode.* Thus it appears as a quasi-particle which can have some internal excitation. This mode can store energy, which moves together with the soliton because the mode is localised around the soliton. But, as this energy is taken from the translational kinetic energy of the soliton, its dynamics can be deeply modified.

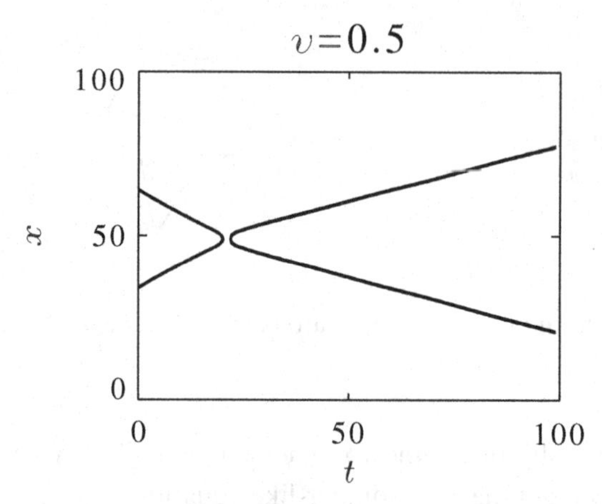

Figure 5.4. Kink–antikink collision in the ϕ^4 model, with a *high* initial velocity v_i. The figure shows the position of the two solitons versus time. After the collision, the solitons emerge with a final velocity v_f which is significantly smaller than their input velocity because some part of their kinetic energy has been transferred into an oscillation of their slope. Although the solitons pass through each other, their trajectories seem to be interrupted at the collision point. This is an artefact which occurs because at the time of the collision the field ϕ is momentarily very small everywhere in the system. The position of the soliton and the antisoliton is not defined at this moment.

For instance the existence of the internal mode is noticeable when the soliton is perturbed by a spatially periodic potential. If the soliton moves at velocity v with respect to such a potential having a spatial period ℓ, it experiences a time periodic perturbation with period $T = \ell/v$. If T coincides with $T_1 = 2\pi/\omega_1$, it resonates with the internal mode frequency. Some of the soliton kinetic energy is transferred to the internal mode. This leads to a selection of some soliton velocities which are more likely than others [65]. But the excitation of the internal mode of a soliton can have much more dramatic effects in the soliton–antisoliton collisions of the ϕ^4 model [37].

We pointed out in Section 2.4 that calling the solutions of the ϕ^4 model 'solitons' is inaccurate. These 'kinks' are not solitons because their energy is not exactly preserved in collisions. It may happen that collisions excite the internal mode of the soliton. The energy which is stored there is taken from the kinetic energy of the solitons so that they emerge from the collision with a velocity v_f smaller than their input velocity v_i as shown in Figure 5.4.

But, as we have shown that a kink and an antikink attract each other, if the transfer of energy to the internal mode is too large, it may happen that the two excitations cannot escape from their attractive potential well. They turn into a bound pair of excitations which oscillate with respect to each other, similar to the sine-Gordon breather, as shown in Figure 5.5. It should however be noticed that the breather is not

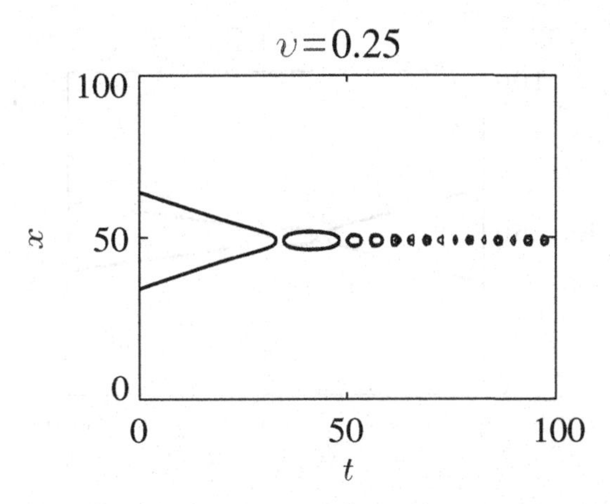

Figure 5.5. Kink–antikink collision in the ϕ^4 model, with a *low* initial velocity v_i. After the collision the two solitons make up a breather, as shown by the oscillations of their positions.

an exact solution of the ϕ^4 model. It decays slowly because it loses energy through the emission of nonlocalised excitations belonging to the continuum spectrum.

Therefore there is a critical velocity v_c below which the transfer of energy to the internal mode is larger than the initial kinetic energy, which leads to the formation of a bound state. However the energy transferred to the internal mode is not lost for the kink–antikink pair because the internal mode is *localised.* If a collision can excite it, another collision can 'de-excite' it. If this occurs, the energy of the internal mode is restored to the kinetic energy of the solitons. The possibility to 'de-excite' the internal mode results directly from the equation which describes the dynamics of the system. It only includes a second-order derivative with respect to time so that it is invariant with a change of t into $-t$. Consequently if the two solitons come to the second collision point in a state where the phases of their internal modes are equal to their values at the outcome of the first collision, the second collision will be the exact time reversal of the first one. It will cancel the excitation of the internal mode and the two solitons will recover enough kinetic energy to escape from their attracting potential well. The bound state will be broken and the solitons will escape to infinity, as shown in Figure 5.6.

The outcome of this low-speed resonant collision is a reflection of the two excitations. This mechanism takes place when the time interval between two collisions is equal to $T_{12} = nT_1 + \delta$, where δ is a phase factor related to the phase of the internal mode after the collision, and T_1 is the period of the internal mode. This mechanism leads to the existence of *resonance windows* in which the solitons are able to escape from each other after the collision although their input velocity is below the critical velocity v_c (Figure 5.7).

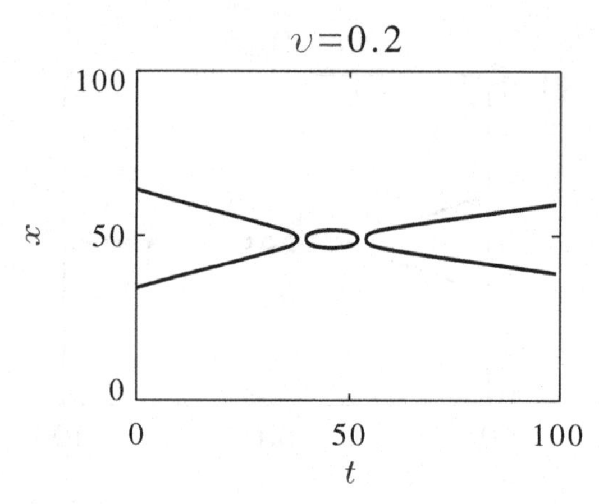

Figure 5.6. Resonant kink–antikink collision in the ϕ^4 model. The two solitons leave the first collision with an excited internal mode while, after the second collision, the internal mode is no longer excited.

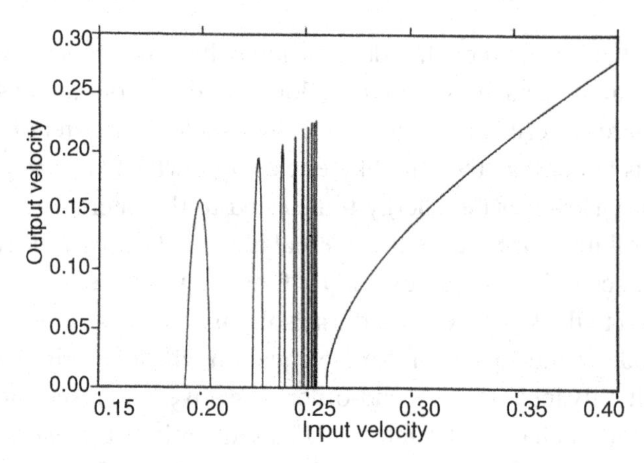

Figure 5.7. Diagram showing the output velocity after the collision of a soliton and an antisoliton in the ϕ^4 model versus their input velocity (adapted from reference [37]).

The same phenomenon exists for the double sine-Gordon (DSG) model [36] in which the internal mode is the oscillation of the distance between the two 'sub-kinks' which make up the DSG soliton. In this case it is even possible to exhibit higher order resonances which occur after two, three, four, ... collisions. In a diagram similar to Figure 5.7, the second-order resonances appear as a series of secondary windows on both sides of the windows of the first-order resonances. Each secondary window can then have a set of third-order resonances on its boundaries, and so on. This leads to a self-similar pattern which can easily be analysed in terms of the excitation of the internal mode. The distribution of the secondary windows with respect to the border of the primary windows is identical, except for a scale factor, to the distribution of the primary windows with respect to the critical velocity v_c. The diagram $v_f = f(v_i)$ looks like a fractal, but this auto-similarity is however not preserved at all scales because, besides the excitation of the internal mode, each collision also excites some modes of the continuum. As they are not localised, they carry away some energy which is lost for the soliton–antisoliton pair.

This example points out the importance of the spectrum of the excitations around the soliton solution for the dynamics of the solitons. But, to study the complex dynamics which arise because the ϕ^4 kinks have an internal mode, the perturbative expansion is not sufficient. The collective coordinate method that we introduce in the next chapter is the appropriate method for this purpose.

6

Collective coordinate method

The previous chapter has shown the power of the perturbative approach, but also its limits: when a soliton is subjected to a disturbance which induces or modifies its motion as a whole, the perturbation may diverge.

But we have seen that solitons are highly stable 'objects' which often behave as quasi-particles. Therefore it seems natural to introduce a coordinate for the position of this quasi-particle. This approach replaces the field, and its amplitude versus space and time corresponding to an infinite number of degrees of freedom, by a single object whose position is defined by one or a few coordinates. They are called *collective coordinates* because they characterise a whole set of degrees of freedom of the physical system.

This method is only meaningful if the properties of the field are such that the 'object' that we characterise by the collective coordinates has a real individuality, even in the presence of perturbations. As we shall see, this hypothesis is well justified for solitons. However the collective coordinate method is a variational method, and like all the methods of this type, it can only give satisfactory results if the class of functions which are used to replace the exact solution is appropriate, i.e. if the collective coordinates are properly chosen.

6.1 sine-Gordon soliton interacting with an impurity: effective Lagrangian method

Let us study for instance the pendulum chain in which the pendulum of index 0 is different from the others: although it has the same moment of inertia as the other pendula, the distance between its centre of mass and the axis is $\ell(1 - \varepsilon)$ instead of ℓ as for the other pendula. The gravitational energy of this particular pendulum is thus different from the others. The Hamiltonian of the chain is

$$H = \sum_n \frac{1}{2} I \left(\frac{\mathrm{d}\theta_n(t')}{\mathrm{d}t'} \right)^2 + \frac{1}{2} C(\theta_{n+1} - \theta_n)^2 + mg\ell\,(1 - \varepsilon\delta_{n0})(1 - \cos\theta_n). \tag{6.1}$$

Using the continuum limit approximation and introducing appropriate dimensionless time and space variables, we get the Hamiltonian

$$H = \int \mathrm{d}x \left[\frac{1}{2}\theta_t^2 + \frac{1}{2}\theta_x^2 + (1 - \varepsilon\delta(x))\,(1 - \cos\theta) \right]. \tag{6.2}$$

Our goal is to calculate the effect of the perturbation $\varepsilon\delta(x)$ on a sine-Gordon soliton which travels toward the defect with a small velocity ($v \ll 1$).

The idea of the method is to assume that the unperturbed sine-Gordon soliton,

$$\theta(x,t) = 4 \arctan \exp\left(\frac{x - vt}{\sqrt{1 - v^2}} \right) \simeq 4 \arctan \exp(x - vt) \quad \text{for} \quad v \ll 1, \tag{6.3}$$

preserves its shape, but that its position is no longer equal to vt, but to a quantity $X(t)$ to be determined. Thus we assume that, in the presence of the perturbation, the solution can be written

$$\theta_X(x,t) = 4 \arctan \exp[x - X(t)], \tag{6.4}$$

where $X(t)$ is the collective coordinate giving the position of the soliton. The method is therefore a variational method because we look for an approximate solution within a particular class of functions (here the soliton solution) depending on some parameters (here $X(t)$) which have to be determined to get an optimal approximation of the exact solution. As the soliton is described with the same functional form as the unperturbed solution, the method cannot describe its possible distortion, contrary to the perturbative approach of Chapter 5.

It could be tempting to choose the trial function

$$\theta(x,t) = 4 \arctan \exp\left(\frac{x - X(t)}{\sqrt{1 - \dot{X}^2}} \right), \tag{6.5}$$

in order to introduce the Lorentz contraction factor. This expression makes the calculation much more difficult and the improvement that it brings is not significant because the method ignores other corrections which can be more important, such as a possible emission of radiations by the soliton when it meets the defect. The perturbative method of Chapter 5 can include the distortion of the soliton and the emission of nonlocalised waves, but it fails to describe a drastic change in the soliton dynamics, such as its reflection by the impurity for instance.

In order to derive the dynamic equation for the collective coordinate, the Hamiltonian formalism is not convenient because it requires the determination of the momentum which is conjugate to the collective coordinate. It is simpler to use the

Lagrangian

$$L = \int \mathrm{d}x \left[\frac{1}{2}\theta_t^2 - \frac{1}{2}\theta_x^2 - (1 - \varepsilon\delta(x))(1 - \cos\theta) \right], \tag{6.6}$$

from which the equation of motion

$$\theta_{tt} - \theta_{xx} + (1 - \varepsilon\delta(x))\sin\theta = 0 \tag{6.7}$$

derives. If we put the ansatz (6.4) into the Lagrangian (6.6), we get

$$\begin{aligned} L &= \int \mathrm{d}x [2\dot{X}^2 \operatorname{sech}^2(x - X) - 2\operatorname{sech}^2(x - X) - 2(1 - \varepsilon\delta(x))\operatorname{sech}^2(x - X)] \\ &= 4\dot{X}^2 - U(X), \end{aligned} \tag{6.8}$$

if we define $U(X) = 8 - 2\varepsilon \operatorname{sech}^2 X$. A dot indicates, as usual in mechanics, the time derivative of the function $X(t)$ which only depends on t.

The expression that we have obtained after performing the integration with respect to x is formally identical to the Lagrangian of a particle moving in the potential $U(X)$. This particle has the mass $m_0 = 8$, which agrees with the soliton mass derived from the calculation of the energy of the sine-Gordon soliton (see Equation (2.26), expressed in dimensionless variables). The Lagrangian (6.8) is called an 'effective Lagrangian' because it is only an approximation of the actual Lagrangian, obtained with a specific assumption for the expression of the field. The collective coordinate method uses it like a true Lagrangian to derive an equation of motion for $X(t)$.

In the absence of any impurity ($\varepsilon = 0$), the potential would be a constant. Thus the soliton would move at constant speed as expected. However, if $\varepsilon > 0$, we find that the impurity generates an attractive potential for the soliton. The dynamic equation which results from the effective Lagrangian is

$$8\ddot{X} + 4\varepsilon \operatorname{sech}^2 X \tanh X = 0. \tag{6.9}$$

If we multiply it by $\dot{X}$ and integrate, we get

$$\dot{X}^2 = \frac{\varepsilon}{2}\operatorname{sech}^2 X + C, \tag{6.10}$$

where C is an integration constant which can be determined from the initial condition that we chose: the soliton is sent toward the impurity with velocity v. Thus, for $X \to -\infty$, $\dot{X} = v$, which gives $C = v^2$. We can separate the variables to get

$$\mathrm{d}t = \frac{\mathrm{d}X}{\sqrt{v^2 + \varepsilon/(2\cosh^2 X)}}. \tag{6.11}$$

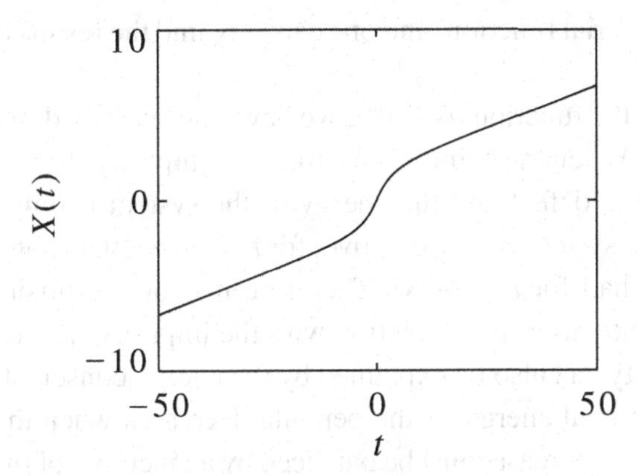

Figure 6.1. Time evolution of the soliton position $X(t)$ given by Equation (6.14) for $v = 0.1$ and $\varepsilon = 0.3$.

Choosing the origin of time to be when $X = 0$, we get

$$t = \int_0^X d\xi \, \frac{\cosh \xi}{\sqrt{v^2 \cosh^2 \xi + \frac{\varepsilon}{2}}}. \tag{6.12}$$

Introducing the constant $\alpha = \sqrt{1 + \varepsilon/(2v^2)}$ and using the change of variable $\alpha u = \sinh \xi$, we obtain

$$t = \frac{1}{v} \int_0^{\sinh X/\alpha} \frac{du}{\sqrt{1+u^2}} = \frac{1}{v} [\operatorname{arcsinh} u]_0^{\sinh X/\alpha} \tag{6.13}$$

which finally leads to the solution

$$X(t) = \operatorname{arcsinh} (\alpha \sinh(vt)) . \tag{6.14}$$

Therefore this calculation determines the position of the soliton versus time in the presence of the defect. For $\varepsilon > 0$, which corresponds to an attractive defect, the solution (6.14) shows that the soliton speeds up when it comes close to the impurity, and then slows down and recovers its initial velocity after it has passed the defect, as shown in Figure 6.1. As the soliton has moved at a speed larger than its speed without the impurity along one part of its trajectory, the only influence of the defect on its dynamics is to induce a phase advance.

However, a study using a direct numerical simulation of the field equation (6.7) shows that, if it has a small initial velocity, the soliton can emerge from its interaction with the defect with a speed smaller than its initial speed, or it can even be *trapped by the impurity*. Therefore the collective coordinate calculation that we made may completely fail in some cases. This is an illustration of the limit of variational methods: the success of the method heavily lies on the expression of the solution that we assume as a trial function, and one can only find the results that this function is able to describe.

It turns out that the function (6.4) that we have chosen only describes the field in terms of a soliton. When the soliton is far from the impurity (for $t \to \pm\infty$), it does not interact with the defect and the energy of the system is only the energy of a free soliton. As the system is conservative, for $t \to +\infty$, the soliton has to recover the energy that it had for $t \to -\infty$. Therefore it is not surprising to find that it preserves its velocity after its interaction with the impurity. The temporary speed-up near the impurity can also be explained by the energy conservation because, for $\varepsilon > 0$, the gravitational energy of the pendula decreases when the soliton arrives near the defect. This decrease must be balanced by an increase of the kinetic energy, in order to preserve the total energy.

The disappointing result that we get does not mean that the collective coordinate method is intrinsically bad, but it emphasises the need to proceed with care in order to use it properly.

6.2 Improving the method with a second collective coordinate

In order to deliver correct results, the collective coordinate method requires a preliminary analysis of the properties of the system in order to detect the modes which could play a role in its dynamics. Numerical simulations of the exact equations of motion are often useful. For instance in our particular case of interest they show that, even when the soliton is not trapped by the defect, its final velocity is smaller than its initial velocity, and moreover a localised oscillation remains at the impurity site. Other essential information can be collected from a study of the linearised equations of motion.

Let us consider the linearised form of Equation (6.7)

$$\theta_{tt} - \theta_{xx} + (1 - \varepsilon\delta(x))\theta = 0. \tag{6.15}$$

It has the oscillatory solutions $\theta(x,t) = \mathrm{e}^{\mathrm{i}\Omega t} f(x)$, where the function $f(x)$ is a solution of

$$-f_{xx} + (1 - \varepsilon\delta(x)) f(x) = \Omega^2 f(x) \tag{6.16}$$

which is again an equation formally equivalent to a Schrödinger equation, with a potential proportional to the Dirac δ function. Such a potential has a bound state

$$f(x) = \frac{A}{2}\,\mathrm{e}^{-\varepsilon|x|/2} \quad \text{corresponding to} \quad \Omega^2 = 1 - \frac{\varepsilon^2}{4}, \tag{6.17}$$

which decays exponentially when one moves away from the impurity. Therefore, besides the soliton solution and the nonlocalised plane waves, Equation (6.7) has the localised solution

$$\theta(x,t) = \frac{A}{2}\,\mathrm{e}^{\mathrm{i}\Omega t}\mathrm{e}^{-\varepsilon|x|/2} + \text{c.c.} = A\cos\Omega t\;\mathrm{e}^{-\varepsilon|x|/2}. \tag{6.18}$$

As this localised mode can store a part of the energy of the soliton, it is essential to include it in the collective coordinates to get a correct result. Its spatial dependence is only determined by the nature of the impurity through the value of the parameter ε, but its amplitude depends on interaction with the soliton. Consequently we choose the following trial solution for the collective coordinate calculation

$$\theta(x,t) = 4\arctan[\exp(x - X(t))] + a(t)\,\mathrm{e}^{-\varepsilon|x|/2}. \tag{6.19}$$

It introduces *two* collective coordinates, the position $X(t)$ of the soliton as before, but also the amplitude $a(t)$ of the localised mode.

Then we proceed as before to derive an effective Lagrangian, using the additional assumptions that $\dot{X}$, a and $\dot{a}$ are small (we assume that $\dot{X}$, a and $\dot{a}$ are of order ε) i.e. we assume that the impurity mode will only be weakly excited and, as before, that the velocity of the soliton is small. The trial function (6.19) is put into the expression of the Lagrangian (6.6) and we only keep the contributions with the two lowest orders in ε. Taking into account a weak excitation of the impurity mode, we can expand the nonlinear term into

$$1 - \cos\theta = 1 - \cos\theta_X + a\,\mathrm{e}^{-\varepsilon|x|/2}\sin\theta_X + \frac{a^2}{2}\mathrm{e}^{-\varepsilon|x|}\cos\theta_X + \mathcal{O}(a^3) \tag{6.20}$$

$$\begin{aligned} &= 2\,\mathrm{sech}^2(x - X) - 2a\mathrm{e}^{-\varepsilon|x|/2}\,\mathrm{sech}(x - X)\tanh(x - X) \\ &\quad + \frac{a^2}{2}\mathrm{e}^{-\varepsilon|x|/2}(1 - 2\,\mathrm{sech}^2(x - X)) + \mathcal{O}(a^3). \end{aligned} \tag{6.21}$$

When the integration with respect to x is performed to get the effective Lagrangian, the interaction between the soliton and the defect appears in two types of terms:

(i) Terms including an interaction *via* the impurity site, for the integrals containing the function $\delta(x)$.

(ii) Direct interaction terms for integrals such as

$$\int \mathrm{d}x\;\mathrm{e}^{-\varepsilon|x|}\,\mathrm{sech}(x - X), \tag{6.22}$$

which can be formally calculated but are small except when the soliton is very close to the impurity ($X \simeq 0$) because they contain the product of two localised functions which are generally not centred at the same point. Moreover these terms are multiplied by $\dot{a}\dot{X}$, or a^2, which are of order ε^2 according to our hypotheses, so that they can be neglected.

Taking into account these approximations, the effective Lagrangian is found to be

$$L = 4\dot{X}^2 - 8 + 2\varepsilon \operatorname{sech}^2 X + \frac{1}{\varepsilon}[\dot{a}^2 - \Omega^2 a^2] + \varepsilon a F(X) + \varepsilon a^2 G(X), \quad (6.23)$$

where we introduce the frequency of the localised mode $\Omega^2 = 1 - \varepsilon^2/4$, and we define $G(X) = -\operatorname{sech}^2 X$ and $F(X) = 2 \operatorname{sech} X \tanh X$. This Lagrangian generalises the expression (6.8) obtained previously because it contains the same contribution for the soliton. But it also includes a contribution for the localised mode alone, and a coupling term which allows the exchange of energy between the soliton and the impurity mode.

The dynamic equations which derive from this Lagrangian for the two collective coordinates are

$$\frac{\mathrm{d}}{\mathrm{d}t}\frac{\partial L}{\partial \dot{X}} - \frac{\partial L}{\partial X} = 0 = 8\ddot{X} + 4\varepsilon \operatorname{sech}^2 X \tanh X - \varepsilon a F'(X) - \varepsilon a^2 G'(X) \quad (6.24)$$

$$\frac{\mathrm{d}}{\mathrm{d}t}\frac{\partial L}{\partial \dot{a}} - \frac{\partial L}{\partial a} = 0 = \frac{1}{\varepsilon}[2\ddot{a} + 2\Omega^2 a] - \varepsilon F(X) - 2\varepsilon a G(X), \quad (6.25)$$

F' and G' being the derivatives of F and G with respect to X. Equations (6.24) and (6.25) are much simpler than the original field equation because they are only two coupled differential equations. This is the appeal of the collective coordinate approach. It reduces a problem with an infinite number of degrees of freedom to a problem with only two degrees of freedom.

However these two equations are still hard to solve analytically because they are nonlinear. They can be easily used to study the behaviour of the system with the following initial conditions: one starts from a large negative X value and $\dot{X}(0) = v$, while $a(0) = \dot{a}(0) = 0$ because the impurity mode is not excited when the soliton is sent from far away. Their numerical simulation is much simpler and faster than a simulation of the field equation because it only has to solve two differential equations instead of a partial differential equation.

But, since we have carried out an analytical calculation to reduce the complexity of the problem, we would indeed like to go on with analytical methods. Rather than deriving the solutions for $X(t)$ and $a(t)$, it is analytically possible to obtain an approximate condition for the trapping of the soliton by computing the amount of energy that it transfers to the localised mode. The steps are as follows:

- In each of the equations we neglect the term which has the highest order in ε, i.e. $\varepsilon a^2 G'(X)$ in the first equation and $\varepsilon a G(X)$ in the second one.
- We then try to evaluate the excitation of the localised mode by assuming that, while the soliton comes toward the defect, $X(t)$ evolves according to the solution (6.14) that we found in the absence of the localised mode.

With these assumptions, it is possible to solve the equation for the amplitude of the localised mode. It is driven by the term $F(X(t)) = f(t)$ which, according to our second hypothesis, is now known,

$$\ddot{a} + \Omega^2 a = \frac{\varepsilon^2}{2} \frac{2 \sinh X(t)}{\cosh^2 X(t)} = \frac{\varepsilon^2 \alpha \sinh vt}{1 + \alpha^2 \sinh^2 vt} = f(t). \tag{6.26}$$

This equation is simpler to solve if we define $\xi = \dot{a} + \mathrm{i}\Omega a$ because it reduces to the first-order equation

$$\dot{\xi} - \mathrm{i}\Omega\xi = f(t). \tag{6.27}$$

The solution of this equation without the right hand side is $\xi(t) = A\,\mathrm{e}^{\mathrm{i}\Omega t}$. The solution of the full equation is then obtained by varying A. We get $\dot{A}\,\mathrm{e}^{\mathrm{i}\Omega t} = f(t)$ so that

$$\xi(t) = \mathrm{e}^{\mathrm{i}\Omega t} \int_{-\infty}^{t} \mathrm{d}\tau\, \mathrm{e}^{-\mathrm{i}\Omega\tau} f(\tau). \tag{6.28}$$

The amount of energy stored in the localised mode can be deduced from the Lagrangian (6.23) if we remember that the potential energy comes with a minus sign in the Lagrangian. The sum of the kinetic and potential energy of the localised mode is therefore

$$E_{\mathrm{loc}} = \frac{1}{\varepsilon}(\dot{a}^2 + \Omega^2 a^2) = \frac{1}{\varepsilon}|\xi|^2. \tag{6.29}$$

As a result the energy which stays in the localised mode after the soliton has passed is

$$E_{\mathrm{loc}}^{\infty} = \lim_{t\to+\infty} E_{\mathrm{loc}} = \frac{1}{\varepsilon}\left| \int_{-\infty}^{+\infty} \mathrm{d}\tau\, \mathrm{e}^{-\mathrm{i}\Omega\tau} f(\tau)\right|^2, \tag{6.30}$$

i.e. the square of the modulus of the Fourier component of $f(\tau)$ at the frequency of the localised mode.

An approximate condition for the trapping of the soliton is obtained by saying that it will be trapped if it transfers all its initial kinetic energy $E_c = 4v^2$ to the localised mode, i.e. $E_{\mathrm{loc}}^{\infty} = 4v^2$. In spite of the approximations that we had to make to reach this condition, it is in good agreement with the results of numerical simulations of the field equations.

Equations (6.24) and (6.25) for the collective coordinates can even predict more subtle effects such as resonances in the diffusion of solitons by the impurity. Similarly to the discussion of the kink–antikink collisions in the ϕ^4 model, after one or two oscillations in the vicinity of the defect, the trapped soliton may get back a large part of the energy stored in the localised mode, and then escape to infinity [64].

We have shown examples of the application of the collective coordinate method to the topological solitons of the sine-Gordon model, but the same method is valid for other soliton equations. The case of topological solitons is favourable because they are extremely stable, but also because their amplitude is fixed so that they can be described, as a first step, by a single collective coordinate, the position of the soliton. For the ϕ^4 model, besides the position (which is associated with the Goldstone mode in the spectrum of the excitations around the soliton), one must also include the amplitude of the localised mode, the shape mode, because it can store energy, as discussed for the collisions.

Generally, in order to select the appropriate collective coordinates, it is very useful to study the spectrum of the localised excitations of the system with and without the soliton. It always includes the translation mode which will be associated with the position variable, but it may include other modes such as the shape mode of the ϕ^4 soliton, or the impurity mode in the case that we studied above. A collective coordinate should be associated with each of them. Moreover, in the case of nontopological solitons such as the KdV or NLS solitons, their amplitude must also be considered as a collective coordinate, as well as a phase factor for the NLS soliton because it is a two-parameter solution.

The effective Lagrangian method that we introduced in this chapter is simple, but is raises delicate questions:

- How is it possible to take into account some degrees of freedom that we have ignored, and in particular the radiation of nonlocalised modes?
- If we attempt to introduce all degrees of freedom, how can we avoid some double counting? The degrees of freedom that we introduce with the collective variables must be subtracted from the other degrees of freedom of the system.

A more complete theory has been developed [29] to rigorously introduce collective variables which can be defined in terms of additional equations that constrain the system. In order to understand why constraints may be necessary, one must think that, if we want to simultaneously describe the position of the soliton and its nonlocalised distortions, we must find a way to select, without any ambiguity, the part which is attributed to a small translation of the soliton, from the part which is attributed to its distortion, in the difference between the perturbed solution and the exact soliton solution. The study of the dynamics of the system in the presence of these constraints is based on the formalism proposed by Dirac to solve Hamilton's equations for systems under constraints. It is possible to write *exact equations* for the dynamics of the collective coordinates together with all the degrees of freedom of the system, including the nonlocalised modes.

7

The inverse scattering transform

The inverse scattering transform, proposed in 1974 for the KdV equation by Gardner, Greene, Kruskal and Miura [74] and further developed in the period 1974–8, can be viewed as the zenith of mathematical soliton theory. It provides a systematic method to get all the solutions of an integrable system. In particular it shows how solitons play the role of 'nonlinear normal modes', exactly like the Fourier modes for a linear equation [12].

A complete presentation of the inverse scattering transform is beyond the scope of this book, but we however shall introduce it for the KdV equation because this fairly simple example is sufficient to understand the main ideas and the beauty of this approach. Then we shall briefly explain how it can be generalised [4].

7.1 Inverse scattering transform for the KdV equation

7.1.1 The principles of the inverse scattering transform

Let us assume that we want to study the time evolution of a spatially localised initial condition $u(x, t = 0)$ which evolves according to the KdV equation which we shall write

$$u_t - 6uu_x + u_{xxx} = 0. \tag{7.1}$$

It should be noticed that the sign of the nonlinear term is the opposite of the sign that we used in Chapter 1, which was the sign that came out naturally from the hydrodynamics of shallow water waves (Equation (1.3)). The sign that we select in this chapter turns out to be convenient for the calculations, but it is easy to switch from one to the other by changing the sign of the unknown function $u = -\phi$.

Before going into the details, let us sketch the main ideas of the inverse scattering transform. The basis of the method is to define an associated linear problem as

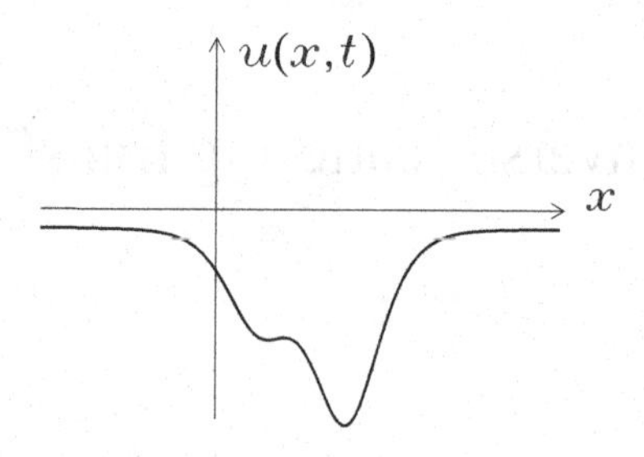

Figure 7.1. Schematic plot of the potential of the Schrödinger equation (7.2).

follows: we consider the Schrödinger equation

$$\left[-\frac{\mathrm{d}^2}{\mathrm{d}x^2} + u(x,t)\right] \psi(x) = \lambda\, \psi(x) \tag{7.2}$$

which defines an eigenvalue problem. The potential u (Figure 7.1) is chosen to be the solution of the KdV equation that we are looking for. Therefore it depends on one parameter, t, the time variable of the KdV equation. As we assume that u is a localised solution ($\lim_{|x|\to\infty} u(x,t) = 0$), the Schrödinger equation (7.2) generally has a spectrum including:

- *Discrete* eigenvalues, denoted by λ_m, associated with spatially localised ψ solutions.
- A *continuous spectrum*, denoted $\lambda_k = k^2$ (where we choose $k \geq 0$), associated with eigenfunctions which behave like a plane wave $\psi \simeq \mathrm{e}^{\pm ikx}$ at infinity.

The potential u can be characterised by the discrete spectrum of the Schrödinger equation (7.2) and by its scattering properties, i.e. the transmission T and reflection R of an incident wave e^{-ikx}. The three quantities λ_m, T and R depend on the potential, and therefore they are functions of its parameter t.

We shall prove a theorem which states that, if $u(x,t)$ evolves according to the KdV equation, then

- The discrete eigenvalues λ_m do not depend on the time parameter t.
- The coefficients R and T can be easily calculated at any time as a function of their initial values calculated at $t = 0$.

The *consequence* is that, if we know the initial condition $u(x, t = 0)$, we can compute λ_m, T and R for $t = 0$ by solving a linear problem. Then, knowing these quantities at any time $t > 0$ using the results of the theorem, we know the scattering properties of the potential, which we wish to determine because it is the solution of the KdV equation. But there is a linear method to solve the inverse scattering problem, which can be used to build up a potential from its scattering properties, so that $u(x, t)$ can be computed *through a sequence of linear steps.*

Solving the KdV equation with the inverse scattering transform can thus be schematised as follows:

$u(x, t = 0)$ —— Solving KdV / nonlinear problem ——→ $u(x, t)$

Solution of the associated linear problem ↓ ... ↑ Inversion of the scattering data

$\lambda_m, R(t = 0), T(t = 0)$ —— Evolution of the scattering data ——→ $\lambda_m, R(t), T(t)$

7.1.2 *Inversion of the scattering data*

Let us consider the Schrödinger problem

$$\psi_{xx} - [u(x) - \lambda]\, \psi(x) = 0 \tag{7.3}$$

with the boundary condition $\lim_{|x|\to\infty} u(x) = 0$. The dependence on the parameter t has not been explicitly indicated to simplify the notation, but it should not be forgotten.

The *bound states* have an energy lower than the value of the potential at infinity. Therefore they correspond to negative eigenvalues that we denote by $\lambda_m = -\kappa_m^2$, assuming $\kappa_m > 0$. The corresponding eigenfunction ψ_m can be normalised, and we henceforth choose it to be of norm unity.

For $|x| \to \infty$, the potential almost vanishes, so that Equation (7.3) simply becomes

$$\frac{\mathrm{d}^2\psi_m}{\mathrm{d}x^2} - \kappa_m^2\, \psi_m(x) = 0. \tag{7.4}$$

This determines the asymptotic behaviour of the wave function for $|x| \to \pm\infty$,

$$\psi_m \sim \mathrm{e}^{\pm\kappa_m x}. \tag{7.5}$$

The inversion of the scattering data determines the potential from the behaviour of the solutions at large distances. For the bound states, this behaviour can be defined by giving the quantities

$$C_m = \lim_{x\to+\infty} \psi_m(x)\, \mathrm{e}^{\kappa_m x}, \tag{7.6}$$

where $\psi_m(x)$ is the normalised eigenfunction. The condition $\lim_{|x|\to\infty} u(x) = 0$ guarantees that this limit exists.

The *continuum states* correspond to positive eigenvalues that we denote by $\lambda_k = k^2$. The corresponding eigenfunctions cannot be normalised because, at infinity, they have an oscillatory behaviour proportional to $e^{\pm ikx}$. If we study the scattering by the potential of a wave coming from $+\infty$, the solution is of the form

$$\psi_k(x) \simeq \begin{cases} e^{-ikx} + R(k)\,e^{ikx} & \text{for } x \to +\infty \\ T(k)\,e^{-ikx} & \text{for } x \to -\infty, \end{cases} \tag{7.7}$$

with a condition which expresses energy conservation $|R|^2 + |T|^2 = 1$.

Knowing the behaviour of ψ when x tends to $+\infty$, i.e. knowing λ_m (or κ_m), C_m and $R(k)$, is sufficient to determine the scattering potential $u(x)$. The mathematical problem of the inversion of the scattering data has been extensively studied because it is of great practical importance:

- In quantum mechanics, the only possibility that we have to determine the interaction potentials between particles is to analyse the results of scattering experiments in which a beam of particles is scattered by a target.
- There are numerous applications in which we try to characterise an object by the waves that it scatters. Radars operate this way, as well as seismic prospecting for oil which uses the scattering of acoustic waves sent into ground by explosions or, in a much more familiar situation, when we look at an object which is illuminated by light!

Here we give only the main result of this theory. A more complete discussion can be found in the book by Drazin and Johnson [4]. In the one-dimensional case which is of interest here, the potential $u(x)$ is the solution of an integral equation called the Gel'fand–Levitan–Marchenko equation

$$u(x) = -2\frac{dK(x,x)}{dx}, \tag{7.8}$$

where the two-variable function K, which is expressed twice as a function of x in Equation (7.8), is a solution of the integral equation

$$K(x,y) + B(x+y) + \int_x^{+\infty} dz\, B(z+y)K(x,z) = 0 \tag{7.9}$$

with

$$B(\xi) = \sum_{m=1}^{N} C_m^2(t)e^{-\kappa_m \xi} + \frac{1}{2\pi}\int_{-\infty}^{+\infty} R(k,t)\,e^{ik\xi}\,dk, \tag{7.10}$$

the index m designating one of the N bound states of the potential.

Equation (7.9) may seem complicated with respect to the KdV equation, but this is a *linear* equation for the unknown function $K(x, y)$. Moreover, in this equation, the variable x plays the role of a parameter because the integration is not carried out with respect to x. Thus Equation (7.9) is not really a two-variable equation, but instead a family of one-variable equations. Furthermore, it is seldom necessary to explicitly solve this equation to get the information that we need on the evolution of an initial condition $u(x, t = 0)$. This is what makes the inverse scattering transform so useful.

7.1.3 Time evolution of the scattering data

In order to apply the inverse scattering transform, we need the time evolution of the scattering data λ_m, C_m, $R(k)$, which enter into the kernel $B(\xi)$ of the integral equation (7.9).

Let us first show that, as previously stated, λ_m does not depend on time if $u(x, t)$ evolves according to the KdV equation. For this we start from the eigenvalue problem associated with the KdV equation, where u is assumed to be a solution of the KdV equation (7.1). Equation (7.2) gives

$$u = \frac{\psi_{xx}}{\psi} + \lambda, \tag{7.11}$$

which is then introduced into the KdV equation (7.1). To perform the calculation, it is important to notice that the eigenvalue λ of the Schrödinger problem does not depend on x but it could depend on t because the potential $u(x, t)$ depends on the 'parameter' t. Computing the space and time derivatives of u, such as $u_t = \psi_{xxt}/\psi - \psi_{xx}\psi_t/\psi^2 + \lambda_t$ for instance, and defining the quantity

$$S(x, t) = \psi_t + \psi_{xxx} - 3(u + \lambda)\psi_x, \tag{7.12}$$

the KdV equation multiplied by ψ^2 becomes

$$\lambda_t \psi^2 + (\psi S_x - \psi_x S)_x = 0, \tag{7.13}$$

which will allow us to determine the time evolution of the scattering data.

▷ *Theorem 1* – If $u(x, t)$ evolves according to the KdV equation and vanishes at infinity, the discrete eigenvalues λ_m of the linear problem (7.2) do not depend on time.

▷ *Proof* – Integrating Equation (7.13) with respect to space, we get

$$\lambda_t \int_{-\infty}^{+\infty} \mathrm{d}x\, \psi^2 + [\psi S_x - \psi_x S]_{-\infty}^{+\infty} = 0. \tag{7.14}$$

As the discrete eigenvalues are associated with normalisable eigenfunctions, the first integral is simply equal to 1. Moreover, because u vanishes at infinity and ψ is a bound state which must therefore also vanish at infinity as well as

ψ_x and its other spatial derivatives, S also tends to 0 at infinity. As a result the integrated terms within the brackets cancel, and we obtain $\lambda_t = 0$. Thus, all the discrete eigenvalues λ_m are simply constants.

This proof cannot be used for the continuum part of the spectrum because the associated eigenfunctions do not vanish at infinity and are not normalisable. But the spectrum of eigenvalues belonging to the continuum does not change if u evolves with the boundary conditions $|u| \to 0$ at infinity: it is always made of all the real numbers $k \geq 0$. If we wish to study one eigenfunction $\psi_k(x)$ and the corresponding reflection coefficient $R(k, t)$, *we select a priori the value of k* which means that we impose $k_t = 0$. Equation (7.13) is thus reduced to

$$(\psi S_x - \psi_x S)_x = 0, \tag{7.15}$$

which shows that the quantity $g(\lambda, t) = (\psi S_x - \psi_x S)$ does not depend on x but only on the value of k that we have selected, and of course on time. But we get

$$(\psi S_x - \psi_x S)_x = \psi S_{xx} - \psi_{xx} S \tag{7.16}$$

$$= \psi \left[S_{xx} - S\,(u - \lambda)\right], \tag{7.17}$$

if we use Equation (7.3). Combining Equations (7.15) and (7.17), we obtain

$$S_{xx} - S\,(u - \lambda) = 0, \tag{7.18}$$

because the function ψ is not identically zero. Consequently we reach the remarkable result that S obeys *the same equation* as the function ψ. Thus, for a given eigenvalue, S and ψ belong to the same eigenspace of the operator $(\partial_{xx} - u(x)\cdot)$. But the discrete eigenvalues of a one-dimensional Schrödinger equation are not degenerate [127], which means that S and ψ must be proportional to each other

$$S_m = A\, \psi_m, \tag{7.19}$$

where A is a factor, which cannot depend on x, but could, a priori, depend on t.

▷ *Theorem 2* – If $u(x, t)$ evolves according to the KdV equation and vanishes at infinity, the scattering data $C_m(t)$ and $R(k, t)$ depend on time according to

$$C_m(t) = C_m(0)\, \mathrm{e}^{4\kappa_m^3 t} \quad \text{and} \quad R(k, t) = R(k, 0)\, \mathrm{e}^{\mathrm{i}8k^3 t}. \tag{7.20}$$

▷ *Proof* – Let us first consider the discrete eigenvalues before studying the continuous spectrum.

- For a discrete eigenvalue, using the relation (7.19) and the definition (7.12) of function S, we get

$$A\psi = \psi_t + \psi_{xxx} - 3(u+\lambda)\psi_x. \tag{7.21}$$

After making the product by ψ and integrating with respect to space, we obtain

$$A \underbrace{\int_{-\infty}^{+\infty} dx\, \psi^2}_{=1} = \frac{1}{2}\frac{d}{dt} \underbrace{\int_{-\infty}^{+\infty} dx\, \psi^2}_{=1} + \int_{-\infty}^{+\infty} dx\, [\psi\psi_{xxx} - 3(u+\lambda)\psi\psi_x] \tag{7.22}$$

$$A = 0 + \int_{-\infty}^{+\infty} dx\, [\psi\psi_{xxx} - 3(\psi_{xx} + 2\lambda\psi)\psi_x] \tag{7.23}$$

$$= \left[\psi\psi_{xx} - 2\psi_x^2 - 3\lambda\psi^2\right]_{-\infty}^{+\infty} = 0, \tag{7.24}$$

where we used $u\psi = \psi_{xx} + \lambda\psi$ and the property that ψ and its derivative vanish at infinity because ψ is a bound state. The result $A = 0$ means that S is identically zero for discrete eigenvalues, which implies

$$\psi_t + \psi_{xxx} - 3(u+\lambda)\psi_x = 0 \quad \text{with} \quad \lambda = -\kappa_m^2. \tag{7.25}$$

As u vanishes at infinity while $\psi_m(x)$ is equivalent to $C_m(t)\,e^{-\kappa_m x}$, we get

$$\left[\frac{dC_m}{dt} - \kappa_m^3 C_m - 3\left(-\kappa_m^2\right)(-\kappa_m C_m)\right] e^{-\kappa_m x} = 0 \tag{7.26}$$

which leads to the first relation $C_m(t) = C_m(0)\,e^{4\kappa_m^3 t}$.

- For the eigenvalues belonging to the continuum, we use the expression $g(\lambda, t) = (\psi S_x - \psi_x S)$ in both limits $\pm\infty$.

  For $x \to -\infty$, using the definition (7.12) of S, $\psi_k(t) = T(k,t)\,e^{-ikx}$ leads to the equation

$$S = \frac{dT}{dt} e^{-ikx} + 4ik^3 T\, e^{-ikx}, \tag{7.27}$$

which gives $(\psi S_x - \psi_x S) = 0$ for $x \to -\infty$, i.e. $g(\lambda, t) = 0$, so that $(\psi S_x - \psi_x S) = 0$ for all x.

  If we now use this equation for $x \to +\infty$ where $\psi_k(x) = e^{-ikx} + R(k,t)\,e^{ikx}$, we get the equation

$$2ik\frac{dR}{dt} + 16k^4 R = 0, \tag{7.28}$$

which leads to the second relation $R(k,t) = R(k,0)\,e^{i8k^3 t}$, and completes the proof of the theorem.

7.1.4 Examples of applications

With Theorems 1 and 2, and the Gel'fand–Levitan–Marchenko equation (7.8), all pieces are in place to determine the time evolution of a given initial state of the KdV equation.

(i) The initial state $u(x, t = 0)$ is treated as a potential for a Schrödinger equation for which we compute:
- The energies of the bound states $\lambda_m = -\kappa_m^2$.
- The behaviour of the bound states at large distances, which determines the parameter C_m.
- The reflection coefficient of this potential for plane waves, $R(k, t)$.

(ii) Then Theorems 1 and 2 give the behaviour of these scattering data at any time t.

(iii) Finally the Gel'fand–Levitan–Marchenko equation (7.8) gives $u(x, t)$.

Equation (7.8) is often hard to solve, but there is however an important case for which we can get an exact solution: it is the case of a separable kernel $B(z + y)$, i.e. a kernel which can be written as a linear combination of the products of a function of z by a function of y.

Assuming that

$$B(z + y) = \sum_{m=1}^{N} F_m(z)G_m(y), \tag{7.29}$$

Equation (7.9) becomes

$$K(x, y) + \sum_{m=1}^{N} F_m(x)G_m(y) + \sum_{m=1}^{N} G_m(y) \int_x^{+\infty} \mathrm{d}z\, F_m(z)K(x, z) = 0. \tag{7.30}$$

We can look for solutions of the form

$$K(x, y) = \sum_{m=1}^{N} L_m(x)G_m(y), \tag{7.31}$$

so that the equation becomes

$$\sum_{m=1}^{N} L_m(x)G_m(y) + \sum_{m=1}^{N} F_m(x)G_m(y) + \sum_{m=1}^{N}\sum_{p=1}^{N} G_m(y)L_p(x) \int_x^{+\infty} \mathrm{d}z\, F_m(z)G_p(z) = 0, \tag{7.32}$$

which has the solution

$$L_m(x) + F_m(x) + \sum_{p=1}^{N} L_p(x) \int_x^{+\infty} dz\, F_m(z) G_p(z) = 0. \tag{7.33}$$

Therefore we get a set of N coupled *algebraic* equations for the unknown L_m, in which x plays the role of a simple parameter.

Such a separable case is obtained for all reflectionless potentials, i.e. the potentials such that $R(k) = 0$, because, in this case, the expression (7.10) for $B(\xi)$ reduces to

$$B(\xi) = \sum_{m=1}^{N} C_m^2\, e^{-\kappa_m \xi}. \tag{7.34}$$

This happens when the initial condition is a multisoliton solution. Then, as shown below with some examples, each discrete eigenvalue λ_m corresponds to a soliton. Its amplitude and position are respectively given by λ_m and C_m. Consequently the inverse scattering transform can be used to identify the solitons which are present in a given initial condition, to determine how many there are and which are their amplitudes and positions.

Single soliton initial condition

We know that the KdV equation (7.1) has the soliton solution

$$u = -A\, \mathrm{sech}^2 \left[\sqrt{\frac{A}{2}} (x - 2At) \right], \tag{7.35}$$

in which we have taken into account the change of variable $\phi = -u$ with respect to the KdV equation (1.3).

Let us consider for instance the initial condition $u(x, t = 0) = -2\, \mathrm{sech}^2 x$, corresponding to amplitude $A = 2$, which leads to the associated linear eigenvalue problem

$$\psi_{xx} + (2\, \mathrm{sech}^2 x + \lambda)\psi = 0. \tag{7.36}$$

It turns out that we recover the Schrödinger equation that we have already met when we studied the perturbations of a sine-Gordon soliton (Equation (5.9)). We saw that, for this particular ratio between the depth and the width of the potential well, there is only one bound state ($N = 1$). It was the Goldstone mode of the sine-Gordon equation,

$$\psi(x) = \frac{1}{\sqrt{2}}\, \mathrm{sech}\, x \quad \text{with} \quad \lambda_1 = -1 \quad \text{i.e.} \quad \kappa_1 = 1. \tag{7.37}$$

The asymptotic behaviour $\psi(x) \sim \sqrt{2}\,\mathrm{e}^{-x}$ gives $C_1(0) = \sqrt{2}$. Moreover, as the potential is reflectionless, we can use the expression (7.34) for $B(\xi)$. Hence we get

$$\kappa_1 = 1, \quad C_m(t) = \sqrt{2}\,\mathrm{e}^{4t} \quad \text{and} \quad B(\xi) = 2\,\mathrm{e}^{8t-\xi}. \tag{7.38}$$

Defining two functions, F_1 and G_1, by $F_1(x) = (2\,\mathrm{e}^{8t})\mathrm{e}^{-x}$ and $G_1(y) = \mathrm{e}^{-y}$, we obtain $K(x, y) = \mathrm{e}^{-y} L_1(x)$ where function $L_1(x)$ is a solution of the equation

$$L_1(x) + 2\,\mathrm{e}^{8t}\mathrm{e}^{-x} + L_1(x) \int_x^{+\infty} \mathrm{d}z\,(2\,\mathrm{e}^{8t})\mathrm{e}^{-z}\,\mathrm{e}^{-z} = 0. \tag{7.39}$$

It leads to

$$L_1(x) = -\frac{2\mathrm{e}^{8t-x}}{1 + \mathrm{e}^{8t-2x}} \quad \text{and} \quad K(x, y) = -2\frac{\mathrm{e}^{8t-x-y}}{1 + \mathrm{e}^{8t-2x}}. \tag{7.40}$$

The solution of the KdV equation is finally deduced from Equation (7.8),

$$u(x, t) = -2\frac{\mathrm{d}K(x, x)}{\mathrm{d}x} \tag{7.41}$$

$$= -8\frac{\mathrm{d}}{\mathrm{d}x}\frac{\mathrm{e}^{8t-2x}}{(1 + \mathrm{e}^{8t-2x})^2} \tag{7.42}$$

$$= -2\,\mathrm{sech}^2(x - 4t) \tag{7.43}$$

which is the soliton with amplitude $A = 2$, as expected. This simple example has shown that the initial condition $u(x, t = 0) = -2\,\mathrm{sech}^2 x$ is indeed a permanent profile soliton solution, and we have determined its speed.

Generalisation

The calculation can be easily generalised to the case of N solitons every time that the potential is reflectionless. One has to solve the system (7.33) of N equations with N unknowns to determine the function $L_m(x)$.

The initial condition $u(x, 0) = -6\,\mathrm{sech}^2 x$ is another example where the analytical calculation is fairly easy. In this case, the potential of the associated Schrödinger equation is deeper with respect to its width than in the previous case. It now has two bound states, which means that this initial condition is a two-soliton solution. The calculation shows that it is given by

$$u(x, t) = -12\frac{2 + 4\cosh(2x + 24t) + \cosh(4x)}{[3\cosh(x - 12t) + \cosh(3x + 12t)]^2}. \tag{7.44}$$

To solve the Schrödinger equation, we can notice that this particular initial condition leads to the same potential as the study of the small oscillations around a ϕ^4 soliton (Equation 5.43). We have seen that this potential has two bound states, the Goldstone mode and the internal mode of the ϕ^4 soliton. Using the results obtained for the ϕ^4 soliton one can easily derive the solution (7.44). This is left as an exercise for the reader.

More generally, the asymptotic expressions of the N soliton solutions are functions of the variables $\xi_m = x - x_0 - 4\kappa_m^2 t$, $m = 1 \dots N$. The discrete eigenvalues $\kappa_m^2 = -\lambda_m$ determine the velocities, and hence the amplitudes, of the solitons while the parameters $C_m^2(0)$ determine the relative positions of these solitons which make up the N-soliton solution. Therefore N-soliton solutions of the KdV equation can be systematically obtained by assuming a reflectionless potential and choosing some values for $C_m(0)$ and κ_m [4, 74].

The calculation for a reflecting potential is seldom explicitly possible, but it is often useless because the determination of the bound states of the initial condition gives *the soliton content of the initial condition*, which is often what we are looking for. Then the subsequent evolution of these solitons can be computed and the contribution of the small-amplitude oscillations can be neglected because they disperse in the system.

Consequently the inverse scattering transform can be used as a true 'soliton detector'. It can immediately explain, without any calculation, why an initial condition which is a dip in the KdV equation describing the hydrodynamics of shallow water waves cannot generate solitons, as we saw in Chapter 1. Such an initial condition with a negative ϕ would give a potential $u = -\phi$ which is always positive, above its limits at infinity. Our knowledge of quantum mechanics tells us that such a potential does not have any bound state.

7.2 The inverse scattering transform: a 'nonlinear Fourier analysis'

The inverse scattering transform can be viewed as a *nonlinear analogue* of the Fourier analysis. Let us examine how the Fourier transform is used to solve a linear partial differential equation such as the linearisation of the KdV equation

$$u_t + \alpha u_x + u_{xxx} = 0. \tag{7.45}$$

Given an initial condition $u(x, t = 0)$, the method first computes its spectrum by a forward Fourier transform, which determines the weight of each component of wavevector k in the initial condition through

$$\widetilde{U}(k) = \int_{-\infty}^{+\infty} \mathrm{d}x \, u(x, 0) \, \mathrm{e}^{-ikx}. \tag{7.46}$$

Then we know the time evolution of each component through the solutions $u_0 \mathrm{e}^{\mathrm{i}[kx - \omega(k)t]}$ of the linear equation that determines the dispersion relation $\omega(k) = \alpha k - k^3$.

The solution at any time can then be built from the formula

$$u(x,t) = \frac{1}{2\pi}\int_{-\infty}^{+\infty} \mathrm{d}k\, \widetilde{U}(k)\, \mathrm{e}^{\mathrm{i}[kx-\omega(k)t]}. \tag{7.47}$$

The key of the method is the separability of the time evolution of components having different wavevectors. Hence their time evolution can be derived without a priori knowing the solution $u(x,t)$.

For the nonlinear KdV equation, the process implemented by the inverse scattering transform is exactly analogous. The spectrum of the initial condition is here made of the scattering data $(\lambda_m, C_m(t), R(k,t))$. Their derivation corresponds to the forward Fourier transform in the linear case. Then the evolution of each component according to the KdV equation is deduced from Theorems 1 and 2, exactly as the linear equation determines the dispersion relation $\omega(k)$, from which the speed of each component is obtained. In the nonlinear case too, the key is that the time evolution of the scattering data can be computed without knowing the full solution $u(x,t)$ because each part of the scattering data evolves independently of the others. The solution $u(x,t)$ at any time is then obtained with the Gel'fand–Levitan–Marchenko equation, which corresponds to the reverse Fourier transform (7.47).

Therefore, in this analysis the solitons play the role of the *nonlinear normal modes* of the system. The inverse scattering transform is a systematic process to compute the time evolution of an initial state according to the KdV equation, this is why it has been used [141] to analyse the solitons which propagate as internal waves in the Andaman sea (see Section 1.6), for instance.

The inverse scattering transform can be used to study the evolution of an initial condition but also to generate N-soliton solutions or to derive an infinite number of conservation laws, which attests that the KdV equation corresponds to a *completely integrable system.* However the major drawback of this method is that it cannot be generalised to all nonlinear equations and moreover, contrary to the Fourier transform for linear equations, it is *specific* to each equation to which it can be applied because the associated linear problem depends on the equation.

7.2.1 A step towards generalisation: the Lax method

The Lax method [110] generalises the formalism of the inverse scattering transform beyond the approach that we have presented for the KdV equation. Let us assume that we want to solve the nonlinear equation $u_t = N(u)$ where N is a nonlinear operator, which does not depend on the variable t, but may contain the variable x and derivatives with respect to x.

The idea of the Lax method is to rewrite the equation as an equation between operators defined in a Hilbert space $\mathcal{H}$ which is different from the functional space $\mathcal{E}$ in which we solve the differential equation.

We write the equation as

$$L_t - [M, L] = 0 \tag{7.48}$$

where L and M are linear operators, defined in $\mathcal{H}$, which may depend on u. For the KdV equation, the operator L is $L = -\partial_{xx} + u(x,t)$, while M will be specified later. The notation $[M, L]$ designates the commutator of the operators as usual, $[M, L] = ML - LM$, and L_t means the derivative of L wherever time appears explicitly. Thus for KdV, $L_t = u_t$.

A scalar product is defined in Hilbert space. We denote it by $\langle\phi|\psi\rangle$ where ϕ and ψ are two elements of $\mathcal{H}$. The operator L is self-adjoint, i.e. $\langle\phi|L\psi\rangle = \langle L\phi|\psi\rangle$ for all ϕ and ψ belonging to the Hilbert space $\mathcal{H}$.

In the same spirit as the previous section, a linear problem

$$L\,\psi = \lambda\,\psi \tag{7.49}$$

is associated with the nonlinear equation, where the eigenvalue λ is a scalar. Let us show that, if the operators L and M are related by Equation (7.48), then the eigenvalues λ do not depend on the variable t. We derive Equation (7.49) with respect to time to get

$$L_t\psi + L\psi_t = \lambda_t\psi + \lambda\psi_t \tag{7.50}$$

$$(ML - LM)\psi + L\psi_t = \lambda_t\psi + \lambda\psi_t. \tag{7.51}$$

Reorganising the terms we have

$$\lambda_t\psi = M\underbrace{L\psi}_{\lambda\psi} - LM\psi + (L-\lambda)\psi_t, \tag{7.52}$$

or

$$\lambda_t\psi = (L-\lambda)(\psi_t - M\psi). \tag{7.53}$$

Computing the scalar product by ψ, and using its linearity, we obtain

$$\langle\psi|\lambda_t\psi\rangle = \lambda_t\langle\psi|\psi\rangle = \langle\psi|(L-\lambda)(\psi_t - M\psi)\rangle \tag{7.54}$$

$$= \langle\underbrace{(L-\lambda)\psi}_{=0}\,|\,(\psi_t - M\psi)\rangle \tag{7.55}$$

because the operator L is self-adjoint, as well as the product by the real eigenvalue λ. Consequently, unless $\langle\psi|\psi\rangle$ is identically equal to 0 which would imply $\psi = 0$ according to the properties of a scalar product, we can conclude that $\lambda_t = 0$.

Relation (7.53) then implies that $(L - \lambda)(\psi_t - M\psi) = 0$, which means that $(\psi_t - M\psi)$ is an eigenfunction of the operator L, associated with the same eigenvalue λ. Thus, for a nondegenerate eigenvalue λ, there exists a constant C such that

$$(\psi_t - M\psi) = C\psi \tag{7.56}$$

i.e. $\psi_t = M'\psi$ if we define $M' = M + C$. But, as C is a constant, L and C commute, hence $[M, L] = [M', L]$. This shows that, if the pair of operators (L, M) verifies Equation (7.48), then the pair (L, M') verifies it too.

Consequently it is possible to find a pair of operators, (L, M), called the *Lax pair*, which verifies the two conditions

$$L_t - [M, L] = 0. \tag{7.57}$$

$$\text{and if} \quad L\psi = \lambda\psi \quad \text{then} \quad \lambda_t = 0 \quad \text{and} \quad \psi_t = M\psi. \tag{7.58}$$

In the steps that we just described, one recognises the path that we followed for the KdV equation, but it is now expressed in a compact formalism, which is easier to generalise. If we select $L = -\partial_{xx} + u(x, t)$, so that the associated linear problem $L\psi = \lambda\psi$ corresponds to a Schrödinger equation allowing us to reconstruct the potential from the Gel'fand–Levitan–Marchenko equation, the last step is to determine M so that Equation (7.48) leads to the nonlinear equation that we wish to solve.

The simplest would be to choose $M = c\partial_x - I$, where I is the identity operator, leading to $[L, M] = (-cu_x\cdot)$. Then Equation (7.48) gives $u_t + cu_x = 0$, which is the linear wave equation. Thus this choice is not a good one because it leads to an equation which is too simple. We must look for a more complex operator.

The operator $M = c\partial_{xx}$ would lead to $[L, M] = -c\,(u_{xx}\cdot + u_x\partial_x)$. As this operator is not reduced to a simple product, Equation (7.48) is not a partial differential equation for u. This shows that the choice of the operator M is not free. It must be such that $[L, M]$ reduces to a multiplicative operator, but, moreover it must lead to an interesting equation!

Further analysis shows that M must be a differential operator of odd order, containing enough free parameters. One can check that the choice

$$M = -4\frac{\partial^3}{\partial x^3} + u\frac{\partial}{\partial x} + \frac{\partial}{\partial x}u \tag{7.59}$$

leads to the KdV equation.

The Lax method opens the door to a generalisation of the inverse scattering transform as it was initially introduced for the KdV equation because it can be used to build a full hierarchy of equations, through the introduction of operators M which can get more and more complicated. The method ensures that they correspond to completely integrable systems, having exact soliton solutions. For instance the equation that follows the KdV equation in the hierarchy is

$$u_t + 30u^2 u_x - 20uu_x - 10uu_{xxx} + u_{xxxxx} = 0. \tag{7.60}$$

The richness of the Lax method is that it can be generalised even further because it separates the space $\mathcal{E}$, in which the equation is solved, from the space $\mathcal{H}$ of the associated linear problem. Thus $\mathcal{H}$ can be extended, for instance by selecting *matrix operators* for L and M.

7.2.2 The AKNS method (Ablowitz–Kaup–Newell–Segur)

Two generalisation schemes have been proposed which go beyond the Lax method. In 1971, V. Zakharov and A. B. Shabat introduced [193] a matrix method which can be used to solve the nonlinear Schrödinger equation. In 1974, M. Ablowitz, D. Kaup, A. Newell and H. Segur (AKNS) also proposed [12] a scheme which starts from a (2×2) matrix operator and can treat a variety of completely integrable nonlinear systems such as Korteweg–deVries, sine-Gordon, nonlinear Schrödinger etc. by an inverse scattering transform.

Studying these extensions is beyond the scope of this book. We shall only give a few ideas on the AKNS scheme which is well described in the original paper [12]. Complements can be found in the book by Drazin and Johnson [4].

Let us consider a linear operator depending on two functions $q(x, t)$ and $r(x, t)$

$$L = \begin{pmatrix} \partial_x & -q(x,t) \\ r(x,t) & -\partial_x \end{pmatrix}, \tag{7.61}$$

which acts in a space of elements $\psi(x) = (\psi_1(x), \psi_2(x))$ which are sets of two functions of the variable x. The eigenvalues ζ of the operator L are defined by

$$L\psi = -\mathrm{i}\zeta\psi. \tag{7.62}$$

The important point is that we *impose* that the eigenvalues ζ do not depend on time, and that the eigenfunctions depend on t according to

$$\psi_t = M\psi, \tag{7.63}$$

where M is a linear operator acting in the same space. We recognise the Lax pair (L, M), whereas the functions ψ are called the *Jost functions*.

The compatibility of the two equations (7.61) and (7.62) restricts the possible choices for the operator M. Equation (7.62) can be written as

$$\frac{\partial \psi}{\partial x} = R\psi \quad \text{with} \quad R = \begin{pmatrix} -\mathrm{i}\zeta & q \\ r & \mathrm{i}\zeta \end{pmatrix}. \tag{7.64}$$

The compatibility between $\psi_x = R\psi$ and $\psi_t = M\psi$, requires the two conditions

$$\psi_{xt} = R_t\psi + R\psi_t = R_t\psi + RM\psi \tag{7.65}$$

and

$$\psi_{tx} = M_x\psi + M\psi_x = M_x\psi + MR\psi, \tag{7.66}$$

which implies the relation

$$R_t - M_x - [M, R] = 0 \tag{7.67}$$

which generalises Equation (7.48) of the Lax method. As it sets conditions on the functions $q(x, t)$ and $r(x, t)$, this equation defines the nonlinear equations that the method can solve.

The scattering data and their inversion process have to be defined. They are obtained by considering the behaviour of ψ when $|x|$ tends to infinity. As in the KdV case, we look for localised solutions so that q and r are assumed to vanish when $|x|$ tends to infinity. In this limit, L is reduced to

$$L \simeq \begin{pmatrix} \partial_x & 0 \\ 0 & -\partial_x \end{pmatrix}. \tag{7.68}$$

The solutions of Equation (7.62) for $|x| \to \pm\infty$, denoted by $\psi_\pm$, are easily determined:

$$\psi_+(x;\zeta) \simeq \mathrm{e}^{-\mathrm{i}\zeta x}\begin{pmatrix} 1 \\ 0 \end{pmatrix} = \begin{pmatrix} \psi_{1+} \\ \psi_{2+} \end{pmatrix} \text{ and } \psi'_+(x;\zeta) \simeq \mathrm{e}^{\mathrm{i}\zeta x}\begin{pmatrix} 0 \\ 1 \end{pmatrix} = \begin{pmatrix} \psi'_{1+} \\ \psi'_{2+} \end{pmatrix} \tag{7.69}$$

$$\psi_-(x;\zeta) \simeq \mathrm{e}^{-\mathrm{i}\zeta x}\begin{pmatrix} 1 \\ 0 \end{pmatrix} = \begin{pmatrix} \psi_{1-} \\ \psi_{2-} \end{pmatrix} \text{ and } \psi'_+(x;\zeta) \simeq \mathrm{e}^{\mathrm{i}\zeta x}\begin{pmatrix} 0 \\ 1 \end{pmatrix} = \begin{pmatrix} \psi'_{1-} \\ \psi'_{2-} \end{pmatrix}. \tag{7.70}$$

However these solutions are not independent from each other because they are linked by Equation (7.62) valid for any x. It means that a solution, which tends to ψ_+ when x tends to $+\infty$, is generally a combination of the four solutions $\psi_{1\pm}$, $\psi_{2\pm}$, for $x \to -\infty$.

Expressing the two boundary conditions in a matrix form yields

$$\Psi_+ = \begin{pmatrix} \psi_{1+} & \psi'_{1+} \\ \psi_{2+} & \psi'_{2+} \end{pmatrix} \quad \text{and} \quad \Psi_- = \begin{pmatrix} \psi_{1-} & \psi'_{1-} \\ \psi_{2-} & \psi'_{2-} \end{pmatrix}. \tag{7.71}$$

Thus there exists a matrix S, called the diffusion matrix, which connects these two matrices through $\Psi_- = S\,\Psi_+$. It generalises the coefficients C_m and $R(k)$ of the KdV equation. Here too, one generally finds two types of eigenvalues ζ, and therefore two types of matrices $S(\zeta)$. The discrete eigenvalues again correspond to the solitons, while the eigenvalues of the continuous spectrum correspond to nonlocalised waves.

The study shows that the time evolution of the scattering data stays very simple and independent of the complete expression of the functions $r(x,t)$ and $q(x,t)$. This separability is essential for the success of the method. The two functions $r(x,t)$ and $q(x,t)$ can be obtained from a Gel'fand–Levitan–Marchenko equation, which deals with (2×2) matrices, but stays formally identical to the equation that we gave for the KdV equation. Its kernel B is again expressed as a function of the scattering data $S(\zeta)$.

This scheme is very general and can be applied to a large variety of integrable systems. For instance, in order to obtain the nonlinear Schrödinger equation

$$\mathrm{i}u_t + u_{xx} + 2|u|^2 u = 0, \tag{7.72}$$

one must start from

$$L = \begin{pmatrix} \partial_x & \mathrm{i}u^*(x,t) \\ -\mathrm{i}u(x,t) & -\partial_x \end{pmatrix}. \tag{7.73}$$

However, the expression of M is complicated [102].

7.2.3 The inverse scattering transform and perturbation theory

We saw that the inverse scattering transform extends well beyond the case of the KdV equation for which it was introduced. Therefore it can be used to study the properties of many physical systems, particularly as a 'soliton detector' in the analysis of experimental or numerical data.

However, we know that a physical system is seldom described by a completely integrable equation. Generally the equation which describes its time evolution contains additional terms, which are small but nevertheless very important, especially because they are responsible for the deviations from integrability.

One attraction of the inverse scattering transform is that it can be used to calculate the role of these perturbations. This can be easily understood from the simple example of the perturbed KdV equation

$$u_t - 6uu_x + u_{xxx} = \varepsilon P(x,t), \tag{7.74}$$

where the perturbation is chosen as $P(x,t) = \gamma(t)u(x,t)$, for instance to study the propagation of waves when the depth of the water varies [102].

Examining the calculations presented above for the KdV equation, one can check that Equation (7.13) was deduced from the relation

$$\psi^2 (\lambda_t - u_t + 6uu_x - u_{xxx}) = -(\psi S_x - \psi_x S)_x, \tag{7.75}$$

in which the KdV equation had been used to cancel the u-dependent term on the left hand side.

Therefore, in the presence of the perturbation, we get a new equation for the eigenvalue λ_m:

$$\psi_m^2 (\lambda_{mt} - \varepsilon P(x, t)) = -(\psi_m S_{mx} - \psi_{mx} S_m)_x. \tag{7.76}$$

As before we integrate with respect to space and use the normalisation of the eigenfunction ψ_m associated with a discrete eigenvalue as well as the conditions $\psi_m \to 0$ and $S_m \to 0$ at infinity. We get

$$\lambda_{mt} - \varepsilon \int_{-\infty}^{+\infty} \psi_m^2 P(x, t) \, \mathrm{d}x = 0. \tag{7.77}$$

Now λ_{mt} no longer vanishes. The discrete eigenvalues drift slowly versus time. But we know that they determine the soliton speed, so that this calculation tells us how the soliton reacts to the perturbation.

This example shows the power of inverse scattering to study perturbations of solitons. The paper [102] is a plentiful reference on this question.

Part III

Examples in solid state and atomic physics

Introduction

In the first part of this book we studied some examples of *macroscopic* solitons. In hydrodynamics or nonlinear electrical lines, it is easy to convince ourselves that solitons are there because we see them or observe their voltage pulse on the screen of an oscilloscope. In optical fibres more apparatus is needed for the observation, but the detection of individual solitons is still possible. In Josephson junctions as well, we have a direct proof of the presence of solitons through their signature in the current–voltage characteristics or thanks to the microwaves that they emit when they move back and forth in the junction.

In *microscopic* physics, such as solid state or macromolecular physics, it is not possible to observe a single soliton. Solitons only show up through their consequences on the properties of the materials, for instance on the dielectric response, on light or neutron scattering, on RMN spectra, on the electronic energy levels measured by absorption spectroscopy, etc. Therefore, at the microscopic level the notion of solitons is more controversial. Very often we should not expect to find the exact solitons that we described in the first part, with their exceptional stability and immunity to collisions. Instead we may have quasi-solitons, as we shall see, for instance, for dislocations in crystals.

However the soliton concept often provides a very useful first-order description to understand some properties of solids and macromolecules. In the third and fourth parts of the book, we shall examine some examples, in increasing order of complexity:

- *Solitons and atomic motions in solids,* which can be properly analysed with classical mechanics.
- *Solitons and magnetic properties of solids.* This is an example where the soliton picture is particularly appropriate and can be accurately tested experimentally. A quantum theory is necessary but it can be reduced to a quasi-classical analysis.

- *Solitons and the dynamics of charge carriers in conducting polymers,* which illustrate a very fruitful cooperation between theoreticians and experimentalists. The quantum description is essential in this case.
- *Solitons in Bose–Einstein condensates,* which provide a highly tunable medium where solitons and nonlinear localisation play a major role. A quantum description is also required for this atomic system.
- *Solitons in biological macromolecules.* This is a topic which is still controversial because it is difficult to obtain experimental evidence, but soliton models, in spite of their simplicity compared to the complexity of biological systems, provide results for energy localisation and transfer in proteins and DNA dynamics which are begining to be experimentally tested.

8

The Fermi–Pasta–Ulam problem

8.1 The physical question

Prior to a study of specific applications in solid state physics, it is interesting to examine this problem of the dynamics of a nonlinear particle lattice because it is at the origin of the 'rediscovery' of the soliton by Zabusky and Kruskal in 1965.

The 'FPU problem', as it is known presently, bears the name of the scientists who raised it.[1] Their goal was to study the thermalisation of a solid [66]. As revealed by S. Ulam later [179], these authors were looking for a theoretical physics problem suitable for investigation using one of the very first computers, the 'MANIAC'. Their work, published in 1955 in a classified Los Alamos Laboratory report, had been completed shortly before the death of Enrico Fermi in 1954. Ulam told later that Fermi did not suspect the importance of this discovery and considered the work 'minor'.

Fermi, Pasta and Ulam decided to study how a crystal evolves toward thermal equilibrium by simulating a chain of particles of mass unity, linked by a quadratic interaction potential, but also by a weak nonlinear interaction. This one-dimensional system is described by the Hamiltonian

$$H = \sum_{i=0}^{N-1} \frac{1}{2} p_i^2 + \sum_{i=0}^{N-1} \frac{1}{2} K (u_{i+1} - u_i)^2 + \frac{K\alpha}{3} \sum_{0}^{N-1} (u_{i+1} - u_i)^3, \qquad (8.1)$$

where u_i is the displacement along a chain of atom i with respect to its equilibrium position, and p_i its momentum. The coefficient $\alpha \ll 1$ measures the strength of the nonlinear contribution in the interaction potential. The two ends of the chain were assumed to be fixed, i.e. $u_0 = u_N = 0$.

The common approach in physics is to think in terms of 'normal modes', related to the displacements through $A_k = \sqrt{\frac{2}{N}} \sum_{i=0}^{N-1} u_i \sin(ik\pi/N)$ and which

[1] Mary Tsingou, who took part in the numerical study is not an author of the report, but her contribution is however recognised by two lines of acknowledgements!

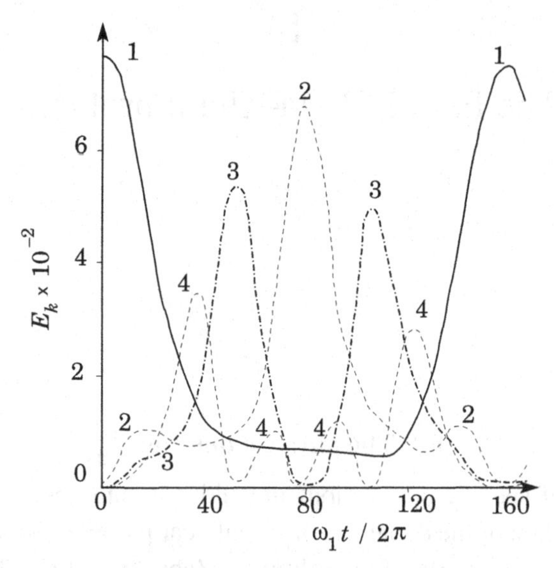

Figure 8.1. FPU recurrence: the plot shows the time evolution of the kinetic and potential energy $E_k = \frac{1}{2}(\dot{A}_k^2 + \omega_k^2 A_k^2)$ of each of the foot lowest modes. Initially, only mode 1 was excited (from reference [66]).

have the frequency $\omega_k^2 = 4K \sin^2(k\pi/2N)$. They provide a basis to rewrite the Hamiltonian (8.1) as

$$H = \frac{1}{2} \sum_k \left(\dot{A}_k^2 + \omega_k^2 A_k^2\right) + \alpha \sum_{k,\ell,m} c_{k\ell m} A_k A_\ell A_m. \tag{8.2}$$

The last term, due to the nonlinear contribution in the potential, leads to a coupling between the modes.

Fermi, Pasta and Ulam thought that, due to this term, energy introduced in a single mode, the mode $k = 1$ in their simulation, would slowly drift to the other modes, until the equipartition of energy predicted by statistical physics would be reached. The beginning of the calculation indeed suggested that it would be the case. Modes 2, 3, ..., were successively excited. However, by accident [128], one day they let the program run long after the steady state had been reached. When they realised their oversight and came back to the computer room, they noticed that the system, after remaining in a steady state for a while, had then departed from it. To their great surprise, after 157 periods of the mode $k = 1$, almost all the energy (all but 3%) was back in the lowest frequency mode, as shown in Figure 8.1.

The initial state seems to be almost perfectly recovered after this recurrence period. Further calculations, performed later with faster computers, showed that the same phenomenon can repeat itself many times, and that a 'super-recurrence' period exists, after which the initial state is recovered to a much higher accuracy. Thus, contrary to the expectations of the authors, the drift of the energy of mode 1 initially excited, toward the modes with a higher wavenumber does not occur.

This highly remarkable result, known as the FPU paradox, shows that nonlinearity is not enough to guarantee the equipartition of energy. To understand it, it is necessary to stop thinking in terms of linear normal modes, and to consider the full nonlinearity intrinsically. It also means that one should stop thinking in Fourier space, and come back to real space. A 'mode' is a localised excitation in Fourier space, but it is fully delocalised in real space. Conversely a soliton is localised in real space, but extended in Fourier space.

8.2 Fermi, Pasta and Ulam: the characters

Born in Rome, ENRICO FERMI (1901–54) has been one of the brightest physicists of the twentieth century, who made major experimental and theoretical discoveries. His name is famous for his contributions to statistical physics, elementary particle physics, the control of nuclear energy, etc. It is thanks to his trip to receive the Nobel prize in Sweden in 1938 that he left Pascist Italy, and emigrated to the United States where he studied atomic fission and set up the first controlled chain reaction in Chicago in 1942. He was naturally called to be one of the leaders of the Manhattan project for the development of nuclear energy and the atomic bomb. We are less familiar with his work on various nonlinear problems at the end of his life, before his premature death from stomach cancer. (Picture: AIP Emilio Segré Visual Archives, Segré Collection.)

JOHN R. PASTA (1918–81) did not immediately have such a bright career. In spite of his interest for physics he had to leave the New York City College during the great depression. He was first a real estate agent and then a police officer in New York from 1941 to 1942. He was then recruited as radar and cryptography specialist by the American army during the second world war. His earnings as a GI allowed him to return to the University, where he gained a PhD in theoretical physics in 1951. He immediately started to work in Los Alamos, on the MANIAC computer which was under construction and testing.

His career went on as a computer expert for the Atomic Energy Commission, where he extensively developed the Mathematics and Computer Division. John R. Pasta then became a physics professor and, in 1964, Dean of the Computer Science Department of the University of Illinois, where he focused on the use of computers to solve applied problems in physics and mathematics. (Picture: Department of Computer Science at the University of Illinois.)

Born in Poland, Stanislaw M. Ulam (1909–84) quickly became interested in mathematics and obtained a PhD in 1933 under the supervision of Banach (1892–1945). Following a first invitation by von Neumann to the prestigious Princeton Institute for Advanced Study in 1935, he visited the United States several times before finally leaving Poland before the start of the second world war. He became an American citizen in 1943, and was invited by von Neumann himself to join the Los Alamos team to prepare the atomic bomb. Among other things he solved the problem of the initiation of fusion in the hydrogen bomb. In collaboration with N. C. Metropolis he invented the Monte Carlo method to solve problems requiring statistical sampling. He stayed in New Mexico until 1965 before his appointment as a mathematics professor at the Colorado University. (Picture: AIP Emilio Segré Visual Archives, Ulam Collection.)

The famous "Fermi–Pasta–Ulam" study actually involved a fourth contributor, Mary Tsingou as indicated on the first page of the FPU Los Alamos report published in 1995:

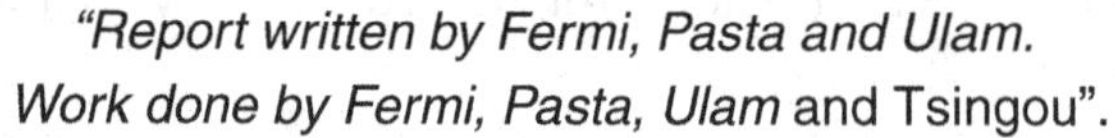

"Report written by Fermi, Pasta and Ulam.
Work done by Fermi, Pasta, Ulam and Tsingou".

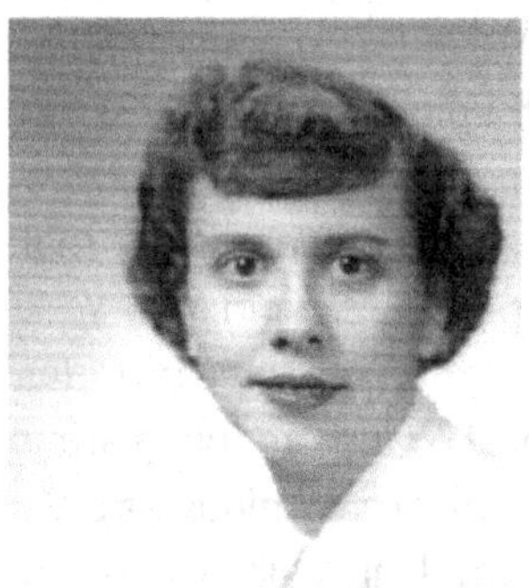

Born in October 14th 1928 at Milwaukee, Wisconsin, Mary Tsingou gained a BSc in 1951 at the University of Wisconsin, and an MSc in mathematics in 1955 at University of Michigan. In 1952, following a suggestion by her mathematics professor, a woman, she applied for a programmer position at Los Alamos National Laboratory. Because of the Korean war, there was a shortage of American young men and staff positions were also offered to young women.

She was initially assigned to the T1 division (T for Theoretical) at Los Alamos National Laboratory but she quickly moved to division T7, led by Nicholas Metropolis, to work on the first ever computer, the MANIAC I. Together with Mary Hunt, she was the first programmer to start exploratory work on it. She remembers it as an easy task because of the very limited possibilities of the computer: 1000 words.

Mary Tsingou mostly worked with John R. Pasta, but also with Stanislaw Ulam. However, she had little contact with Enrico Fermi, at that time professor in Chicago, who visited Los Alamos only for short periods, mostly during the summer. However, she knew Nella, Fermi's daugher, much better who didn't want to stay with her parents during their visits to Los Alamos. Both girls, in their early 20s, slept in the same dormitory.

It appears that the only reason Mary Tsingo is not an author of the FPU report, is that she was not involved in the writing. It was the custom of the time ... but did not apply to Fermi since, as noted in S. Ulam's biographical book, he died before the writing of the paper.

In 1958, Mary Tsingou married Joseph Menzel who was also working at Los Alamos for the Protective Force of the Atomic Energy Commission. She stayed her whole life in this small city and worked successively on different problems, always with computers. She became one of the early experts in Fortran (FORmula TRANslator) invented by IBM in 1955, and was assigned to help researchers in the laboratory.

After her seminal programming work on the MANIAC, in the beginning of the seventies she came back to the FPU problem with Jim Tuck looking for recurrences. But she also considered numerical solutions of Schrödinger equation, the mixing of two fluids of different densities, with J. Von Neumann, and other problems. Finally, in the eighties during Ronald Reagan's presidency, she was deeply involved in the Star Wars project calculations.

Retired in 1991, Mary Tsingou Menzel is still living at Los Alamos, very close to the place where the FPU problem was designed and studied: it is time for a proper recognition of her work.

The FPU numerical experiment has been performed on the MANIAC (Mathematical Analyser, Numerical Integrator And Computer) built in 1952 for the Manhattan project, which was used in the development of Mike, the first hydrogen bomb. Richard Feynman (1918–87) and Nicholas C. Metropolis (1915–99), exasperated by the slow and noisy mechanical calculators which were nevertheless necessary for the design of the bombs, promoted its construction. The name of the computer, MANIAC, had been chosen by the project director, N. Metropolis, who hoped that it would stop the rising fashion of naming computers by acronyms.

The effect has been exactly the opposite, although, according to Metropolis [128], George Gamow (1904–68) was instrumental in rendering this and other computer names ridiculous when he dubbed the MANIAC 'Metropolis And von Neumann Install Awful Computer'. The MANIAC was able to perform about 10^4 operations per second, which must be compared to the 10^8 operations per second of any personal computer today.

It was Fermi who had the genius to propose that computers could be used to study a problem or test a physical idea by simulation, instead of simply performing standard calculus. He proposed to check the prediction of statistical physics on the system now called FPU. The discovery that resulted from this study has not only been at the origin of the soliton concept and of many features of chaotic phenomena, but it also introduced the concept of *numerical experiment.* This has led to a complete revolution in the investigation of some physical phenomena, since it is now possible to study collisions of black holes or quantum phenomena which are beyond any experimental or analytical investigation.

The remarkable discovery of FPU was made in 1953. In the same year J. D. Watson and F. Crick discovered the double helix structure of DNA, which led to another revolution, in the world of biology. It is interesting to note that the numerical simulations of FPU were performed by Mary Tsingou, who is not recognised as an author of this discovery. Similarly the experimental contribution of Rosalind Franklin (1920–58) to the discovery of the double helix has been greatly minimised [61]. Fermi, who qualified the FPU result as 'minor' intended to present it in an honorary conference of the American Mathematical Society. He became seriously ill before the conference and this work stayed buried for several years in a classified report of the Los Alamos Laboratory. It was Ulam who presented the results to the scientific community in several lectures.

8.3 The solution of the FPU problem

The solution of the FPU paradox was found ten years later by Zabusky and Kruskal in terms of solitons [190]. They studied the equations of motion which derive from Hamiltonian (8.1)

$$\ddot{u}_i = K(u_{i+1} + u_{i-1} - 2u_i) + K\alpha[(u_{i+1} - u_i)^2 - (u_i - u_{i-1})^2]. \tag{8.3}$$

Let us look for a solution which has a small amplitude with respect to the lattice spacing, denoted by a. We define v such that $u = a\varepsilon v$ with $\varepsilon \ll 1$. Dividing

Equation (8.3) by εa, we get

$$\ddot{v}_i = \frac{c^2}{a^2}(v_{i+1} + v_{i-1} - 2v_i) + \frac{c^2 \alpha \varepsilon}{a}[(v_{i+1} - v_i)^2 - (v_i - v_{i-1})^2], \tag{8.4}$$

in which we introduce the speed of sound $c = a\sqrt{K}$.

Fermi, Pasta and Ulam were considering the low-wavenumber modes since they were exciting the mode with the lowest wavenumber $k = 1$, which corresponds to a wavelength equal to twice the system size. As numerical simulations showed that recurrence occurs before any large-wavenumber mode had been excited, we can restrict the investigation to the long-wavelength modes, which amounts to studying the system in the continuum limit approximation. We shall introduce the dimensionless variable $X = \varepsilon x/a$, where the factor $\varepsilon \ll 1$ expresses the hypothesis that v varies slowly when the space variable along the chain changes by a. In order to estimate orders of magnitude, it is also convenient to introduce the dimensionless variable $\theta = ct/a$.

Now the calculation is standard. We put the expression $v_{i\pm 1} = v(x_i \pm a) = v(X \pm \varepsilon)$, expressed in terms of its Taylor expansion, into Equation (8.4) and we get

$$\frac{c^2}{a^2}\frac{\partial^2 v}{\partial \theta^2} = \frac{c^2}{a^2}\left(\varepsilon^2 \frac{\partial^2 v}{\partial X^2} + \frac{\varepsilon^4}{12}\frac{\partial^4 v}{\partial X^4}\right) + \frac{c^2 \alpha \varepsilon}{a}\left[\left(\varepsilon \frac{\partial v}{\partial X} + \frac{\varepsilon^2}{2}\frac{\partial^2 v}{\partial X^2} + \cdots\right)^2 - \left(\varepsilon \frac{\partial v}{\partial X} - \frac{\varepsilon^2}{2}\frac{\partial^2 v}{\partial X^2} + \cdots\right)^2\right]. \tag{8.5}$$

Keeping terms up to order ε^4 leads to

$$\frac{\partial^2 v}{\partial \theta^2} = \varepsilon^2 \frac{\partial^2 v}{\partial X^2} + \frac{\varepsilon^4}{12}\frac{\partial^4 v}{\partial X^4} + \alpha a \varepsilon^4 \frac{\partial}{\partial X}\left[\left(\frac{\partial v}{\partial X}\right)^2\right]. \tag{8.6}$$

This equation contains weakly dispersive and weakly nonlinear terms. If they were absent, we would get a permanent profile solution in the frame moving at the speed of sound, which is equal to ε using our units. Thus we can assume that, due to the dispersive and nonlinear terms, the solution will slowly evolve in the mobile frame. To formalise this, we change to the moving frame by defining $\xi = X - \varepsilon\theta$ and we introduce the 'slow time' $\tau = \varepsilon^3\theta$. The choice of the exponents of ε is directed by the balance between nonlinearity and dispersion, as discussed earlier, in Chapter 1. Using these new variables in Equation (8.4), we can check that the frame change cancels the terms at order ε^2, while the next terms, at order ε^4, lead to the equation

$$2\frac{\partial^2 v}{\partial \xi \partial \tau} + \frac{1}{12}\frac{\partial^4 v}{\partial \xi^4} + \alpha a \frac{\partial}{\partial \xi}\left[\left(\frac{\partial v}{\partial \xi}\right)^2\right] = 0. \tag{8.7}$$

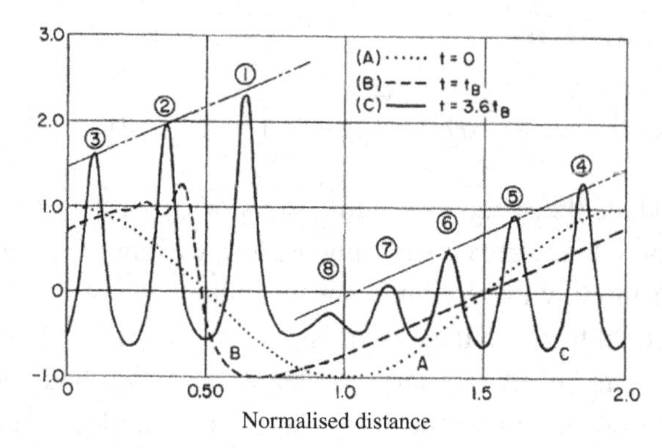

Figure 8.2. Time evolution of the FPU initial condition, plotted in real space, at different instants. (Adapted from reference [190].)

Defining $w = v_\xi$, we get the KdV equation

$$w_\tau + \frac{1}{24} w_{\xi\xi\xi} + \alpha a \, w w_\xi = 0. \tag{8.8}$$

The derivation of this soliton equation is the basis for the explanation of the FPU paradox proposed by Zabusky and Kruskal. As shown in Figure 8.2, the sinusoidal initial condition (dotted line) first tends to make sharp fronts (dashed line), which is a consequence of the nonlinearity of the KdV equation, as discussed in Chapter 1. Then the initial excitation breaks into a series of pulses, which are solitons. Figure 8.2 even shows the linear relation (1.5) between the excess of the velocity of a soliton over the sound velocity and its amplitude. This relation explains why the maxima of the solitons, which are equidistant from each other, line up in a straight line. The solitons that emerge from the initial condition preserve their shapes and velocities. In their motion in the finite system with periodic boundary conditions,[2] from time to time, they come back to the positions that they had initially, restoring the initial condition. This is why recurrence is observed.

Figures 8.2 and 8.3 immediately draw attention to the peculiarities of the propagation of waves in a nonlinear medium. The FPU paradox would probably not have been a mystery for more than ten years, if, before Zabusky and Kruskal, somebody had had the idea to look at the dynamics of the nonlinear lattice as a function of the space coordinate. But physicists were so used to analysing linearised problems, which naturally lead to a description in Fourier space, and only adding nonlinearity afterwards as a coupling between the modes, that the observation of solitons emerging from the sinusoidal initial condition had not occurred.

[2] The calculation performed by Zabusky and Kruskal used periodic boundary conditions, while the FPU calculation used fixed boundaries. This does not change the analysis because either the solitons pass through the boundaries and enter on the opposite side, or they are totally reflected.

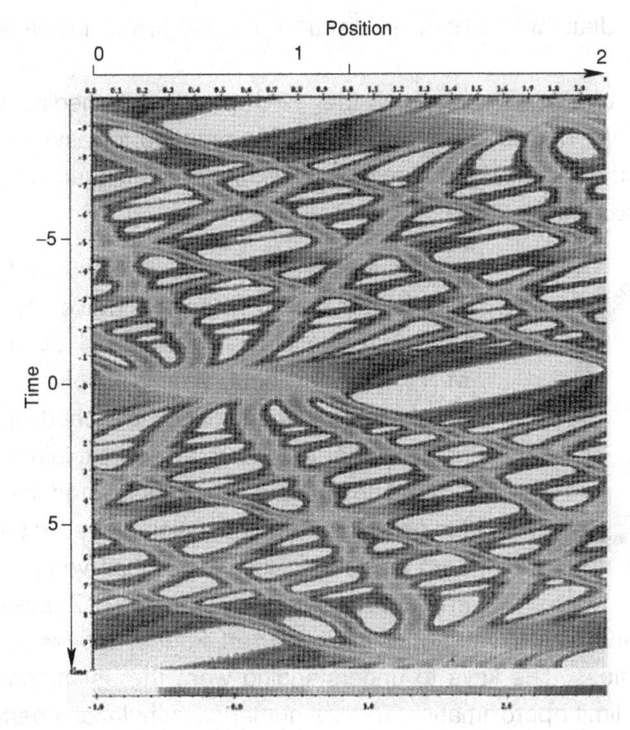

Figure 8.3. Space–time evolution of the initial condition $\cos(\pi x)$ in a system with periodic boundary conditions, obeying the Korteweg–de Vries equation. Time and space are respectively plotted vertically and horizontally. The amplitude of the local velocity is shown by a grey scale, as shown at the bottom of the figure. (Courtesy of N. J. Zabusky.)

8.4 Kruskal and Zabusky: the pioneers

While he was a teenager, MARTIN D. KRUSKAL, born in 1925 in New Rochelle, met Richard Courant (1888–1972) who became his PhD supervisor at the New York University a few years later. He became a professor in the prestigious Princeton University where he stayed for about forty years. His research was first devoted to pure mathematics but he quickly became interested in the emerging plasma physics, which is also the physics of stars or thermonuclear generators.

In 1954, M. D. Kruskal visited Los Alamos for six weeks within a research programme on controlled fusion, the Matterhorn project.

He met S. M. Ulam with whom he discussed many topics in mathematics and physics.

The results of what we now call the 'FPU' problem immediately caught his attention because he recognised an interesting problem for asymptotic analysis. However he did not consider it immediately within the continuum limit approximation. (Picture: M. D. Kruskal)

In January 1960, NORMAN J. ZABUSKY (1929–), who was a post-doc at Princeton University following his PhD at Caltech, was fascinated by a seminar by M. D. Kruskal about 'A few unsolved mathematical problems'. He immediately switched research topic and together they were able to explain the FPU paradox in 1965 because they discovered the link between the problem of the relaxation of a solid toward equilibrium and the solitary wave observed by J. S. Russell in a Scottish canal. (Picture: N. J. Zabusky)

Their first attempts to study the interaction of Fourier modes in the spirit of FPU were fruitless. The keys to understanding were the asymptotic analysis in the continuum limit approximation and the numerical simulations performed by N. Zabusky, which he called 'KdV cinema'. Unable to get assistance from the Los Alamos group, he decided to develop his own code with the valuable help of Gary Deem [192]. The simulations allowed him to follow the time evolution of the system in real space and not in Fourier space, so that he could identify the nonlinear localised waves, clearly visible in Figure 8.3. Using periodic boundary conditions instead of the fixed boundary conditions of Fermi, Pasta and Ulam, turned out to be another important advance because it exhibited the progressive waves more clearly. Moreover, M. D. Kruskal and N. J. Zabusky understood the importance of dispersive effects in the discrete lattice, which had been overlooked by Fermi, Pasta and Ulam.

It is interesting to notice that the KdV equation, which results from the asymptotic analysis presented above, was not taught by mathematicians at that time and, for instance, it was not mentioned in the book by Courant and Hilbert [43], which was already an authoritative reference. This is surprising because the section on partial differential equations, starting from first-order equations, continued with second-order equations such as the wave equation and the heat equation, before jumping to fourth order equations such as the biharmonic or the telegraphist's equations. Third-order equations such as the KdV equation were completely ignored, probably due to a lack of applications. During a meeting on plasmas, Cliff Gardner and George Morikawa attracted the attention of Kruskal and Zabusky over the work of Korteweg and de Vries [104], well known by hydrodynamicists and listed in a footnote in the book by H. Lamb [107]. This turned out to be the clue!

It is also Zabusky and Kruskal who coined the word 'soliton', which they introduced by putting it, for the first time, in the title of their paper [190]. In order to stress the quasi-particle properties of the solitary waves they thought of the suffix 'on', as in electron, proton, boson, etc. Zabusky first proposed *solitron* as an abbreviation for 'solita-**r**-y wave' before noticing that it was the name of a company in the United States. They finally opted for *soliton*, which is now famous!

M. D. Kruskal discovered the general method for solving the KdV equation [74] and then devoted his research to the strange 'surreal' numbers which generalise the familiar real numbers and have important applications for asymptotic analysis, a topic at the root of his research all through his career. N. Zabusky later obtained important results in hydrodynamics, again using with acuity numerical simulations and visualisation of the results, an aspect that he considers crucial and calls 'visiometrics'.

8.5 FPU and the Japanese School

In the meantime the Japanese School was, independently, very active in this field. As discussed, the original FPU report [66], unpublished, was known by very few people, most of them within America. This is why Nobuhiko Saito (1919–) together with his PhD student Hajime Hirooka (1939–) were developing in Japan, closely related research work, in complete ignorance of what had been done in Los Alamos ten years earlier.

As explained by the title of Hirooka's PhD Thesis, 'The approach to thermal equilibrium in a nonlinear lattice', the motivation was to explore the mechanism of ergodicity. As the mechanics of collisions in a gas were too complicated, they thought that an anharmonic lattice would be a more appropriate system; in the end it appeared that its dynamics were difficult to solve. This is why, in 1964 they started to perform numerical simulations on a small computer provided by NEC; it took all night to compute the evolution for only five lattice sites over several periods of the lowest mode! In 1965, they switched to the new supercomputer provided by the Hitachi Co. to Tokyo University, installed under a national plan for all Japanese universities and research institutes.

Saito and Hirooka considered a one-dimensional anharmonic lattice with quadratic and quartic potentials between neighbouring particles, and with both ends fixed. The initial excitation was slightly different since all particles were at rest, while a constant force was applied to the first particle. They also prepared a similar system, with only harmonic potentials, and analytically calculated several quantities of interest, in particular the long-time averages of the squares of the velocities of the particles $\langle \dot{u}_i^2 \rangle$ and the correlations $\langle \dot{u}_i \dot{u}_{i+1} \rangle$ of the velocities of neighbouring particles. The long-time averages of $\langle \dot{u}_i^2 \rangle$ were the same for all particles, but the correlation functions did not vanish in the long term, contrary to what is expected at thermodynamic equilibrium. However, the results for the anharmonic lattice vibration obtained with the computer were almost the same as the harmonic case, and did not show any tendency to approach the Maxwellian distribution of velocities, contrary to expectations.

It is interesting to notice that they hesitated to publish these results, because they were not familiar with the computer calculation and were afraid of having introduced some errors. However, in the meantime, they found numerical experiments similar to the original FPU one in several papers on nonlinear oscillations, in particular, those of Ford [69, 70]. These papers also convinced Morikazu Toda (1917–) to study solitons and to introduce a lattice with exponential interactions, nowadays called the Toda lattice [176].

Soon after, they also learned that the original FPU report [66] was reproduced in the collected papers of Fermi. Very excited, Saito and Toda read the paper at the library of Tokyo University of Education (now Tsukuba University) and shared their knowledge about this topic. As they had found results similar to the findings of Fermi, Pasta and Ulam, they finally decided to publish their calculations [164, 165]. Afterwards, they considered the FPU's simpler initial conditions and found the induction phenomenon and the occurrence of the random character of lattice vibrations [88].

Finally, they discovered the seminal Zabusky–Kruskal paper [190] soon after the publication of their papers in 1967. Norman Zabusky went to Kyoto in 1968 for the International Conference of Statistical Mechanics, where he showed his movies and in particular the KdV cinema, which had 'the power to communicate unbiased information in a credible manner, much beyond the power of words, graphs and equations' [191].

At that time, M. Toda gave [177] the first analytical estimate of the recurrence time, based on exact solutions of the Toda lattice. Introducing $T_1 = 2N/\sqrt{K}$, the period of the first mode, he got

$$T_R = \frac{3}{\pi^{3/2}\sqrt{2}} \frac{N^{3/2}}{\sqrt{a\alpha}}\, T_1 \simeq 0.38 \frac{N^{3/2}}{\sqrt{a\alpha}}\, T_1. \tag{8.9}$$

which compares well with the first empirical estimate made by Zabusky [191]. Toda's result is a little larger than Zabusky's formula: the pre-factor is 0.31 instead of 0.38 in Equation (8.9). The discrepancy is due to the fact that when solitons pass through each other, they accelerate due to the compression of the lattice.

9

A simple model for dislocations in crystals

9.1 Plastic deformation of crystals

Let us examine the deformation of a crystal under sheer stress (Figure 9.1(a)) due to a force $\vec{F}$ applied on its upper surface while its bottom surface is held fixed.

For a weak force the deformation is elastic but, beyond some threshold, the crystal undergoes a permanent plastic deformation, which persists even after the force has been suppressed. On a microscopic scale, this deformation occurs because an atomic plane has slipped with respect to another as schematised in Figure 9.1(b). Let us use typical data on the mechanical properties of a solid to evaluate the order of magnitude of the force creating such a deformation, or rather of the sheer stress $\sigma = F/S$, where S is the area over which the force is applied.

The microscopic structure of the crystal is schematised in Figure 9.2. The stress σ induces a relative motion x of the two atomic layers which are on both sides of the slip plane. This leads to a reorganisation of the atomic bonds, some becoming longer while others shorten. They recover their initial structure when the displacement x has reached the value of the lattice spacing a.[1] Thus the function $\sigma(x)$ must have the periodicity of the crystal. In order to get the order of magnitude of the critical stress σ_c which leads to a plastic deformation, we can consider the simplest periodic function

$$\sigma = \sigma_c \sin \frac{2\pi x}{a}. \tag{9.1}$$

In order to evaluate σ_c, we can use the theory of elasticity, valid for small displacements. For small x, σ can be approximated by

$$\sigma \approx \sigma_c \frac{2\pi x}{a}. \tag{9.2}$$

[1] Strictly speaking this is only true in the bulk of the crystal. Near the boundaries some bonds are broken. However, boundary effects can be neglected because the lattice spacing is of the order of an Angström (10^{-10} m). Even for a very small crystal having a size of 0.1 mm, boundary effects only concern about one millionth of the bonds along each line of atoms.

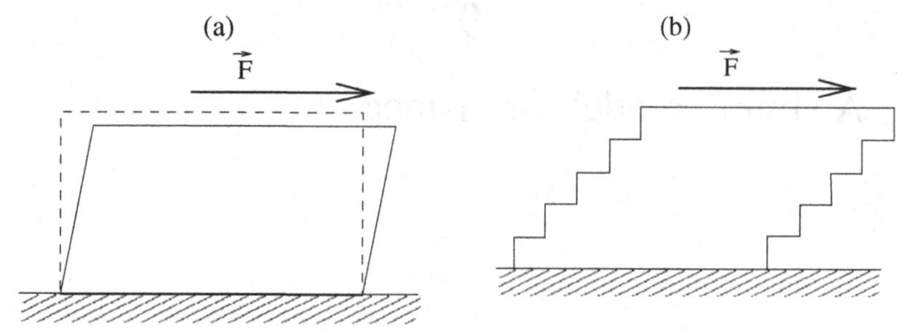

Figure 9.1. Plastic deformation of a crystal under sheer stress: (a) macroscopic view, (b) schematic microscopic view.

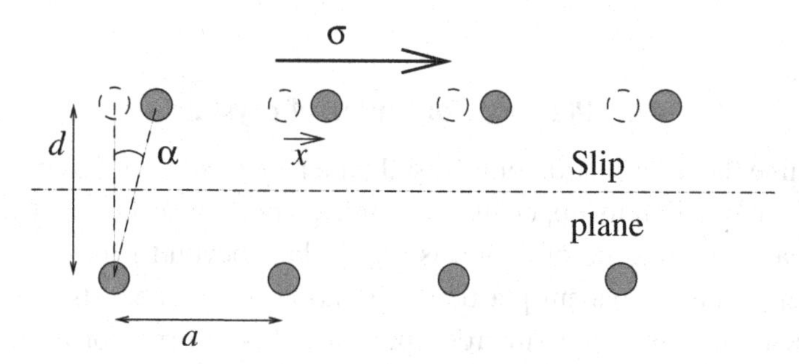

Figure 9.2. Microscopic view of a crystal under sheer stress.

The sheer stress modulus G of a solid distorted with an angle α (Figure 9.2) is defined by

$$\alpha = \frac{\sigma}{G} \quad \text{with} \quad \alpha \approx \frac{x}{d}, \tag{9.3}$$

where d is the interatomic distance in the direction perpendicular to the slip plane. Therefore the elasticity theory leads to $\sigma \approx Gx/d$. Comparing with (9.2), we get

$$\sigma_c \approx \frac{aG}{2\pi d}. \tag{9.4}$$

The sheer stress modulus of a solid such as aluminium is $G = 2.5 \times 10^{10}$ N m^{-2}, whereas measurements of σ_c give 10^6 to 10^7 N m^{-2}, depending on the purity of the material. As a and d are of the same order of magnitude, our estimate of σ_c appears to be wrong by several orders of magnitude. Such a disagreement cannot be attributed to the approximations that we have made. They could at most alter the result by one order of magnitude. We must *question the basic hypothesis of the calculation.*

We assumed that an atomic layer was sliding as a whole with respect to the next one. In fact one could imagine another process for sliding the layer, with a much

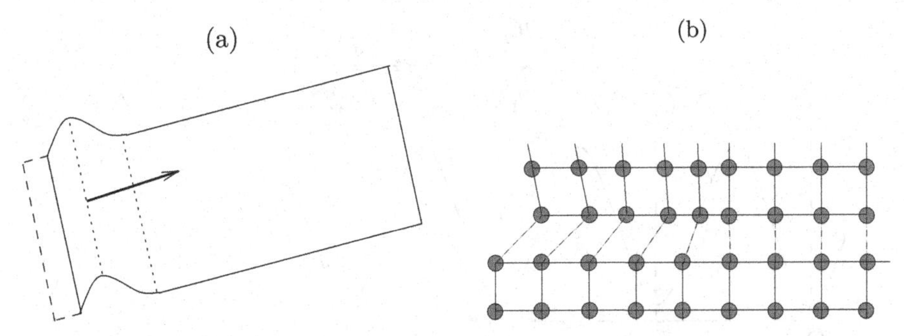

Figure 9.3. (a) How to move a carpet with minimal effort by creating a fold on one side. (b) Schematic plot of an 'edge' dislocation in a crystal. It is created by the motion of one end by one lattice spacing.

smaller stress. A simple analogy helps to understand how it can be done. Assume that we want to slide a carpet on the floor. We could simply pull on one end. This is the principle of the calculation that we made to evaluate σ_c. But we could also make a fold along one side (Figure 9.3(a)) and push it to the other side. When the fold reached the opposite side, the carpet would have been moved by the displacement that created the fold initially. But the force required to move the fold would be much smaller than the force that would have to be applied to pull the carpet as a whole. For the plastic deformation of a solid, there is a process which is exactly analogous to the fold in the carpet: the *creation of a dislocation*. Instead of moving the whole atomic layer, we can move only one of its sides by one lattice spacing. This generates a defect in the crystal packing because, along one line, the atomic density is higher than in the perfect crystal. This defect is called an *edge dislocation* [99]. On a simplified figure such as Figure 9.3(b), the distortion may seem very large. Actually, when one looks at the crystal packing at a larger scale, the defect may be hardly visible (see an example in Figure 5 of reference [99]). The motion of the dislocation from one side of the crystal to the other generates an overall plastic distortion resulting from local distortions only. The creation of the dislocation requires much less effort than the displacement of an atomic layer as a whole. Moreover crystals are seldom perfect, and generally they already contain a lot of dislocations. In order to distort them it is sufficient to move these pre-existing dislocations, so that the sheer stress modulus can be considerably smaller than the modulus given by the motion of a whole atomic layer.[2] The edge dislocation that we described

[2] Plumbers know this phenomenon very well. When they want to bend a copper pipe, they select a pipe which has been cold-drawn, and hence contains a lot of dislocations, while they use pipes which have been annealed when they want them to stay straight and resist bending. Similarly, when we bend an iron wire for the first time it is a bit difficult, but then, after a few bending–straightening cycles, it bends very easily because we have created a large number of dislocations.

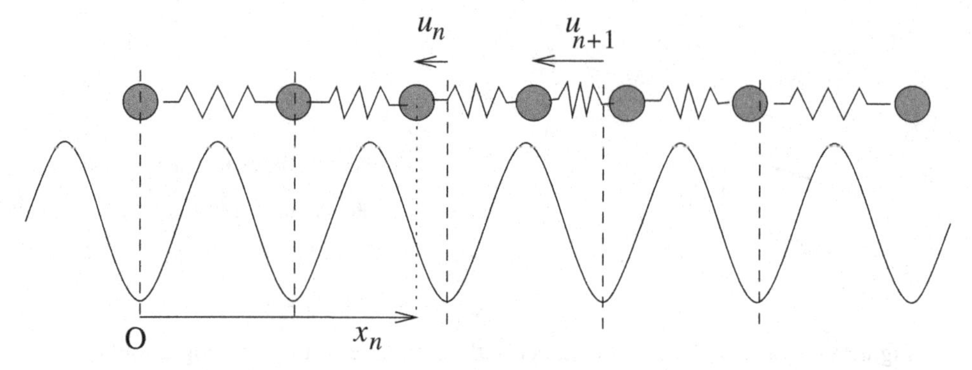

Figure 9.4. The Frenkel–Kontorova (FK) model of an edge dislocation.

is the simplest one, but there are other possible geometries for creating defects in the crystal packing. All of them are distortions which are localised along lines or surfaces.

9.2 A one-dimensional model: the Frenkel–Kontorova model

A model containing the essentials of the physics of a dislocation was proposed in 1939 by Frenkel and Kontorova (FK) [2, 72]. It describes the dynamics of a line of atoms above the slip plane (Figure 9.4).

The position of the atom of index n is measured with respect to a fixed origin, chosen to be at the equilibrium position of one particular atom in a perfect crystal (see Figure 9.4). It can also be given by its displacement with respect to its equilibrium position, $u_n = x_n - na$, where a is the lattice spacing along the line of atoms that we are studying. An atom is subjected to the potential $V(u_n)$ created by the atoms which are below the slip plane. This 'substrate potential' has the periodicity of the lattice, and the FK model chooses the simplest periodic function

$$V(u_n) = V_0 \left(1 - \cos \frac{2\pi u_n}{a}\right). \tag{9.5}$$

The model must also take into account the interactions of the atoms along the line.[3] The substrate potential $V(u_n)$ cannot ignore nonlinearity because we want to describe motions which may be as large as the period of the potential. However, a harmonic approximation is possible for the interaction potential between atoms $n-1$ and n because it depends on the *relative* displacement of two neighbouring atoms, which is small with respect to the lattice spacing, even in the core of a

[3] Or more precisely the interaction between the atomic planes orthogonal to the slip plane which are above the slip plane.

dislocation. It is written as

$$W(u_{n-1}, u_n) = \frac{C}{2}(u_n - u_{n-1})^2, \tag{9.6}$$

so that the Hamiltonian of the model is

$$H = \sum_n \frac{p_n^2}{2m} + W(u_{n-1}, u_n) + V(u_n) \tag{9.7}$$

$$= \sum_n \frac{p_n^2}{2m} + \frac{C}{2}(u_n - u_{n-1})^2 + V_0\left(1 - \cos\frac{2\pi u_n}{a}\right), \tag{9.8}$$

where m is the mass of an atom and $p_n = m\, du_n/dt$ its momentum.

This model contains the basic ingredients for soliton solutions, the nonlinearity of the substrate potential and the cooperativity coming from the interatomic interactions. We can show that indeed solitons exist in this system, ... almost.

Son of a notorious activist of the pre-revolutionary period, YAKOV ILYICH FRENKEL (1894–1952) came to the University of Saint Petersburg during the Russian revolution. After his first paper, devoted to the differential gear in cars(!), the beginning of his career was chaotic, in particular because he was accused of being a 'Bolshevik agitator' and sent to prison. As a young researcher he was supported by Paul Ehrenfest (1880–1933) who, being at the heart of the events which revolutionised the physics of the twentieth century, played a large role in helping young Russian theoreticians to reach international level in the difficult post-revolution period. (Picture: AIP Emilio Segré Visual Archives, Frenkel Collection.)

He obtained a grant of the Rockefeller Foundation in Europe, which allowed him to spend 1926 in Germany, France and England, where he met many physicists: Einstein, Brillouin, Langevin... and then in 1931 he spent several months in the United States; two post-doctoral stays in a foreign country, well before it became common!

His major contributions were in solid state and liquid physics but he also contributed significantly to classical electrodynamics and nuclear physics. His approach was characterised by the introduction of original and simple physical models, using minimal mathematical tools. The model for plasticity that he developed with T. Kontorova was the first attempt in solid state physics to describe the dynamics of a two-dimensional defect with a one-dimensional model of a discrete chain. This approach was very successful and became common in solid state physics of course but also in other areas of physics. Frenkel published the first complete course in theoretical physics in the Soviet Union and, with V. A. Fock (1898–1974), he was the first to teach quantum mechanics there. A talented painter and violin player [73], he died suddenly in 1952.

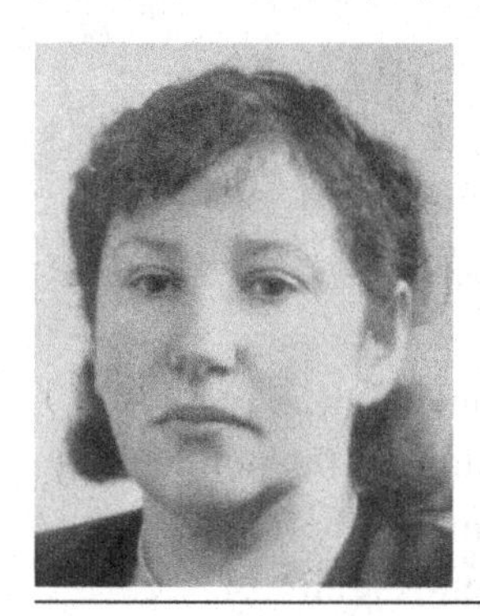

TATIANA KONTOROVA (1911–1977) in Saint Petersburg, studied at the Polytechnic State University of Saint Petersburg. She gained her Ph.D. in 1939, thanks to her work on dislocations with Y. I. Frenkel. She was a researcher and a teacher in the same institute, and wrote about 50 scientific papers. She also translated the book by William Shockley, *Electronic Theory of Semiconductors*, into Russian. (Picture: Ioffe Institute).

9.3 Continuum limit approximation: the sine-Gordon equation

The equations of motion of the atoms, which derive from Hamiltonian (9.8), are

$$m\frac{\mathrm{d}^2 u_n}{\mathrm{d}t^2} = C(u_{n+1} + u_{n-1} - 2u_n) - \frac{2\pi V_0}{a}\sin\frac{2\pi u_n}{a}. \tag{9.9}$$

As in the Fermi–Pasta–Ulam problem, we get a set of coupled nonlinear differential equations. This is a common situation in solid state physics. This system, although it has a simple form *does not have any known analytical solution.* As in the FPU case, approximations are required to solve it. We don't want to linearise the equations because our goal is to study the large atomic displacements in the heart of a dislocation, but, as in the FPU case, we can use a continuum limit approximation by replacing the set of discrete variables $u_n(t)$ by the continuous function $u(x,t)$ so that $u_n(t) = u(x = na, t)$. Expanding $u_{n\pm1}(t)$ around $u_n(t)$ we get

$$u_{n\pm1}(t) = u[(n \pm 1)a, t] = u(na,t) \pm a\frac{\partial u}{\partial x}(na,t) + \frac{a^2}{2}\frac{\partial^2 u}{\partial x^2}(na,t) + \cdots \tag{9.10}$$

If we truncate the expansion at order 2, the set of equations (9.9) becomes

$$\frac{\partial^2 u}{\partial t^2} - \frac{Ca^2}{m}\frac{\partial^2 u}{\partial x^2} + \frac{2\pi V_0}{ma}\sin\frac{2\pi u}{a} = 0. \tag{9.11}$$

Defining $\theta(x,t) = 2\pi u(x,t)/a$, $c_0^2 = Ca^2/m$, $\omega_0^2 = 4\pi^2 V_0/(ma^2)$, we obtain the sine-Gordon equation

$$\frac{\partial^2 \theta}{\partial t^2} - c_0^2\frac{\partial^2 \theta}{\partial x^2} + \omega_0^2 \sin\theta = 0. \tag{9.12}$$

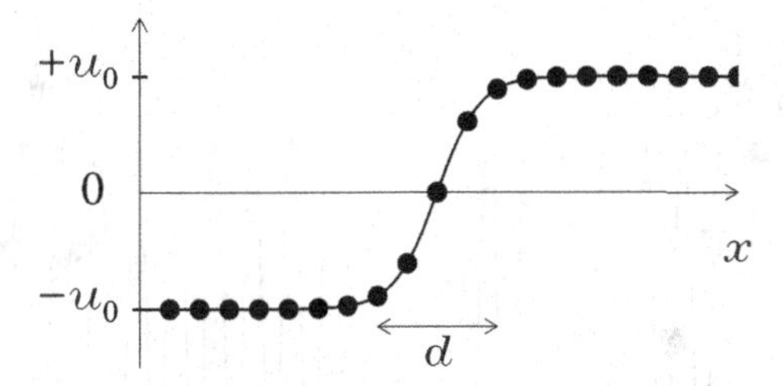

Figure 9.5. The circles show the positions of the atoms in a dislocation, while the curve shows a continuum limit approximation of these positions. The width of the soliton is only about two lattice spacings.

This approach suggests therefore that the solutions of the SG equation (9.12) could describe the dislocations that we discussed above.

9.4 Are dislocations solitons?

The result that we just obtained has to be questioned because, if dislocations were solitons, once they had been created, they would move at constant speed with a permanent profile. Of course our experience tells us that a deformation initiated in a material does not grow spontaneously once the stress that created it is suppressed. There is certainly something wrong in the model.

Actually, it is not the model which is wrong, but how we solved it. The continuum limit approximation is not justified for the dynamics of dislocations. In actual crystals, dislocations are not very extended with respect to the lattice spacing a, as suggested by Figure 9.5.

Therefore the continuum limit approximation can only give a rather crude picture. Dislocations are not solitons in the strict sense, but it does not mean that the soliton concept is useless to describe excitations in highly discrete systems. It provides a good starting point, but it must be completed. The lattice brings in new features, such as a trapping of the soliton or the emission of small amplitude radiations.

In order to understand this, let us examine the positions of the particles with respect to the substrate potential $V(u)$, depending on the strength of their coupling. When the coupling is strong, the soliton is broad, which means that the transition of the particles from one well to the next evolves very slowly, from zero to one lattice spacing. Thus there are particles at all levels of the substrate potential, even at its maximum. In such a quasi-continuum case, it is easy to understand why the dislocation (or the soliton) can move freely. Let us look at Figure 9.6(a) and assume that we want to move the soliton from its present position marked by the solid arrow in Figure 9.6(a) toward a new position marked by the dotted arrow on the figure. When the displacement is completed, the particles will sit at positions

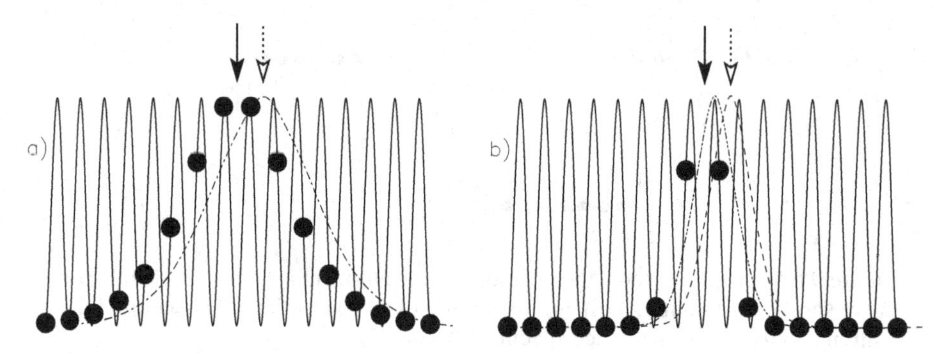

Figure 9.6. Position of harmonically coupled particles subjected to a periodic substrate potential. Figures (a) and (b) show, respectively, a wide and narrow defect. The particles in the right side of the chain have been translated by one lattice spacing. In case (b), the atoms situated near the centre of the dislocation are far from the top of the potential because the strength of the interaction potential is not large enough to maintain them there. In order to move the defect by one lattice spacing, those atoms will have to temporarily climb to the potential maximum, which costs energy.

corresponding to the intersections between the potential curve (solid line) and the dash–dotted curve. During the progressive translation of the curve, from a position centred under the solid arrow to a position under the dotted arrow, observing how the intersection points move tells us that some particles climb the substrate potentials, while others move downward. There is no energy cost because the potential energy increases are balanced by the decays for each intermediate position of the soliton. In the continuum version of the model, which leads to the sine-Gordon equation, the physical system is invariant with *any* translation. The free motion of the soliton is simply a manifestation of this translational symmetry, which gives rise to the Goldstone mode, as we saw in Chapter 5.

When the coupling is weak, as is the case for dislocations, the defect is highly localised, i.e. the width of the dislocation is only of the order of one or two lattice spacings. Although some particles are lifted from the bottom of the potential well as shown in Figure 9.6(b), in the minimum energy configuration it would cost too much potential energy to retain particles at the top of the potential maximum, with respect to the small gain in elastic energy that it would provide, so that there are no particles near the top. If we translate the defect, when its centre reaches the position between the two arrows in Figure 9.6(b) some particles will have climbed to a maximum. In this intermediate position, the defect has a higher energy than when it is centred under one of the arrows. Therefore there is a barrier to translation, which is called the Peierls–Nabarro (PN) barrier in dislocation theory, and which has been re-discovered in the theory of discrete solitons (which should be called quasi-solitons although we have kept the denomination soliton for convenience).

In our qualitative discussion to show the origin of the Peierls–Nabarro barrier, we only considered the variations of the substrate potential energy during the translation. To be complete, we should also consider the variation of the elastic interaction between the particles. But this does not change the result: the translation of a narrow defect does not occur at constant potential energy.

The case of a dislocation, extended over a few sites, is intermediate between an individual particle, which has to climb the full height of the substrate potential in order to move in the periodic potential because it cannot borrow energy from another particle moving down in the potential, and the continuum limit in which the soliton moves freely. In such an intermediate case, the defect moves as a collective atomic excitation over an effective potential which has the period of the lattice, but an amplitude significantly smaller[4] than the amplitude of the substrate potential. This explains why the critical sheer stress for the plastic deformation of a crystal is several orders of magnitude smaller than the sheer stress necessary to slide one whole plane over another. However the critical sheer stress does not vanish because dislocations are not free. Even when dislocations already exist in a material, the stress must exceed some threshold in order to move them and create a plastic deformation.

When a dislocation moves, it radiates lattice vibrations, which are responsible for the heating that we can feel if we impose large plastic deformations on a material, thereby moving many dislocations. It is possible to understand this radiation of phonons if we notice that, when a narrow defect moves, a given atom is first taken to the maximum of the substrate potential before falling into its next minimum. When it reaches the minimum, it tends to oscillate instead of stopping exactly at the minimum. This does not occur when the coupling is strong because the atom is maintained by a strong elastic coupling with its neighbours. The radiation of extended modes takes energy away from the defect, which tends to slow down as shown in Figure 9.7. A sharp decay of the velocity of the narrow kink, associated with a strong emission of phonons, is observed first. Subsequently the velocity stabilises and stays almost constant, at a value which does not depend on the initial velocity.

It is possible to give a simple analysis of these numerical observations. In order to compute the emitted radiation, we can use the linearised equations of motion and treat the effect of the soliton as if it were an external excitation, causing motion of the particles above the substrate potential. This effect is represented by adding a force $f_n(t)$ to the linearised equations of motion in order to describe the action of

[4] It can be shown that the PN barrier decreases exponentially with the width of the soliton.

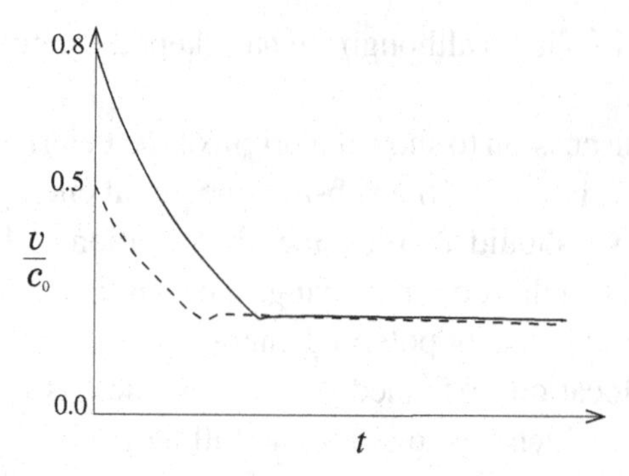

Figure 9.7. Time evolution of the velocity of a narrow soliton-like excitation in the Frenkel–Kontorova model. The solid line corresponds to an initial velocity of 0.8 and the dashed line to an initial velocity of 0.5, in units of c_0.

the soliton on the particles. The equations become

$$\ddot{u}_n - C(u_{n+1} + u_{n-1} - 2u_n) + \omega_0^2 u_n = f_n(t). \tag{9.13}$$

To understand the phenomena qualitatively, it is not necessary to know the expression of $f_n(t)$. The important point is that, since the soliton moves at velocity v in a lattice of periodicity a, it is legitimate to assume that $f_{n+1}(t) = f_n(t - a/v)$, or, more generally $f_n(t) = f_0(t - na/v)$. The force can excite plane waves of the form $\cos(\omega t - qn) = \cos\omega(t - qn/\omega)$ if the resonance condition $q/\omega = a/v$ is verified. The intersection of this 'soliton dispersion relation' $\omega = qv/a$ with the dispersion relation of the linear waves in the lattice shows that the resonant emission of radiation stops at the velocity v for which a stabilisation is observed [151]. Radiation through harmonics remains possible, but it is much weaker.

9.5 Applications

The Frenkel–Kontorova model can be used to understand the plastic deformation of materials. Cooperative effects are necessary to explain why crystals exhibit plasticity, unlike glasses which do not have any long-range order.

The model also explains why the more perfect a crystal is, the less plastic it is. By definition a perfect crystal does not have dislocations. They must be created before the crystal can be deformed. On the other hand, in an imperfect crystal it is sufficient to put pre-existing dislocations in motion. Moreover, as dislocations are quasi-solitons, initiating their motion is rather easy although the barrier of the Peierls–Nabarro potential has to be overcome. The perfect pure silicon crystals made by the electronic industry are extremely brittle because they do not include the dislocations that would allow them to deform rather than break.

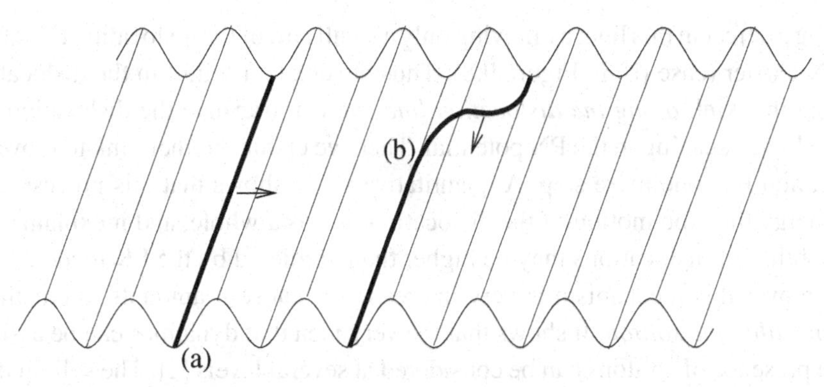

Figure 9.8. Schematic picture of the motion of a dislocation line (thick line) in the two-dimensional Peierls–Nabarro potential. It must be stressed that this potential *is not the substrate potential applied to the atoms* but a potential for the dislocations. We can think of moving the straight dislocation line as a whole (case (a)), but it is also possible to create a kink in the dislocation (case (b)) and then move it in the transverse direction.

Soliton properties can be used to establish quantitative predictions on dislocations. For instance the energy to create a dislocation is directly deduced from the soliton energy calculated in Chapter 2. The friction coefficient for the motion of a dislocation can be computed from the vibrations radiated by a quasi-soliton. A maximum strain rate can also be estimated from the limiting velocity of the dislocations, which is the speed c_0.

Studying the FK model for the dislocation, we restricted our investigation to one dimension. Actually, when we plastically deform a sample by a sheer stress, we move a *dislocation line* as schematised in Figure 9.8. The dislocation of the FK model is a point on this line but, to better understand how it moves, it is necessary to consider the slip plane parallel to the sheer stress. When we look from this two-dimensional viewpoint, the Peierls–Nabarro potential due to discreteness takes the shape of a 'corrugated iron sheet', with its undulations orthogonal to the plane in which we plotted the FK model (Figure 9.4). When it is in equilibrium in the minimum of the PN potential, the dislocation line lies along the bottom of one of these potential valleys. Working with the FK model, we implicitly assumed that the dynamics of the crystal was unchanged by any translation orthogonal to the plane of Figure 9.4. In this framework, moving the quasi-soliton of the FK model amounts to moving the dislocation line as a whole in the slip plane (case (a) in Figure 9.8). But, in the crystal, nothing forces the dislocation line to stay straight. This motion is energetically costly because all parts of the dislocation line must simultaneously overcome a maximum of the PN potential. The 'trick' of introducing a dislocation to avoid the motion of a plane of atoms as a whole can also be used to move the dislocation itself! Instead of moving the dislocation line as a whole, we can imagine creating a defect in this line, by moving only a small part of the dislocation line above the PN barrier (case (b) in Figure 9.8). Thus we create a 'kink' in the dislocation. Moving this kink *along the dislocation line*, we can translate the dislocation line by one lattice spacing in the PN potential. Then we create another kink to move the dislocation line one more step. A quantitative study shows that this process costs less energy than the motion of the dislocation line as a whole, and it explains why the mobility of dislocations may be higher than predicted by the FK model.

Moreover this mechanism is very interesting because it amounts to creating *a soliton within the soliton*. It shows that the very idea that dynamics can be assisted by the presence of solitons can be considered at several levels [2]. The soliton itself can become an object which may exhibit dynamics involving solitons! The spirit is similar to the idea that leads to 'second sound': we move from the physical system to a system of collective objects (the solitons for dislocations or the phonons for second sound). These objects themselves can then show some properties of the original physical system.

10

Ferroelectric domain walls

This chapter introduces a second example where the atomic positions in a crystal are well described by an equation which has quasi-soliton solutions. The existence of these solitons can explain some experimental properties of dielectric materials which resisted theoretical analysis for a long time.

10.1 Ferroelectric materials

A ferroelectric is a material which has a permanent electric dipole moment, which persists even in the absence of any external field. The dipole moment per unit volume is the *polarisation* $\vec{P}$ of the ferroelectric. It can be reversed by the application of a strong external electric field in the opposite direction. The ferroelectric polarisation depends on temperature and vanishes above a critical temperature T_c. For temperatures higher than T_c the material is in a state called paraelectric, without any spontaneous polarisation, as shown in Figure 10.1.

Two types of ferroelectrics can be distinguished. They are called displacive and order–disorder ferroelectrics.

10.1.1 A displacive ferroelectric: barium titanate

Barium titanate $BaTiO_3$ is the prototype of a displacive ferroelectric crystal, which has an ordered structure which changes symmetry at the ferroelectric–paraelectric transition. In the high-temperature paraelectric phase, the crystal structure is cubic. The titanium ion Ti^{4+} is in the centre of the unit cell, surrounded by a regular octahedron of oxygen ions O^{2-}. When the temperature is lowered below $T_c =$ 393 K the crystal becomes ferroelectric because the cell distorts to a tetragonal structure. The oxygen octahedron extends along the c axis, and instead of one equilibrium position for the titanium in the cell centre, the Ti^{4+} ion has two possible

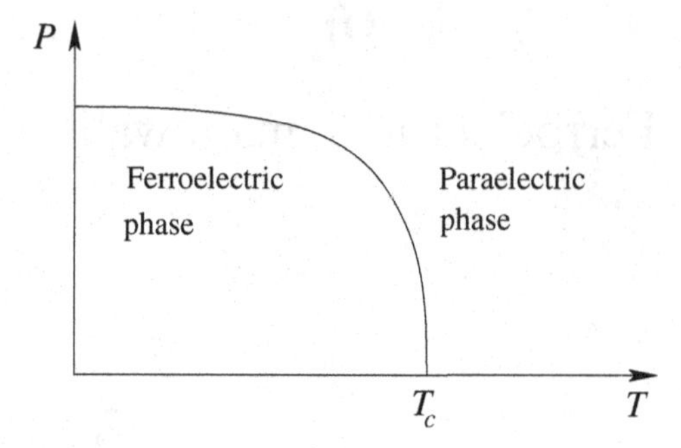

Figure 10.1. Variation of the spontaneous polarisation P of a ferroelectric material versus temperature.

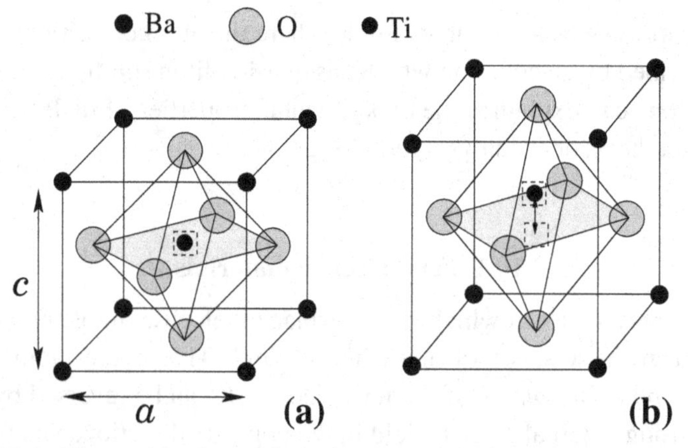

Figure 10.2. Figure (a) shows the structure of barium titanate in the high-temperature phase while Figure (b) shows the structure below the critical temperature. The dashed-line squares indicate the possible positions of the Ti^{4+} ion. In the paraelectric phase (a) a single position is available. It is located in the median plane of the cell. In the ferroelectric phase, there are two possible positions, on both sides of the median plane. The distortion of the tetragonal phase with respect to the cubic phase has been exaggerated to make the difference between the two structures clearly visible.

equilibrium positions, shifted up or down with respect to the median plane of the cell (Figure 10.2). The centre of the positive charges (Ti^{4+} and Ba^{2+}) no longer coincides with the centre of the negative charges (O^{2-}). The shifts along the c axis, $\pm u_0$, are associated to two states of spontaneous polarisation $\pm \overrightarrow{P_s}$.

The structures of the two phases of barium titanate indicate that the crystalline potential in which the titanium ion is evolving has the shape shown in Figure 10.3, depending on the phase. In the ferroelectric phase, the titanium is sufficiently mobile within the distorted oxygen octahedron to allow a strong electric field to switch it from one minimum to the other, which reverses the polarisation.

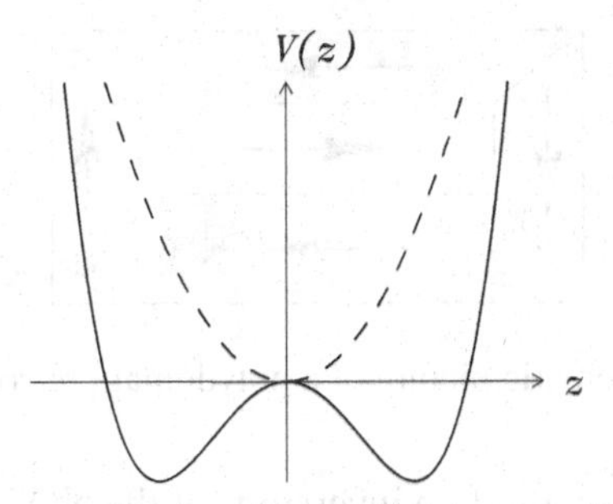

Figure 10.3. Plot of the crystalline potential for the titanium ions of barium titanate in the paraelectric phase (dashed line) and in the ferroelectric phase (solid line).

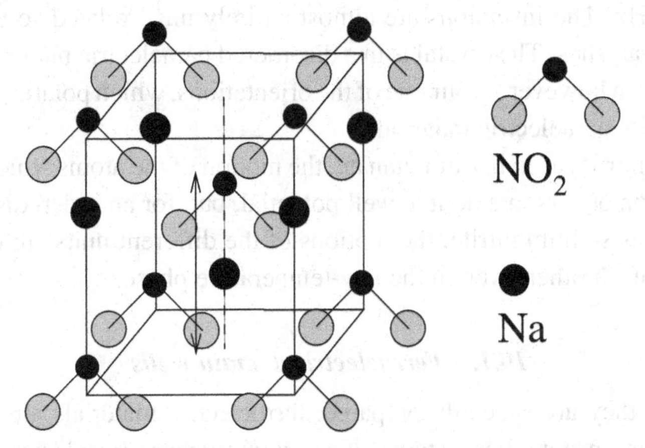

Figure 10.4. Structure of $NaNO_2$ in its ferroelectric phase at low temperature ($T < T_c$).

The ferro–para phase change occurs through a *global and coordinated displacement* of the ions. The ingredients for solitons are therefore present due to the double-well potential in which the titanium ions are sitting, and their coupling which is strong enough to lead to cooperative effects.

10.1.2 Order–disorder ferroelectric: sodium nitrite

The polarisation of sodium nitrite $NaNO_2$ comes from the intrinsic dipole moments of the NO_2^- groups. At low temperature, the dipolar interactions favour a phase where all the NO_2^- groups have the same orientation. Their dipole moments sum up to generate a polarisation in the material, even in the absence of an external field. This is the ordered ferroelectric phase (Figure 10.4). When the temperature rises, the thermal motions grow and, for $T > T_c$, the polarisation of the NO_2^- groups can be reversed by a motion of the nitrogen atom which is able to cross the imaginary line joining the two oxygens. The inversion of the NO_2^- group is similar to the inversion of the NH_3 molecule, better known because it is at the basis of masers. In

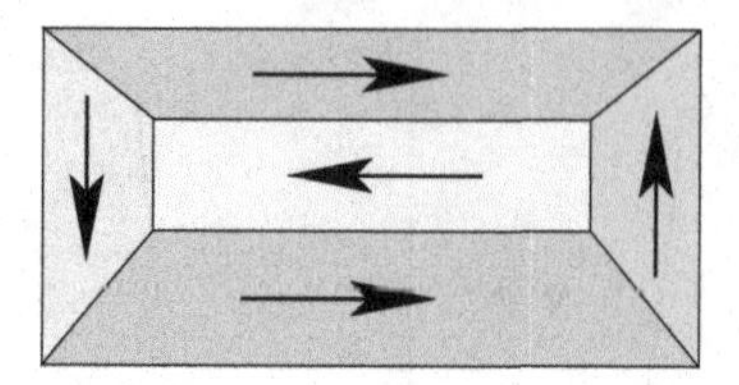

Figure 10.5. Schematic picture of a polydomain ferroelectric sample.

the high-temperature phase, random inversions of the NO_2^- groups occur in sodium nitrite. The lattice of the sodium ions is also slightly modified with respect to the low-temperature phase. The frequency of inversion of NO_2^- groups is very high, around 10^{10} Hz. The inversions are almost entirely uncorrelated so that the total polarisation vanishes. The crystal is in a disordered paraelectric phase. An external electric field can however favour one of the orientations, which polarises the crystal, as is usual for a paraelectric material.

In sodium nitrite, as in barium titanate, the motion of the atoms which determine the polarisation occurs in a double well potential, but, for an order–disorder ferroelectric such as sodium nitrite, the motions of the different units are only weakly correlated to each other, even in the low-temperature phase.

10.1.3 Ferroelectric domain walls

Except when they are specially prepared, ferroelectric materials are seldom in a single polarisation state. For instance, if a barium titanate crystal is cooled from its cubic paraelectric phase, when the tetragonal distortion appears it can take place along any of the axes of the cube, and for each axis in any of the two possible orientations. As a result the material is generally divided into *domains* of uniform polarisation, separated by *domain walls* which are narrow regions in which the polarisation changes from the orientation of one domain to the orientation of the neighbouring domain (Figure 10.5). Their width is typically a few lattice spacings. In a material such as barium titanate in which the symmetry of the paraelectric phase allows multiple directions for the ferroelectric polarisation, some domain walls may separate two domains in which the polarisations are orthogonal to each other. More often, domain walls separate domains which have opposite polarisations.

10.2 A one-dimensional ferroelectric model

It is possible to propose a model for a ferroelectric crystal in the same spirit as the Frenkel–Kontorova model for a dislocation, i.e. a one-dimensional model which only includes the essential features relevant for the problem of interest. This model was proposed quasi-simultaneously by Serge Aubry on the one hand [19] and by

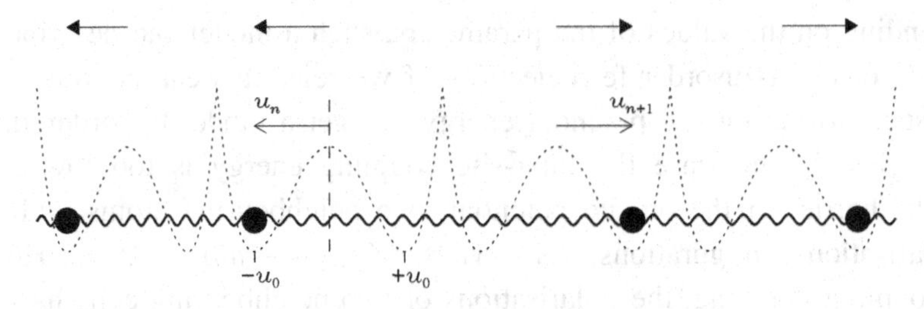

Figure 10.6. Schematic picture of the one-dimensional ferroelectric model.

Jim Krumhansl and Robert Schrieffer [106] on the other. It allows the calculation of the structure of the domain wall and the analysis of some physical properties of the materials, such as their dielectric response.

The model considers a chain of coupled particles, corresponding to the titanium ions in the case of barium titanate, each one being in a double well potential which plays the role of the substrate potential in the FK model. For a ferroelectric crystal the potential represents the effect of the other atoms in the cell. In the simplest version of the model, this substrate potential is assumed to be *fixed*. More elaborate models can include its deformation to take into account the lattice distortions which accompany the motion of the titanium ions. At the simplest level of description which we have chosen, it would be meaningless to try to quantitatively describe the potential in which the titanium is evolving. It is essential to preserve its main feature, the double-well shape, but it is possible to choose an expression which is qualitatively correct while still being simple enough to allow analytical calculations. We have already met such a potential with the ϕ^4 model. For the ferroelectric model we shall write it as

$$V(u) = V_0 \left(1 - \frac{u^2}{u_0^2}\right)^2 . \tag{10.1}$$

The model is plotted in Figure 10.6. It considers nearest-neighbour coupling only and its Hamiltonian is

$$H = \sum_i \frac{1}{2} m \dot{u}_i^2 + \frac{C}{2} (u_{i+1} - u_i)^2 + V_0 \left(1 - \frac{u_i^2}{u_0^2}\right)^2 , \tag{10.2}$$

which includes the kinetic energy term, a harmonic coupling energy and the substrate potential energy. As in the Frenkel–Kontorova model for dislocations, the coupling term involves the *relative* displacement of two neighbouring ions, which stays small, so that a harmonic coupling potential can be chosen. In this expression the variable u_i denotes the displacement of the ions with respect to the middle point of their two equilibrium positions, i.e. the median plane of the barium titanate unit cell for instance.

Depending on the values of the parameters, such a model can describe either displacive or order–disorder ferroelectrics. If we select a weak coupling energy with respect to the on-site potential energy, we get a model for order–disorder ferroelectrics. In this case the inter-site coupling energy is too low to push above the barrier of the on-site potential, two neighbouring atoms with opposite polarisation configurations, $\pm u_0$: $V_0 \gg \frac{1}{2}C(-u_0 - u_0)^2 = 2Cu_0^2$. However, if the coupling is strong, the polarisations of two neighbouring cells have to be very similar and the polarisation evolves slowly in space. A polarisation reversal over one lattice spacing, as in the order–disorder case, is not possible and the physics is dominated by long-wavelength modes for the atomic displacements. This is the case for the displacive ferroelectrics, which we shall investigate more thoroughly.

For a strong coupling, we can use the continuum limit approximation which replaces the set of discrete functions $u_i(t)$ by the two-variable field $u(x,t)$, a transformation which is now familiar because we have met it several times. The Hamiltonian (10.2) is replaced by the continuous integral

$$H = \int \frac{\mathrm{d}x}{\ell}\left[\frac{1}{2}mu_t^2 + \frac{1}{2}mc_0^2u_x^2 + V_0\left(1 - \frac{u^2}{u_0^2}\right)^2\right], \tag{10.3}$$

where ℓ denotes the lattice spacing and $c_0^2 = C\ell^2/m$ is a constant, homogeneous to the square of a velocity. Its physical meaning will appear later. Going from the discrete to the continuous sum is achieved by noting that one term of the discrete sum (10.2) gives the energy of a unit cell. The energy per unit length is thus obtained by dividing one term of the sum by ℓ, and replacing the finite difference by its Taylor expansion at the lowest nonvanishing order. Integrating this energy density with respect to space gives the Hamiltonian (10.3).

10.3 Structure of the domain walls in the continuum limit approximation

The equation of motion of the atoms is readily obtained from Hamiltonian (10.3) as

$$u_{tt} - c_0^2u_{xx} - \omega_0^2u\left(1 - \frac{u^2}{u_0^2}\right) = 0, \tag{10.4}$$

where we define $\omega_0^2 = 4V_0/mu_0^2$, which is the frequency at which particles of mass m would oscillate near one minimum of the substrate potential.

10.3.1 Small-amplitude solutions: phonons

The ground states of the system are obtained when the ions are in the potential minima at $u = \pm u_0$. Let us first consider small oscillations around the bottom of the well. We look for a solution $u = u_0 + \varepsilon\eta$ with $\varepsilon \ll 1$. Keeping only the first order terms in ε, we get

$$\varepsilon\frac{\partial^2\eta}{\partial t^2} - c_0^2\varepsilon\frac{\partial^2\eta}{\partial x^2} - \omega_0^2(u_0+\varepsilon\eta)\left(1-\frac{(u_0+\varepsilon\eta)^2}{u_0^2}\right) = 0$$

i.e.

$$\varepsilon\left[\frac{\partial^2\eta}{\partial t^2} - c_0^2\frac{\partial^2\eta}{\partial x^2} + 2\omega_0^2\eta\right] = 0 + \mathcal{O}(\varepsilon^2). \quad (10.5)$$

The solutions of Equation (10.5) are the plane waves $\eta = \eta_0\, \mathrm{e}^{\mathrm{i}(\omega_q t - qx)} + \text{c.c.}$ having the dispersion relation $\omega_q^2 = 2\omega_0^2 + c_0^2 q^2$. They are the phonon modes of the ferroelectric crystal. This expression gives the meaning of c_0, which is the phase velocity of the phonons in the limit of large wavevectors.

10.3.2 Large-amplitude solutions: the domain walls

Except for some constant factors, Equation (10.4) is the ϕ^4 equation of motion that we introduced in Chapter 2. It is easy to derive its permanent profile solution, as we did for the sine-Gordon equation. We look for u as a function of $z = x - vt$. The equation simplifies into

$$-v^2u_{zz} + c_0^2u_{zz} + \omega_0^2u\left(1-\frac{u^2}{u_0^2}\right) = 0. \quad (10.6)$$

As usual, we multiply it by $\mathrm{d}u/\mathrm{d}z$ and integrate with respect to z to obtain

$$(c_0^2 - v^2)\frac{1}{2}\left(\frac{\mathrm{d}u}{\mathrm{d}z}\right)^2 - \frac{1}{4}\omega_0^2u_0^2\left(1-\frac{u^2}{u_0^2}\right)^2 = K, \quad (10.7)$$

where K is a constant. Far from the domain wall ($|z| \to +\infty$), the polarisation takes one of its two equilibrium values $u = \pm u_0$ with $\mathrm{d}u/\mathrm{d}z = 0$, so that the constant K must vanish. Hence Expression (10.7) can be written as

$$\frac{1}{u_0^2}\left(\frac{\mathrm{d}u}{\mathrm{d}z}\right)^2 = \frac{\omega_0^2}{2(c_0^2 - v^2)}\left(1-\frac{u^2}{u_0^2}\right)^2, \quad (10.8)$$

which shows that the solution must be such that $v^2 < c_0^2$. The solution is finally obtained by a new integration with respect to z,

$$u = \pm u_0 \tanh\frac{\omega_0}{\sqrt{2}c_0\sqrt{1 - v^2/c_0^2}}(z - z_0), \quad (10.9)$$

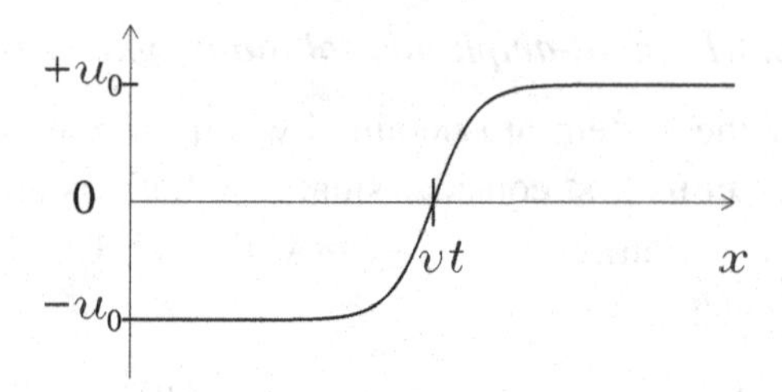

Figure 10.7. Plot of the solution (10.9) (with a + sign) describing a ferroelectric domain wall.

where z_0 is an integration constant which sets the position of the centre of the 'kink' (see Figure 10.7). Apart from the constants, we recover the solution (2.84) that we had given for the ϕ^4 kink, which describes a smooth variation between the two positions $\pm u_0$ in a small region of space. Thus the solution of Equation (10.4) is a wall which separates two domains of opposite polarisation. The energy of the material is the same in the two polarisation states, which are degenerate ground states of the system. Their existence suggests that we should find *topological solitons* in such a system, and this is indeed confirmed by the calculation.

The width L of the domain wall is

$$L = \frac{\sqrt{2}c_0}{\omega_0}\sqrt{1 - \frac{v^2}{c_0^2}}, \tag{10.10}$$

which includes the Lorentz contraction factor with respect to the speed c_0, that we have already met for nonlinear Klein–Gordon equations such as the SG equation.

As this result has been obtained in the continuum limit approximation, it is only valid if the width of the kink at rest, $L_0 = \sqrt{2}c_0/\omega_0 = \ell\sqrt{Cu_0^2/2V_0}$, is much larger than the lattice spacing ℓ, i.e. if the inter-site coupling energy $4Cu_0^2$ is large with respect to the barrier of the on-site substrate potential V_0. It is the strong coupling hypothesis that we made for displacive ferroelectrics, which is not valid for order–disorder ferroelectrics. Actually, even for displacive ferroelectrics such as barium titanate, the domain walls have a width of only a few lattice spacings. Therefore, as for the dislocations, their potential energy depends on their position within the unit cell. They are trapped by a Peierls–Nabarro potential. A uniform electric field, which applies a force on all the atoms, can move them. For instance a field in the negative direction, which tends to move the atoms from the site $+u_0$ toward the site $-u_0$ tends to increase the size of the domain $u = -u_0$, i.e. it tends to move the wall plotted in Figure 10.7 in the positive direction. This agrees with the experimental observation that exposing a ferroelectric to a strong electric field tends to increase the size of the domains which are polarised in the direction of the field. Like the moving dislocations, the ferroelectric domain walls, which are narrow, radiate small-amplitude atomic oscillations when they move. This leads to an effective friction force which tends to oppose their motion.

We calculated the solution corresponding to the constant $K = 0$, which is a single domain wall. Other values of K lead to multikink solutions, describing a lattice of domain walls. As for the sine-Gordon equation, a two-wall solution cannot simply be obtained by summing two single-wall solutions because the equation is nonlinear. However, in practice, as the solution exponentially decays toward the equilibrium values $\pm u_0$ away from its centre, the interaction between two walls decreases very quickly with the distance between their centres. As soon as the walls are separated by more than five to ten times their width, the multi-wall solution is almost identical to a superposition of single walls.

10.3.3 Energy of a domain wall

The energy of a domain wall is an important parameter because the structure of the material is determined by the balance between the wall energy and the energy that can be gained by putting dipoles in antiparallel configurations thanks to the existence of domains.

It can be calculated from Hamiltonian (10.3) by replacing the space and time derivatives by their expressions as a function of u_z, which leads to

$$H = \frac{m}{2\ell} \int \mathrm{d}x \left[(v^2 + c_0^2) u_z^2 + \frac{1}{2} \omega_0^2 u_0^2 \left(1 - \frac{u^2}{u_0^2} \right)^2 \right]. \tag{10.11}$$

Using Expression (10.8), we obtain

$$H = \frac{m}{4\ell} \omega_0^2 u_0^2 \int \mathrm{d}x \left[\frac{v^2 + c_0^2}{c_0^2 - v^2} + \frac{c_0^2 - v^2}{c_0^2 - v^2} \right] \left(1 - \frac{u^2}{u_0^2} \right)^2 \tag{10.12}$$

$$= \frac{m}{2\ell} \omega_0^2 u_0^2 \frac{c_0^2}{c_0^2 - v^2} \int_{-\infty}^{+\infty} \mathrm{d}x \left(1 - \frac{u^2}{u_0^2} \right)^2 . \tag{10.13}$$

If we take into account the soliton solution, H becomes

$$H = \frac{m}{2\ell} \omega_0^2 u_0^2 \frac{c_0^2}{c_0^2 - v^2} \int_{-\infty}^{+\infty} \mathrm{d}z \left(\operatorname{sech}^2 \frac{z - z_0}{L} \right)^2 \tag{10.14}$$

$$= \frac{m}{2\ell} \omega_0^2 u_0^2 \frac{c_0^2}{c_0^2 - v^2} L \underbrace{\left[-\frac{1}{3} \tanh^3 Z + \tanh Z \right]_{-\infty}^{+\infty}}_{=4/3} \tag{10.15}$$

$$= \frac{m}{2\ell} \omega_0^2 u_0^2 \frac{c_0^2}{c_0^2 - v^2} \frac{\sqrt{2} c_0 \sqrt{1 - v^2/c_0^2}}{\omega_0} \frac{4}{3} \tag{10.16}$$

where we introduce $Z = (z - z_0)/L$. The result can be simplified into

$$H = \frac{Mc_0^2}{\sqrt{1 - v^2/c_0^2}} \qquad \text{where} \qquad M = \frac{2\sqrt{2}}{3}\frac{\omega_0 u_0^2}{c_0 \ell} m. \tag{10.17}$$

Therefore, as is usual for solitons, we find that the wall energy is invariant by translation since it does not depend on the position z_0 of the centre.[1] Moreover we recover an expression equivalent to the relativistic energy of a particle of mass M. This confirms the pseudo-particle character of solitons, which must have a velocity lower than c_0 as found in the derivation of the solution. We can also notice that their width L includes a 'relativistic' contraction factor.

Solitons describing ferroelectric domain walls are microscopic solitons which can however still be directly observed. At low temperature the domain walls are well separated from each other. The domains of a ferroelectric crystal can be visualised under a microscope and their motion under the effect of an external electric field can be observed.

10.4 Dielectric response of a ferroelectric material

Ferroelectric materials are used in many devices due to the properties of their dielectric response function, which describes their polarisation under an external electric field. This polarisation comes from atomic displacements in the crystal and, in spite of its simplicity, the one-dimensional model that we introduced can predict the main features of the dielectric response of a ferroelectric and show how solitons are determinant in some properties of the material.

Let us consider the system in some particular microscopic state, as shown for instance in Figure 10.6, and assume that it is excited by an oscillatory external field $E(t)$ at frequency ω (in this one-dimensional model, only the component of the field along the chain is relevant). The field will drive the charged particles into motion but, depending on its frequency, the response of the system may be drastically different. We can distinguish two types of response:

(i) A strong response around the frequency of oscillation of the particles around their equilibrium positions $\pm u_0$. This response is due to the phonons described in Section 10.3.1. Their frequencies are around 10^{13} Hz, corresponding to electromagnetic waves in the far-infrared range, which have a wavelength much larger than the unit cell parameter, so that a field in this frequency

[1] Of course this is only true if discreteness effects can be neglected. Otherwise the energy includes a Peierls–Nabarro term, which depends on the position of the wall within a unit cell. In this case the energy is only invariant by discrete translations representing an integer number of unit cells.

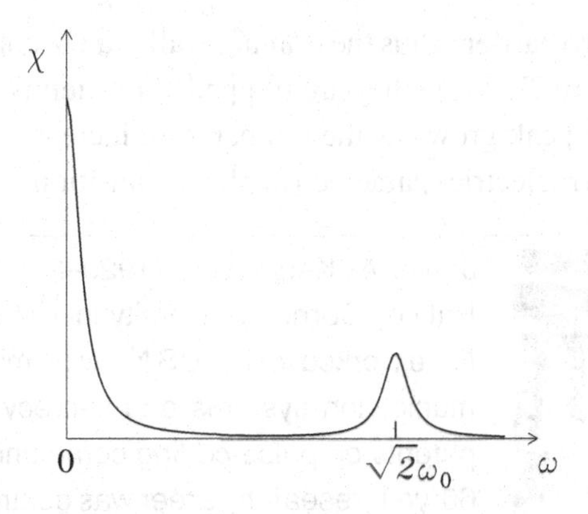

Figure 10.8. Schematic plot of the dielectric response (susceptibility χ) of a ferroelectric material versus the external driving frequency ω.

range only excites the long-wavelength phonons. According to the phonon dispersion curve, their frequency is $\sqrt{2}\omega_0$, explaining the peak at this frequency in Figure 10.8.

(ii) If the system has mobile solitons, we can expect another strong component of the response in the very low frequency range because the field will induce a motion of the solitons. The lower the frequency ω, the higher the amplitude of the displacement of a soliton before its motion reverses its direction when the field changes sign. The large-amplitude motions of the solitons, extending the size of the domains polarised in the direction of the field, may lead to a very large polarisation. This generates a very strong 'central peak' in the dielectric susceptibility in the vicinity of the frequency $\omega = 0$ (Figure 10.8). This central peak, which had been observed experimentally, had been without a satisfactory explanation for a long time. Tentative explanations involved static defects. It is likely that such defects contribute but they cannot explain the shape of the central peak versus ω. Both solitons and antisolitons, corresponding to the plus or minus sign in the solution (10.9), contribute similarly to the central peak. Under an external field, they move in opposite directions, but their motion is always such that it tends to increase the size of the domains which are polarised in the direction of the field.

This fairly simple model for ferroelectric domain walls, seems however to suffer from an internal contradiction: we said that domain walls are trapped by discreteness effects, and now we invoke their mobility to explain the central peak! How can this contradiction be resolved? The solution comes from the effect of temperature which has been completely ignored up to now. Thermal fluctuations can lead to fluctuations in the energy of the solitons which are of the same order of magnitude as the Peierls–Nabarro barrier. Thus the domain walls can become very mobile near the critical temperature T_c, while they are trapped at low temperature. Experiments show that the central peak grows as the temperature increases, and particularly in the vicinity of the ferroelectric–paraelectric phase transition.

JAMES A. KRUMHANSL (1920–2004) was awarded his PhD by Cornell University in 1943. During World War II, he worked for the US Navy on microwave pulse communication systems and secrecy systems, receiving patents on pulse coding communications circuits. His 60-year research career was centred on the properties of solid materials, movements of atoms within solids and other topics of condensed-matter physics, but his research interests also included communication and information systems, applied mathematics, nonlinear science and molecular biological physics. He played a major role in two technical events which turned out to have a strong influence: the development of the first Xerox copy machine and the creation of the first internet network over the United States, which was the first stage of what is now the World Wide Web, which is deeply changing our society. He liked to tell that it was difficult to convince the directors of the Xerox company that the copy machine could have a commercial future until the development team made a copy of a US dollar bill, using a green ink that had exactly the colour of the dollar: the directors were convinced that the copy machine could actually 'make money'! And it did. (Faculty Biography Collection. Courtesy of the Division of Rare and Manuscript Collections, Cornell University Library.)

In the solitons' community, he is particularly known for his work on martensitic materials and for the development of the transfer integral formalism for nonlinear field equations. This later work was done in close collaboration with J. R. Schrieffer (1931–), 1972 Nobel prize winner for his contribution to the BCS microscopic theory of superconductivity. He has been a tireless advocate of the role of nonlinear excitations in physics.

In parallel to his scientific activities, he had very important and broad responsibilities in the science administration. He was president of the American Physical Society, served on the board of several scientific journals and was an assistant director of the National Science Foundation. But he became most widely recognised for his testimony before Congress in 1987, when he questioned the financing of a superconducting supercollider.

10.5 Thermodynamics of a nonlinear system

10.5.1 The correlation function

When we wrote Equation (10.4) to derive the soliton solution, we treated the system as if it were isolated, while it is actually in contact with a thermal bath at temperature T. The equation that we solved did not take temperature into account, and this is why we can say that we studied the system at zero temperature.

To get results that can be compared with experimental data, it is essential to take into account the thermal bath which leads to fluctuations in the atomic motions. The response to an external field which is measured in a macroscopic experiment, for instance during the recording of a Raman spectrum, is not determined by the instantaneous positions of the atoms, but by *statistical averages* of these positions.

Contrary to the case of the macroscopic systems that we studied in Part I, at the microscopic level, in a solid or a macromolecule, thermal fluctuations involve energies which are of the same order of magnitude as the energies associated with the solitons or their dynamics. Their role cannot be ignored. For instance thermal fluctuations may lead to a motion of the solitons, moving them forward or backward when a favourable fluctuation helps a particle to overcome a potential barrier. This explains qualitatively why domain walls, trapped by discreteness effects at low temperature may become highly mobile at high temperature. In the presence of thermal fluctuations, the solitons have a diffusive motion very similar to the motion of a Brownian particle, under the effect of the fluctuations of the particles of the lattice.

Linear response theory shows that the dielectric response of a system to an external field can be calculated as a function of the displacement–displacement correlation function

$$C_{uu}(x, t, x_0, t_0) = \langle u(x_0, t_0)\, u(x_0 + x, t_0 + t)\rangle, \tag{10.18}$$

where we denote by $\langle . \rangle$ the statistical average over a large number of identical systems, at the same temperature T, but with different fluctuations. For a homogeneous system which is invariant to space and time translations, the statistical average does not depend on x_0 and t_0, but only on x and t. Thus it will be denoted by $C_{uu}(x, t)$.

The dielectric response, as well as neutron or light scattering data, can be expressed as a function of the dynamic structure factor $S(q, \omega)$, which is the Fourier transform of the correlation function

$$S(q, \omega) = \int_{-\infty}^{+\infty} \mathrm{d}x \int_{-\infty}^{+\infty} \mathrm{d}t\, C_{uu}(x, t)\, \mathrm{e}^{\mathrm{i}(\omega t - qx)} \tag{10.19}$$

$$= \int_{-\infty}^{+\infty} \mathrm{d}x \int_{-\infty}^{+\infty} \mathrm{d}t\, \langle u(0, 0)\, u(x, t)\rangle \mathrm{e}^{\mathrm{i}(\omega t - qx)}, \tag{10.20}$$

where we set $t_0 = 0$ and $x_0 = 0$ for convenience since the correlation function does not depend on them. Therefore the correlation function (or its Fourier transform) contains the core of the information that we want for a ferroelectric crystal in contact with a thermal bath at temperature T. The problem is now to compute $\langle u(0,0)\, u(x,t)\rangle$. We shall examine several approaches.

10.5.2 The soliton gas model

The first method is only approximate, but nevertheless very powerful. It is based on the existence of the solitons. When we discussed the dielectric response from a qualitative point of view, we noticed that the fluctuations, which enter into the correlation function $\langle u(0,0)\, u(x,t)\rangle$ have two possible origins:

(i) The small vibrations around the equilibrium positions $\pm u_0$.
(ii) The motion of the solitons.

Contribution (i) can be easily computed by assuming that the vibrations are small enough to allow a harmonic approximation.

The calculation of contribution (ii) seems much more difficult, but we can take advantage of the exceptional stability of solitons, which can be treated as 'quasi-particles' forming a 'soliton gas' whose thermodynamic properties can be determined because we know how to compute the energy of the solitons as a function of their velocities. As the gas is one-dimensional, we must take into account the topological constraint mentioned in Section 2.4.1, which comes from the shape of the potential having two minima only, so that a soliton must necessarily be followed by an antisoliton. They cannot pass through each other but can only be reflected in collisions. We must consider a gas of particles which can be individually identified, i.e. a classical gas. The average density of solitons at temperature T is obtained from the Boltzmann factor $\exp(-E/k_B T)$. The energy–velocity relation (10.17) can then be used to derive the velocity distribution at a given temperature.

Let us show that these results are sufficient to allow the calculation of the correlation functions by examining the case of the product $u_k u_p$ of two displacements at the same time t_0. This approach does not take into account the exact shape of the solitons. They are assumed to be so narrow that the particles only occupy the minima $u = \pm u_0$. This is of course only an approximation, but Figure 10.9, drawn for a realistic case, shows that it is rather well verified. This approximation is especially good at low temperature, for a low soliton density, because then there are few particles in the soliton core, where they are in an intermediate position between one potential minimum and the next. Therefore the only possible values for the product $u_k u_p$ are:

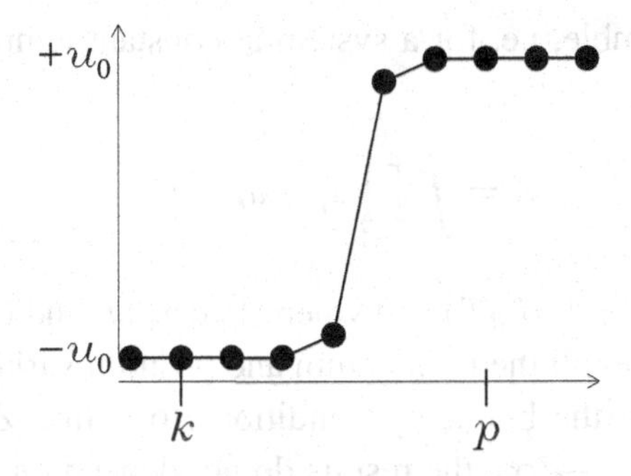

Figure 10.9. Position of the particles in a configuration having one soliton between sites k and p.

- u_0^2 if there are no solitons between sites k and p.
- $-u_0^2$ if there is a single soliton between sites k and p.
- u_0^2 if there are two solitons etc.

As it is possible to calculate the probability of having a soliton along the length L which separates sites k and p, from the energy of a soliton and the temperature, it is possible to assign a probability to each of the values listed above so that the average $\langle u_k u_p \rangle$ can be determined. A similar reasoning can be used to calculate the correlation function $\langle u_k(0) u_p(t) \rangle$ from the correlation $\langle u_k(0) u_p(0) \rangle$. The idea is that the value of $u_p(t)$ is equal to $u_p(0)$ unless a soliton has passed at site p during the time interval t. If this happens $u_p(t)$ has been switched to $-u_p(0)$. To get $u_p(t)$ it is thus sufficient to calculate the probability that a soliton passes over a given site during the time interval t, which can be done from the soliton density and average speed.

These examples show how the correlation functions can be deduced from the picture of a soliton gas. They also point out the power of the soliton concept to completely renew the approach to the statistical physics of a nonlinear system.

10.5.3 The transfer integral method

For a one-dimensional system with short-range interactions, such as the ferroelectric model that we are studying here, it is possible to determine the thermodynamic properties with a method which does not involve the approximations made in the soliton gas calculation, and which is elegant and very general. It allows the calculation of the partition function of the system, from which all thermodynamic properties can be derived. For the ϕ^4 model applied to ferroelectrics it has been developed by Aubry [19] and Krumhansl and Schrieffer [106].

Let us consider a chain of N particles. Its Hamiltonian H is a function of the $2N$ variables (p_n, u_n), p_n being the canonical momentum conjugate to the position u_n.

In the canonical ensemble, i.e. for a system at constant temperature, its *partition function* is given by

$$Z = \int \prod_{n=1}^{N} \mathrm{d}p_n \mathrm{d}u_n \, \mathrm{e}^{-\beta H}, \tag{10.21}$$

where we introduce $\beta = 1/(k_B T)$ and where $\prod \mathrm{d}p_n \mathrm{d}u_n$ indicates that the integral must be carried out over all the momentum and position variables of the chain.

We need to specify the boundary conditions to define Z completely. In the thermodynamic limit $N \to \infty$, the results do not depend on these boundary conditions, so that we can select those which are the most convenient for the calculation. We shall impose periodic boundary conditions by assuming that the displacement of particle $N + 1$ is identical to that of particle 1. In the calculation this is achieved by introducing the Dirac distribution $\delta(u_{N+1} - u_1)$ in the partition function via $\int \mathrm{d}u_{N+1} \, \delta(u_{N+1} - u_1) = 1$. Taking into account the expression (10.2) of the Hamiltonian, the partition function becomes

$$Z = \int \prod_{n=1}^{N} \mathrm{d}p_n \mathrm{e}^{-\beta p_n^2/2m} \int \prod_{n=1}^{N+1} \mathrm{d}u_n \, \mathrm{e}^{-\beta \sum_{n=1}^{N} [C/2(u_{n+1}-u_n)^2 + V(u_n)]} \times \delta(u_{N+1} - u_1). \tag{10.22}$$

The kinetic part is readily integrated and gives $(2\pi m k_B T)^{N/2}$ while the multiple integral over the variable u_n, which is the configurational part that we shall denote by Z_u, seems very hard to calculate. Actually, thanks to the one-dimensional character of the model and to a coupling restricted to nearest neighbours, the calculation is possible through the introduction of an auxiliary operator called the *transfer operator*.

If we define $R(u_n, u_{n+1}) = C(u_{n+1} - u_n)^2/2 + V(u_n)$, the configurational part Z_u of (10.22) becomes

$$Z_u = \int du_{N+1} \prod_{n=1}^{N} \mathrm{d}u_n \, \mathrm{e}^{-\beta R(u_n, u_{n+1})} \, \delta(u_{N+1} - u_1). \tag{10.23}$$

The transfer operator method introduces an operator $T(y, y')$, acting on a space of functions $f(y)$, defined by

$$T(y, y') f(y') = \int_{-\infty}^{+\infty} \mathrm{d}y' \, \mathrm{e}^{-\beta R(y,y')} \, f(y') = \int_{-\infty}^{+\infty} \mathrm{d}y' \, K(y, y') \, f(y'), \tag{10.24}$$

where $K(y, y') = \mathrm{e}^{-\beta R(y,y')}$ is the kernel of the integral operator. The method can be viewed as a generalisation of the transfer matrix method used to solve the

Ising model, which is familiar in statistical physics courses. The action of the operator on a function of the variable y' is another function, depending on the variable y.

We can then define the eigenfunctions $\varphi_q(y)$ of the transfer operator, associated with the eigenvalues λ_q, by

$$T(y, y')\,\varphi_q(y') = \int_{-\infty}^{+\infty} \mathrm{d}y'\; K(y, y')\,\varphi_q(y') = \lambda_q \varphi_q(y). \qquad (10.25)$$

The set of the eigenfunctions of the transfer operator defines an orthonormal basis for the functional space,[2] which can be used to expand $\delta(u_{N+1} - u_1)$ as

$$\delta(u_{N+1} - u_1) = \sum_q \varphi_q^*(u_{N+1})\varphi_q(u_1), \qquad (10.26)$$

where we introduce a simplified notation for the sum over the index q which generally involves both discrete eigenvalues and a continuum.

The definition of the transfer operator and the Expansion (10.26) allow us to rewrite the configurational part Z_u of the partition function as

$$Z_u = \int \mathrm{d}u_{N+1} \cdots \mathrm{d}u_1 \left[\sum_q \varphi_q^*(u_{N+1})\varphi_q(u_1)\right] K(u_{N+1}, u_N) \cdots K(u_2, u_1). \qquad (10.27)$$

[2] Strictly speaking this is only true for symmetric operators $T(y, y') = T(y', y)$. The operator that we have defined is not symmetric because $R(y, y') = C/2(y' - y)^2 + V(y)$ is not symmetric with respect to the change of y into y'. If we want to keep the asymmetrical operator, which is convenient for some of the calculations, we should define two types of eigenfunctions, the 'right' eigenfunctions defined by $T(y, y')\,\varphi_q(y') = \lambda_q \varphi_q(y)$ and the 'left' functions defined by $T(y, y')\,\varphi_q^*(y) = \lambda_q \varphi_q^*(y')$. The orthogonality condition of the functions is $\int \varphi_q^*(y)\varphi_{q'}(y)\mathrm{d}y = \delta(q - q')$.

To perform numerical calculations with the transfer operator, it is however, convenient to define a symmetrical operator because it becomes possible to use programs which deal with symmetrical matrices after the problem has been put in a discrete form by discretising space. The operator can be symmetrised by introducing $R_1(y, y') = C/2(y' - y)^2 + 1/2V(y) + 1/2V(y')$. This symmetrised form is introduced into the equation that defines the eigenfunctions of the operator by

$$\int \mathrm{d}y' \exp[-\beta R(y, y')]\varphi_q(y') = \int \mathrm{d}y' \exp[-\beta R_1(y, y')]\; \exp[-\beta V(y)/2] \exp[+\beta V(y')/2]\; \varphi_q(y') = \lambda_q \varphi_q(y).$$

Multiplying this equation by $\exp[+\beta V(y)/2]$, we get a new eigenvalue equation for a symmetric transfer operator

$$\int \mathrm{d}y' \exp[-\beta R_1(y, y')]\; \phi_q(y') = \lambda_q \phi_q(y)$$

which has a new set of eigenfunctions $\phi_q(y) = \exp[+\beta V(y)/2]\; \varphi_q(y)$, but the same eigenvalues as the original operator.

If we integrate first with respect to u_1, we get

$$Z_u = \sum_q \int \mathrm{d}u_{N+1} \cdots \mathrm{d}u_2\, \varphi_q^*(u_{N+1}) K(u_{N+1}, u_N) \cdots$$

$$K(u_3, u_2) \underbrace{\int \mathrm{d}u_1 K(u_2, u_1) \varphi_q(u_1)}_{=\lambda_q \varphi_q(u_2)} \tag{10.28}$$

$$= \sum_q \lambda_q \int \mathrm{d}u_{N+1} \cdots \mathrm{d}u_2 \varphi_q^*(u_{N+1}) K(u_{N+1}, u_N) \cdots K(u_3, u_2)\, \varphi_q(u_2). \tag{10.29}$$

With a sequence of integrations with respect to u_2, then $u_3 \cdots$ up to u_N we proceed along the indices of the chain (hence the name 'transfer integral') and we get

$$Z_u = \sum_q \lambda_q^N \int \mathrm{d}u_{N+1}\, \varphi_q^*(u_{N+1}) \varphi_q(u_{N+1}) \tag{10.30}$$

$$= \sum_q \lambda_q^N, \tag{10.31}$$

if we use the normalisation condition of the eigenfunctions $\int \varphi_q^*(y) \varphi_q(y) \mathrm{d}y = 1$.

In the thermodynamic limit, i.e. for $N \to \infty$, the sum is dominated by the largest eigenvalue λ_0 of the transfer operator. Putting together the kinetic part and the configurational part, we get the partition function of the model

$$Z = (\sqrt{2\pi m k_B T})^N \lambda_0^N \tag{10.32}$$

which yields its *free energy*

$$F = -k_B T \ln Z \tag{10.33}$$

$$= -\frac{N k_B T}{2} \ln (2\pi m k_B T) + N \varepsilon_0, \tag{10.34}$$

using the notation $\lambda_q = \mathrm{e}^{-\beta \varepsilon_q}$ for the eigenvalues of the transfer operator.

A similar calculation can be performed to calculate the mean value $\langle u \rangle$, which is the *order parameter* for the ferroelectric–paraelectric transition; it is finite in the ferroelectric phase and vanishes in the paraelectric phase. Due to the periodic boundary conditions, all sites of the lattice are equivalent so that

$$\langle u \rangle = \langle u_n \rangle = \langle u_{N+1} \rangle \tag{10.35}$$

$$= \frac{1}{Z} \int \prod_{n=1}^{N} \mathrm{d}p_n \mathrm{d}u_n\, u_{N+1}\, \mathrm{e}^{-\beta H}. \tag{10.36}$$

In this expression, the integrals over the momenta p_n appear both in the numerator and within Z in the denominator and can be simplified. The calculation of the configurational part of the numerator is exactly identical to the one that we made for Z_u, except for the extra factor u_{N+1}. It can thus be performed similarly and gives

$$\langle u \rangle = \frac{1}{Z_u} \sum_q \lambda_q^N \int \mathrm{d}u_{N+1}\, \varphi_q^*(u_{N+1}) \varphi_q(u_{N+1})\, u_{N+1} \qquad (10.37)$$

$$= \frac{\sum_q \lambda_q^N \int_{-\infty}^{+\infty} \mathrm{d}u\, |\varphi_q(u)|^2\, u}{\sum_q \lambda_q^N}. \qquad (10.38)$$

In the thermodynamic limit, the sums are dominated by the term containing the largest eigenvalue λ_0, so that only

$$\langle u \rangle = \int_{-\infty}^{+\infty} \mathrm{d}u\, \varphi_0^*(u)\, \varphi_0(u)\, u \qquad (10.39)$$

remains. This expression gives the meaning of the eigenfunction of the transfer operator which is associated with the largest eigenvalue because it shows that $\varphi_0^*(u)\varphi_0(u)$ is the statistical weight that should be applied to each value of u to compute the average $\langle u \rangle$.

The same method can be used to calculate the static correlation functions $\langle u_k u_p \rangle$. Their analytical expression contains a sum which involves all the eigenvalues of the transfer operator, even in the thermodynamic limit. In practice, as it is seldom possible to compute the full spectrum of the transfer operator, the analytical calculation cannot be carried to the end, and only approximate expressions can be obtained. This highlights the appeal of the soliton gas approach, which relies on knowledge of the nonlinear excitations of the system.

10.5.4 Determination of the spectrum of the transfer operator

In spite of the limitations that we just mentioned, the transfer operator method is popular because it leads to exact results for the thermodynamics of the one-dimensional model, provided that we can compute the largest eigenvalue of the transfer operator (or the smallest value of ε_q if we use the notation $\lambda_q = \exp[-\beta\varepsilon_q]$) and the corresponding eigenfunction φ_0. This calculation can always be carried out numerically, but analytically it is only possible in the strong coupling case, i.e. in the continuum limit approximation.

The goal is to solve the eigenvalue equation (10.25), which becomes

$$I = \int_{-\infty}^{+\infty} \mathrm{d}y' \, \varphi(y') \, \mathrm{e}^{-\beta[C/2(y'-y)^2 + V(y)]} = \mathrm{e}^{-\beta\varepsilon} \, \varphi(y), \tag{10.40}$$

when the kernel has been replaced by its expression. In this equation we omit the index q of the eigenfunction. When the coupling is strong, the term $C(y'-y)^2$ grows quickly as soon as $y' \neq y$. As it appears in the exponential with a minus sign, it implies that the contribution to the integral of the values of y' which are far from y is very small. We can use this feature and perform a Taylor expansion of $\varphi(y')$ around the value $y' = y$. It leads to

$$I = \mathrm{e}^{-\beta V(y)} \int_{-\infty}^{+\infty} \mathrm{d}y' \left[\sum_{n=0}^{+\infty} \frac{\varphi^{(n)}(y)}{n!} (y'-y)^n \right] \mathrm{e}^{-\beta C/2(y'-y)^2} \tag{10.41}$$

$$= \mathrm{e}^{-\beta V(y)} \sum_{n=0}^{+\infty} \frac{\varphi^{(n)}(y)}{n!} \int_{-\infty}^{+\infty} \mathrm{d}y' \, (y'-y)^n \, \mathrm{e}^{-\beta C/2(y'-y)^2} \tag{10.42}$$

$$= \mathrm{e}^{-\beta V(y)} \sum_{n=0}^{+\infty} \frac{\varphi^{(2n)}(y)}{(2n)!} \int_{-\infty}^{+\infty} \mathrm{d}y' \, (y'-y)^{2n} \, \mathrm{e}^{-\beta C/2(y'-y)^2} \tag{10.43}$$

because the integrals of the odd functions vanish. Computing the Gaussian integrals, we get

$$I = \mathrm{e}^{-\beta V(y)} \sum_{n=0}^{+\infty} \frac{\varphi^{(2n)}(y)}{(2n)!} \sqrt{\frac{2\pi}{\beta C}} \frac{(2n-1)(2n-3)\cdots 3 \cdot 1}{(\beta C)^n} \tag{10.44}$$

$$= \mathrm{e}^{-\beta V(y)} \sqrt{\frac{2\pi}{\beta C}} \sum_{n=0}^{+\infty} \frac{\varphi^{(2n)}(y)}{(2n)(2n-2)\cdots 4 \cdot 2} \frac{1}{(\beta C)^n} \tag{10.45}$$

$$= \mathrm{e}^{-\beta V(y)} \sqrt{\frac{2\pi}{\beta C}} \sum_{n=0}^{+\infty} \frac{1}{n!(2\beta C)^n} \frac{\mathrm{d}^{2n}\varphi(y)}{\mathrm{d}y^{2n}}. \tag{10.46}$$

The sum is an operator acting on $\varphi(y)$ which turns out to be the expansion of the exponential of an operator, so that we obtain

$$I = \mathrm{e}^{-\beta V(y)} \sqrt{\frac{2\pi}{\beta C}} \, \mathrm{e}^{1/2\beta C \, \mathrm{d}^2/\mathrm{d}y^2} \, f(y). \tag{10.47}$$

If we define s_0 such that $\exp(-\beta s_0) = \sqrt{2\pi/(\beta C)}$ and if we combine Equations (10.40) and (10.47), we obtain the equation

$$I = \mathrm{e}^{-\beta[-1/2\beta^2 C \, \mathrm{d}^2/\mathrm{d}y^2 + s_0 + V(y)]} \varphi(y) \simeq \mathrm{e}^{-\beta\varepsilon} \, \varphi(y), \tag{10.48}$$

where the commutator of the operators that we combine in the exponential has been neglected.[3] The eigenfunction $\varphi(y)$ is thus a solution of the equation

$$-\frac{1}{2\beta^2 C}\frac{\mathrm{d}^2 f(y)}{\mathrm{d}y^2} + V(y)f(y) = (\varepsilon - s_0)\, f(y), \tag{10.49}$$

which is formally identical to the Schrödinger equation for a particle in the substrate potential $V(u)$, with a pre-factor of the spatial derivative equal to $1/(\beta^2 C)$ instead of the usual $\hbar^2/m$.

Thus, in the strong coupling case, where the continuum limit approximation is valid, the derivation of the thermodynamic properties of the one-dimensional ferroelectric model is formally equivalent to looking for the ground state of a quantum particle having mass $m^* = \hbar^2 C/[(k_B T)^2]$ and subjected to a double-well potential.

Although the exact solution of this problem is not known for the nonlinear potential $V(y)$ of the ferroelectric model, it is easy to make some predictions about the results, with a basic knowledge of quantum mechanics.

Without any calculation, we can make some qualitative predictions. At low temperature, the effective mass m^* of the particle is very high. Thus we expect its ground state to be very close to the minimum of the potential $V(y)$, i.e. much lower than the potential barrier V_0 (Figure 10.10). In this case the 'wave function' $\varphi_0(y)$ will have maxima in the vicinity of the potential minima, and will be negligible in the region $y \simeq 0$. But we have shown that $|\varphi_0|^2$ gives the weight of each value of u in the calculation of $\langle u \rangle$. This indicates that, at low temperature, only the values of u in the vicinity of $\pm u_0$ will contribute significantly to the statistical average. This is in agreement with our expectations for the ferroelectric phase, where the material has a dielectric polarisation which can be either in one direction or in the other.

Conversely, at high temperature, the effective mass m^* is now very small. We can expect the energy level of the particle to be much higher than the barrier V_0. The corresponding eigenfunction will still show some slight maxima around the values $y = \pm u_0$ but it will also be large for all the y values in the range between points A and B in Figure 10.10, and in particular around $u = 0$. The statistical meaning of $|\varphi_0|^2$ tells us that, in many unit cells, the material will be in a state where the atoms are not near the minima of the potential. Taking into account the possible excitations of the system, which are the phonons and the solitons, the transfer integral calculation predicts a high thermal motion of the particles, but also the creation of many domain walls when the material is at high temperature.

[3] When two operators C and D commute with their commutator $[C, D]$, the equality $\mathrm{e}^C\, \mathrm{e}^D = \mathrm{e}^{C+D}\, \mathrm{e}^{[C,D]/2}$ holds.

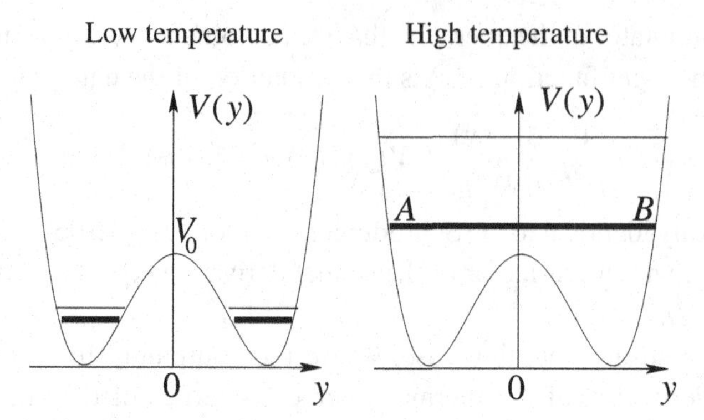

Figure 10.10. Qualitative behaviour of the energy levels of the pseudo-Schrödinger problem associated with the determination of the eigenvalues of the transfer operator of the one-dimensional ferroelectric model, at low and high temperatures. The thick lines indicate the energy level which enters into the expression of the partition function.

However the calculation of the partition function does not exhibit any qualitative change of behaviour at a particular critical temperature. It does not predict the ferroelectric–paraelectric phase transition but only a gradual decrease of the permanent polarisation as temperature increases. This is not surprising because we should not expect this one-dimensional model with short-range interactions to have a phase transition[4] according to standard results in statistical physics [109].

Although the exact calculation of the energy levels of the particle in the potential $V(y)$ cannot be made, it is nevertheless possible to obtain *quantitative results* for the free energy of the model, using some approximation methods of quantum mechanics, especially in the low temperature range [106]. In the temperature range of the ferroelectric phase, the ground state energy can be calculated in two steps. As we predict that it will be very close to the minimum of the potential because m^* is very large, a first approximation of the ground state $E_0 = \varepsilon_0 - s_0$ is obtained from a harmonic expansion of $V(y)$ around one of its minima. Moreover we know that, in such a double-well potential, the ground state is split into two levels which are close to each other due to tunnelling. Indeed if the barrier V_0 were infinitely high, the two wells would be completely separated and we would have two identical states for a particle situated in one well or the other. For a finite value of V_0, a quantum particle can tunnel from one well to the other, inducing a coupling between the wells, which lifts the degeneracy of the two states [41]. The single energy level E_0 is replaced by a pair of levels $E_0 \pm t_0$, where $2t_0$ is the 'tunnel splitting', which

[4] We shall see, however, in Chapter 16 that another one-dimensional system which can be studied within the same formalism does show a phase transition.

can be calculated by the WKB approximation of quantum mechanics [106]. Only the lowest level enters into the expression of the partition function and of the free energy (10.33), with $\varepsilon_0 = s_0 + E_0 - t_0$.

Thus the approximation methods of quantum mechanics can provide a quantitative value for the free energy of the one-dimensional ferroelectric model, to a good accuracy. However, the transfer integral method gives a *value* for F but it does not give its *physical meaning*. It does not tell us which properties of the system are responsible for a given contribution to the free energy. A comparison between the transfer integral results and those given by other methods can answer this question. An alternative method to determine the thermodynamic properties of a system is to search for all possible excitations of the system, and to evaluate their contribution to the free energy. We have seen that the ferroelectric model has two classes of excitations, phonons and solitons. The contribution of the phonons can be calculated in the harmonic approximation, assuming that the particles stay near the minimum of the potential and expanding it to second order. This calculation shows that the contribution of the phonons corresponds to the E_0 term given by the transfer operator [106]. This is not surprising since E_0 was indeed obtained from a harmonic approximation of the potential $V(y)$. Calculating the soliton contribution to the thermodynamics seems more delicate, but the 'soliton gas' description gives the clue. The soliton contribution can be evaluated as the free energy of an ideal gas of particles for which we know the relation between energy and velocity. The result shows that it gives the tunnel contribution t_0 of the transfer operator method [106]. Here too this is understandable a posteriori: tunnel splitting occurs because the potential has two energetically degenerate minima, and it is also the existence of these minima which allows the existence of solitons in the system.

10.5.5 Conclusion

The transfer integral method is a very general method in statistical physics, which can be used for any lattice having a nonlinear on-site potential and a harmonic coupling between nearest neighbours. In the continuum limit, it reduces the study of the thermodynamics of a nonlinear lattice to the solution of a fictitious quantum problem for a particle in the substrate potential. If we know how to solve it we get an exact result for the nonlinear system in the continuum limit. Even if the quantum problem cannot be solved exactly, the approximation methods of quantum mechanics can give satisfactory results.

Another example of the transfer integral method will be examined in Chapter 16. However, the drawback of this method is that it does not naturally lead to a physical understanding of the quantitative results that it provides. It is useful to complete it by other approaches. For a system having soliton solutions, we can take advantage of the quasi-particle properties of the solitons. This is a fruitful point of view because, beyond the understanding of the transfer integral results, it can also be used to derive results on the dynamics of the system, such as the calculation of the dynamic correlation functions allowing the analysis of spectroscopic or neutron diffraction experimental data, which is not provided by thermodynamics.

11
Incommensurate phases

11.1 Examples in solid state physics

Incommensurate phases exist in systems where several length scales compete, leading to *frustration*.

Mono-atomic layers adsorbed on crystals, such as krypton on graphite, provide a simple example which has been extensively studied experimentally. The structure of the adsorbate depends on the ratio between the lattice spacing of graphite (denoted by b) and the equilibrium distance a that the krypton atoms would select if they were not constrained by their interaction with the graphite surface. The value of a depends on temperature and on the pressure in the krypton gas phase, which determines the average density of atoms adsorbed on the surface.

At low temperature and krypton pressure, the atoms sit in the potential minima created by the regular graphite lattice, forming 'commensurate' phases which are pinned by the lattice. If the krypton pressure increases, the krypton atoms adsorbed on the surface get closer to each other and organise themselves in a phase which has a lattice spacing which is not an integer multiple of the graphite period. Such a phase is called an 'incommensurate' phase. Varying temperature and pressure, it is possible to observe a large variety of phases, making complex bi-dimensional structures.

This example, which is the simplest to describe and understand, is the system that we shall study in this chapter. However the ideas which emerge from this study are also valid for numerous physical systems in which two length scales compete:

- Two-dimensional structures, analogous to those formed by adsorbed gases, can also appear when some metals like caesium are intercalated within the graphite layers.
- Three-dimensional incommensurate structures are made by the mercury atoms in $Hg_{3-\delta}AsF_6$. They make up a three-dimensional lattice which is incommensurate with the tetragonal lattice of AsF_6.

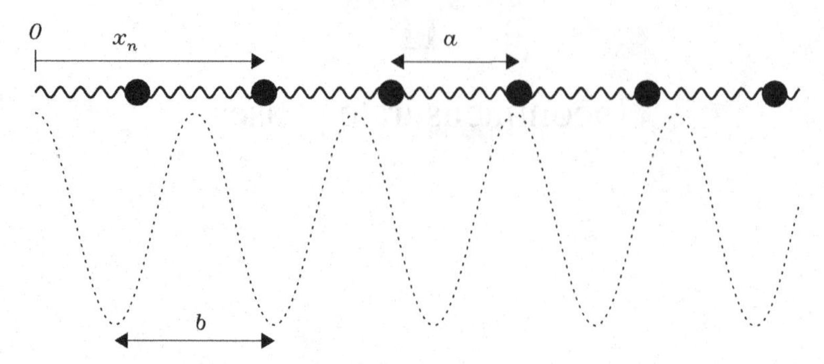

Figure 11.1. Schematic plot of the Frenkel–Kontorova model for an incommensurate phase: the substrate potential has the period b, and the equilibrium length of the bond connecting the adsorbed atoms is a.

- In some magnetic compounds containing rare earth ions, and in some ferroelectrics ($NaNO_2$, $RbLiSO_4$, $(NH_4)_2BeF_4$), the spin–spin or dipole–dipole interactions lead to an expression of the free energy which contains first and second derivatives of the magnetisation or polarisation. As a result the system can minimise its free energy by selecting a modulated structure for the magnetisation or polarisation. Generally the modulation wavelength depends on temperature, and it is not commensurate with the period of the underlying crystal lattice.
- In charge-density wave conductors, the electronic density is modulated with a period incommensurate with the crystal lattice.

11.2 A one-dimensional model for incommensurate phases: the Frenkel–Kontorova model

The Frenkel–Kontorova model (Figure 11.1) that we introduced as a dislocation model in Chapter 9 can also be adapted to describe incommensurate phases if we take into account the existence of two different periods in the system. Its general structure, with a lattice of coupled atoms subjected to a substrate potential, is perfectly suitable to describe a mono-atomic layer adsorbed on a crystal.

The sinusoidal substrate potential describes the interaction energy between the adsorbed atoms and the crystal. It has the period b of the crystal lattice. The potential describing the interaction between the atoms can be chosen to be harmonic because we assume that the distance between them does not deviate very much from its equilibrium value a. For dislocations we had $a = b$, but here the ratio a/b can take any value, and it is not necessarily an integer number.

In such a system, two terms of the energy compete: the interaction energy between the atoms is minimised by putting the particles at distance a from each other, while the substrate potential energy is minimal if the particles are at distance b from each other (or a multiple of b) so that they can sit in the minima of the potential. We cannot say a priori where the equilibrium positions of the atoms will be located.

Therefore, contrary to the dislocation case, we cannot choose the deviation from these equilibrium positions as the appropriate coordinates of the atoms. Instead we choose an origin O, fixed with respect to the substrate potential, and the position of atom n is defined by its coordinate x_n with respect to this origin (Figure 11.1).

Initially, we shall only study the equilibrium structures of this system, because they already show a large variety of phenomena, even if the dynamics are not considered. Thus, instead of starting from the Hamiltonian as we did for dislocations, we can consider the potential energy of the model only. It is given by

$$U = \sum_n \frac{\lambda}{2}(x_{n+1} - x_n - a)^2 + V(x_n), \tag{11.1}$$

if we denote the harmonic coupling constant by λ. For equilibrium states, the kinetic energy is, of course, zero. It is often convenient to rewrite this expression in terms of $u_n = x_n - na$ which measures the deviation with respect to the state that the harmonic interaction potential would impose if it were acting alone. It leads to

$$U = \sum_n \frac{\lambda}{2}(u_{n+1} - u_n)^2 + V(u_n + na). \tag{11.2}$$

The equilibrium structure of the system is obtained by minimising its potential energy, which yields a set of coupled nonlinear equations,

$$\lambda(u_{n+1} + u_{n-1} - 2u_n) - V'(u_n + na) = 0. \tag{11.3}$$

11.3 Commensurate phases

These are the simplest phases and their structure is easy to obtain. If the two distances a and b are equal, or if a is equal to Mb (M being an integer), there is an obvious solution which minimises the energy by putting all the particles exactly at the bottom of substrate potential wells. The lattice of particles is organised into a phase which has a period commensurate with the period of the substrate potential.

This is still true if the ratio between a and b is a rational number, $a = (M/N)\,b$, because, in this case, the system has the period $Na = Mb$. The structure of the phase can be a bit more difficult to determine if the two integers M and N are large, but it can nevertheless be calculated exactly. As the system is periodic, with period Mb, we have

$$x_{n+N} = x_n + Mb = x_n + Na \quad \text{i.e.} \quad u_n = u_{n+N}, \tag{11.4}$$

which is valid for all n. This implies that the structure is fully known if we calculate N unknowns from (11.3), which can be written for $n = 1, \ldots, N$ for instance. The first equation includes $u_0 = u_N$ and the last one includes $u_{N+1} = u_1$, so that the system of equations is closed (and finite), and thus it can in principle be solved.

In practice, due to the nonlinear term introduced by the derivative of the substrate potential, a numerical solution is often necessary.

11.4 The commensurate–incommensurate transition

When the periods a and b are close to each other but not equal, incommensurate phases may exist and their structure *cannot be obtained* by a perturbative approach because the vicinity of the two periods leads to a phenomenon similar to resonance. Reasoning in the spatial domain as we are used to doing in the time domain, we can notice that the network of interacting atoms, which has an intrinsic period a, is excited by a force which has the period $b \simeq a$, deriving from the substrate potential. We know that driving a system near its 'eigenfrequency' (spatial frequency here) can lead to a very large response, even for a small driving force. This causes a failure of the perturbative approach. In this case we shall show that localised soliton-like excitations appear in the system, *the discommensurations.*

Before we look for the structure of these discommensurations, let us show that, if the ratio a/b increases gradually from unity, above a critical value, the structure of the system changes qualitatively. For this value, a commensurate–incommensurate transition occurs. We can show that it should be expected by examining two limiting cases.

Let us assume that $a = b + \delta$, where δ is very small with respect to b, and that the ratio b/a is irrational.

As discussed above, we expect competition between the substrate potential energy and the elastic coupling energy. Let us evaluate the energy of a set of N_0 particles in two extreme cases:

(i) If the substrate potential wins, all the particles tend to sit in its minima. The phase is commensurate with the substrate. Each of the bonds between neighbouring particles has the length b and therefore it brings the contribution $\frac{1}{2}\lambda(b-a)^2 = \frac{1}{2}\lambda\delta^2$ to the total energy, while the substrate energy potential of each particle is equal to the minimum $V_{\min}$ of the potential. The energy of a set of N_0 particles in the commensurate phase is

$$U_{\text{com}} = N_0\left[\frac{1}{2}\lambda\delta^2 + V_{\min}\right]. \tag{11.5}$$

(ii) Conversely, if the harmonic coupling potential wins, the particles settle in positions which are separated by distance a in order to cancel the elastic coupling energy. They can no longer be in the minima of the substrate potential, and as a/b is irrational, in the limit of large N_0, they can occupy any position with respect to the period of the potential. The energy of this perfectly incommensurate phase is therefore

$$U_{\text{inc}} = N_0\langle V(x)\rangle, \tag{11.6}$$

where $\langle V(x) \rangle$ is the mean value of the function $V(x)$.

Above a value δ_c of the difference between a and b, the commensurate phase has more energy per particle than the incommensurate phase. An incommensurate phase appears for $\delta > \delta_c$. The value of δ_c for which the commensurate and incommensurate phases have the same energy is

$$\delta_c \simeq \sqrt{\frac{2(\langle V(x) \rangle - V_{\min})}{\lambda}}. \tag{11.7}$$

This result, obtained from crude estimates, gives a good approximation of the actual critical value, which can be determined numerically.

However, a more accurate value of δ_c can be obtained if we compute the energy of a discommensuration, and then determine the value of δ for which the formation of a discommensuration decreases the total energy of the system.

11.5 Structure of the incommensurate phase

Let us look for a solution of the set of equations (11.3) for a strong coupling $\lambda b^2 \gg |V|$ allowing a continuum limit approximation. The variables u_n are replaced by the continuous field $u(n)$, where n is now treated as a continuous space variable. As we are looking for the *equilibrium* (static) structure, the time dependence is ignored.

If we define the field $\varphi(n) = u_n + n\delta$ which measures the displacement of the particles with respect to the perfectly commensurate case, we have

$$x_n = na + u_n = n(b + \delta) + u_n = nb + \varphi(n). \tag{11.8}$$

Taking into account the periodicity b of the potential V and of its derivative, we get $V'(na + u_n) = V'(nb + \varphi(n)) = V'(\varphi(n))$. This allows us to write the set of equations (11.3) as

$$\varphi(n+1) + \varphi(n-1) - 2\varphi(n) = \frac{1}{\lambda} V'(\varphi(n)), \tag{11.9}$$

which becomes

$$\frac{\mathrm{d}^2\varphi}{\mathrm{d}n^2} = \frac{1}{\lambda} V'(\varphi), \tag{11.10}$$

in the continuum limit approximation. As usual this equation is solved by multiplying it by $(\mathrm{d}\varphi/\mathrm{d}n)$ and integrating with respect to n to get

$$\frac{1}{2}\lambda \left(\frac{\mathrm{d}\varphi}{\mathrm{d}n}\right)^2 = V(\varphi) + \mathcal{E}. \tag{11.11}$$

The integration constant $\mathcal{E}$ must be determined. Since we are looking for discommensurations, we must search for localised defects in the commensurate phase. It means that, far from the discommensuration, the system should be in a perfectly

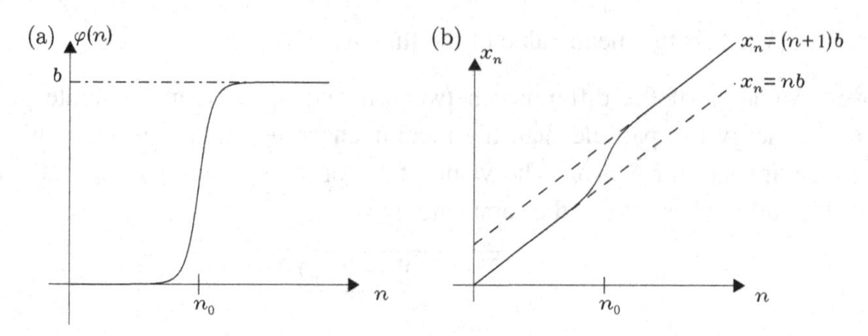

Figure 11.2. a) Variation of $\varphi(n)$ in the vicinity of a discommensuration. b) Position of the atoms in the presence of this discommensuration. The dashed lines correspond to the commensurate phases $x_n = nb$ and $x_n = (n+1)b$.

commensurate phase, i.e. with the particles at the minima of the potential V so that $\varphi = 0$. Since, in this limit, we also have $(\mathrm{d}\varphi/\mathrm{d}n) = 0$ and $V(\varphi) = V_{\min}$, it leads to $\mathcal{E} = -V_{\min}$ and Equation (11.11) becomes

$$\frac{1}{2}\lambda\left(\frac{\mathrm{d}\varphi}{\mathrm{d}n}\right)^2 = V(\varphi) - V_{\min}. \tag{11.12}$$

Given a particular expression for the potential, such as the sinusoidal potential of the Frenkel–Kontorova model, $V(\varphi) = V_0[1 - \cos(2\pi\varphi/b)]$, the solution of Equation (11.12) can be obtained by a separation of the variables, as we did when we solved the sine-Gordon equation in Chapter 2 (see Equation (2.20)). For the sinusoidal potential, in terms of the parameters that we introduced for the Frenkel–Kontorova model of an incommensurate phase, the sine-Gordon soliton solution gives

$$\varphi(n) = \frac{2b}{\pi}\arctan\left\{\exp\left[2\pi\sqrt{\frac{V_0}{\lambda b^2}}(n - n_0)\right]\right\}, \tag{11.13}$$

which is plotted in Figure 11.2.

As shown in this figure, φ vanishes for $n \ll n_0$, and we simply have $x_n = nb$ in this region, i.e. a commensurate phase, until the discommensuration is met. Near this defect, the elastic energy which had been accumulated in all the harmonic coupling terms, compressed to length b instead of their equilibrium distance a, is abruptly released by a sharp rise of x_n in a small region. The system then recovers commensurate structures, and thus starts again to build up the elastic energy of the compressed bonds.

11.6 Calculation of δ_c

In order to determine the critical value of δ above which the first discommensuration appears, it is not necessary to solve for the variable $\varphi(n)$. The energy of a discommensuration can be obtained from the calculation of the potential energy of the system,

$$U = \sum_n \frac{\lambda}{2}(u_{n+1} - u_n)^2 + V(na + u_n). \tag{11.14}$$

In the continuum limit approximation, it becomes

$$U = \int \mathrm{d}n \left[\frac{\lambda}{2} \left(\frac{\mathrm{d}\varphi}{\mathrm{d}n} - \delta \right)^2 + V(\varphi) \right] \tag{11.15}$$

since $u_n = -n\delta + \varphi(n)$. As $\varphi(n)$ vanishes exactly in the commensurate phase, the potential energy U_{com} of a commensurate phase is

$$U_{\text{com}} = \int \mathrm{d}n \left[\frac{\lambda}{2}\delta^2 + V_{\text{min}} \right], \tag{11.16}$$

which is proportional to the length of the system because the function which is integrated is a constant. The energy of a discommensuration is the difference between the energy U of the system in the presence of this discommensuration and the energy U_{com} of the perfectly commensurate phase

$$E_{\text{disc}}(\delta) = U - U_{\text{com}} \tag{11.17}$$

$$= \int \mathrm{d}n \left[\frac{\lambda}{2} \left(\left(\frac{\mathrm{d}\varphi}{\mathrm{d}n} \right)^2 - 2\delta \frac{\mathrm{d}\varphi}{\mathrm{d}n} \right) + V(\varphi) - V_{\text{min}} \right]. \tag{11.18}$$

Using Equation (11.12), this can be simplified into

$$E_{\text{disc}}(\delta) = \int_{-\infty}^{+\infty} \mathrm{d}n \; 2[V(\varphi) - V_{\text{min}}] - \lambda\delta \int_{-\infty}^{+\infty} \mathrm{d}n \frac{\mathrm{d}\varphi}{\mathrm{d}n}. \tag{11.19}$$

The first integral, that we shall denote by $E_{\text{disc}}(0)$, does not depend on δ. It can be calculated without knowing the solution $\varphi(n)$ for a discommensuration, as we did in Chapter 2 to compute the energy of a sine-Gordon soliton. We get

$$E_{\text{disc}}(0) = \int_{-\infty}^{+\infty} \mathrm{d}n \; 2[V(\varphi) - V_{\text{min}}] \tag{11.20}$$

$$= \int_{-\infty}^{+\infty} \mathrm{d}n \sqrt{2\lambda(V(\varphi) - V_{\text{min}})} \underbrace{\sqrt{\frac{2}{\lambda}(V(\varphi) - V_{\text{min}})}}_{=\mathrm{d}\varphi/\mathrm{d}n} \tag{11.21}$$

$$= \int_{\varphi=0}^{\varphi=b} \sqrt{2\lambda(V(\varphi) - V_{\text{min}})} \, \mathrm{d}\varphi. \tag{11.22}$$

Its value only depends on the particular function chosen for the substrate potential. For a sinusoidal substrate potential $V(\varphi) = V_0[1 - \cos(2\pi\varphi/b)]$, it is given by

$$E_{\text{disc}}(0) = \frac{4b}{\pi}\sqrt{\lambda V_0}. \tag{11.23}$$

For a discommensuration, the second integral of Equation (11.19) is equal to $\varphi(+\infty) - \varphi(-\infty) = b$, so that the energy of a discommensuration is given by

$$E_{\text{disc}}(\delta) = E_{\text{disc}}(0) - \lambda\delta b. \tag{11.24}$$

The first term is the energy of the discommensuration in a particular case $\delta = 0$, i.e. $a = b$. It is the case that we considered in Chapter 2 on topological solitons and for the Frenkel–Kontorova model for dislocations discussed in Chapter 9: for the sinusoidal potential, $E_{\text{disc}}(0)$ is simply the energy of a sine-Gordon soliton.

But Equation (11.24) shows that, if $\delta \neq 0$, the energy of the discommensuration includes a second contribution which tends to *reduce* its value below the energy of a soliton. If the mismatch $\delta = a - b$ between the equilibrium length of the bonds and the period of the potential is large enough, the right term becomes larger than $E_{\text{disc}}(0)$: *the system can reduce its energy by creating discommensurations.* The physical explanation is that a discommensuration releases the elastic energy accumulated in the bonds compressed with respect to their equilibrium length in the commensurate phase.

The value δ_c for which the energy of the discommensuration vanishes is the critical value beyond which the commensurate phase is no longer the phase with the lowest energy. It corresponds to the *commensurate–incommensurate transition*, and is equal to

$$\delta_c = \frac{E_{\text{disc}}(0)}{\lambda b} = \frac{4b\sqrt{\lambda V_0}}{\pi\lambda b} = \frac{4}{\pi}\sqrt{\frac{V_0}{\lambda}}, \tag{11.25}$$

while Formula (11.7), obtained by a crude method, gave $\delta_c = 2\sqrt{V_0/\lambda}$, which is very close.

As Formula (11.24) shows that the energy of a discommensuration becomes negative for $\delta > \delta_c$, in this range of δ, we can imagine that the system will create more and more discommensurations in order to lower its energy as much as possible. This is not true due to the interaction energy between the discommensurations. As their number grows this energy rises, and the equilibrium structure of the incommensurate phase is determined by a balance between the energy gained by creating a discommensuration and the energy cost of its interactions with the other discommensurations. Therefore the phase is made of a lattice of discommensurations. Their distance is such that the average growth of x_n versus n follows $x = n\,a$ although the system is in a commensurate phase as much as possible, except in

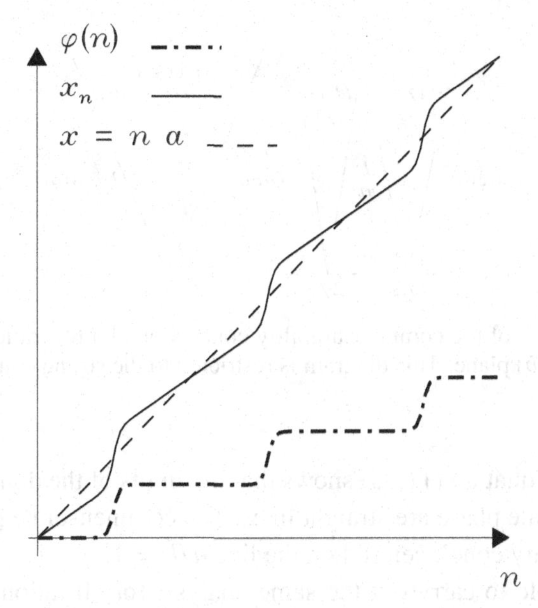

Figure 11.3. Variation of $\varphi(n)$ and x_n versus n for an incommensurate phase made of a lattice of discommensurations. The thin dashed line is a plot of $x = na$.

the regions of discommensurations which are local defects in the commensurate phase (Figure 11.3). This solution makes the optimal compromise between the coupling which tends to impose the solution $x = n\,a$ and the substrate potential which tends to select a commensurate phase. The corresponding variation of $\varphi(n)$ is a multisoliton of the nonlinear equation (11.10).

11.7 Phase diagram

Since the model showed the role of two fundamental parameters:

(i) The ratio a/b (or the value of δ) which determines the degree of frustration,
(ii) The ratio of the amplitude of the substrate potential to the coupling constant λ,

we can look for its phase diagram, i.e. determine the domains in the parameter space where the system is in a commensurate phase and the domains in which it is in an incommensurate phase.

We already know this diagram in the vicinity of $a/b = 1$, because Expression (11.25) determines the maximum value of δ for which the system stays commensurate. In our analysis we implicitly assumed that $a > b$ ($\delta > 0$) to make the discussion easier, but all results stay valid for $a < b$ ($\delta < 0$). In this case, the discommensuration tends to bring the particles closer to each other. It is an antisoliton. If we choose to plot the phase diagram by putting a/b along the horizontal axis and $\sqrt{V_0/\lambda}$ along

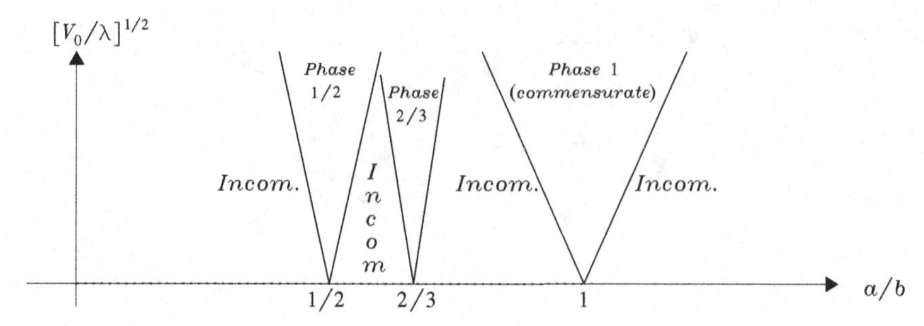

Figure 11.4. Plot of the commensurability cones around a few rational values in the $(\sqrt{V_0/\lambda}, a/b)$ plane. This diagram is restricted to the strong coupling limit.

the vertical axis, Equation (11.25) shows that the limits of the domain of existence of the commensurate phase are straight lines. The commensurate phase lies within a 'commensurability cone', centred on the line $a/b = 1$.

But it is possible to carry out the same analysis for all rational values $a/b = M/N$, where M and N are integers. For these rational a/b ratios a critical frustration depending on M/N exists, above which the commensurate phase is replaced by an incommensurate phase. The loss of stability of the commensurate phase also occurs through the formation of discommensurations, which have a more complex structure than the one we studied because the commensurate phase from which they emerge is itself more complex. When $\sqrt{V_0/\lambda}$ is small enough to ensure the validity of the continuum limit approximation, the variation of δ_c is again proportional to $\sqrt{V_0/\lambda}$ so that the phase diagram must be completed by other 'commensurability cones' as shown in Figure 11.4. Outside of these cones, the system is in an incommensurate phase.

However, Figure 11.4 clearly shows that this reasoning is incomplete for two reasons. First, we should in principle consider commensurability cones around all rational numbers, i.e. 'almost everywhere' in the figure. Second, the limits of the cones, as we have described them, meet each other so that, in some domains, we cannot specify which phase the system should select.

In fact, for large values of M and N, the commensurability cones become extremely narrow but, more importantly, the diagram plotted in Figure 11.4 is only valid for very small values of the parameter $\sqrt{V_0/\lambda}$. When V_0 increases, the strong coupling hypothesis has to be dropped. The straight lines which draw the boundary of the cones are curved by discreteness effects. The full phase diagram, which involves a numerical solution was determined by Serge Aubry [21]. It is plotted in Figure 11.5. The domains which do not contain the indication of a ratio are the domains of incommensurate phases. They are separated by narrow domains of complex commensurability (such as 3/7, 4/9, etc.). It should be noticed that, when

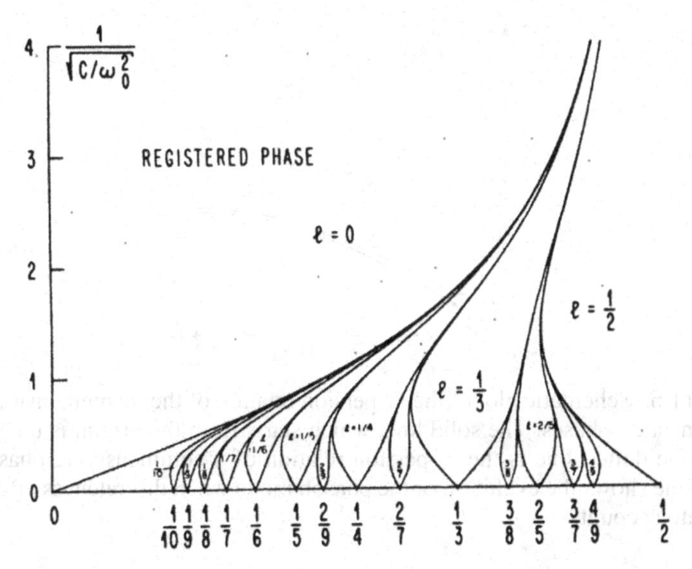

Figure 11.5. Full phase diagram showing the commensurability cones in the $(\sqrt{V_0/\lambda}, a/b)$ plane (from reference [21]).

V_0 increases, the diagram gets simpler, and only phases with a simple commensurability ratio subsist. This is why experiments only detect such 'simple' phases.

11.8 Dynamics of the incommensurate phase

Up to now we have studied equilibrium structures, but it is of course interesting to investigate the dynamics of the commensurate and incommensurate phases. The Hamiltonian of the system is obtained by adding a kinetic energy contribution to the potential energy (11.1). This leads to equations of motion which are the dynamic counterparts of Equation (11.9),

$$m\ddot{\varphi}_n - \lambda[\varphi(n+1) + \varphi(n-1) - 2\varphi(n)] + V'(\varphi(n)) = 0, \qquad (11.26)$$

where m is the mass of the atoms.

In the continuum limit approximation, the situation is rather simple:

- Any commensurate phase is pinned to the substrate potential because the atoms are in potential minima. The phase can vibrate around its equilibrium structure, without any overall translation.
- Conversely any incommensurate phase is free to slide with respect to the substrate potential because it is made of solitons, which are freely mobile if the continuum limit approximation is valid. The physical explanation of this behaviour, which may seem surprising at first, is analogous to the one that we gave for dislocations:

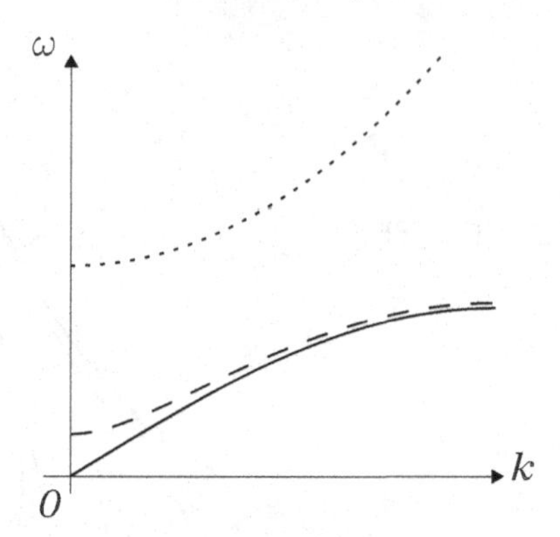

Figure 11.6. Schematic plot of the dispersion relation of the incommensurate and commensurate phases. The solid line, which starts from the origin, is the phason mode. The dotted line is the dispersion relation of a commensurate phase. The dashed line shows the evolution of the phason mode when discreteness effects are taken into account.

in the incommensurate phase, there are particles at every level of the substrate potential, even on its maxima. While some particles need energy to climb over the barriers of the substrate potential, the incommensurate phase gets this energy from other particles which are moving down in this potential.

To study the dynamics of the incommensurate phase, we can look for a solution of the continuum version of Equation (11.26) with the form

$$\varphi(n,t) = \varphi_0(n) + \Delta\varphi(n,t), \tag{11.27}$$

where $\varphi_0(n)$ is the multisoliton solution which describes the incommensurate phase that we are studying, and $\Delta\varphi$ a small deviation from this solution. The equation is linearised by only keeping the first-order terms in $\Delta\varphi$. This shows that the incommensurate phase has a vibrational mode whose dispersion relation $\omega(k)$ passes through the origin. The $k = 0$ solution, for which $\omega = 0$, is the Goldstone mode of the multisoliton solution. It describes an overall sliding of the incommensurate phase with respect to the substrate. This mode is called the 'phason mode'. Its dispersion relation for $k \neq 0$ can be obtained by a collective coordinate method. Each discommensuration is treated as a quasi-particle which, in the continuum limit, is perfectly free to move over the substrate but interacts with neighbouring quasi-particles.

However, the phason mode is never observed because it is affected by discreteness effects. Serge Aubry [21] demonstrated what is now called the 'Aubry transition' or 'transition by breaking of analyticity': if V_0 is below a critical value V_{0c}, the phason mode exists but, above V_{0c}, the incommensurate phase is pinned to the lattice. Instead of vanishing, the frequency of the small displacements of the incommensurate phase tends to a nonzero value when k vanishes, which is called the *Peierls–Nabarro frequency* because the pinning of the incommensurate phase is analogous to the pinning of a dislocation by discreteness effects.

The pinned incommensurate phase has remarkable properties, such as its response to a pulling force. When the force gradually increases, the stretching of the atomic lattice increases by steps along a curve called the 'devil's staircase' because it has an infinity of steps of different sizes. Examining a step in detail shows that it is in fact made of smaller steps, and so on. Of course experiments can only detect the largest steps, but this behaviour has nevertheless been observed experimentally by X-ray structural studies of an incommensurate phase under pressure. This surprising property occurs because all the solitons which make the multisoliton solution pinned by the Peierls–Nabarro potential are not in the same position with respect to the discretisation potential. When the force increases, some of them are unpinned first, and others follow later in a sequence of local unpinning transitions which generate the 'devil's staircase'.

11.9 Formation of the discommensurations

11.9.1 Boundary effects in a finite system

Let us now consider a system of length L with free boundary conditions. We shall again choose the simplest case $a = b + \delta$ and assume that the continuum limit approximation is valid. We shall look for the limit of stability of a commensurate phase, but we shall take into account the boundary effects which were neglected in our previous analysis which assumed that the system was infinite.

Besides Equation (11.10), the minimisation of the potential energy U must take into account the boundaries. Let us for instance consider the first particle of the lattice, that we shall label by index 0. It is linked to a single neighbour, particle 1. In the strong coupling approximation where the positions of the particles are mostly determined by the coupling terms, the first 'spring', linking particles 0 and 1, minimises its energy by selecting a length equal to its equilibrium length, which is possible because one of its ends is free (see Figure 11.1 and assume that the atomic chain is interrupted on the left side). Thus

$$x_1 - x_0 = a = b + \delta \tag{11.28}$$

$$= b + \varphi(1) - \varphi(0) \quad \text{i.e.} \quad \varphi(1) - \varphi(0) = \delta. \tag{11.29}$$

In the continuum limit, the conditions expressing that boundaries 0 and L are free and therefore

$$\left(\frac{\mathrm{d}\varphi}{\mathrm{d}n}\right)_{x=0} = \left(\frac{\mathrm{d}\varphi}{\mathrm{d}n}\right)_{x=L} = \delta. \tag{11.30}$$

Equation (11.12) is still valid because we look for the stability of a phase which is commensurate in the bulk of the material. If we combine it with Equation (11.30) for the particular cases $n = 0$ or $n = L$, we get

$$\frac{\lambda}{2}\,\delta^2 = V(\varphi_{\text{ext}}) - V_{\text{min}}. \tag{11.31}$$

Consequently the value φ_{ext}, of φ at the extremity of the lattice, will vary in order to satisfy this condition, which determines the positions of the particles at the two ends of the lattice. However, this is only possible if δ does not exceed the limiting value δ_{cL} defined by

$$\frac{\lambda}{2}\,\delta_{cL}^2 = V_{\text{max}} - V_{\text{min}}. \tag{11.32}$$

Comparing with Expression (11.7), we immediately notice that

$$\delta_{cL} > \delta_c. \tag{11.33}$$

Thus, for a finite medium, if we gradually increase the difference δ between the equilibrium length of the springs and the period of the substrate potential, the incommensurate phase becomes unstable in the bulk while the particles at the ends are still in stable positions. But discommensurations cannot appear in the bulk of the material, even if their energy, once they are formed, is negative (i.e. if $\delta > \delta_c$). To create a discommensuration in the bulk, it would be necessary to move a very large number of atoms (an infinity if the system were infinite) above a barrier of the substrate potential.[1] This can also be understood from Figure 11.2: in order to create the discommensuration, all the atoms on the right of the discommensuration must be moved by b. Consequently discommensurations can only emerge from the boundaries. In the whole range $\delta_c < \delta < \delta_{cL}$, the commensurate phase is *metastable*. The study of the finite system is therefore essential to determine the real domain of existence of the commensurate phase.

11.9.2 Action of an external force

Let us now examine the finite system of the previous section under the effect of an external force F applied to all the atoms (it could be an electric field for a phase made of ions, or a sheer stress). To take this force into account in the calculations, we can replace the substrate potential $V(\varphi)$ by the effective potential $\widetilde{V}(\varphi) = V(\varphi) - F\varphi$, plotted in Figure 11.7. Indeed a commensurate phase can only exist if $\widetilde{V}(\varphi)$ has

[1] This is the discrete counterpart of the conservation of the topological charge in the sine-Gordon model (see Chapter 2).

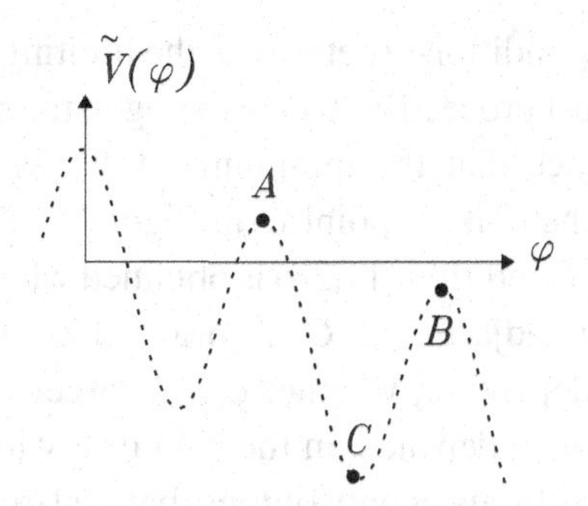

Figure 11.7. Plot of the effective potential $\widetilde{V}(\varphi)$ in the presence of an external force F.

minima. As $\widetilde{V}'(\varphi) = V'(\varphi) - F$, the condition

$$F < F_{\max} = \max[V'(\varphi)] \tag{11.34}$$

must be verified.

Changing the potential induces a change of the critical value of δ for which the ends become unstable. Let us study again the case where the bulk of the material is in a commensurate phase and let us look for the conditions that φ must obey at the boundaries. Equations (11.10) and (11.11), which come from the minimisation of the potential energy of the system, are still valid provided that they are written in terms of the effective potential $\widetilde{V}$. Therefore we have

$$\frac{1}{2}\lambda\left(\frac{\mathrm{d}\varphi}{\mathrm{d}n}\right)^2 = \widetilde{V}(\varphi) + \mathcal{E}. \tag{11.35}$$

To determine the constant $\mathcal{E}$, we can consider the bulk of the material, which is assumed to be in the commensurate phase: the particles are therefore in the minima of the substrate potential, separated by b which corresponds to $\mathrm{d}\varphi/\mathrm{d}n = 0$. Equation (11.35) applied in the bulk of the material therefore gives

$$\mathcal{E} = -\widetilde{V}_{\min}(\varphi). \tag{11.36}$$

Applying Equation (11.35) at the boundaries, we get

$$\frac{1}{2}\lambda\left(\frac{\mathrm{d}\varphi_{\text{ext}}}{\mathrm{d}n}\right)^2 = \widetilde{V}(\varphi_{\text{ext}}) - \widetilde{V}_{\min}. \tag{11.37}$$

The reasoning from the previous section to determine the condition that a free boundary imposes is still valid, so that we still have $\mathrm{d}\varphi_{\text{ext}}/\mathrm{d}n = \delta$ which, as above, leads to an equation which must be obeyed by φ_{ext},

$$\frac{1}{2}\lambda\delta^2 = \widetilde{V}(\varphi_{\text{ext}}) - \widetilde{V}_{\min}. \tag{11.38}$$

In order to exploit this condition to determine the limiting value of δ for which the boundary is stable, we must proceed with care owing to the shape of the potential $\widetilde{V}$. Let us assume for instance that the minimum of $\widetilde{V}$, which corresponds to the commensurate phase in the bulk, is point C in Figure 11.7. The largest value of δ which is compatible with Condition (11.38) is obtained when $\widetilde{V}(\varphi_{\text{ext}})$ is a maximum of the potential, which is adjacent to C. A priori it could be either point B or point A in Figure 11.7, depending whether φ_{ext} is larger or smaller than φ in the bulk. The appropriate choice depends on the sign of δ which determines whether discommensurations are solitons or antisolitons, but it also depends on the free end that we consider because, in order to create a discommensuration of a given type, the displacements that we have to impose to the end atoms have opposite signs on both ends of the lattice. As a result, whatever the sign of δ, there will always be one of the two ends for which the maximum that determines the stability is point B in Figure 11.7, i.e. the maximum adjacent to C for which the value of $\widetilde{V}$ is the closest to $\widetilde{V}(C)$.

The limit of stability at one of the two ends is thus reached when δ takes the limiting value defined by

$$\frac{\lambda}{2}\delta_{cL}^2 = \widetilde{V}_{\text{max}}(B) - \widetilde{V}_{\text{min}}(C). \tag{11.39}$$

As the difference $\widetilde{V}_{\text{max}}(B) - \widetilde{V}_{\text{min}}(C)$ decreases when the force increases, the value of δ_{cL} decays when F grows. Consequently, for a given value of δ, we can define a critical force $F_c(\delta)$ above which the incommensurate phase becomes unstable at one of the two ends of the material. For $F > F_c(\delta)$ a discommensuration is generated at the unstable end and it travels inside the material. When it is far enough inside, the vicinity of the boundary returns to the same state as it was before the creation of the discommensuration. The process will repeat and new discommensurations will be created until their density in the bulk has reached the value which is predicted by a study of the bulk for the parameter δ which is appropriate for the material.

Thus the boundary appears as a source of solitons. Their creation rate depends on the difference between the applied force ($F > F_c$) and the critical force F_c. This mechanism plays a role, for instance, in the diffusion of atoms belonging to adsorbed layers.

11.10 Conclusion

We only gave a very partial view of incommensurate phases, which could be developed in various directions. Studying models in more than one dimension is very important to correctly describe surfaces. This raises new questions: How are the dislocations oriented? Can they cross each other? What is their role in thermal fluctuations? Some of these questions are answered in a review paper by Jacques Villain [183].

There is also a very interesting link with dynamical systems and chaos. Equation (11.9), which determines the structure of the system, can also be written as a mapping [22]

$$p_{n+1} = p_n + \frac{1}{\lambda} V'(\varphi_n) \tag{11.40}$$

$$\varphi_{n+1} = p_n + \varphi_n + \frac{1}{\lambda} V'(\varphi_n) \quad (\text{modulo } b) \tag{11.41}$$

if we define $p_n = \varphi_n - \varphi_{n-1}$. This is a nonlinear map due to the derivative of the potential. The trajectories in the phase space $\{p, \varphi\}$ correspond to different physical situations:

- Discrete cycles which run along a finite number of points before coming back to the starting point correspond to commensurate phases. The period of the phase, in units of the lattice spacing b, is given by the number of points in the cycle.
- Incommensurate phases are associated with sets of points drawing a continuous line because the mapping never comes back exactly to a point visited earlier.
- There are also chaotic regions, which correspond to *disordered phases*.

If we consider the simple example of the sinusoidal potential,

$$V(\varphi) = V_0[1 - \cos(2\pi\varphi/b)], \tag{11.42}$$

we get the *standard map*

$$p_{n+1} = p_n + \frac{2\pi V_0}{\lambda b} \sin\left(\frac{2\pi\varphi_n}{b}\right) \tag{11.43}$$

$$\varphi_{n+1} = p_n + \varphi_n + \frac{2\pi V_0}{\lambda b} \sin\left(\frac{2\pi\varphi_n}{b}\right) \quad (\text{modulo } b), \tag{11.44}$$

which becomes chaotic for $V_0/(\lambda b^2) > \alpha_c \simeq 0.8$. Below this critical value, the application generates trajectories on a torus. They correspond to the equilibrium ground state. This is the range where the phason modes exist. For higher values of $V_0/(\lambda b^2)$ the trajectories lie on a Cantor set. Discreteness effects dominate and the incommensurate phase is pinned to the lattice.

Dynamical system theory can be used to recover some of the results that we discussed in terms of solitons, but open problems subsist, in particular describing amorphous materials such as glasses.

12

Solitons in magnetic systems

Magnetic systems have been extensively studied for two reasons:

(i) They are excellent model systems for phase transitions, so that they allow very complete theoretical and experimental investigations of this fundamental problem in statistical physics.
(ii) They have important applications, in particular for high density data recordings, which require high-performance materials and have therefore prompted detailed investigations.

Actual materials are indeed three-dimensional, although two-dimensional magnetic films are made for data recording, but, as discussed below, in some magnetic solids, the magnetic interactions are almost perfectly one-dimensional. Magnetic materials are one of the best examples of systems where the theory of solitons at the atomic scale can be very accurately tested experimentally [130].

12.1 Ferromagnetism and antiferromagnetism

In a solid, the magnetic moment of the atoms has two origins:

(i) The spin of the electrons and nuclei which is a quantum intrinsic angular momentum of the particles associated with a magnetic moment.
(ii) The orbital angular momentum of the electrons coming from the circulation of the electrons around the nucleus, which is also quantised and associated with a magnetic moment.

The total magnetic moment $\overrightarrow{M}$ is related to the total angular momentum $\overrightarrow{J}$ by $\overrightarrow{M} = \gamma \overrightarrow{J}$ where γ is the gyromagnetic ratio. For the materials that we consider here, the magnetic moment is due to unpaired electrons and it is given by $\overrightarrow{M} = \gamma \overrightarrow{S}$, where $\overrightarrow{S}$ is the total spin of an atom.

Understanding how magnetic moments interact in a solid is one of the toughest problems in solid state physics and it is far from being fully solved [18, 114]. There are however well established basic results. The leading term in the interaction is not the dipolar coupling term between the magnetic moments as one might believe at first. A much larger contribution comes from the *electrostatic* interactions of the electrons. Quantum particles are truly indistinguishable, and this imposes symmetry conditions on the quantum state of the system which is a function of the orbital and spin variables of the particles. As a result the energy of a system of two interacting electrons with spins $\overrightarrow{S_1}$ and $\overrightarrow{S_2}$ depends on their spins through these symmetry conditions and can be written as

$$U = -\mathcal{J}\,\overrightarrow{S_1} \cdot \overrightarrow{S_2}, \tag{12.1}$$

which is called the Heisenberg exchange Hamiltonian. The constant $\mathcal{J}$, called the *exchange integral*, is very difficult to calculate in a solid. It is determined by the overlap of the electronic clouds of the atoms, and therefore it depends very strongly on their distance. The exchange integral decays exponentially with the interatomic distance which implies that:

- Nearest neighbour interactions dominate.
- In an anisotropic solid where the distances between the magnetic atoms are smaller in one direction than in the others, the magnetic interactions are almost one-dimensional. TMMC,[1] ($(CH_3)_4N^+\ MnCl_3^-$), is an excellent example. The distances between the Mn^{2+} ions which carry the magnetic moments are 3.25 Å along one axis and 9.15 Å in the directions orthogonal to it. This leads to a ratio of 10^{-5} between the exchange integrals in the transverse direction and along the axis. For TMMC it is really possible to speak of 'spin chains'. Another example which has been extensively studied is $CsNiF_3$ where the ratio between the transverse and longitudinal exchange integrals is about 10^{-3}.

Magnetic materials show:

- *Cooperative effects* because neighbouring spins interact with each other.
- *Nonlinear effects* because the interaction proportional to $\overrightarrow{S_1} \cdot \overrightarrow{S_2}$ depends on the cosine of the angle $\theta = (\overrightarrow{S_1}, \overrightarrow{S_2})$.

Therefore nonlinear excitations such as solitons can be expected in these systems. This is confirmed by the theoretical analysis presented in this chapter, which has been verified by numerous experiments.

The type of magnetic order which is observed in a material depends on the sign of the exchange integral $\mathcal{J}$:

[1] Tetramethylammonium manganese tetrachloride

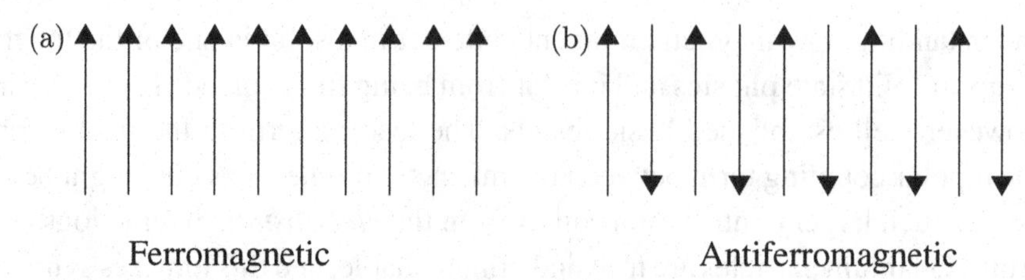

Figure 12.1. Schematic plot of the organisation of the magnetic moments in a ferromagnetic material (a) and in an antiferromagnetic material (b). Each arrow corresponds to an individual magnetic moment in the material.

- If $\mathcal{J} > 0$, the smallest interaction energy U is obtained when the two spins $\overrightarrow{S_1}$ and $\overrightarrow{S_2}$ point in the same direction. Thus the system tends to self-organise with a ferromagnetic order in which all spins are parallel and point in the same direction, at least as long as the temperature is low enough to prevent thermal fluctuations from perturbing this order. The sum of the magnetic moments is not zero and the material exhibits a permanent magnetic moment due to the superposition of individual moments (Figure 12.1(a)). This is the case for $CsNiF_3$. Its chains of Ni^{2+} are the prototype for ferromagnetic spin chains.
- Conversely, when $\mathcal{J} < 0$ the interaction energy U is minimal when two neighbouring spins point in opposite directions (Figure 12.1(b)). The material shows antiferromagnetic order and its total magnetic moment is zero. This is the case for TMMC in which the magnetic interaction is mediated by the electronic shells of the chlorine.

12.2 Equations for the dynamics of a spin chain

Let us consider first a spin chain with a ferromagnetic interaction ($\mathcal{J} > 0$), making a one-dimensional lattice in the z direction, and subjected to an external magnetic field in the direction x orthogonal to the chain. Its Hamiltonian is

$$H = -\mathcal{J} \sum_n \vec{S}_n \cdot \vec{S}_{n+1} + \sum_n A \left(S_n^z\right)^2 - \gamma B_x \sum_n S_n^x. \tag{12.2}$$

Besides the first term, which is the interaction term discussed above, there are two additional contributions:

(i) An anisotropy term, proportional to the parameter $A > 0$, which is such that any increase in the z component of the spin increases the energy of the chain. This terms favours the motion of the spins in the xy plane, orthogonal to the chain. This is a common situation for one-dimensional magnetic systems.

(ii) A term due to the external magnetic field $\vec{B}$ which tends to align the spins in the x direction ($\gamma > 0$).

Owing to these different contributions, in the ground state of the chain all spins are parallel and point along x. The chain is in a ferromagnetic state as shown in Figure 12.1(a). This system reminds us of the pendulum chain studied in Chapter 2, with the important difference that the coupling between adjacent sites is no longer harmonic. However the analogy suggests the existence of two types of excitations:

(i) Small-amplitude excitations, corresponding to an oscillation of the spins around their equilibrium positions. These are the spin waves, also called *magnons*.
(ii) Topological excitations coming from the degeneracy of the ground state because any 2π rotation of the spins around the z axis does not change the energy of the system. These excitations will be 2π rotations of the spins, connecting two different realisations of the ground state in two regions of the chain.

The analogy is qualitatively correct but we shall however discover that the spin chain is more complex than the pendulum chain because it can have excitations which are intermediate between topological and nontopological solitons. In order to establish this result, we must write the dynamic equations for the spins, which are quantum quantities. Their dynamics must therefore be deduced from the dynamic laws of quantum mechanics.

The spin quantum observable is associated with three operators S^x, S^y and S^z, corresponding to the three components x, y and z. We denote the component by an exponent rather than an index, which is more usual, to save the index for the particle number. The x component of the spin operator of the nth particle is denoted by S_n^x. The scalar product of the Hamiltonian (12.2) must be understood as a product of operators: $S_n^x S_{n+1}^x + S_n^y S_{n+1}^y + S_n^z S_{n+1}^z$.

Operators associated with different sites are defined in different state spaces, so that they commute. Conversely the spin operators of a given site do not commute. They follow the usual commutation relations of an angular momentum,

$$\left[S_n^x, S_n^y\right] = S_n^x S_n^y - S_n^y S_n^x = i\hbar S_n^z, \tag{12.3}$$

and the relations which can be derived by a circular permutation of the components x, y, z. Using the Heisenberg picture of quantum mechanics, the operators are time-dependent, and their time evolution is given by

$$i\hbar \frac{d\vec{S}_n}{dt} = [\vec{S}_n, H]. \tag{12.4}$$

Using the commutation relations, we get the equations governing the time evolution of the spin operators:

$$\frac{dS_n^x}{dt} = \mathcal{J}\left\{S_n^y\left(S_{n-1}^z + S_{n+1}^z\right) - \left(S_{n-1}^y + S_{n+1}^y\right)S_n^z\right\} - A\left(S_n^z S_n^y + S_n^y S_n^z\right) \tag{12.5}$$

$$\frac{dS_n^y}{dt} = \mathcal{J}\left\{S_n^z\left(S_{n-1}^x + S_{n+1}^x\right) - \left(S_{n-1}^z + S_{n+1}^z\right)S_n^x\right\} + A\left(S_n^z S_n^x + S_n^x S_n^z\right) + \gamma B_x S_n^z \tag{12.6}$$

$$\frac{dS_n^z}{dt} = \mathcal{J}\left\{S_n^x\left(S_{n-1}^y + S_{n+1}^y\right) - \left(S_{n-1}^x + S_{n+1}^x\right)S_n^y\right\} - \gamma B_x S_n^y. \tag{12.7}$$

A full quantum treatment of these equations is difficult but it is possible to take a quasi-classical limit, which can be viewed as an approximation which leads to a set of equations for the quantum mean values of the components of the spin. Thus these components become numbers rather than operators, and the spin can be treated as a *vector*. Let us introduce the vector $\vec{\sigma}$ defined by

$$\vec{S}_n = \hbar\sqrt{S(S+1)}\,\vec{\sigma}_n, \tag{12.8}$$

in which we introduce the norm of the spin, determined by the eigenvalue $\hbar^2 S(S+1)$ of the operator $\vec{S}^2 = (S^x)^2 + (S^y)^2 + (S^z)^2$, S being the spin quantum number. With this definition, the vector $\vec{\sigma}$ is a unitary vector. The classical limit ignores the quantisation of the spin components, which are treated as scalars that commute with each other. This approximation is best if S is large because quantum mechanics tells us that the possible values for a spin component such as S^z lie between $-S\hbar$ and $+S\hbar$, and differ by the quantum $\hbar$. If S gets bigger the number of possible values grows and approaches a classical behaviour with an infinite number of allowed values. Written in terms of $\vec{\sigma}$, the commutation relations between the spin components (12.3) become

$$[\sigma^x, \sigma^y] = \sigma^x\sigma^y - \sigma^y\sigma^x = i\sigma^z \frac{1}{\sqrt{S(S+1)}}, \tag{12.9}$$

showing that, in the limit of large S, the commutator $[\sigma^x, \sigma^y]$ tends to zero. In this limit the variables σ^x, σ^y, σ^z commute, as they would do if they were classical variables.

In practice very large values of S are not necessary to ensure that the classical approximation for the spin dynamics gives satisfactory results. For instance for Mn^{2+} which has $S = 5/2$, experiments show that this approximation, which seems very crude, actually gives very good results.

With this approximation, the discrete set of equations for the spin components are deduced from Equations (12.5) to (12.7) by the change of scale which introduces $\vec{\sigma}$, and by making no distinction between products of two spin components which

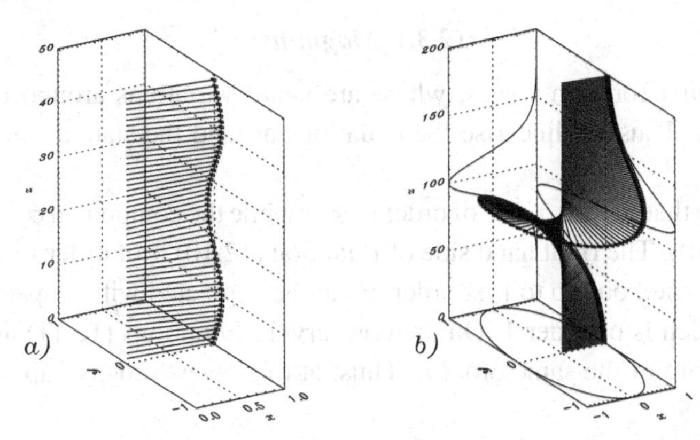

Figure 12.2. Plot of a magnon (a) and a soliton (b) in a spin chain having the Hamiltonian (12.2). Figure (b) also shows the projection of the solution on the three planes defined by the x, y, z axes.

only differ in the order of the factors. We get

$$\frac{d\sigma_n^x}{dt} = S\mathcal{J}\left\{\sigma_n^y\left(\sigma_{n-1}^z + \sigma_{n+1}^z\right) - \left(\sigma_{n-1}^y + \sigma_{n+1}^y\right)\sigma_n^z\right\} - 2SA\sigma_n^y\sigma_n^z \tag{12.10}$$

$$\frac{d\sigma_n^y}{dt} = S\mathcal{J}\left\{\sigma_n^z\left(\sigma_{n-1}^x + \sigma_{n+1}^x\right) - \left(\sigma_{n-1}^z + \sigma_{n+1}^z\right)\sigma_n^x\right\} + 2SA\sigma_n^x\sigma_n^z + \gamma B_x\sigma_n^z \tag{12.11}$$

$$\frac{d\sigma_n^z}{dt} = S\mathcal{J}\left\{\sigma_n^x\left(\sigma_{n-1}^y + \sigma_{n+1}^y\right) - \left(\sigma_{n-1}^x + \sigma_{n+1}^x\right)\sigma_n^y\right\} - \gamma B_x\sigma_n^y, \tag{12.12}$$

with $S = \hbar\sqrt{\mathcal{S}(\mathcal{S}+1)}$.

These equations are intrinsically nonlinear, which could be predicted because Hamiltonian (12.2) contains an interaction term which is a scalar product.

12.3 Magnons and solitons

Currently no exact solution of the set of nonlinear equations that we derived for the variables $\sigma_n^x(t)$, $\sigma_n^y(t)$, $\sigma_n^z(t)$, is known. To study the spin dynamics we shall have to make further approximations. They depend on the solution that we are looking for (Figure 12.2), which may be troubling because it seems as if we are biasing the equations to get the solutions that we wish to obtain. However we have seen in Chapter 4 that this is legitimate for nonlinear systems which may have several classes of solutions, very different from each other. In mathematical terms, this can be viewed as a *selection of the basis* to express the solutions. For instance, when we look for plane wave solutions, we restrict ourselves to a search for nonlocalised solutions.

12.3.1 Magnons

Let us look first for spin waves, which are small variations around the equilibrium position. Thus we linearise the equations around the state $\sigma_n^x = 1$, $\sigma_n^y = 0$, $\sigma_n^z = 0$.

We assume that σ_n^y and σ_n^z are of order $\varepsilon \ll 1$ while the component σ_n^x is of order 1 (Figure 12.2(a)). The right hand side of Equation (12.10) is of order ε^2 so that, in a calculation carried out up to first order, it can be neglected with respect to the left hand side which is of order 1. On the contrary, in Equations (12.11) and (12.12), all the terms are of the same order ε. Thus, at the lowest order of approximation, we get

$$\frac{\mathrm{d}\sigma_n^x}{\mathrm{d}t} = 0 \tag{12.13}$$

$$\frac{\mathrm{d}\sigma_n^y}{\mathrm{d}t} = S\mathcal{J}\left(2\sigma_n^z - \sigma_{n-1}^z - \sigma_{n+1}^z\right) + 2SA\sigma_n^z + \gamma B_x \sigma_n^z \tag{12.14}$$

$$\frac{\mathrm{d}\sigma_n^z}{\mathrm{d}t} = S\mathcal{J}\left(\sigma_{n-1}^y + \sigma_{n+1}^y - 2\sigma_n^y\right) \qquad\qquad - \gamma B_x \sigma_n^y . \tag{12.15}$$

This system, which is linear (because we chose to linearise the original equations) has 'plane-wave' solutions, with a wavevector q and a frequency ω, given by

$$\sigma_n^y = Y\,\mathrm{e}^{\mathrm{i}(qna-\omega t)} \quad \text{and} \quad \sigma_n^z = Z\,\mathrm{e}^{\mathrm{i}(qna-\omega t)}, \tag{12.16}$$

where a is the lattice spacing and Y and Z are the amplitudes of the components of the wave. Putting these expressions in the wave equations, and simplifying by $\mathrm{e}^{\mathrm{i}(qna-\omega t)}$, we get a homogeneous system

$$-\mathrm{i}\omega Y = \left(S\mathcal{J}[2 - 2\cos(qa)] + \gamma B_x + 2SA\right)\, Z \tag{12.17}$$

$$-\mathrm{i}\omega Z = \left(S\mathcal{J}[2\cos(qa) - 2] - \gamma B_x\right) \qquad Y, \tag{12.18}$$

which has a nonvanishing solution only if its determinant vanishes, i.e.

$$\omega^2 = [2S\mathcal{J}(1 - \cos(qa)) + \gamma B_x]\,[2S\mathcal{J}(1 - \cos(qa)) + \gamma B_x + 2AS]\,. \tag{12.19}$$

This equation is the magnon dispersion relation. It becomes much simpler if the anisotropy coefficient A vanishes. In such a case, we recover a dispersion relation which has the familiar expression

$$\omega = \pm\left[\gamma B_x + 4S\mathcal{J}\sin^2\left(\frac{qa}{2}\right)\right], \tag{12.20}$$

with a gap for frequencies between 0 and γB_x, controlled by the external magnetic field.

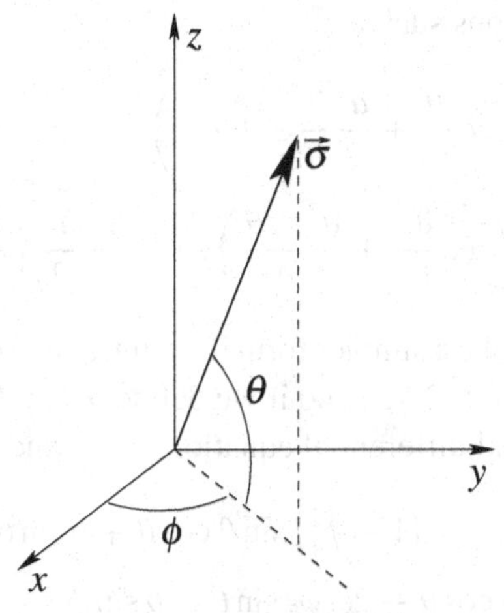

Figure 12.3. Definition of the angles θ and ϕ which characterise the unitary vector associated with each spin.

12.3.2 Solitons

Let us now look for another type of solution, which can involve large-amplitude motions (Figure 12.2(b)). Linearisation should be avoided in this case, but we can use the continuum limit approximation. As $\vec{\sigma}$ is a unitary vector, its three components are not independent from each other, which means that two variables are enough to fully characterise the magnetic moment at a given site. We can introduce [122, 187] the two angular variables shown in Figure 12.3:

- θ which is the angle between the spin and the xy plane. This angle is expected to be small if the anisotropy is strong.
- ϕ which characterises the rotation of the spin in the xy plane.

The reference state, which is a ground state of the system, is $\phi = \theta = 0$. The three components of the unitary vector $\vec{\sigma}$ are given in terms of these angular parameters by

$$\sigma_n^x = \cos\theta_n \cos\phi_n \tag{12.21}$$

$$\sigma_n^y = \cos\theta_n \sin\phi_n \tag{12.22}$$

$$\sigma_n^z = \sin\theta_n. \tag{12.23}$$

Let us put these expressions in the equations of motion (12.10)–(12.12), written in the continuum limit by introducing $\theta(z, t)$ and $\phi(z, t)$ and expressing the quantities at sites $n \pm 1$ by expansions such as

$$\cos\theta_{n\pm1} = \cos\left(\theta_n \pm a\frac{\partial\theta}{\partial z} + \frac{a^2}{2}\frac{\partial^2\theta}{\partial z^2} + \cdots\right) \tag{12.24}$$

$$\simeq \cos\theta_n - \left(\pm a\frac{\partial\theta}{\partial z} + \frac{a^2}{2}\frac{\partial^2\theta}{\partial z^2}\right)\sin\theta_n - \frac{a^2}{2}\left(\frac{\partial\theta}{\partial z}\right)^2\cos\theta_n. \tag{12.25}$$

The resulting equations take a simpler form if we introduce dimensionless variables $\xi = z\sqrt{2A/\mathcal{J}a^2}$ and $\tau = t\,(2AS)$ and if we define $b = \gamma B_x/(2AS)$. This leads to the following set of partial differential equations for θ and ϕ

$$\cos\theta\;\phi_\tau = -\theta_{\xi\xi} + \left(1 - \phi_\xi^2\right)\sin\theta\cos\theta + b\sin\theta\cos\phi \tag{12.26}$$

$$\theta_\tau = \phi_{\xi\xi}\cos\theta - 2\theta_\xi\phi_\xi\sin\theta - b\sin\phi. \tag{12.27}$$

This set of equations is still too complex to be solved exactly, and we must go ahead with other simplifying assumptions. We shall consider high anisotropies A so that the spin dynamics take place in the xy plane or in its immediate vicinity because excursions far from this plane cost a lot of energy. In this case θ is small and we can assume that it is of order $\varepsilon \ll 1$.

Moreover, as we look for soliton solutions, we can search for permanent profile solutions which are functions of the variable $\xi - v\tau$. Assuming that the velocity can be of order 1, it implies that ξ and τ are of the same order of magnitude. The continuum limit approximation imposes a slow variation with respect to ξ, i.e. the assumption that first-order derivatives with respect to ξ are of order ε, so that $\partial/\partial\tau$ must also be of order ε.

Finally we shall assume that b is of order ε^2, which sets limits on the acceptable values of the field and anisotropy parameters. This approximation is perfectly coherent with the classical limit which assumes $S \gg 1$ since $b \propto 1/S$.

Within these hypotheses, keeping only the leading terms in Equations (12.26) and (12.27) leads to

$$\phi_\tau = \theta \quad \text{noting that, at order } \varepsilon, \quad \sin\theta \simeq \theta \tag{12.28}$$

$$\theta_\tau = \phi_{\xi\xi} - b\sin\phi. \tag{12.29}$$

If we combine the two equations, we get the second-order differential equation

$$\phi_{\tau\tau} - \phi_{\xi\xi} + b\sin\phi = 0. \tag{12.30}$$

We recognise the sine-Gordon equation for the variable ϕ.

Therefore we have succeeded in deriving an equation which has soliton solutions, at the expense of making a series of approximations. This is a typical situation when a real system is studied, and it appears in many other physical situations.

The treatment of nonlinear phenomena requires specific methods because the usual linear approximation cannot be used. It is done in several steps:

- A qualitative analysis which exhibits the different classes of solutions: small-amplitude spin waves or topological solitons.
- The second step is to simplify the equations with assumptions which are adapted to the type of solution that we are looking for.
- The last step, which should not be forgotten, is to check the validity of the results which have been obtained with these simplified equations. This is what we have to do now.

12.4 Validity of the sine-Gordon approximation

There are two alternatives to check the validity of the approximations: a numerical simulation of the original equations (prior to any simplifying assumption) or a comparison with experiments.

12.4.1 Orders of magnitude

The soliton solution of the sine-Gordon equation (12.30) is

$$\phi_{SG} = 4 \arctan \exp\left[\frac{\sqrt{b}(\xi - v\tau)}{\sqrt{1-v^2}}\right] \tag{12.31}$$

which gives the value of θ thanks to Equation (12.28),

$$\theta_{SG} = -2\frac{v\sqrt{b}}{\sqrt{1-v^2}} \operatorname{sech}\left[\frac{\sqrt{b}(\xi - v\tau)}{\sqrt{1-v^2}}\right]. \tag{12.32}$$

This solution is a 2π rotation of the variable ϕ, accompanied by a local distortion of θ (Figures 12.4 and 12.5). In the core of the soliton, the spins move out of the xy plane, which does not happen for the pendulum chain.

First we can check that the orders of magnitude are coherent with our assumptions:

- As ϕ varies from 0 to 2π according to Equation (12.31), ϕ is indeed of order 1.
- The derivative of ϕ with respect to the spatial variable ξ yields a factor $\sqrt{b}$; as we have assumed that b is of order ε^2, we check that $\partial\phi/\partial\xi$ is of order ε in agreement with our assumptions.
- The solution θ, which also includes a factor $\sqrt{b}$, is indeed of order ε.
- The derivative of θ with respect to the space variable ξ is proportional to b so that it is of order ε^2 as expected.

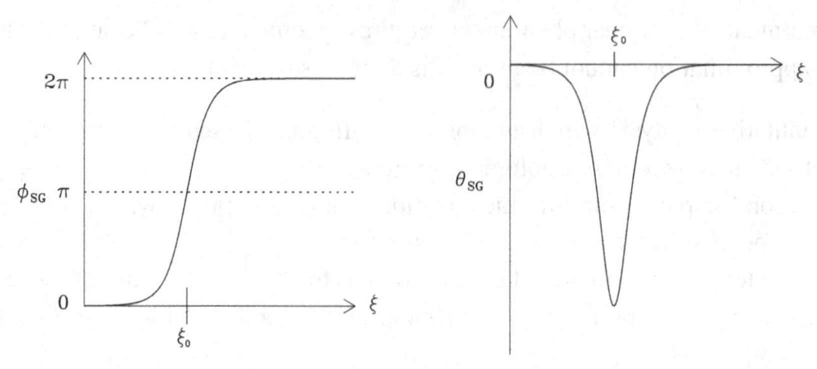

Figure 12.4. Plot of the solutions ϕ_{SG} and θ_{SG} of the coupled-spin chain, centred around the site ξ_0.

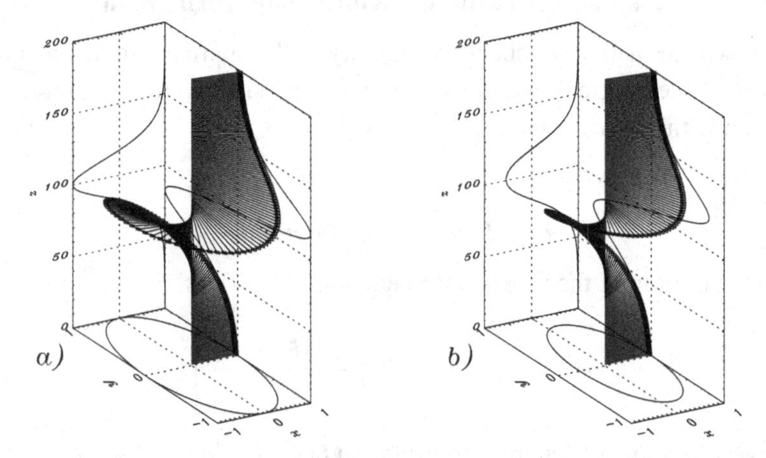

Figure 12.5. Three-dimensional plots of the static (a) and dynamic (b) solutions for the spin chain, with their projections on the three planes of the x, y, z frame. The projection on the xy plane shows that, in the centre of the dynamic solution, a spin does not take a position which is exactly opposed to its position at rest, as it does for the static solution. For this dynamic solution, the variation of the vector $\vec{\sigma}$ when we move from one side of the solution to the other is not a full rotation around the axis of the chain.

It should be noticed that, to preserve this coherence, the factor $1/\sqrt{1-v^2}$ which appears in θ should not become large. Thus the approximations can be expected to be better at low velocity.

Moreover it is possible to check that the *static* sine-Gordon solution for the variable ϕ, which is therefore such that $\phi_{\xi\xi} = b\sin\phi$ and is associated with $\theta = 0$, is an exact solution of the set of equations (12.26) and (12.27), obtained in the continuum limit approximation, but *without* the approximations that led us to the sine-Gordon equation. As soon as the velocity is different from zero, the solution derived from the sine-Gordon equation is no longer an exact solution for the spin chain. We can determine to what extent it is still an acceptable solution by introducing it as an initial condition in numerical simulations of the discrete set of equations (12.10), (12.11) and (12.12).

12.4.2 Numerical simulations

To check the validity of the solutions, we should simulate equations which describe the system with the minimum of approximations. It would be ideal to use the quantum equations, but they are not tractable, even at the numerical level. However, Equations (12.10)–(12.12) are perfectly suitable for a numerical check because they appear only as the first level of approximation of the exact quantum equations, and, moreover a discrete set of equations is more readily solved numerically than the partial differential equations that we derived from them. These partial differential equations are preferable for an analytical solution, but, if we wanted to simulate them we would have to discretise space again. It is thus more natural and simpler to start directly from the discrete equations.

It should however be stressed that such a test of the validity of the solution does not allow us to conclude in favour of the existence of solitons in a one-dimensional magnetic system because the equations that we simulate are only an approximation of the quantum equations, and the Hamiltonian (12.2) from which we derived the quantum equations is itself an incomplete picture of reality. For instance it neglects second-neighbour interactions, or the coupling of the spins with lattice distortions, which may modify the exchange integral. Only experiments can decide, but simulations are nevertheless very useful in practice because, before launching a complex experiment, it is necessary to determine the optimal conditions for its success. Numerical tests are appropriate for this purpose and moreover they provide data which can help in the analysis of the experimental results. The only unambiguous answer that simulation could give is a negative one: if it turned out that the solitons do not survive when they are tested in original equations, it would be hopeless to expect them in the actual magnetic material.

While numerical simulations cannot replace experiments, they cannot exempt us from the theoretical analysis that we carried out either. Looking numerically for solitons in a system as complex as the spin chain without a good starting point would have little chance of success.

The numerical calculations show that the sine-Gordon solutions are very good solutions when b and v are small, as we predicted. But simulations detect *additional branches of solutions.*

Figure 12.6 shows that, for a given value of the energy, two values of the velocity are possible in each direction [160, 188]. The out-of-plane deviations θ of two solutions with opposite velocities have opposite signs. In some velocity ranges, it is even possible to find three values for the energy, which correspond to three different solutions.

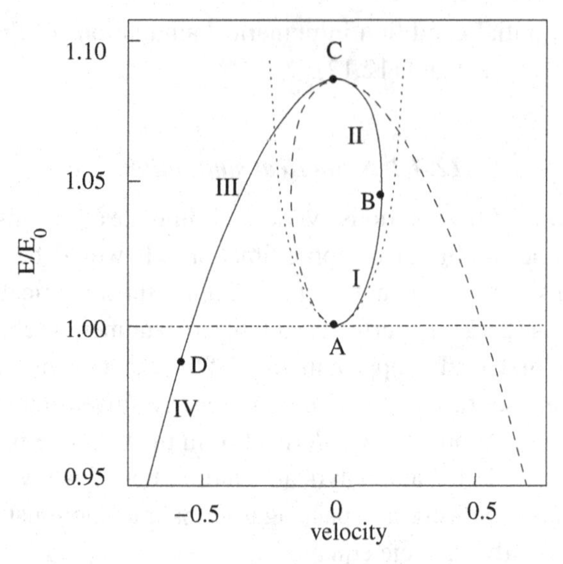

Figure 12.6. Plot of the ratio of the energy E of the different solutions of the spin chain to the energy E_0 of a static sine-Gordon soliton (adapted from reference [160]). The solid line and the dashed line correspond to the case where the out-of-plane deviation θ is negative and positive, respectively. The dotted lines for $v > 0$ or $v < 0$ show the energy–velocity relation for a sine-Gordon soliton.

The various branches of solutions which are observed in Figure 12.6 correspond to solutions which are different from each other, but nevertheless have the same qualitative shape. The variable ϕ evolves like a kink, which travels together with a pulse of the variable θ, as plotted in Figure 12.4. To simplify the discussion of the different solutions, let us restrict our attention to solutions with a negative out-of-plane deviation θ, which means that their velocity is positive (see Equation (12.32)). Figure 12.6 shows that there is a first branch, marked by I, between points A and B where the energy grows when the velocity increases, which is very similar to the sine-Gordon solution. On the other hand branches marked II, III, IV, exhibit very large deviations with respect to the SG soliton, for instance because the energy decreases when the velocity increases. Moreover a detailed study shows that, between points A and D in the figure, the out-of-plane deviation θ grows from 0 to $\pi/2$. At this point the nature of the soliton has deeply changed with respect to the sine-Gordon soliton. Instead of a rotation of the spins around the axis of the chain, their motion takes place in a plane containing the axis, and the soliton has lost its topological character. The gradual shift between topological solitons and nontopological localised excitations, which is exhibited by the numerical simulations [160], shows that the border between the different classes of solitons that we introduced in Part I is not tight.

Figure 12.6 shows the existence of two static solutions: the SG solution corresponding to $\theta = 0$ that we have already discussed (point A in Figure 12.6), but also another solution with a higher energy, at point C. For this second solution the angle θ is no longer zero, and the rotation of the spins is achieved through a deviation out of the xy plane.

Therefore the numerical simulations reveal a world much richer than the one that we discovered from the analytical study. However, in spite of all the approximations that we had to make, analytical calculations led to qualitatively correct solutions, and moreover they had the merit of pointing out the possible existence of localised solutions in a spin chain, completely different from the magnons.

12.4.3 Experimental observations

Experimental observations confirm the presence of solitons and magnons in spin chains, and thus they validate the analytical and numerical studies. They are carried out by nuclear magnetic resonance and inelastic neutron scattering.

Nuclear magnetic resonance (NMR)

The energy of an atom of spin $\vec{S}$ in a static magnetic field $\vec{B}$ parallel to the z axis is

$$U = -\gamma B_z S_z. \tag{12.33}$$

Since the component S_z of the spin is quantised (for instance $\pm 1/2\,\hbar$, $\pm 3/2\,\hbar$ and $\pm 5/2\,\hbar$ for the Mn^{2+} ion), this leads to a set of energy levels separated by $\gamma \hbar B_z$.

A magnetic field which oscillates at a frequency that we denote by ω_0 may induce a transition between these levels if $\hbar\omega_0 = \gamma\hbar B_z$. In a frequency scan, such a transition appears as a peak in the energy absorbed by the sample.

In a solid which contains an assembly of atoms with spin, the width of the NMR absorption peak is proportional to the inverse of the time (denoted by T_2) during which two individual spins are in phase. Here we meet again the idea, introduced in Chapter 10 on ferroelectrics, that the correlation function determines the response.

A soliton passing over a site temporarily reverses its spin, thereby reducing T_2. This is one of the methods used to detect the presence of solitons in magnetic systems.

Inelastic neutron scattering

This second experimental method is highly used because it provides a direct measurement of the dynamic structure factor $\widetilde{\mathbf{S}}_{\perp}(q, \omega)$ since the neutron scattering cross-section is proportional to this factor given by

$$\widetilde{\mathbf{S}}_{\perp}(q,\omega) = \int d^3\vec{r} \int dt \, \langle S_{\perp}(0,0) S_{\perp}(\vec{r},t)\rangle \, e^{i(\omega t - \vec{q}.\vec{r})}, \qquad (12.34)$$

where $S_{\perp}$ is the spin component which is orthogonal to the momentum transferred between the neutron and the spin lattice. A measurement of the dynamic structure factor is therefore a measurement of the spatial correlation function of the spins.

As discussed in Chapter 10, the presence of solitons leads to a characteristic 'signature' in the structure factor $\widetilde{\mathbf{S}}(q,\omega)$, the central peak which appears for low values of the wavevector q and frequency ω.

The thermodynamics of the soliton gas for the sine-Gordon model at temperature T shows that the width of the central peak is:

- Proportional to q, the other variables T and B being fixed.
- Proportional to T, the other variables q and B being fixed.
- Proportional to $\sqrt{B}$, the other variables q and T being fixed.

Moreover the amplitude of the peak is an exponential function of T and $B^{-1/2}$.

For a one-dimensional magnetic compound, these results are rather well verified experimentally. From the known microscopic parameters of $CsNiF_3$, the calculation gives an energy E_0 of the solitons equal to $E_0/k_B = 34$ K. Neutron scattering measurements lead to $E_0/k_B = 28$ K. Although the agreement is only crude, pointing out the limits of the sine-Gordon model, it nevertheless gives a remarkable proof of the value of the soliton concept because the central peak can by no means be explained in terms of the linear spin waves.

Recently experiments using electron spin resonance [17] have detected another type of soliton, the breather, in copper benzoate, which has a spin quantum number which is only 1/2, so that the quasi-classical approximation is not valid. These breathers have been described by a quantum version of the sine-Gordon model, which shows that the validity of the classical limit is not a prerequisite for the existence of solitons in a magnetic chain.

In magnetic systems the experimental results are in much better agreement with theory than in ferroelectrics because magnetic solitons, which describe the domain walls, are broad with respect to the lattice spacing, at least as long as the spin quantum number is high enough to allow a quasi-classical approximation. Thus they are generally not trapped by discreteness and the soliton-gas picture is good.

However the width of the magnetic domain walls depends on temperature because the substrate potential of the sine-Gordon equation is actually an effective potential. Its amplitude decreases when temperature increases because the thermal fluctuations of the spins tend to prevent their alignment along the external magnetic field. At very low temperature these fluctuations are very small and the effect of the external field is felt much more strongly by the spins. An increase of the amplitude

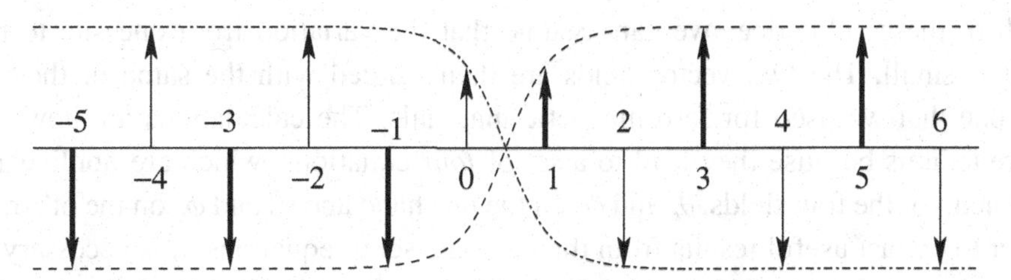

Figure 12.7. Schematic plot of a topological soliton in an antiferromagnetic chain. Spins of the odd sites have been marked by a thick line to make their identification easier, but all the atoms in the lattice carry the same spin. The dashed lines show the variation of the two fields associated with the even and odd spins in a continuum limit approximation for an antiferromagnetic material.

of the substrate potential leads to a decay in the soliton width (see Equation (2.22)). Thus, at very low temperature, we must expect the role of pinning by discreteness to be larger, which is even more enhanced because the solitons are less likely to be thermally activated over the maxima of the Peierls–Nabarro potential. This is confirmed by experiments.

In thin crystalline films of yttrium-iron garnet (YIG), sub-micron probes using the Hall effect have been used to measure the tiny variations of the magnetic field which are associated with the motion of magnetic domain walls. Very small motions of the domain walls can be detected [139]. These experiments have shown that the domain walls move by discrete steps, which have a size exactly equal to the lattice spacing in the direction of the motion of the domain wall (1.75 nm). These experiments have provided a direct observation of the Peierls–Nabarro potential that we discussed for dislocations in Chapter 9. Magnetic systems are appropriate for this observation because the motion of the domain walls induces variations in the local magnetic field, which can be measured accurately.

12.5 Solitons in antiferromagnetic spin chains

The case of antiferromagnetic spin chains is peculiar because, as shown in Figure 12.7, a soliton leads to a π rotation of the spins. Thus a spin reversal persists after a soliton has passed, and this is why antiferromagnetic solitons are much easier to detect in neutron scattering or NMR experiments than their ferromagnetic counterparts which reverse the spin only temporarily, while they are at a particular site [32].

From a theoretical point of view, this case is interesting too because it requires a special treatment of the continuum limit approximation. As shown in Figure 12.7, in such a medium, the spin is reversed from one site to the next, which seems to preclude any continuum limit approximation. The trick is to consider *two* coupled fields, a field for the even sites $S_e(x, t)$ and a field for the odd sites $S_o(x, t)$. In each of these sublattices we can assume that the variation from one site to the next is small. The two vector fields are then studied with the same method as the one that we used for ferromagnetic materials. The calculations are however more tedious because they lead to a set of *four* equations, which are nonlinearly coupled, for the four fields, θ_e and ϕ_e on the one hand and θ_o and ϕ_o on the other. In order to extract useful results from this intricate set of equations, it is necessary to determine the field variables which have an order of magnitude very different from the others to partially decouple the equations [129]. Theoretical studies [189], very well confirmed by experiments, have exhibited a large variety of excitations.

12.6 Conclusion

Magnetic materials provide very interesting examples, both theoretical and applied, in which mobile solitons exist in a microscopic system. They allow accurate measurements which are in good agreement with theoretical predictions, especially in the low temperature range, where thermal fluctuations are weak.

13
Solitons in conducting polymers

The application of the soliton concept in polymer physics is a spectacular example of a very fruitful interdisciplinary approach. Solitons were first suggested in this context by theoretical physicists, experts in quantum mechanics and field theory [86]. Chemists worked hard to synthesise materials with the requested characteristics before experimental physicists could study the properties of these solitons and confirm the theoretical predictions. This multi-step process culminated in the Nobel prize being awarded in October 2000 to Alan J. Heeger, Alan G. MacDiarmid and Hideki Shirakawa for their studies on the electric conductance of polymers, which is due to solitons.

13.1 Materials

13.1.1 Polyacetylene

Polyacetylene is the simplest conducting polymer, and it became the prototype of these materials. It is a chain of CH units where the electronic orbitals of the carbon are in the sp^2 hybridisation state, which allows the formation of three chemical bonds in a plane (called σ bonds), at 120° from each other. In addition it carries an extra electron, called the π electron, belonging to a p orbital, orthogonal to this plane, allowing the carbon atom to make a fourth bond. A covalent chemical bond is established when two atoms share two electrons. In a $(CH)_n$ chain, a simple bond between the atoms leads to three bonds per carbon atom, involving three of the four electrons available on a carbon atom for bonding. The fourth electron is involved in a double bond with one neighbouring carbon atom. Therefore if a chain of C atoms is bound by alternating single and double bonds, each C atom also being bound to a hydrogen atom, all the bonding possibilities of the atom are exhausted (one possible bond for H and four for C). As double bonds prevent rotation around the C–C bond, there are two different possibilities for building the chain, as shown in Figure 13.1: the *cis* and *trans* configurations. *Trans*-polyacetylene is a zig-zag

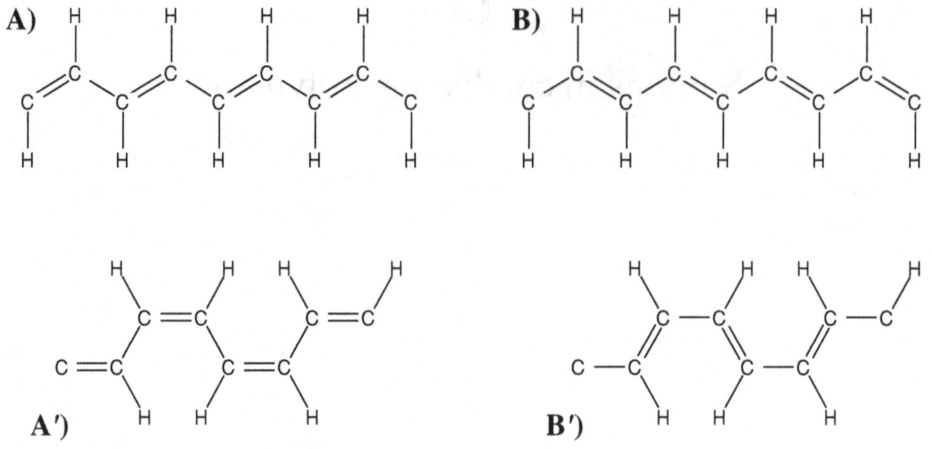

Figure 13.1. Plot of two *trans* configurations (A and B) and two *cis* configurations (A′ and B′) of polyacetylene.

carbon chain, while, in the *cis* configuration, every second C–C bond is parallel to the axis of the chain. The configurations plotted in Figure 13.1 are only schematic because actually double bonds are slightly shorter than single bonds.

For each geometry of the carbon chain corresponding to the *trans* or *cis* configuration, there are moreover two possibilities for selecting the bonds which are double bonds, as shown in Figure 13.1. In the *cis* configuration, these two possibilities, denoted by A′ and B′ in the figure, lead to two slightly different states because, in one case the double bonds are parallel to the axis of the chain, and in the other they are not. The energies of these two *cis* configurations are similar but not exactly equal.

On the other hand, in the *trans* geometry of the carbon chain, the two states denoted by A and B in Figure 13.1 are exactly identical since they are the image of each other in a mirror plane. Thus they have the same energy and it may appear superfluous to mention the existence of the two states. This would be true if *these two states could not coexist within the same molecule.* If this happens, they are connected by a defect in the bond alternation, as shown in Figure 13.2. The carbon atom sitting at the centre of the defect has an unpaired electron, but it is still neutral.

Therefore Figure 13.2 shows that *trans*-polyacetylene is a physical system which can exist in two energetically degenerate states, connected by a spatially localised 'defect'. This situation is *exactly the situation that we described when we introduced topological solitons.* This suggests that the 'defect' could actually be a topological soliton, and, if this is true, one could think that it might be mobile along the molecule. This idea led to a Nobel prize in chemistry . . . after a large interdisciplinary effort to validate it theoretically and experimentally.

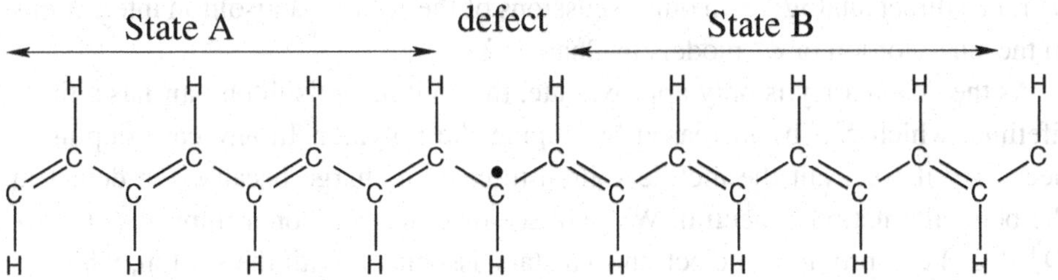

State A defect State B

Figure 13.2. Defect connecting the two possible states of *trans*-polyacetylene in a single polymer chain.

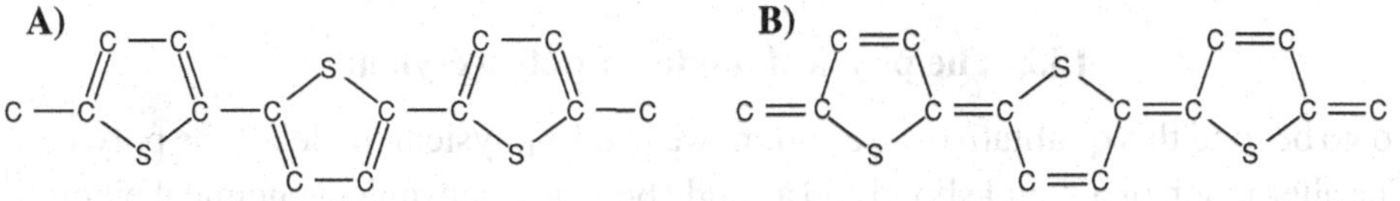

Figure 13.3. The two bonding structures of polythiophene, A and B.

This discovery originates from ideas of chemists in the 1960s, but as they thought that the defect was static, they did not give it the importance that it deserved. Indeed Figure 13.2 suggests that the defect extends over one unit of the chain only, so that discreteness effects would pin this localised structure. A more complete study shows that it is not the case. The defect extends over about seven units, and thus it may be highly mobile. As discussed below, this allows *trans*-polyacetylene to be turned into a very good electric conductor.

13.1.2 Other conducting polymers

This remarkable property of *trans*-polyacetylene comes from the alternation of single and double bonds forming a conjugated π electron system, which means that the π system extends over several neighbouring atoms. This is sometimes described in chemistry as the possibility of swapping the double and single bonds. Many other conjugated polymers may exhibit the same kind of property, allowing solitons. One of the most studied is polythiophene, which also allows two possibilities for setting double bonds as shown in Figure 13.3. This case is similar to *cis*-polyacetylene because the two states are only nearly degenerate.

In this case where the energy levels are different, with $E_B > E_A$, most of the chain is in the ground state A, but a small part may however be in the B state. The energy of the system increases when the size of the B region increases, which means that the two transition regions between the A and B states (the 'defects' in the bond alternation) are not free from each other. They behave like a soliton–antisoliton pair, in a direct analogy with our discussions of the soliton–antisoliton interactions in the sine-Gordon or ϕ^4 models in Chapter 2.

As the degeneracy is only approximate, the soliton–antisoliton pair has a finite lifetime, which can be increased by doping the polymer. In any case doping is necessary if we want the mobile soliton to carry a charge because the defect in the bond alternation is neutral. We shall see that electron donor impurities ($D \rightarrow D^+ + e^-$) can provide an electron to a state associated with the soliton, which is mobile with it. This has been used to synthesise conducting polyacetylenes which have an electric conductivity equal to one tenth of that of copper, one of the best electric conductors.

13.2 The physical model of polyacetylene

To go beyond this qualitative description, we need a physical model of the polymer. The alternation of a double bond and a single bond is a convenient chemical picture, but it does not allow a quantitative study, especially because it cannot describe the spatial extent of the defect in the alternation. A complete treatment of this problem is very complex, but Su, Schrieffer and Heeger (SSH) proposed a picture which contains the basic features which control the physics of the polymer [171]. The elements which must be part of a model are:

- The π electrons, which make the double bonds, because it is their switching from one bond to another which leads to the existence of the two states A and B;
- The distortions of the carbon chains, because the single and double bonds do not have the same length so that a swing of the π electrons from one bond to another is coupled with a deformation of the lattice.

The SSH Hamiltonian includes two parts, one for the dynamics of the carbon chain, linked by the σ bonds between the carbon atoms, and the other one for the π electrons,

$$H = H_\sigma + H_\pi . \tag{13.1}$$

The first step is to establish a satisfactory physical model of the dimerised chain, i.e. a chain which has a unit cell made of one single and one double bond, in the absence of any soliton. Then the defect in the bond alternation will be studied.

13.2.1 *Dynamics of the atoms*

The goal of SSH was to establish the simplest possible model, and thus they only considered the component of the lattice distortion along the direction of the chain. In their model, it is treated at the classical level. Let us denote by U_n the displacement of the nth CH group along the axis of the molecule, with respect to its equilibrium position in a regular structure where all bonds have the same length. The simplest Hamiltonian for the carbon chain is

$$H_\sigma = \sum_n \frac{P_n^2}{2m} + \frac{1}{2}K(U_n - U_{n+1})^2 \tag{13.2}$$

where P_n is the momentum of the nth CH group, m its mass, and K the coupling constant between the groups. The lattice distortion due to the double bonds is small since it is about 0.04 Å which should be compared to the length of the projection of a C–C bond on the chain axis, which is $a = 1.22$ Å. The harmonic expression chosen for the coupling energy is therefore perfectly justified.

13.2.2 The electronic Hamiltonian

Owing to the mass of the CH group, a classical description is satisfactory, but this is not true for the electrons. The simplest way to write their Hamiltonian is to use the second quantisation. We introduce a creation operator $c_n^\dagger$ and an annihilation operator c_n for an electron at site n. The operator $c_n^\dagger c_n$ is the number operator for the electrons at site n, while the operator $c_{n+1}^\dagger c_n$ describes the annihilation of an electron at site n and the creation of an electron at site $n+1$, i.e. the transfer of an electron from site n to site $n+1$. Similarly the operator $c_n^\dagger c_{n+1}$ describes an electron jump from site $n+1$ to site n.

To complete the electronic operator, these transfer operators must be multiplied by a factor $t_{n+1,n}$ which measures the probability of the jump (it is also called the overlap integral because it depends on the overlap of the electron eigenfunctions at sites n and $n+1$). The *coupling* between the atomic and electronic motions is included in the Hamiltonian by a dependence of $t_{n+1,n}$ upon the distance between the atoms, as

$$t_{n+1,n} = -[t_0 - \alpha(U_{n+1} - U_n)], \tag{13.3}$$

where α is a positive parameter. The negative sign in front of the overlap integral arises because the system lowers its energy by delocalising the electrons. With this choice for the interaction term, the probability of transfering an electron when a bond is extended by u with respect to equilibrium is $t_0 - 2\alpha u$, while it becomes $t_0 + 2\alpha u$ when the bond is shortened by u. Thus, it correctly leads to an increased transfer rate for the shortest bonds.

The Hamiltonian for the π electron is therefore

$$H_\pi = -\sum_n [t_0 - \alpha(U_{n+1} - U_n)]\big(c_{n+1}^\dagger c_n c_n^\dagger c_{n+1}\big). \tag{13.4}$$

It should be noticed that this Hamiltonian does not include a $c_n^\dagger c_n$ term, which would describe the energy of an electron at site n. Adding such a term would simply amount to changing the reference energy level for the electronic energy. This Hamiltonian implements what is generally called a 'tight-binding' description of the electronic state because it describes the π electrons at each site and their transfers in terms of electronic states which are localised at a site. It does not include the Coulomb interactions between the π electrons.

Each Hamiltonian H_σ and H_π has been written in a linearised approximation. However *nonlinearity is taken into account in the coupling between the two contributions* to the Hamiltonian. Starting from the two parts (13.2) and (13.4) of the total Hamiltonian which describe the coupling between the electrons and the ions, we have to proceed in three steps of increasing complexity: the static study without a soliton, then with it and finally a dynamic study.

13.3 The ground state of polyacetylene

From the chemical structure, when we look for the ground state, we expect a sequence of single and double bonds, i.e. an alternation of long and short bonds in the geometry of the carbon chain. Therefore we look for the distortion U_n of the lattice with respect to a regular lattice under the form

$$U_n = (-1)^n\, u \quad \text{where } u \text{ does not depend on } n. \tag{13.5}$$

With this hypothesis, the total Hamiltonian becomes

$$H = H_\sigma + H_\pi \tag{13.6}$$

$$= \sum_n \frac{P_n^2}{2m} + 2Ku^2 - [t_0 + \alpha(-1)^n 2u]\left(c_{n+1}^\dagger c_n + c_n^\dagger c_{n+1}\right). \tag{13.7}$$

Moreover the ground state, i.e. the state with the lowest possible energy, is a static state in which the lattice does not have any kinetic energy. Thus its Hamiltonian is simply

$$H = 2NKu^2 - \sum_n [t_0 + \alpha(-1)^n 2u]\left(c_{n+1}^\dagger c_n + c_n^\dagger c_{n+1}\right) \tag{13.8}$$

$$= 2NKu^2 + H_\pi, \tag{13.9}$$

which includes the energy of the static distortion of a chain of N CH groups and the corresponding electronic energy H_π.

To compute the electronic energy we have to determine the energy of the stationary states of the electrons in an atomic lattice, i.e. to find the eigenstates of the electronic Hamiltonian. We know that isolated atoms have well defined discrete energy levels. When the atoms are part of a lattice, their interactions induce a splitting of each of these levels, which leads to the *electronic energy bands* of the quantum theory of solids. Each band corresponds to a set of electronic states, which are labelled by a wavevector, which, in the one-dimensional case, is only a scalar k. The ground state of the system is obtained when the electrons occupy the lowest of these states, taking into account the Pauli exclusion principle which prevents two electrons from being in the same quantum state. The levels are thus filled in increasing order, until the last occupied level, called the Fermi level, denoted by E_F.

This calculation, which is standard in the quantum theory of solids, is generally performed with a fixed atomic lattice of period a. If a lattice distortion leads to a dimerisation with long and short bonds, the band structure is modified. In the middle of the allowed energy band, a forbidden band appears, creating a gap in the allowed energies. In order to understand the physics of polyacetylene, it is essential to see precisely how the calculation of the allowed electronic states is made. Therefore we shall first present a reminder of band theory, starting from the simple case of a regular lattice and then gradually introducing the features which characterise polyacetylene.

13.3.1 A reminder of band theory

Let us consider the Hamiltonian of a homogeneous one-dimensional electronic system,

$$H_h = \sum_n E_0 c_n^\dagger c_n - t_0 \left(c_{n+1}^\dagger c_n + c_n^\dagger c_{n+1} \right), \tag{13.10}$$

where we have included the on-site energy of an electron E_0, which is not in the SSH Hamiltonian, in order to clearly exhibit all contributions to the energy. It corresponds to the Hamiltonian H_π when α is equal to zero. To determine the electronic states, we must specify the boundary conditions. We shall choose the standard boundary conditions of solid state physics: we shall study a lattice with N sites and periodic boundary conditions.

Case $t_0 = 0$

Let us start with the simplest case, $t_0 = 0$, because it will give us hints for studying the case $t_0 \neq 0$. In the absence of any electronic transfer between sites, the Hamiltonian reduces to $H_0 = \sum_n E_0 c_n^\dagger c_n$ and its eigenstates are states localised at a single site. The state of the complete system can be labelled by specifying the number of electrons on each site, such as $|00...010...0\rangle$. If we restrict our attention to states with a single electron, we can use the short notation $|n\rangle$ to specify the site

n where the electron is located. The creation and annihilation operators act as usual on such a state

$$c_n|0\ldots010\ldots0\rangle = |0\ldots000\ldots0\rangle \tag{13.11}$$

$$c_n^\dagger|0\ldots000\ldots0\rangle = |0\ldots010\ldots0\rangle, \tag{13.12}$$

so that

$$c_n^\dagger c_n|0\ldots010\ldots0\rangle = |0\ldots010\ldots0\rangle \quad \text{i.e.} \quad c_n^\dagger c_n|n\rangle = |n\rangle, \tag{13.13}$$

while $c_p^\dagger c_p|n\rangle = 0$ for any integer $p \neq n$. Therefore, for any state $|n\rangle$ we have $H_0|n\rangle = E_0|n\rangle$. The N states $|n\rangle$ are eigenstates of the Hamiltonian, forming a base for the eigenspace associated with the eigenvalue E_0. However this is *not a convenient base because it does not have the symmetries of the system.*

The chain is invariant by the symmetry operator $\mathcal{T}(a)$ which corresponds to the translation along one lattice spacing a. This implies that H_h and $\mathcal{T}(a)$ commute, so that they have a common basis of eigenstates. This is also true for H_0 which is a particular case of the Hamiltonian H_h. The state $|n\rangle$ is an eigenstate of H_0 but it is not an eigenstate of $\mathcal{T}(a)$ because the translation operator moves the electron from site n to site $n+1$, which means that $\mathcal{T}(a)|n\rangle = |n+1\rangle$. An eigenstate of the translation operator can be obtained by a linear combination of the different states $|n\rangle$ as

$$|\chi_k\rangle = \frac{1}{\sqrt{N}} \sum_{p=1}^{N} \mathrm{e}^{ikpa}\,|p\rangle. \tag{13.14}$$

When N is odd, which is more convenient for the calculations, the index k takes the N values $[0, \pm1, \pm2, \ldots, \pm(N-1)/2] \times 2\pi/(Na)$, symmetrical with respect to zero. The kets $|\chi_k\rangle$ appear as the discrete Fourier transform of the set of kets $|n\rangle$.

Using the periodic boundary conditions, it is possible to check that $|\chi_k\rangle$ is indeed an eigenket of $\mathcal{T}(a)$ because

$$\mathcal{T}(a)|\chi_k\rangle = \mathcal{T}(a)\,\frac{1}{\sqrt{N}} \sum_{p=1}^{N} \mathrm{e}^{ikpa}\,|p\rangle = \frac{1}{\sqrt{N}} \sum_{p=1}^{N} \mathrm{e}^{ikpa}\,|p+1\rangle \tag{13.15}$$

$$= \frac{1}{\sqrt{N}} \sum_{p'=2}^{N+1} \mathrm{e}^{ik(p'-1)a}\,|p'\rangle = \mathrm{e}^{-ika}|\chi_k\rangle. \tag{13.16}$$

Moreover, as $|\chi_k\rangle$ is a linear combination of the eigenstates of H_0 which are all associated with the same eigenvalue E_0, $|\chi_k\rangle$ is also an eigenstate of H_0 for this eigenvalue:

$$H_0\,|\chi_k\rangle = \sum_n E_0 c_n^\dagger c_n \left(\frac{1}{\sqrt{N}}\sum_{p=1}^{N} \mathrm{e}^{ikpa}\,|p\rangle\right) \tag{13.17}$$

$$= E_0\,\frac{1}{\sqrt{N}}\sum_{p=1}^{N}\mathrm{e}^{ikpa}\sum_n c_n^\dagger c_n|p\rangle = E_0\,\frac{1}{\sqrt{N}}\sum_{p=1}^{N}\mathrm{e}^{ikpa}\,\underbrace{\sum_n \delta_{np}|p\rangle}_{|p\rangle}$$

$$= E_0\,|\chi_k\rangle. \tag{13.18}$$

Therefore we have derived a set of eigenstates for the Hamiltonian H_0, $\{|\chi_k\rangle\}$, which also obey the translational symmetry of the system. This set will be very useful for $t_0 \neq 0$.

Case $t_0 \neq 0$

Coming back to Hamiltonian (13.10), we immediately notice that the eigenstates $|n\rangle$ of H_0 are no longer eigenstates of H_h. However, as H_h is invariant by the translation $\mathcal{T}(a)$, H_h and $\mathcal{T}(a)$ must have a common basis of eigenstates. It turns out that this is again the basis $|\chi_k\rangle$, and we shall check it by computing the action of H_h on the state $|\chi_k\rangle$ in order to determine the energy $E(k)$ of this eigenstate.

As the first term of H_h is H_0 itself, we have

$$\sum_n E_0 c_n^\dagger c_n \left(\frac{1}{\sqrt{N}}\sum_{p=1}^{N}\mathrm{e}^{ikpa}\,|p\rangle\right) = E_0\,|\chi_k\rangle, \tag{13.19}$$

while the two other terms of H_h lead to

$$\sum_n c_{n+1}^\dagger c_n\,|\chi_k\rangle = \sum_n c_{n+1}^\dagger c_n \left(\frac{1}{\sqrt{N}}\sum_{p=1}^{N}\mathrm{e}^{ikpa}\,|p\rangle\right) \tag{13.20}$$

$$= \frac{1}{\sqrt{N}}\sum_{p=1}^{N}\mathrm{e}^{ikpa}|p+1\rangle = \frac{1}{\sqrt{N}}\sum_{p'=1}^{N}\mathrm{e}^{ik(p'-1)a}\,|p'\rangle \tag{13.21}$$

$$= \mathrm{e}^{-ika}\,|\chi_k\rangle \tag{13.22}$$

and

$$\sum_n c_n^\dagger c_{n+1}\,|\chi_k\rangle = \mathrm{e}^{ika}\,|\chi_k\rangle. \tag{13.23}$$

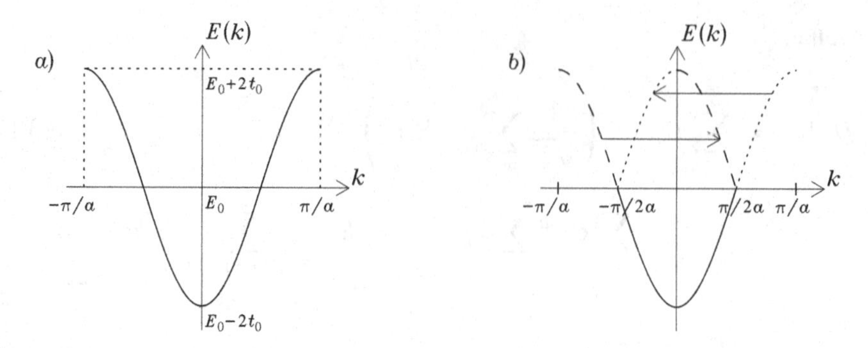

Figure 13.4. (a) Plot of the energy bands of a lattice of period a. (b) If we assume that the chain is a system of period $2a$, we get the reduced Brillouin zone picture. The arrows show the branches of the $E(k)$ curve which have been moved by the folding of the Brillouin zone.

Putting together these three results, we get

$$H_h\,|\chi_k\rangle = E_0\,|\chi_k\rangle - t_0(\mathrm{e}^{-ika} + \mathrm{e}^{ika})|\chi_k\rangle \tag{13.24}$$

$$= [E_0 - 2t_0\cos ka]|\chi_k\rangle = E(k)\,|\chi_k\rangle. \tag{13.25}$$

Consequently, when k varies within the range $[-\pi/a, \pi/a]$, which is called the *Brillouin zone* of the one-dimensional lattice of period a, the energy $E(k)$ takes its values in the range $[E_0 - 2t_0, E_0 + 2t_0]$ (Figure 13.4). When N is very large, the set of eigenvalues $E(k)$ tends to form a continuum, the band of the allowed electronic energies. Moreover this calculation shows that the term $E_0 c_n^\dagger c_n$ in the Hamiltonian only leads to a shift of all the energy levels. Thus it is possible to choose E_0 as the origin of the energies without loss of generality. It is this choice which has been made in the SSH Hamiltonian.

The folding of the Brillouin zone

We know that polyacetylene tends to dimerise into a lattice where the unit cell contains two CH groups. Therefore, in preparation for the study of polyacetylene, it is useful to view the homogeneous chain ($t_{n,n+1} = t_0$) as a system having the period $2a$. Doubling the unit cell amounts to dividing by two the values of k which define the limits of the Brillouin zone. But, as the physical system has not been modified, this new picture ought to give the same values as before for the electronic energies. This is the case if we 'fold' the band structure on the interval $[-\pi/(2a), \pi/(2a)]$ in a scheme called the 'reduced Brillouin zone' scheme.

Choosing henceforth $E_0 = 0$, the expression for the electronic energy is

$$E(k) = -2t_0\cos ka. \tag{13.26}$$

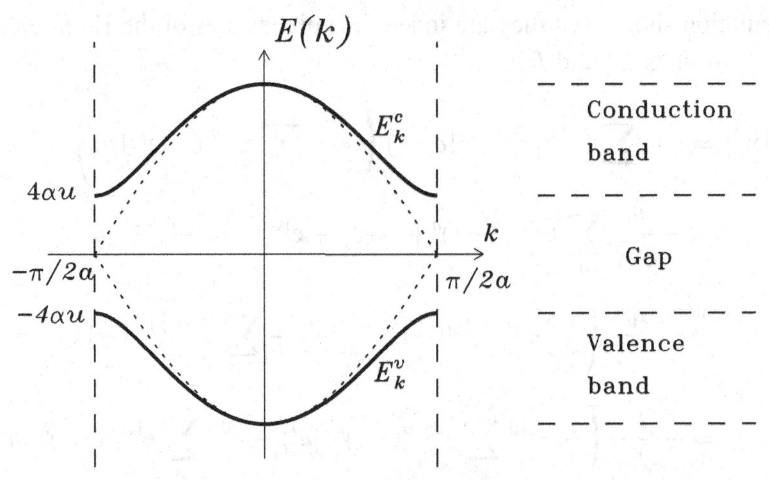

Figure 13.5. Emergence of a gap in the presence of a lattice distortion. The dotted lines are the band structures (13.27) and (13.28) for the homogeneous Hamiltonian H_h, in the reduced Brillouin zone. The solid lines correspond to Equation (13.36) for the Hamiltonian H_π and its negative counterpart E^c_k.

It is simple to check that the transformations $k' = k - \pi/a$ and $E_{k'} = -E_k$ bring the part of the curve $E(k)$ that was in the domain $[\pi/(2a), \pi/a]$ when we were considering the lattice of period a into the domain $[-\pi/(2a), \pi/(2a)]$ without changing the values of the energy. Similarly, it is possible to fold in the reduced Brillouin zone the part of the curve $E(k)$ that was in the range $[-\pi/a, -\pi/(2a)]$ (Figure 13.4(b)).

This transformation replaces the single equation (13.26) by *two* branches,

$$E^v_k = -2t_0 \cos ka \tag{13.27}$$

$$E^c_k = 2t_0 \cos ka, \tag{13.28}$$

with k now belonging to the reduced Brillouin zone $[-\pi/(2a), \pi/(2a)]$ (Figure 13.4(b)). The exponents v and c refer to the valence and conduction band of the material, which are defined in Figure 13.5.

To complete the transformation to the reduced Brillouin zone, we can introduce the eigenstates associated with the two branches,

$$|\chi^v_k\rangle = \frac{1}{\sqrt{N}} \sum_{p=1}^{N} e^{ikpa} |p\rangle, \tag{13.29}$$

$$|\chi^c_k\rangle = \frac{1}{\sqrt{N}} \sum_{p=1}^{N} e^{ikpa} (-1)^p |p\rangle. \tag{13.30}$$

The calculation shows that they are indeed the eigenstates of the Hamiltonian H_h for the eigenvalues E_k^v and E_k^c:

$$H_h\,|\chi_k^c\rangle = -t_0 \sum_n \left(c_{n+1}^\dagger c_n + c_n^\dagger c_{n+1}\right)\left(\frac{1}{\sqrt{N}}\sum_p \mathrm{e}^{\mathrm{i}kpa}(-1)^p\,|p\rangle\right) \tag{13.31}$$

$$= -\frac{t_0}{\sqrt{N}}\sum_n \left(\mathrm{e}^{\mathrm{i}kna}(-1)^n\,|n+1\rangle + \mathrm{e}^{\mathrm{i}k(n+1)a}(-1)^{n+1}\,|n\rangle\right) \tag{13.32}$$

$$= -\frac{t_0}{\sqrt{N}}\left(\sum_{n'} \mathrm{e}^{\mathrm{i}k(n'-1)a}(-1)^{n'-1}\,|n'\rangle + \sum_n \mathrm{e}^{\mathrm{i}k(n+1)a}(-1)^{n+1}\,|n\rangle\right)$$

$$= -\frac{t_0}{\sqrt{N}}\left(-\mathrm{e}^{-\mathrm{i}ka}\sum_{n'} \mathrm{e}^{\mathrm{i}kn'a}(-1)^{n'}\,|n'\rangle - \mathrm{e}^{\mathrm{i}ka}\sum_n \mathrm{e}^{\mathrm{i}kna}(-1)^n\,|n\rangle\right)$$

$$= -\frac{t_0}{\sqrt{N}}(-2\cos ka)\sum_n \mathrm{e}^{\mathrm{i}kna}(-1)^n\,|n\rangle \tag{13.33}$$

$$= 2t_0 \cos ka\,|\chi_k^c\rangle = E_k^c\,|\chi_k^c\rangle. \tag{13.34}$$

A similar calculation confirms that

$$H_h\,|\chi_k^v\rangle = -2t_0 \cos ka\,|\chi_k^v\rangle = E_k^v|\chi_k^v\rangle. \tag{13.35}$$

13.3.2 The band structure of polyacetylene

Let us now examine the Hamiltonian H_π of polyacetylene when α is not equal to zero. We select a value of u, which is *fixed*, and we want to determine the band structure when the overlap integral has two alternate values $t_0 \pm 2\alpha u$ instead of the unique value t_0. The calculation is more complex [171] but the result is simply a deformation of the two branches that we obtained for $\alpha = 0$. Equation (13.27) becomes

$$E_k^v = -\sqrt{4t_0^2 \cos^2 ka + 16\alpha^2 u^2 \sin^2 ka}. \tag{13.36}$$

and the second branch of the curve is given by $E_k^c = -E_k^v$. The existence of two different values for the overlap integral has lifted the degeneracy at the boundaries of the Brillouin zone (Figure 13.5). The energy band that was spanning the full range from $-2t_0$ to $2t_0$ is now divided in two bands, separated by the gap $8\alpha u$. The lower band is called the *valence band* while the upper one is the *conduction band.*

Since we consider a chain of N –CH groups, i.e. $N/2$ unit cells of size $2a$, the index k takes $N/2$ values, so that the curve relative to each energy band corresponds to $N/2$ energy levels. Each of the eigenstates $|\chi_k\rangle$ corresponds to two quantum states because an electron has two possible spin states, so that each energy band contains N quantum states. The chain has N π-electrons, and when they are distributed in the available states according to the Pauli exclusion principle, they exactly fill the valence band. Referring to the total number of states in the band without dimerisation, polyacetylene is sometimes called a 'half-filled band system'.

This distribution of the electronic states explains why polyacetylene tends to spontaneously dimerise into an alternation of long and short bonds because, as the width of the gap is proportional to the distortion u, when u increases, the energies of the electrons in the valence band decrease, particularly for the states lying in the vicinity of the Brillouin zone boundary. The lowering of the valence band energy levels is exactly compensated by an increase of the levels in the conduction band, but this is irrelevant for the energy of the system because these states are not occupied. When only the valence band is filled, the one dimensional lattice can therefore lower its electronic energy by creating a distortion $u \neq 0$, which is known as the Peierls distortion. It explains why polyacetylene is found in a dimerised state where a long and a short bond alternate but, moreover, it shows that *polyacetylene is an insulator* because its valence band is completely filled.

However, the electronic energy gain due to the distortion is not the only energy change introduced by the distortion. We must not forget that the energy of polyacetylene also includes the contribution H_σ. Distorting the lattice has an energetic cost in the σ bonds that connect the carbon atoms, according to Equation (13.9). The equilibrium structure of polyacetylene is achieved for a nonzero value of u such that the electronic energy gain is exactly balanced by the energetic cost of the lattice distortion. This is the condition that determines the dimerised state. This result is a particular case of Peierls' theorem [145] which states that a one-dimensional metal would be unstable against a lattice distortion creating an energy gap above the top occupied state, turning it into an insulator. This theorem forbids metallic electric conduction in a one-dimensional system.

The energy $E_0(u)$ of the ground state of polyacetylene is the sum of the energies E_k^v of the valence band, each state being occupied by two electrons, and the energy of the distortion of the σ bonds. It is equal to

$$E_0(u) = 2KNu^2 + 2\sum_k E_k^v \tag{13.37}$$

$$= 2KNu^2 - 2\sum_k \sqrt{4t_0^2 \cos^2 ka + 16\alpha^2 u^2 \sin^2 ka}. \tag{13.38}$$

For a very large N, the sum can be replaced by an integral taking into account that successive values of k differ by $2\pi/(Na)$ because we have $(N/2)$ k values in the range $-\pi/(2a)$ and $\pi/(2a)$. It gives

$$E_0(u) = 2KNu^2 - 2\int_{-\pi/2a}^{\pi/2a} \mathrm{d}k \frac{Na}{2\pi} \sqrt{4t_0^2 \cos^2 ka + 16\alpha^2 u^2 \sin^2 ka}. \tag{13.39}$$

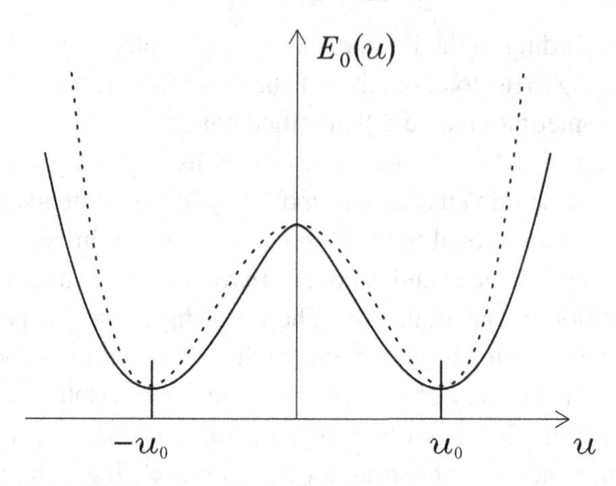

Figure 13.6. The solid line shows the ground state energy $E_0(u)$ of polyacetylene as a function of the lattice distortion u measuring the displacements of the CH groups with respect to the regular reference lattice. The dotted line is the potential (13.44) which allows an analytical study of the defect between the two ground states of polyacetylene.

Introducing the variable $z = 2\alpha u/t_0$ and the elliptic integral

$$E_\ell(m) = \int_0^{\pi/2} \mathrm{d}\theta (1 - m\sin^2\theta)^{1/2}, \tag{13.40}$$

Equation (13.39) can be written in the more compact form

$$E_0(u) = N\left(\frac{Kt_0^2}{2\alpha^2}z^2 - \frac{4t_0}{\pi}E_\ell(1 - z^2)\right). \tag{13.41}$$

A more explicit expression in terms of simple functions can be obtained by expanding the elliptic integral $E_\ell(1 - z^2)$ in the vicinity of $z = 0$,

$$E_0(u) \simeq N\left(\frac{Kt_0^2}{2\alpha^2}z^2 - \frac{4t_0}{\pi}\left[1 + \frac{1}{2}\left(\ln\frac{4}{z} - \frac{1}{2}\right)z^2 + \mathcal{O}(z^4)\right]\right). \tag{13.42}$$

Figure 13.6 shows that the ground state has two minima for $u = \pm u_0$. Therefore this nonperturbative calculation shows that the ground state of polyacetylene has to be dimerised since Figure 13.6 shows that the homogeneous state $z = 0$ is unstable. The distortion with respect to the regular lattice of period a is $U_n = (-1)^n u$ according to Equation (13.5) so that the minima of u, which have opposite signs, differ by the switching of the long and short bonds. They correspond to the two states of *trans*-polyacetylene denoted by A and B in Figure 13.1.

It is interesting to consider some numerical values for polyacetylene. Using the parameters proposed by SSH [171], $\alpha = 4.1$ eVÅ^{-1}, $K = 21$ eVÅ^{-2} and $t_0 = 2.5$ eV, we get a distortion $u_0 \simeq 0.04$ Å, leading to a variation of the length of the C–C bonds equal to $\pm\sqrt{3}u_0 = 0.073$ Å if we take into account the angle of 60° between the bonds and the axis of the chain, along which u is measured. The energy gap is equal to 1.3 eV, which is much higher than the energy $k_B T$ of the thermal fluctuations at room temperature which is about 0.025 eV. This is too high to allow the thermal excitation of charge carriers into the conduction band. This is why polyacetylene in its ground state is an electric insulator.

13.4 The excited state of polyacetylene: the soliton solution

13.4.1 Method

As the study of the ground state has found two degenerate minima, topological solitons interpolating between these two ground states can be expected. It is possible to determine the structure of this interface and its influence on the electronic states following the steps proposed by SSH [171]. A polymer which is in state A on one side and in state B on the other side is studied. The interface between these two regions is assumed to extend over about 50 unit cells and the aim of the investigation is to determine its structure. The Hamiltonian is decomposed into $H = H_1 + V$ where H_1 corresponds to the A and B domains, which are in a homogeneous ground state, while V describes an extended perturbation. In order to specify V, an assumption is made for the general shape of the lattice distortion within the interface: u_n, defined by $U_n = (-1)^n u_n$, is assumed to vary according to the formula

$$u_n = -u_0 \tanh \frac{n}{\ell}, \tag{13.43}$$

where ℓ is a parameter. This hypothesis, which is qualitatively correct owing to the boundary conditions for the interface, determines $t_{n+1,n}$ within the interface. Therefore the quantum problem is completely defined for the regions A and B, as well as inside the interface, because the overlap integral is specified everywhere. The ground state of the quantum problem, with given boundary conditions in the regions A and B can therefore be computed, using a combination of analytical and numerical methods. The last step is to use a variational method to look for the value of the parameter ℓ which leads to the minimal energy for the system. A value of $\ell \approx 7$ unit cells is found, and the corresponding energy of the soliton is found to be $E_{sol} \simeq 0.42$ eV.

It is however possible to perform an analytical study, which can be used to analyse several experimental results, if we assume that the energy $E_0(u)$, that we established for the polyacetylene chain, is also *valid in the core of the interface.* This is obviously an approximation. The value of $E_0(u)$ has been derived by assuming a particular value of u and it is valid for any u, not only the values $u = \pm u_0$ which minimise the energy, but *it was derived with the assumption that the distortion is the same everywhere along the polymer.* Strictly speaking it should not be applied within the interface between the A and B states because u varies in this region. The approximation that we shall use is therefore better when u_n varies only slightly from one site to the next, i.e. it is analogous to the continuum limit approximation. We have to check a posteriori that the soliton is wide enough to justify its validity.

Figure 13.6 shows that, if we adjust the height of the barrier between the two wells by selecting a value of the parameter V_0 equal to the barrier obtained by the quantum calculation, the function

$$V(u) = V_0 \left(1 - \frac{u^2}{u_0^2}\right)^2 \tag{13.44}$$

gives a satisfactory approximation of the function $E_0(u)$ in the range $[-u_0, u_0]$ in which u evolves across the interface.

If we subtract the energy of the σ bonds, we get an approximate expression for the energy of the π electrons in a unit cell,

$$E_\pi(u) = E_0(u) - 2Ku^2 = V_0 \left(1 - \frac{u^2}{u_0^2}\right)^2 - 2Ku^2. \tag{13.45}$$

In the same spirit as the calculation performed by SSH, we shall assume that we can use this expression *derived from a calculation performed on a homogeneous chain* to write an approximate Hamiltonian *for a polymer which is no longer in the homogeneous ground state.* Thus we assume that the expression (13.45) of $E_\pi(u)$ stays valid locally in a system where u_n varies along the lattice. Adding the kinetic energy of the CH groups, and the coupling energy between neighbouring groups due to the σ bonds, we obtain the Hamiltonian

$$H = \sum_n \frac{1}{2} m \left(\frac{\mathrm{d}u_n}{\mathrm{d}t}\right)^2 + \frac{1}{2} K (u_n + u_{n+1})^2 + \left[V_0 \left(1 - \frac{u_n^2}{u_0^2}\right)^2 - 2Ku_n^2 \right]. \tag{13.46}$$

The + sign in the coupling term $(1/2)K(u_n + u_{n+1})^2$, which may look unusual, comes from the transformation $U_n = (-1)^n u_n$.

The appeal of this formula is that it provides an expression of the energy of the polymer only in terms of the classical variables of the model, the positions u_n of the CH groups. However the quantum effects *have not been ignored* because they were included in the calculation of $E_0(u)$, from which we derived the energy per unit cell associated with the π electrons.

13.4.2 The soliton solution

Let us define $\phi = u/u_0$ and change to time and space dimensionless variables $\tau = \omega_0 t$ and $\xi = sx$ with $\omega_0^2 = 4V_0/(mu_0^2)$ and $s = 4V_0/(Ku_0^2a^2)$. Moreover we make a continuum limit approximation, which is consistent with the condition for the validity of the expression for the π electron energy. The Hamiltonian (13.46) becomes

$$H = A\int_{-\infty}^{+\infty} \mathrm{d}\xi \left[\frac{1}{2}\left(\frac{\partial\phi}{\partial\tau}\right)^2 + \frac{1}{2}\left(\frac{\partial\phi}{\partial\xi}\right)^2 + B\phi\frac{\partial\phi}{\partial\xi} + \frac{1}{4}\left(1-\phi^2\right)^2\right], \tag{13.47}$$

$$= A\int_{-\infty}^{+\infty} \mathrm{d}\xi\ \mathcal{H}, \tag{13.48}$$

where $A = 2\sqrt{V_0 K u_0^2}$ and $B = \sqrt{Ku_0^2/V_0}$ are two constants. We can immediately notice that the term $B\phi\phi_\xi$ can be written as an exact spatial derivative $B\left(\phi^2\right)_\xi/2$. Therefore its spatial integration gives a value which only depends on the boundary conditions at infinity. As we look for solutions where the system is in one of its ground states $\phi = \pm 1$ at infinity, the term $B\phi\phi_\xi$ can be ignored because its contribution to the Hamiltonian is only a constant. We shall consider henceforth the Hamiltonian density

$$\mathcal{H}' = \frac{1}{2}\left(\frac{\partial\phi}{\partial\tau}\right)^2 + \frac{1}{2}\left(\frac{\partial\phi}{\partial\xi}\right)^2 + \frac{1}{4}(1-\phi^2)^2. \tag{13.49}$$

Denoting by Π the momentum, conjugate of the variable ϕ, we can write the partial differential equation for ϕ which derives from this Hamiltonian. The Hamilton equations for a continuous medium give

$$\frac{\partial\phi}{\partial\tau} = \frac{\partial\mathcal{H}'}{\partial\Pi} = \Pi \tag{13.50}$$

$$\frac{\partial\Pi}{\partial\tau} = -\frac{\partial\mathcal{H}'}{\partial\phi} + \frac{\partial}{\partial\xi}\frac{\partial\mathcal{H}'}{\partial\phi_\xi} = \phi(1-\phi^2) + \frac{\partial^2\phi}{\partial\xi^2}. \tag{13.51}$$

If we combine them together we get the equation

$$\frac{\partial^2\phi}{\partial\tau^2} - \frac{\partial^2\phi}{\partial\xi^2} - \phi(1-\phi^2) = 0, \tag{13.52}$$

which is the dynamic equation of the ϕ^4 model that we introduced in Section 2.4. Its constant-profile kink solution is

$$\phi(\xi,\tau) = \pm\tanh\frac{\xi - v\tau}{\sqrt{2(1-v^2)}}. \tag{13.53}$$

With the parameters $u_0 = 0.04$ Å, $K = 920$ eVÅ^{-2}, $V_0 = 0.015$ eV and $m = 13 \times 1.67 \cdot 10^{-27}$ kg, we get a width for the static soliton solution $\sqrt{2}/s \approx 7$ unit cells, which is in agreement with the numerical solution of SSH. For such a width the continuum limit approximation is valid. It should however be noticed that, in order to reach this result, we assumed a coupling constant K which is much higher than the value used in the numerical solution of SSH (21 eVÅ^{-2}) which would have given a width of only one unit cell with the model that we introduced. This model, which is interesting because it allows analytical calculations, requires the fitting of one parameter to give acceptable results.

If we introduce the solution (13.53) into the Hamiltonian (13.48) with the Hamiltonian density (13.49), we can calculate the energy of the static soliton,

$$E_{sol} = A \int_{-\infty}^{+\infty} d\xi \left[\frac{1}{2} \left(\frac{1}{\sqrt{2}} \operatorname{sech}^2 \frac{\xi}{\sqrt{2}} \right)^2 + \frac{1}{4} \left(1 - \tanh^2 \frac{\xi}{\sqrt{2}} \right)^2 \right] \tag{13.54}$$

$$= A \int_{-\infty}^{+\infty} \sqrt{2}\, dX \left[\frac{1}{4} \operatorname{sech}^4 X + \frac{1}{4} (1 - \tanh^2 X)^2 \right] \tag{13.55}$$

$$= \frac{A}{\sqrt{2}} \underbrace{\int_{-\infty}^{+\infty} dX \operatorname{sech}^4 X}_{4/3} = \frac{2\sqrt{2}}{3} A. \tag{13.56}$$

Using the numerical values that we introduced above, we get $A = 0.29$ eV and a soliton energy $E_{sol} = 0.28$ eV, which is of the same order of magnitude as the energy of 0.42 eV obtained in the SSH calculation.

If we treat the soliton as a free quasi-particle, with mass M, moving in a medium where the square of the sound velocity is $c_0^2 = Ka^2/m$, its rest energy $Mc_0^2 = 0.28$ eV is much larger than the thermal energy which is of the order of $k_B T \simeq 0.025$ eV at room temperature. Therefore the solitons cannot be thermally created. But if they exist in the lattice, thermal fluctuations can give them some kinetic energy. A simple *estimate* of the mean square velocity of the solitons at room temperature can be made. Assuming that their thermal energy is small with respect to their rest energy, we can use the nonrelativistic expression $Mv^2/2$ of their kinetic energy. We get

$$\frac{v^2}{c_0^2} \simeq \frac{k_B T}{E_{sol}} = \frac{0.025}{0.28} \quad \text{i.e.} \quad v \simeq 0.3\, c_0. \tag{13.57}$$

With the numerical values given above, $c_0 \simeq 10^5$ m s^{-1}, so that an estimate of the thermal velocity of the solitons is $3 \cdot 10^4$ m s^{-1}. This very high value partly comes from the large K that we had to introduce in our simplified analytical model to get a correct soliton width, but is mainly due to the very small effective mass of the carriers, $M = E_{sol}/c_0^2 \simeq 4.6 \cdot 10^{-30}$ kg, i.e. about five electron mass. The quantum calculation of SSH gives six electron mass, so that both approaches are in rather good agreement and confirm that the soliton is indeed a quantum object. This small effective mass is consistent with the small distortion u_0 of the carbon lattice in the core of the soliton. The actual velocity of the carriers is much lower than our crude estimate based on a free particle description. Discreteness effects, although they are very weak are not completely negligible, but the most important factor that reduces the carrier velocity is their diffusion by phonons and impurities, such as the donor centres introduced when polyacetylene is doped to give a charge to the solitons.

13.5 Mechanism of electric conduction in conducting polyacetylene

13.5.1 The principle

The SSH study and the model that we introduced show that polyacetylene has excited states which are approximately described by topological solitons and have a width of seven unit cells. This width, significantly larger than the unit cell, validates the use of the expression of $E_0(u)$ obtained in a homogeneous system to derive the π electron energy. For such a width, discreteness effects are weak enough to allow the solitons to be highly mobile.

Therefore solitons would be very good candidates to play the role of charge carriers and explain the electric conductivity of polyacetylene ... if they were charged. This is not the case as Figure 13.2 suggests: each double bond involves two π electrons, each neighbouring carbon atom providing one. When there is a defect in the single–double bond alternation, the carbon atom at the defect site keeps its π electron, which is not involved in a double bond, and the system stays neutral.

However the quantum solution obtained by SSH shows that the electronic states are modified by the soliton. The distortion of the –CH lattice is associated with the appearance of an electronic energy state which is localised around the soliton centre, contrary to the other states of the energy bands which are extended states. The energy of this new state is *in the middle of the energy gap* between the valence and conduction bands (Figure 13.7(a)). The presence of this extra state must be balanced by a decrease of the electronic density of states in the band, which essentially occurs near the Brillouin zone boundaries.

This mid-gap energy level associated with the soliton is the state occupied by the lonely π electron of Figure 13.2. As the spin of this electron is not paired with the spin of another electron, the soliton appears as *a neutral quasi-particle, which has a spin* 1/2. The existence of this extra state is important because it can be detected experimentally by photo-excitation, providing an indirect proof of the existence of the soliton.

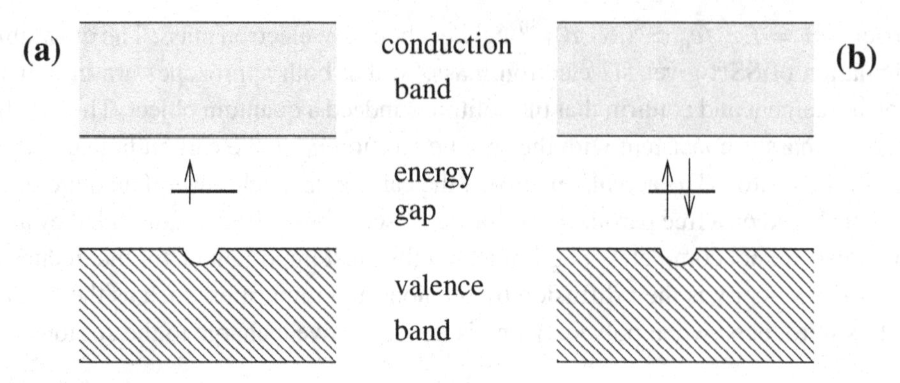

Figure 13.7. Electronic energy band structure for (a) a neutral soliton and (b) a soliton charged by an electron provided by a donor.

Although it is intrinsically neutral, the soliton can nevertheless play a role in the electric conductivity of polyacetylene because the localised electronic state due to the soliton can be populated by a second electron, provided by doping. The principle of this doping is to introduce electron donors to the material, that we shall denote by D. They are intercalated within the polymer chains in the solid, and they lose one electron, which goes to the mid-gap electronic states (Figure 13.7(b)) rather than to the valence states which have a higher energy. The donors stay as D^+ ions, fixed in the polymer matrix.

The charged solitons are *pseudo-particles with a charge* $-e$ *and spin zero* because the mid-gap state is now populated by two electrons which have antiparallel paired spins. Thus, the collective soliton mode allows us to conceive 'particles' with an unusual spin–charge combination. In a nonhydrogenated carbon chain, where single and triple bonds alternate, similar mechanisms can lead to pseudo-particles with the fractional charge $-e/3$.

Using appropriate dopants [86], polyacetylene can be transformed from a very good insulator to a good electric conductor, with conductance of about one tenth of the conductance of copper.

However the charged soliton is not fully free because it interacts with the D^+ ion that stays at some distance d from the polymer chain after the donor atom has lost its electron. The soliton can oscillate in the attractive potential of the ion, forming an oscillatory dipole which has an oscillation frequency lying in range of infrared electromagnetic waves. This is why infrared spectroscopy could be used to study the properties of solitons (Figure 13.8).

It should be noticed that, although it can move like a quasi-particle and has the analytical expression of a soliton, the interface between the two states of *trans*-polyacetylene is not exactly a soliton because it is not preserved in collisions with

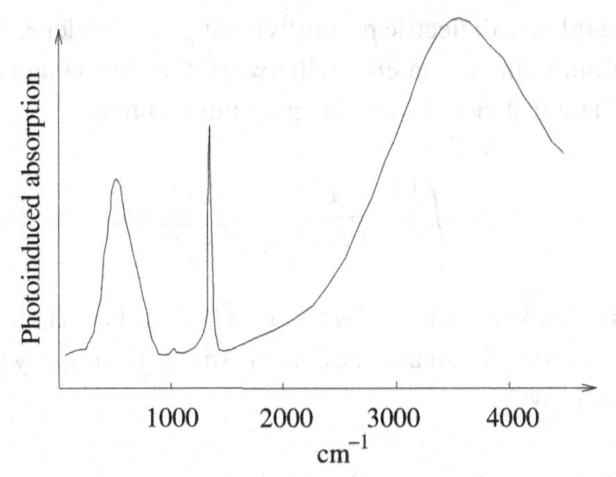

Figure 13.8. Absorption spectrum of *trans*-$(CH)_x$ in the infrared range (from reference [167]). The broad absorption band around 3560 cm^{-1} is due to an electronic transition. The other bands are due to solitons. The mode around 500 cm^{-1} is the translation mode, while the mode around 1300 cm^{-1} is the soliton internal mode. In this experiment, in order to get a soliton density high enough for good detection, the solitons were created by photo-excitation of polyacetylene.

another interface. However, as shown below, the soliton picture turns out to be very useful to understand some of the experimental results that have contributed to our knowledge of the nature of the charge carriers in polyacetylene.

13.5.2 Dynamics of the charged soliton

The charge density $\rho(x)$ of the doped soliton is determined by the wavefunction $\phi_0(x)$ of the electronic state localised around the soliton through

$$\rho(x) = -e|\phi_0(x)|^2. \tag{13.58}$$

The function $|\phi_0(x)|^2$, obtained in the SSH calculation, has the shape of a peak centred on the soliton site and having a spatial extent corresponding to the size of the lattice distortion described by the soliton solution $u(x) = u_0\,\phi(s\,x)$, where ϕ is given by Equation (13.53). The derivative of this solution with respect to x, which exhibits a maximum at $x = 0$, is $u_x = s\,(u_0/\sqrt{2})\,\text{sech}^2(s\,x/\sqrt{2})$. It has the same shape and the same spatial extent as $|\phi_0(x)|^2$ and it even coincides with the approximate expression of $|\phi_0(x)|^2$ obtained by SSH [171]. We shall therefore assume that the charge density associated with the soliton is given by

$$\rho(x) = -\frac{e}{2}\frac{1}{u_0}\frac{\partial u}{\partial x} = -\frac{es}{2}\frac{\partial \phi}{\partial x}. \tag{13.59}$$

Introducing the relative dielectric permittivity of polyacetylene, $\varepsilon_r = 2.3$, we can evaluate the Coulomb interaction energy between the soliton and a donor located at position $x = 0$ at a distance d from the polymer chain as

$$U = \int_{-\infty}^{+\infty} \frac{e}{4\pi\varepsilon_0\varepsilon_r} \frac{\rho(x)}{\sqrt{d^2 + x^2}} \mathrm{d}x. \tag{13.60}$$

This expression is clearly dominated by the $x = 0$ contribution, in particular when the distance d between the donor and the chain is small. To simplify the calculations, let us approximate it by

$$U' = \frac{e\rho(0)}{4\pi\varepsilon_0\varepsilon_r} \frac{a}{d} \tag{13.61}$$

so that

$$U' = \frac{e}{4\pi\varepsilon_0\varepsilon_r} \frac{a}{d} \frac{(-es)}{2} \phi_x(0) = -\frac{e^2 s}{8\pi\varepsilon_0\varepsilon_r} \frac{a}{d} \phi_x(0), \tag{13.62}$$

which amounts to saying that the interaction is determined by the part of the soliton charge which is inside the unit cell where the impurity is located.

This brings an additional contribution to the potential energy, which has to be added to the Hamiltonian (13.48) of the polyacetylene chain if we want to study the soliton–impurity interaction. To analyse the dynamics of the soliton we can use the collective coordinate method by writing the soliton solution as

$$\phi(\xi, \tau) = \tanh\left[\frac{\xi - X(\tau)}{\sqrt{2}}\right], \tag{13.63}$$

where $X(\tau)$ denotes the position of the soliton.

Let us write the Lagrangian which can be deduced from Hamiltonian (13.46) with the Hamiltonian density (13.49), taking into account the additional potential term (13.61). We get

$$L = A \int_{-\infty}^{+\infty} \mathrm{d}\xi \left[\frac{1}{2}\left(\frac{\partial\phi}{\partial\tau}\right)^2 - \frac{1}{2}\left(\frac{\partial\phi}{\partial\xi}\right)^2 - \frac{1}{4}\left(1 - \phi^2\right)^2\right] + \frac{e^2 sa}{8\pi\varepsilon_0\varepsilon_r d} \phi_x(0). \tag{13.64}$$

Putting Expression (13.63), which introduces the collective coordinate, into this Lagrangian leads to

$$L = A\left(\frac{1}{2}\frac{\dot{X}(\tau)^2}{\sqrt{2}} - \frac{1}{2\sqrt{2}} - \frac{\sqrt{2}}{4}\right)\underbrace{\int_{-\infty}^{+\infty}\frac{\mathrm{d}\xi}{\sqrt{2}}\,\mathrm{sech}^4\left[\frac{\xi - X(\tau)}{\sqrt{2}}\right]}_{4/3}$$

$$+\frac{e^2 sa}{8\sqrt{2}\pi\varepsilon_0\varepsilon_r d}\,\mathrm{sech}^2\frac{X(\tau)}{\sqrt{2}} \tag{13.65}$$

$$= \frac{1}{2}\left(\frac{2\sqrt{2}A}{3}\right)\dot{X}^2(\tau) - V_{\text{eff}}(X) \tag{13.66}$$

with the effective potential

$$V_{\text{eff}}(X) = \frac{4}{3\sqrt{2}} - C\,\mathrm{sech}^2\frac{X(\tau)}{\sqrt{2}} \tag{13.67}$$

and the constant $C = e^2 sa/(8\sqrt{2}\pi\varepsilon_0\varepsilon_r d)$. The presence of the donor leads therefore to an attractive potential for the soliton given by $V_{\text{eff}}(X)$. This calculation confirms the value of the soliton mass that we obtained from the energy of the soliton at rest (Equation (13.56)).

The Lagrange equation which results from the effective Lagrangian (13.66) gives an equation for the dynamics of $X(\tau)$,

$$\frac{\mathrm{d}}{\mathrm{d}\tau}\frac{\partial L}{\partial \dot{X}} - \frac{\partial L}{\partial X} = \left(\frac{2\sqrt{2}A}{3}\right)\ddot{X} + \frac{2}{\sqrt{2}}C\,\mathrm{sech}^2\frac{X}{\sqrt{2}}\tanh\frac{X}{\sqrt{2}} = 0. \tag{13.68}$$

If we linearise this equation around the equilibrium position of the soliton in the effective potential of the impurity, we get

$$\left(\frac{2\sqrt{2}A}{3}\right)\ddot{X} + CX = 0, \tag{13.69}$$

which shows that the frequency at which the soliton oscillates in the potential of the donor is $\Omega_0^2 = 3C/2\sqrt{2}A$. Coming back to the dimensional variables, we get the square of this frequency as

$$\Omega^2 = \Omega_0^2\,\omega_0^2 = \frac{4V_0}{Mu_0^2}\frac{3e^2}{32\pi\varepsilon_0\varepsilon_r d}\frac{1}{Ku_0^2} \tag{13.70}$$

$$\Omega \simeq 0.122\cdot 10^{15}\ \text{Hz}, \tag{13.71}$$

if we choose $d = 3$ Å which is a reasonable value for the distance between a donor ion and a polyacetylene chain. The corresponding wavenumber

$$\sigma = \frac{\Omega}{2\pi c} 100 = 651.5 \text{ cm}^{-1}, \tag{13.72}$$

(where c is the speed of light in vacuum) is in rather good agreement with experiments as shown in Figure 13.8. Experiments observe a broad mode, which is not surprising because the exact frequency of the contribution of each soliton depends on the distance d between the donor and the polymer chain. Thus all solitons do not resonate at the same frequency and the experimental width results from the superposition of many modes, with slightly different frequencies.

13.6 An experimental test for the presence of solitons

The observation of the resonance associated with the translation of the soliton is not a stringent test of the role of solitons in carrying charge in polyacetylene because any kind of mobile charge in the presence of the donor ion D^+ would have given a similar response. The calculation of the frequency of oscillation is also not sufficient because the result is highly sensitive to the specific properties of a sample such as the distances between the D^+ ions and the $(CH)_x$ chains. Further studies have been performed to test the unusual charge–spin combination of the solitons ($Q = -e$, $S = 0$). However, the most sensitive test of the validity of the charge transport by a collective mode in polyacetylene has been provided by the study of the internal excitation mode of the solitons.

13.6.1 The oscillation of the soliton slope

We showed in Chapter 5 that ϕ^4 solitons have an internal mode which is an oscillation of their slope around the centre. Its frequency can be deduced from linearisation around the soliton solution, but we shall show here that it can also be derived by the collective coordinate method in order to illustrate another application of this method.

Since the localised mode is an intrinsic property of the soliton, which does not depend on the presence of the impurity, we shall start from the Lagrangian in the absence of the defect,

$$L = A \int_{-\infty}^{+\infty} d\xi \left[\frac{1}{2} \left(\frac{\partial \phi}{\partial \tau} \right)^2 - \frac{1}{2} \left(\frac{\partial \phi}{\partial \xi} \right)^2 - \frac{1}{4} \left(1 - \phi^2 \right)^2 \right], \tag{13.73}$$

and look for a solution as

$$\phi(\xi, \tau) = \tanh(\alpha(\tau)\, \xi). \tag{13.74}$$

Here we did not include a coordinate for the position of the centre of the soliton because we are only investigating the internal mode. Equation (13.74) corresponds to a soliton that does not move, but which has a slope which varies with time. Putting this solution into the Lagrangian we get

$$L = A \int_{-\infty}^{+\infty} d\xi \left[\frac{1}{2} \dot{\alpha}^2 \xi^2 \operatorname{sech}^4(\alpha\xi) - \frac{1}{2} \alpha^2 \operatorname{sech}^4(\alpha\xi) - \frac{1}{4} \operatorname{sech}^4(\alpha\xi) \right] \tag{13.75}$$

$$= A \left[\frac{1}{2} \frac{\dot{\alpha}^2}{\alpha^3} \underbrace{\int_{-\infty}^{+\infty} du\, u^2 \operatorname{sech}^4 u}_{\pi^2/9-2/3} - \frac{1}{2} \frac{\alpha^2}{\alpha} \underbrace{\int_{-\infty}^{+\infty} du \operatorname{sech}^4 u}_{4/3} - \frac{1}{4} \frac{1}{\alpha} \underbrace{\int_{-\infty}^{+\infty} du \operatorname{sech}^4 u}_{4/3} \right]$$

$$= A \left[\frac{\mu}{2} \frac{\dot{\alpha}^2}{\alpha^3} - \frac{1}{3\alpha}(2\alpha^2 + 1) \right], \tag{13.76}$$

in which we define $\mu = \pi^2/9 - 2/3$. This expression leads to the Lagrange equation for the time evolution of $\alpha(\tau)$

$$\frac{d}{d\tau} \frac{\partial L}{\partial \dot{\alpha}} - \frac{\partial L}{\partial \alpha} = A \left[\mu \frac{d}{d\tau} \left(\frac{\dot{\alpha}}{\alpha^3} \right) + \frac{3\mu}{2} \frac{\dot{\alpha}^2}{\alpha^4} + \frac{1}{3} \left(2 - \frac{1}{\alpha^2} \right) \right] = 0 \tag{13.77}$$

i.e.
$$\mu \left(\frac{\ddot{\alpha}}{\alpha^3} - \frac{3\mu}{2} \frac{\dot{\alpha}^2}{\alpha^4} \right) + \frac{1}{3} \left(2 - \frac{1}{\alpha^2} \right) = 0. \tag{13.78}$$

Its static solution is such that $(2 - 1/\alpha^2) = 0$, which gives $\alpha = 1/\sqrt{2}$. Therefore we recover the slope of the ϕ^4 which results from Equation (13.53).

13.6.2 Linearised solution

Let us linearise Equation (13.78) around the value corresponding to a static solution. Introducing $\alpha = (1 + \theta)/\sqrt{2}$, we get

$$\mu \frac{\ddot{\theta}/\sqrt{2}}{[(1+\theta)/\sqrt{2}]^3} + \mathcal{O}(\dot{\theta}^2) + \frac{1}{3} \left(2 - \frac{1}{(1+\theta)^2/2} \right) = 0. \tag{13.79}$$

where $\mathcal{O}(\dot{\theta}^2)$ is a nonlinear term of the order of $\dot{\theta}^2$. The terms which are linear with respect to θ and its derivatives reduce to

$$\ddot{\theta} + \frac{2}{3\mu} \theta = 0. \tag{13.80}$$

This equation for a harmonic oscillator indicates that the soliton has an internal mode which is an oscillation of its slope at frequency Ω_i, given by $\Omega_i^2 = 2/(3\mu) \simeq 1.55$. This value of Ω_i is very close to the frequency of the internal mode of the ϕ^4

soliton deduced from a linearisation of the fluctuations around the soliton, given by $\Omega_1^2 = 3/2$ in dimensionless variables. This shows that the two methods, the collective coordinate approach and direct linearisation around the soliton, are in good agreement.

13.6.3 Observation of the internal mode of the soliton

The expression of the charge density of the soliton $\rho(x) \propto \partial u/\partial x$, and the soliton solution (13.74) give the expression of $\rho(x)$ as $\rho(x) \propto \alpha(\tau)\ \mathrm{sech}^2[\alpha(\tau)\, s\, x]$. Therefore, even when the slope of the soliton oscillates, its charge density stays symmetrical with respect to its centre. The oscillation of the slope does not generate an oscillating dipole moment that could be observed by infrared spectroscopy. However, the internal mode from the soliton has been detected in experiments [167, 182] because a *mobile* soliton with an internal oscillation has an infrared response which contains a contribution from the internal mode. It explains the peak at about $1300\,\mathrm{cm}^{-1}$ in Figure 13.8.

The infrared response of the internal mode comes from its coupling to the translation mode, which is directly active in infrared spectroscopy because it corresponds to the motion of an electric charge which makes an oscillatory dipole with the D^+ ion. It is possible to understand qualitatively why this is the case by examining the dynamics of the soliton in the potential of the defect. When the soliton is on the right side (or the left side) of the defect, the right part (or the left part) of the soliton feels a larger attraction to the defect. The difference in attraction on the two sides of the soliton tends to modify its slope, leading to an exchange of energy between the translation mode and the internal mode of the soliton. If we excite the translation mode by an infrared field, we must therefore also expect a response at the frequency of the internal mode. This is confirmed by experiments, and, since the internal mode is an intrinsic feature of the soliton which would not exist with an ordinary electric charge, these experiments *clearly confirm the role of solitons* as electric charge carriers in conducting polyacetylene. Moreover the frequency of the internal mode is a characteristic of the soliton, which does not depend on its distance from the defect. Therefore we can expect its frequency to be more sharply defined than the frequency of the translational mode. This is confirmed by experiments: the peak at $1300\,\mathrm{cm}^{-1}$ attributed to the internal mode is very sharp (Figure 13.8). From the frequency of the internal mode expressed in its dimensional form, $\omega_i = \omega_0\sqrt{3/2}$, we get the wavenumber $\sigma_i = \omega_i/c/2\pi/100 = 1094\,\mathrm{cm}^{-1}$, close to the experimental value of $1300\,\mathrm{cm}^{-1}$. The frequency of the internal mode of a soliton is very sensitive to its exact shape. The approximation that we made when we replaced the exact potential by a ϕ^4 potential (Figure 13.6) is therefore less suitable for computing the frequency of the internal mode than for studying its shape or energy. Thus we

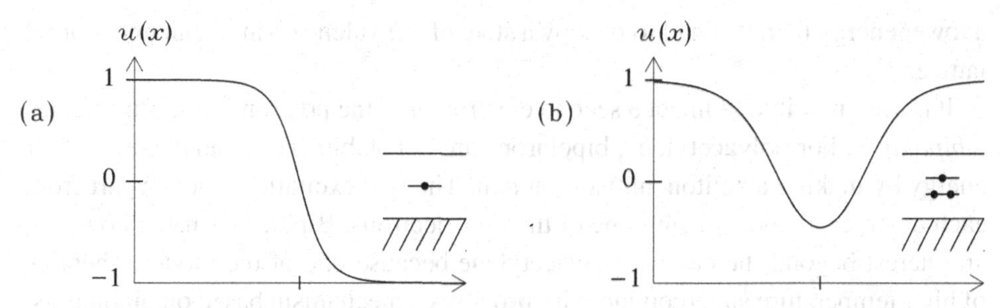

Figure 13.9. Comparison between (a) a soliton and (b) a polaron. Their shapes and their electronic structures are shown (the hatched regions in the insets indicate the valence band).

cannot expect to get a quantitatively correct result for the internal mode with our simple analytical model.

In order to perform an analytical study of the coupling between the translation mode and the internal mode to justify the infrared response of the internal mode, it is possible to use the collective coordinate approach in the same spirit as the study made by D. K. Campbell *et al.* [37] for soliton interactions in the ϕ^4 model. The idea is to look for a solution of the form

$$\phi(\xi, \tau) = \tanh[\alpha(\tau)(\xi - X(\tau))], \tag{13.81}$$

which leads to an effective Lagrangian that yields two coupled equations for $\alpha(\tau)$ and $X(\tau)$. The possibility to study the coupling between the translation and the internal modes is an attraction of the collective coordinate method, compared to the linearisation around the soliton, which finds the translation (Goldstone) mode and the internal mode as independent excitations of the soliton.

13.7 The other nonlinear excitations of polyacetylene

The topological solitons that we have described are not the only localised excitations which can exist in polyacetylene. Numerical simulations have shown that the injection of an electron into an homogeneous polymer chain, in the absence of a soliton, leads to the formation of a nontopological excitation, the *polaron*, which is similar to a bound pair of a soliton and an antisoliton having a slope different from that of the topological soliton.

Figure 13.9 compares a soliton and a polaron by showing their distortion u_n, related to the atomic displacement by $U_n = (-1)^n u_n$. The injected electron deforms the lattice in its vicinity, creating a minimum in the function $u(x)$. This leads to the formation of a localised electronic state, occupied by the injected electron which has a lower energy than if it had to occupy a state of the valence band in an undistorted lattice.

It is even possible to inject a second electron onto the polaron level. This creates a *bipolaron*. For polyacetylene, bipolarons are not stable. They can decrease their energy by making a soliton–antisoliton pair. The two excitations move apart from each other, each one bringing one of the two electrons. Bipolarons have, however, an interest beyond the case of polyacetylene because one of the modern theories of high-temperature superconductivity proposes a mechanism based on bipolarons, which seems consistent with experimental results. In this nonconventional view, proposed by Serge Aubry, the bipolaron is stabilised by discreteness effects. This suggestion is drastically different from the Bardeen–Cooper–Schrieffer (BCS) theory in which two electrons of a pair have the same momentum, without having the same spatial position. In this example we again notice the change of viewpoint brought by the physics of nonlinear excitations that we stressed for the Fermi–Pasta–Ulam problem: conventional approaches work in reciprocal space, i.e. the Fourier transform of real space, while nonlinear theory implies a localisation in real space.

14
Solitons in Bose–Einstein condensates

14.1 Introduction

Sometimes it can take a while for experiment to catch up with theory in physics. Predicted in the 1920s, it would be 70 years before the actual creation of the first Bose–Einstein condensate (BEC) in the laboratory. That achievement established an entirely new branch of atomic physics that continues to provide a treasure-trove to scientists who want to study this strange and extremely small world of quantum physics. Moreover it opens many new possibilities for the experimental study of solitons. As we shall see in this chapter, BEC provides a highly tunable medium where a variety of nonlinear excitations and solitons can be created and studied in different regimes.

The story of Bose–Einstein condensation starts in 1924 when a young Bengali physicist, Satyendra Nâth Bose (1894–1974), sent to Einstein a manuscript where he showed that one can recover Planck's law for black body radiation by treating the photons as a gas of indistinguishable particles [31]. Albert Einstein (1879–1955) then generalised Bose's argument to the case of material particles [60] and, in particular, he demonstrated the condensation property. When an ideal gas obeying Bose statistics is cooled below the critical temperature $k_B T_c = 3.31 n^{2/3}\hbar^2/m$ where n is the density of the gas and m the mass of a particle, a macroscopic fraction of the particles accumulates in a *single* quantum state, the ground state of the box.

The phenomenon predicted by Einstein remained controversial and somewhat mysterious until 1937, when the superfluidity of liquid helium was discovered. Fritz London (1900–54) then noticed that the temperature of the superfluid transition, $T_s = 2.2$ K, is remarkably close to the temperature of the Bose–Einstein condensation for an ideal gas with the same density as liquid helium, $T_c = 3.2$ K, and he made the connection between the two phenomena [119]. However, this connection is not obvious since the superfluidity occurs because of the interaction between particles, while Einstein was dealing with an ideal gas. More quantitatively, the condensate in liquid helium exists and can be detected experimentally, but it never contains more than 10% of the particles of the fluid, while all particles should be condensed at $T = 0$ for an ideal gas.

Since 1970, research into systems closer to Einstein's theory has been very active. Several groups around the world started searching for BECs with a combination of laser and magnetic cooling apparatus. Except for atomic hydrogen, all experiments with gaseous condensates start with laser cooling, a technique developed by C. Cohen-Tannoudji (1933–), S. Chu (1948–) and W. D. Philipps (1948–), the three 1997 Nobel prize winners. The gas of atoms at room temperature is first slowed down and captured in a trap created by laser light. This cools the atoms to about one-ten-millionth of a degree above absolute zero – still far too hot to produce a BEC.

Once the gas is trapped, the lasers are turned off and the atoms are held in place by a magnetic field. The atoms are further cooled in the magnetic trap by selecting the hottest atoms and kicking them out of the trap: this is evaporative cooling, developed by D. Kleppner (1932–) for hydrogen. Then comes the tricky part [45]: trapping a sufficiently high density of atoms at temperatures that are cold enough to produce a BEC. It is worth noting that, under these conditions, the equilibrium configuration of the system would be the solid phase. Thus, in order to observe BEC, one has to maintain the system in a metastable gas phase for a sufficiently long time. One tries to keep the collision rate at the centre of the trap constant during the evaporation process. In a harmonic trap, the collision rate scales as N/T where N is the number of atoms. On the other hand, the phase space density at the centre of the trap scales as N/T^3 and it must be multiplied typically by 10^6 to reach the BEC threshold. Therefore a typical evaporation process consists of dividing the number of atoms and the temperature by 10^3, which leads to the desired gain. Starting with 10^9 atoms at 100 μK, one ends with 10^6 atoms at 100 nK.

The last step of the process is the visualisation of the atom cloud, which is usually performed by illuminating the remaining atom cloud with a flash of resonant light, and imaging the shadow of the atom cloud onto a CCD camera; this gives access to the spatial distribution of the atoms. Alternatively, if one releases the atoms from the trap and waits before transmitting the flash, one obtains the velocity distribution. Both methods are however inherently destructive because absorbed photons heat the atoms in the first technique, while atoms are released from the trap in the second one. This is why, nondestructive *in situ* imaging, relying on dispersion rather than absorption, was designed after the first BEC was reported.

The world's first BEC, achieved [16] by Eric Cornell (1961–), Carl Wieman (1951–) and their collaborators on 5 June 1995 in Boulder, was formed inside a carrot-sized glass cell, and made visible by a video camera; it measured only about 20 microns in diameter. The result was a BEC of about 2,000 rubidium atoms that lasted for 15 to 20 seconds. Shortly thereafter, Wolfgang Ketterle (1957–) also achieved [49] a BEC in his laboratory at MIT with sodium vapour. The 2001 Nobel prize in physics was awarded to E. A. Cornell, W. Ketterle and C. E. Wieman for this achievement.

Today, scientists can produce condensates with much greater numbers of atoms that can last for several minutes, and they continue to yield intriguing new insights into this unusual form of matter. Bose–Einstein condensation has already been found experimentally for many atomic species (atomic hydrogen, metastable helium, lithium, sodium, potassium, rubidium, caesium and ytterbium) and recently for Li_2 molecules.

14.2 Theoretical description of a condensate

In this section we shall discuss a simple approach to describe a pure Bose–Einstein condensate ($T = 0$), taking into account the atom–atom interactions in a mean-field approach (see references [40, 44, 111] where more rigorous approaches, second quantisation and BBGKY hierarchy or Bogoliubov formalism, are presented in detail) at variance with the ideal gas model, initially proposed by Einstein. In a real condensate, atom–atom interactions cannot be neglected despite the very dilute nature of these gases. Indeed, the combination of BEC and harmonic trapping greatly enhances the effects of the atom–atom interactions on important measurable quantities.

14.2.1 The Hartree approximation

Considering a system of N spinless bosons characterised by spatial coordinates $\mathbf{r}_n$, the starting point is the N-body Hamiltonian

$$H = \sum_{n=1}^{N} \left(\frac{\mathbf{p}_n^2}{2m} + V(\mathbf{r}_n) \right) + \frac{1}{2} \sum_{n=1}^{N} \sum_{j \neq n}^{N} W(\mathbf{r}_n - \mathbf{r}_j). \tag{14.1}$$

The first term corresponds to the kinetic energy, V standing for the trapping potential and W for the two-body interaction potential. We look for the ground state of the system within the Hartree approximation, i.e. we consider for the many-body wave function Ψ_N, the ansatz

$$\Psi_N(\mathbf{r}_1, \mathbf{r}_2, \ldots, \mathbf{r}_N, t) = \psi(\mathbf{r}_1, t)\, \psi(\mathbf{r}_2, t)\, \ldots\, \psi(\mathbf{r}_N, t), \tag{14.2}$$

where ψ is some single-particle wave function, normalised to unity, to be determined. This approximation reproduces the basic ingredients of Einstein's model, assuming that all atoms occupy the same quantum state, but the important modification is that the state $\psi(\mathbf{r})$ is now determined using a self-consistent method, instead of being simply the ground state of the one-body Hamiltonian

$$H_1 = \frac{\mathbf{p}^2}{2m} + V(\mathbf{r}) \tag{14.3}$$

for an ideal gas. It is important to emphasise that the description of the N-body system by the factorised wave function (14.2) amounts to neglecting all correlations between particles. As we shall see, this is a very good approximation for the weakly interacting Bose–Einstein condensates, which provides quantitative predictions for the static, dynamic and thermodynamic properties of these trapped gases.

The Hartree approximation (14.2) explicitly assumes that all N particles are in the same macroscopic state; this is exact for *simple BECs* at $T = 0$ to which our presentation is limited. A more rigorous approach [111], with a time-dependent macroscopic number N_0 of particles in the condensate can be also considered. Much more complicated cases, with more than one state with a macroscopic number of particles, called *fragmented BECs*, are also possible but will not be discussed here.

14.2.2 The two-body interaction potential

To proceed further, we also need to consider a simplified version of the two-body potential $W(\mathbf{r})$, which is valid at very low temperature. The problem of a collision between two particles with mass m_1 and m_2, interacting with a potential $W(\mathbf{r}_1 - \mathbf{r}_2)$, is equivalent [45, 108] to the problem of the scattering of a particle with reduced mass $\mu = m_1 m_2/(m_1 + m_2)$ by the potential $W(\mathbf{r})$. The solution of the scattering problem amounts to looking for the eigenfunctions of the one-body Hamiltonian $\mathbf{p}^2/(2\mu) + W(\mathbf{r})$ with a positive energy $E_k = \hbar^2 k^2/(2\mu)$.

At low energy, for a potential $W(\mathbf{r})$ decreasing faster than r^{-3}, one can show that the scattering becomes isotropic. In addition, when k tends to zero, the scattering amplitude usually tends to a finite value $-a$. This quantity which has the dimension of a length, is called the *scattering length.* Assuming for simplicity that W is a spherically symmetrical potential, the calculation of a only requires the knowledge of the zero-energy eigenstates of the Hamiltonian H_1. For usual potentials between alkali atoms with a van der Waals attraction varying as $-c_6/r^6$, one has to precisely know the potential W to determine the scattering length. One finds that a strongly depends on the parameters of the problem, such as the c_6 coefficient; even its sign can be modified!

The critical dependence of the scattering length on the parameters of the two-body interaction potential $W(\mathbf{r})$ opens a way to manipulate the effective interaction between the atoms in the condensate in real time. By varying, for example, the external magnetic field, and taking advantage of a *Feshbach resonance*,[1] one can tune the scattering length all the way from a large and positive scattering length (strongly repulsive effective interactions) to a large and negative scattering length (strongly attractive effective interactions) [93]. This possibility opens applications of paramount importance for the existence of solitary waves, as we shall see in this chapter.

Once the scattering length has been calculated or experimentally measured, one takes advantage of the fact that two potentials having the same scattering length a lead to the same properties for the cold N-body system. Consequently, one replaces the real potential W by the much simpler contact potential[2]

$$W(\mathbf{r}_i - \mathbf{r}_j) \longrightarrow g\delta(\mathbf{r}_i - \mathbf{r}_j), \tag{14.4}$$

by introducing the effective interaction constant $g = 4\pi\hbar^2 a/m$.

The Hamiltonian for N interacting particles in the trapping potential V can thus be written as

$$H = \sum_{n=1}^{N}\left(-\frac{\hbar^2}{2m}\nabla_n^2 + V(\mathbf{r}_n)\right) + \frac{g}{2}\sum_{n=1}^{N}\sum_{j\neq n}^{N}\delta(\mathbf{r}_i - \mathbf{r}_j), \tag{14.5}$$

where ∇_n is the gradient relative to $\mathbf{r}_n$.

14.2.3 The Gross–Pitaevskii equation

Contrary to intuition, introducing Hamiltonian (14.5) into the time-dependent Schrödinger equation is not the appropriate method to derive the dynamic equation because of the mean-field property; several subtle points arise. It is much more appropriate and direct to use a variational approach as follows.

Recognising that Hamiltonian (14.5) is very close to the NLS Hamiltonian (3.76), it is straightforward to introduce the total Lagrangian associated with Hamiltonian (14.5). It reads

$$L = \int \prod_{k=1}^{N} \mathrm{d}\mathbf{r}_k \left[\frac{i\hbar}{2}\left(\Psi_N^*\Psi_{N,t} - \Psi_N\Psi_{N,t}\right) - \sum_{n=1}^{N}\left(\frac{\hbar^2}{2m}|\nabla_n\Psi_N|^2 + V(\mathbf{r}_n)|\Psi_N|^2 + \frac{g}{2}\sum_{j\neq n}^{N}\delta(\mathbf{r}_n - \mathbf{r}_j)|\Psi_N|^2\right)\right]. \tag{14.6}$$

[1] Feshbach resonance is a scattering resonance in which pairs of free atoms are tuned, via the Zeeman effect, into resonance with a vibrational state of the diatomic molecule.

[2] Rather than a pure $\delta(\mathbf{r})$ potential which is singular with respect to scattering, one uses the well behaved pseudo-potential W_p defined by $W_\mathrm{p}(\mathbf{r})\psi(\mathbf{r}) = (4\pi\hbar^2 a/m)\,\delta(\mathbf{r})\,\partial(r\psi(\mathbf{r}))/\partial r$ [45, 90].

Substituting the Hartree ansatz (14.2) in Equation (14.6), we get several terms. Using the notation $\psi_n = \psi(\mathbf{r}_n, t)$, the first one is

$$\int \prod_{k=1}^{N} d\mathbf{r}_k \Psi_N^* \frac{\partial \Psi_N}{\partial t} = \int \prod_{k=1}^{N} d\mathbf{r}_k \left(\prod_{j=1}^{N} \psi_j^* \right) \left(\sum_{\ell=1}^{N} \frac{\partial \psi_\ell}{\partial t} \prod_{k \neq \ell}^{N} \psi_k \right) \tag{14.7}$$

$$= \sum_{\ell=1}^{N} \left(\int d\mathbf{r}_\ell\, \psi_\ell^* \frac{\partial \psi_\ell}{\partial t} \right) \left(\prod_{k \neq \ell}^{N} \underbrace{\int d\mathbf{r}_k \psi_k^* \psi_k}_{=1} \right) \tag{14.8}$$

$$= N \int d\mathbf{r}\, \psi^*(\mathbf{r}, t) \frac{\partial \psi(\mathbf{r}, t)}{\partial t}, \tag{14.9}$$

whereas the second one is simply the complex conjugate of Equation (14.9).
The third term of Equation (14.6) leads to

$$\int \prod_{k=1}^{N} d\mathbf{r}_k \sum_{n=1}^{N} |\nabla_n \Psi_N|^2 = \sum_{n=1}^{N} \left(\int d\mathbf{r}_n\ |\nabla_n \psi(\mathbf{r}_i, t)|^2 \right) \left(\prod_{k \neq n}^{N} \underbrace{\int d\mathbf{r}_k \psi_k^* \psi_k}_{=1} \right)$$

$$= N \int d\mathbf{r}\ |\nabla \psi(\mathbf{r}, t)|^2, \tag{14.10}$$

and the fourth to

$$\int \prod_{k=1}^{N} d\mathbf{r}_k \sum_{n=1}^{N} V(\mathbf{r}_n)\, |\Psi_N|^2 = N \int d\mathbf{r}\, V(\mathbf{r})\, |\psi(\mathbf{r}, t)|^2. \tag{14.11}$$

Finally, the interaction term is

$$\int \prod_{k=1}^{N} d\mathbf{r}_k \sum_{n=1}^{N} \sum_{j \neq n}^{N} \delta(\mathbf{r}_n - \mathbf{r}_j)\, |\Psi_N|^2 = N(N-1) \int d\mathbf{r}\ |\psi(\mathbf{r}, t)|^4. \tag{14.12}$$

Collecting all terms, one obtains

$$L = N \int d\mathbf{r} \left[\frac{i\hbar}{2} [\psi^* \psi_t - \psi \psi_t^*] - \frac{\hbar^2}{2m} |\nabla \psi|^2 - V(\mathbf{r})\, |\psi(\mathbf{r}, t)|^2 - \frac{g}{2}(N-1)\, |\psi(\mathbf{r}, t)|^4 \right]. \tag{14.13}$$

The stationarity condition of the action with respect to ψ^* gives

$$\frac{\delta L}{\delta \psi^*} = 0 = N\biggl(i\hbar\psi_t + \frac{\hbar^2}{2m}\nabla^2\psi(\mathbf{r},t) - V(\mathbf{r})\,\psi(\mathbf{r},t)$$
$$-g(N-1)\,|\psi(\mathbf{r},t)|^2\,\psi(\mathbf{r},t)\biggr). \tag{14.14}$$

Thus, we obtain

$$i\hbar\frac{\partial\psi(\mathbf{r},t)}{\partial t} = \left[-\frac{\hbar^2}{2m}\nabla^2 + V(\mathbf{r}) + g(N-1)\,|\psi(\mathbf{r},t)|^2\right]\psi(\mathbf{r},t). \tag{14.15}$$

The physical interpretation of this equation is simple. A particle with wave function $\psi(\mathbf{r},t)$ evolves in the external potential $V(\mathbf{r})$ plus the mean-field potential created by the remaining particles. Note the occurrence of a factor $(N-1)$ rather than N, ensuring that the interaction term disappears for $N=1$. However, in the remainder of the chapter we will ignore the difference between $N-1$ and N. This assumption is perfectly valid since, typically, in actual BEC experiments N is at least 10^5. The mean field is thus proportional to the scattering length a through g, and to the local density $N|\psi(\mathbf{r},t)|^2$. The equation describing the dynamic evolution of the BEC can thus be written as

$$i\hbar\frac{\partial\psi(\mathbf{r},t)}{\partial t} = \left[-\frac{\hbar^2}{2m}\nabla^2 + V(\mathbf{r}) + gN\,|\psi(\mathbf{r},t)|^2\right]\psi(\mathbf{r},t). \tag{14.16}$$

Writing, as it is conventional, the above equation (14.16) in terms of the order parameter, the *condensate wave function* $\Phi(\mathbf{r},t) \equiv \sqrt{N}\psi(\mathbf{r},t)$, we get

$$i\hbar\frac{\partial\Phi(\mathbf{r},t)}{\partial t} = \left[-\frac{\hbar^2}{2m}\nabla^2 + V(\mathbf{r}) + g\,|\Phi(\mathbf{r},t)|^2\right]\Phi(\mathbf{r},t). \tag{14.17}$$

Derived independently by Gross and Pitaevskii in 1961, Equations (14.16) and (14.17) are both called[3] $T=0$ time-dependent Gross–Pitaevskii (GP). A similar nonlinear equation has also been considered for the theory of superfluid helium near the λ-point [76], however, the terms of this equation have a different physical meaning.

Equations (14.16) and (14.17) are used to describe the ground state of a condensate as well as its excitations. They have the form of three-dimensional nonlinear Schrödinger equations, the nonlinearity coming from the mean-field term. In the absence of interaction ($g=0$), both equations reduce to the usual Schrödinger equation for the single particle Hamiltonian (14.3). However, if we temporarily neglect the external potential V but consider nonzero values for g, our experience

[3] In cases for which the condensate wavenumber $N(t)$ is not conserved in time, it is clear that both equations are not equivalent; Equation (14.17) implies Equation (14.16) but not vice versa.

with the NLS equation (Chapter 3) immediately suggests that the behaviour would be very different depending on the sign of the interaction parameter, which can be tuned from positive to negative values by modifying the magnetic field. We can expect both types of excitation: bright ($g < 0$) and grey ($g > 0$) solitons. This will be confirmed in Section 14.5.

The predictions derived from this equation are in excellent agreement with experiments dealing with a quasi-pure condensate. At $T = 0$, the validity criterion is $na^3 \ll 1$, which expresses that the fluid must be dilute. For typical gaseous condensates $na^3 \sim 10^{-5}$, so that the mean-field approximation is excellent.

14.2.4 Lagrangian and Hamiltonian densities

The Lagrange equations (see Appendix B) applied to the Lagrangian density

$$\mathcal{L} = \frac{i\hbar}{2}[\Phi^* \Phi_t - \Phi\Phi_t^*] - \frac{\hbar^2}{2m}|\nabla\Phi|^2 - V(\mathbf{r})|\Phi(\mathbf{r}, t)|^2 - \frac{g}{2}|\Phi(\mathbf{r}, t)|^4, \quad (14.18)$$

lead to Equation (14.17).

As shown in Chapter 3, it is then straightforward to derive the Hamiltonian density $\mathcal{H}$. Introducing the canonical momenta, conjugate variables of Φ and Φ^*,

$$p_\Phi = \frac{\partial \mathcal{L}}{\partial \Phi_t} = +\frac{i\hbar}{2}\Phi^* \quad \text{and} \quad p_{\Phi^*} = \frac{\partial \mathcal{L}}{\partial \Phi_t^*} = -\frac{i\hbar}{2}\Phi, \quad (14.19)$$

we get the Hamiltonian density

$$\mathcal{H} = \sum_\ell p_{\Phi^\ell}\frac{d\Phi^\ell}{dt} - \mathcal{L} \quad (14.20)$$

$$= \frac{\hbar^2}{2m}|\nabla\Phi|^2 + V(\mathbf{r})|\Phi(\mathbf{r}, t)|^2 + \frac{g}{2}|\Phi(\mathbf{r}, t)|^4, \quad (14.21)$$

where Φ^ℓ denotes Φ and Φ^*. Integrating over spatial coordinates, we finally obtain the so-called Gross–Pitaevskii energy

$$E = \int d\mathbf{r}\left[\frac{\hbar^2}{2m}|\nabla\Phi|^2 + V(\mathbf{r})|\Phi(\mathbf{r}, t)|^2 + \frac{g}{2}|\Phi(\mathbf{r}, t)|^4\right], \quad (14.22)$$

which will be of later use.

14.2.5 The ground state of the Gross–Pitaevskii equation

The ground state can be easily obtained within the above mean-field theory. For this, the condensate wave function can be written as

$$\Phi(\mathbf{r}, t) = \phi(\mathbf{r})\, e^{-i\mu t/\hbar}, \quad (14.23)$$

where ϕ is real and normalised so that the space integral of its squared modulus is N. Because of the nonlinearity, μ is in general not the energy per particle but the chemical potential (hence the notation) [111].

The time-dependent GP equation (14.17) becomes in that simplest case

$$\left(-\frac{\hbar^2}{2m}\nabla^2 + V(\mathbf{r}) + g\phi^2(\mathbf{r})\right)\phi(\mathbf{r}) = \mu\,\phi(\mathbf{r}). \tag{14.24}$$

Alternatively one may introduce the transformation $\psi(\mathbf{r},t) = \psi(\mathbf{r})\,\mathrm{e}^{-\mathrm{i}\mu t/\hbar}$ into Equation (14.16), which leads to

$$\left(-\frac{\hbar^2}{2m}\nabla^2 + V(\mathbf{r}) + Ng\psi^2(\mathbf{r})\right)\psi(\mathbf{r}) = \mu\,\psi(\mathbf{r}). \tag{14.25}$$

Note that in the present case, the time-independent Gross–Pitaevskii equation (14.24) is strictly equivalent to the one-particle equation (14.25).

14.3 Magnetic traps

To go further we have to specify the external trapping potential V. The device takes advantage of the magnetic interaction energy $-\vec{\mu}\cdot\vec{B}$ between the atomic magnetic moment $\vec{\mu}$ and the magnetic field $\vec{B}$. An important feature characterising magnetic traps for alkali atoms is that the confining potential can be safely approximated with the quadratic form

$$V(\mathbf{r}) = \frac{m}{2}\left(\omega_x^2 x^2 + \omega_y^2 y^2 + \omega_z^2 z^2\right). \tag{14.26}$$

The shape of the confining field determines, of course, the symmetry of the problem. Spherical or axially symmetric traps can be used, for instance. The first two experiments with rubidium [16] and sodium [49] were carried out with axial symmetry. In this case, one defines an axial coordinate z and a radial coordinate $\rho = (x^2 + y^2)^{1/2}$, with associated frequencies ω_z and $\omega = \omega_x = \omega_y$. The ratio between the axial and radial frequencies $\lambda = \omega_z/\omega$ fixes the asymmetry of the trapping potential which can be rewritten as

$$V(\mathbf{r}) = \frac{m\omega^2}{2}(x^2 + y^2 + \lambda^2 z^2). \tag{14.27}$$

If the confinement is tight in two directions and relatively weak in the third ($\lambda < 1$), this gives rise to cigar-shaped (Figure 14.1(a)), rather than spherical, trapped clouds. On the other hand, $\lambda > 1$ leads to disk-shaped condensates (Figure 14.1(b)). The first solution is the most commonly used nowadays.

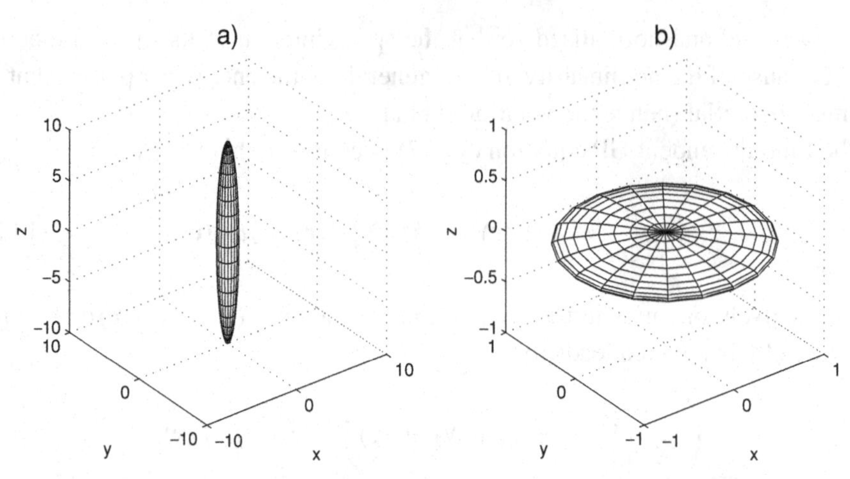

Figure 14.1. Representation of the spatial distribution $V(\mathbf{r}) = 1$ where V is the potential given by Equation (14.27): (a) corresponding to $\lambda = 0.1$ gives rise to a cigar-shaped condensate whereas (b) with $\lambda = 10$ is a disk-shaped one.

14.4 Dynamic properties

The study of elementary excitations is a task of primary importance for quantum many-body theories. In the case of Bose fluids, it plays a crucial role in understanding the properties of superfluids, and consequently, after the experimental realisation of BECs in trapped Bose gases, excitations in these systems have been intensively studied. Measurements of the frequency of the lowest modes have quickly become available and the propagation of wave packets has also been observed.

14.4.1 Linearisation of the Gross–Pitaevskii equation

In the low-temperature limit that we are considering, the properties of the excitations do not depend on temperature and the spectrum of excited states can be found from the frequencies of the linearised GP equation. Let us study the small deviations $\delta\Phi(\mathbf{r}, t)$ from a reference solution $\Phi_0(\mathbf{r}, t)$. Introducing the ansatz

$$\Phi(\mathbf{r}, t) = \Phi_0(\mathbf{r}, t) + \delta\Phi(\mathbf{r}, t) \tag{14.28}$$

into Equation (14.17), due to the nonlinear term, we get a pair of coupled equations for $\delta\Phi(\mathbf{r}, t)$ and its complex conjugate $\delta\Phi^*(\mathbf{r}, t)$

$$\begin{aligned} i\hbar\frac{\partial\delta\Phi(\mathbf{r}, t)}{\partial t} &= -\frac{\hbar^2}{2m}\nabla^2\delta\Phi(\mathbf{r}, t) + V(\mathbf{r})\,\delta\Phi(\mathbf{r}, t) \\ &\quad + 2g\,|\Phi_0(\mathbf{r}, t)|^2\,\delta\Phi(\mathbf{r}, t) + g\Phi_0^2(\mathbf{r}, t)\delta\Phi^*(\mathbf{r}, t). \end{aligned} \tag{14.29}$$

Note the factor 2 in front of the term proportional to $g\delta\Phi(\mathbf{r},t)$. The equation for $\delta\Phi^*(\mathbf{r},t)$ is the complex conjugate of Equation (14.29).

The behaviour of small deviations from the GP ground state solution is of particular interest. This case corresponds to the reference solution

$$\Phi_0(\mathbf{r},t) = \phi(\mathbf{r})\,\mathrm{e}^{-\mathrm{i}\mu t/\hbar} \tag{14.30}$$

with ϕ real and the solution of the time-independent GP equation (14.24). Writing the perturbation as

$$\delta\Phi(\mathbf{r},t) = [u(\mathbf{r})\,\mathrm{e}^{-\mathrm{i}\omega t} + v^*(\mathbf{r})\,\mathrm{e}^{+\mathrm{i}\omega t}]\,\mathrm{e}^{-\mathrm{i}\mu t/\hbar}, \tag{14.31}$$

gives the Bogoliubov–de Gennes equations

$$\hbar\omega\, u(\mathbf{r}) = (H_1 - \mu + 2g\phi^2(\mathbf{r}))u(\mathbf{r}) + g\phi^2(\mathbf{r})\,v(\mathbf{r}) \tag{14.32}$$

$$-\hbar\omega\, v(\mathbf{r}) = (H_1 - \mu + 2g\phi^2(\mathbf{r}))v(\mathbf{r}) + g\phi^2(\mathbf{r})\,u(\mathbf{r}), \tag{14.33}$$

where H_1 is the single particle Hamiltonian (14.3).

14.4.2 Dispersion relation for a condensate in a box

Let us consider the case where atoms are trapped in a cubic box of size L, and assume periodic boundary conditions. In the simplest case of a weakly interacting gas without external potential, $V = 0$, the time-independent equation (14.24) has the homogeneous solution $\phi = \sqrt{n}$, where $n = N/L^3$ is the density of the gas, if $\mu = gn$. It is then clear that the functions u and v have the form of plane waves $u(\mathbf{r}) = A\,\exp(\mathrm{i}\mathbf{q}\cdot\mathbf{r})$ and $v(\mathbf{r}) = B\,\exp(\mathrm{i}\mathbf{q}\cdot\mathbf{r})$.

Let us first consider the case of an effective *repulsive interaction* ($a > 0$ which leads to $g > 0$). Explicit solutions of Equations (14.32) and (14.33) then yield the dispersion relation

$$\omega(q) = \sqrt{c_s^2 q^2 + \frac{\hbar^2 q^4}{4m^2}}, \tag{14.34}$$

if one introduces the hydrodynamic speed of sound $c_s = \sqrt{ng/m}$ and the modulus of the wavevector $q = |\mathbf{q}|$. This is the Bogoliubov spectrum for a dilute Bose gas, as it was first derived by Nikolai Bogoliubov (1909–92). In the limiting case of the ideal Bose gas ($g = 0$), the spectrum is a parabola. In the general interacting case, it has the sound-wave form $c_s q$ in the low momenta domain ($\hbar q \ll 2mc_s$), but reduces to the free-particle frequency $\hbar q^2/(2m)$ for high momenta (Figure 14.2).

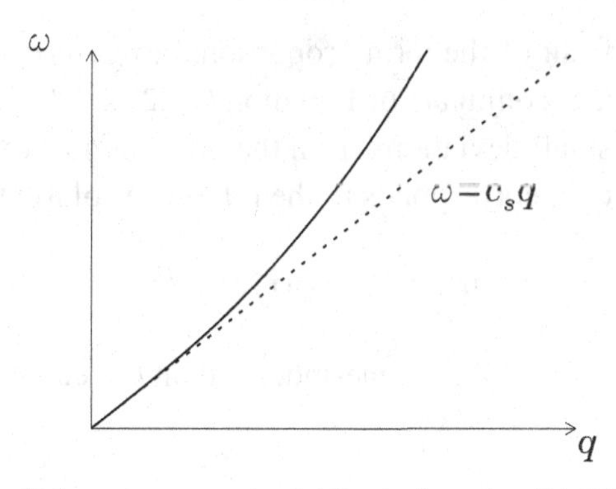

Figure 14.2. Bogoliubov spectrum (solid line) given by (14.34) in the case of an effective repulsive interaction ($a > 0$).

In the case of an effective *attractive interaction* ($a < 0$) between particles, the dynamic stability condition is satisfied if and only if

$$-\frac{n|g|}{m}q^2 + \frac{\hbar^2 q^4}{4m^2} \geq 0 \quad \text{for all} \quad q > 0. \tag{14.35}$$

In the thermodynamic limit of an infinite number of condensate atoms with a fixed mean density n, the stability condition cannot be satisfied as q can be arbitrarily close to zero in an infinite box. However experiments are performed in a finite container. In a cubic box of size L, the components of the wavevector $\mathbf{q}$ of an atom are integer multiples of $2\pi/L$. The smallest nonzero modulus of wavevector that can be achieved is therefore $2\pi/L$, by taking only one nonzero component. The dynamic stability condition (14.35) can then be rewritten as

$$\frac{\hbar^2}{4m}\left(\frac{2\pi}{L}\right)^2 \geq \frac{N}{L^3}\frac{4\pi\hbar^2|a|}{m}, \tag{14.36}$$

by using the explicit form of the effective interaction constant g and the density of the gas n. In terms of the scattering length, one obtains

$$\frac{N|a|}{L} \leq \frac{\pi}{4}. \tag{14.37}$$

This shows that condensates with a negative scattering length a can contain a limited number of atoms proportional to the size of the condensate. The physical meaning of Equation (14.36) is that the energy gap in the spectrum of a particle in the box between the ground state and the first excited states should be larger than the mean-field energy per particle: stabilisation of low wavenumbers is thus provided by the discrete spectrum of the atoms in the trapping potential. This condition can thus be qualitatively extended to the case of an isotropic harmonic trap, $\hbar\omega > |g|N/\ell^3$ where ω is the oscillation frequency of the atoms in the trap and $\ell = \sqrt{\hbar/(m\omega)}$ is the typical extension of the ground state of the trap.

14.4.3 Gaussian solutions for a condensate in a trap

When the external potential V is not negligible, the three-dimensional problem can be restated as a variational problem corresponding to the minimisation of the action related to the Lagrangian density (14.18), using the collective coordinate method. As explained in Chapter 6, the choice of the trial function is crucial. Guided by the exact result in the absence of interactions ($g = 0$), we choose the Gaussian ansatz

$$\Phi(x, y, z, t) = A(t) \prod_{\eta=x,y,z} \exp\left[-\frac{\eta^2}{2\ell^2\sigma_\eta^2} + i\eta^2\beta_\eta(t) \right], \tag{14.38}$$

which is precisely the ground state of the linear Schrödinger equation. We again use the harmonic oscillator length $\ell = \sqrt{\hbar/(m\omega)}$ as a characteristic width for the Gaussian. For simplicity, we have assumed in the ansatz (14.38) that the condensate, initially in a steady state, is only excited by a temporal variation of the trap frequencies so that no oscillation of the centre-of-mass motion of the condensate takes place. A more general ansatz has been considered [146] but it leads to very lengthy calculations.

The variational parameters appearing in the ansatz (14.38), the amplitude A, the three widths σ_η and the curvatures β_η^{-2}, will be determined by the collective coordinate method. Inserting this ansatz into (14.18), one calculates the effective Lagrangian $L = \int \mathrm{d}\mathbf{r}\, \mathcal{L}$. Then, using Lagrange equations for each variational parameter, one derives [146] the different equations for the evolutions of the parameters.

The first one, $\pi^{3/2}|A(t)|^2\ell^3\sigma_x(t)\sigma_y(t)\sigma_z(t) = N$, concerns the particle number conservation and shows that if the amplitude increases, the widths have to decrease. One also gets $\beta_\eta = -m\dot{\sigma}_\eta/(2\hbar^2\sigma_\eta)$ for the curvatures. Finally, the widths of the condensate satisfy

$$\ddot{\sigma}_x + \omega_x^2\sigma_x = \frac{\hbar^2}{\ell^4 m^2\sigma_x^3} + \sqrt{\frac{2}{\pi}}\frac{a\hbar^2 N}{m^2\ell^5\sigma_x^2\sigma_y\sigma_z}, \tag{14.39}$$

and two similar equations for σ_y and σ_z, obtained by cyclic permutation of indices. In addition to the time-dependent term, one recognises the attractive term due to the trapping potential and the dispersive one proportional to σ_η^{-3} (originally the kinetic term) which tends to spread the packet. Finally, there is the nonlinear interaction term which tends to either spread or compress the condensate if a is positive or negative.

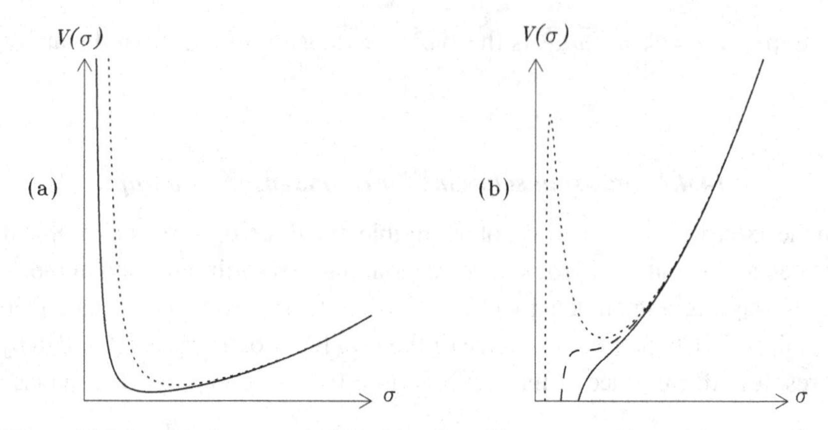

Figure 14.3. In (a), the potential $V(\sigma)$ is plotted for cases with positive scattering lengths: the solid line is for $P = 1$ and the dotted one for $P = 10$. In (b), cases with negative scattering lengths are presented: the solid line is for $P = -1$, the dotted one for $P = -0.2$ and the dashed one for $P_c = -4/5^{5/4}$.

Consequently, in the context of this approximation, one has to solve the system of three ordinary differential equations (14.39), whose equilibrium points correspond to stationary states of the condensate.

In the spherical case $\lambda = 1$, Equation (14.39) can be reduced to

$$\ddot{\sigma} + \omega^2 \sigma = \frac{\hbar^2}{\ell^4 m^2 \sigma^3} + \sqrt{\frac{2}{\pi}} \frac{a\hbar^2 N}{m^2 \ell^5 \sigma^4}, \tag{14.40}$$

which corresponds to the motion of a particle in the potential

$$V(\sigma) = \frac{\omega^2}{2}\left[\sigma^2 + \frac{1}{\sigma^2} + \frac{2P}{3\sigma^3}\right]. \tag{14.41}$$

by introducing $P = (aN/\ell)\sqrt{2/\pi}$, which measures the effect of the interactions on the condensate density. This equation can be formally integrated but the representation of this potential as shown in Figure 14.3 is sufficient to understand the main properties of the condensate.

For a *positive* scattering length a ($P > 0$), in the limit $\sigma \to 0$, the potential energy is dominated by the positively diverging repulsive interaction ($\simeq 1/\sigma^3$), whereas for large values of σ, the trapping potential dominates. As shown in Figure 14.3(a), there is only one stable equilibrium point and consequently only anharmonic oscillations around the bottom of the potential well $V(\sigma)$ may be expected.

On the other hand, for a *negative* scattering length, one can have two possible behaviours depending on the value of the parameter P. For large values of σ, the trapping potential still dominates but the interaction energy is now negatively diverging in the small-width domain, so that $\sigma = 0$ is always the global minimum of the potential energy.[4] For $P > P_c$, there are two equilibrium points, one stable and one unstable as shown by the dotted line in Figure 14.3(b). Both points coincide at a critical value P_c (see the dashed line in the picture) which is given by the following two equations:

$$\frac{\partial V}{\partial \sigma} = 0 \quad \Rightarrow \quad \sigma^5 = \sigma + P_c \tag{14.42}$$

$$\frac{\partial^2 V}{\partial \sigma^2} = 0 \quad \Rightarrow \quad \sigma^5 = -3\sigma - 4P_c, \tag{14.43}$$

which lead to $P_c = -4/5^{5/4}$.

As shown by the solid line in Figure 14.3(b), for $P < P_c$, there are no equilibrium points and all initial conditions will converge toward a vanishing width: the condensate will collapse in a finite time. Using the definition of the parameter P, the condition to avoid the collapse can be rewritten as

$$\frac{N|a|}{\ell} < \frac{\sqrt{8\pi}}{5^{5/4}} \simeq 0.67. \tag{14.44}$$

This is a condition very similar to (14.37) obtained by considering the stability of linear solutions of the GP equation. The present method shows that there is one equilibrium point for $P > P_c$, but the figure emphasises that this is only a metastable position; consequently Condition (14.44) is necessary, but not sufficient, to avoid collapse, since the initial condition is also important. Even if $P > P_c$, a too large initial condensate would collapse by overcoming the energy barrier.

The above analysis can be extended [147] to cylindrical traps ($\lambda \neq 1$) and one finds that $|P_c|$ is a decreasing function of λ. This means that one can have more particles in the condensate before a collapse takes place if it is cigar-shaped ($\lambda \ll 1$) rather than disk-shaped ($\lambda \gg 1$). This is because the GP equation of a cigar-shaped condensate is closer to a one-dimensional NLS equation, for which collapse does not exist, whereas, for the disk-shaped condensate, GP is closer to a two-dimensional NLS, where collapse is possible (see Section 3.7).

Experiments are in good agreement with these results and have allowed the exploration of the dynamic response (the collapse) of the condensate to a sudden shift of the scattering length a to a value more negative than the critical one [56]. This is in particular possible in the vicinity of the Feshbach resonance. The level of control allows one to see how all characteristics of the collapse depend on the magnitude of a, on the initial number and density of condensate atoms, and on the initial size and shape of the BEC before the transition to instability.

[4] The GP equation does not apply of course for a too small width, since the regime of validity of the Born approximation, required to use the mean-field approximation [40], is $a \ll \sigma$.

14.5 Soliton solutions

14.5.1 From 3D to 1D Gross–Pitaevskii equation

Let us consider the three-dimensional equation (14.17) for a cigar-shaped condensate confined in a cylindrically symmetric parabolic trap, i.e. the case in which the trapping potential in z is much weaker than the trapping potential in x and y, i.e. $\lambda \ll 1$. This external potential suggests mapping the three-dimensional GP equation into an effective one-dimensional equation, which would greatly simplify the analysis.

The solution of this difficult nonlinear problem can be very accurately approximated by assuming that it is possible [147] to factor the condensate wave function as

$$\Phi(\rho, z, t) = \sqrt{\frac{N}{\ell^3}}\, f(\rho) b(s, \tau), \tag{14.45}$$

where we introduce the change of variables $\tau = \omega t$, $\rho = (x^2 + y^2)^{1/2}/\ell$, $s = z/\ell$, $\ell = \sqrt{\hbar/(m\omega)}$ being the harmonic oscillator length corresponding to the radial confinement frequency ω. The physical justification is that, in the perpendicular direction ρ, the trapping and the nonlinear term tend to compress the wave packet, competing against the linear dispersion provided by the kinetic-energy term. On the other hand, the trapping force in the s-direction is much smaller so that, along this axis, there is only a competition between the nonlinear attraction and the dispersion.

Introducing the ansatz (14.45) into Equation (14.17), one gets

$$\mathrm{i}\frac{\partial (fb)}{\partial \tau} = \left[-\frac{1}{2}\nabla_\perp^2 - \frac{1}{2}\frac{\partial^2}{\partial s^2} + \frac{1}{2}(\rho^2 + \lambda^2 s^2) - 2Q\,|fb|^2 \right] fb, \tag{14.46}$$

where $\nabla_\perp^2 = \dfrac{1}{\rho}\dfrac{\partial}{\partial \rho}\left(\rho \dfrac{\partial}{\partial \rho}\right)$ and $Q = -2\pi a N/\ell$.

Temporarily neglecting the nonlinear term (i.e. taking the particular case $Q = 0$), Equation (14.46) can be rewritten as

$$\frac{1}{b(s,\tau)}\left[\mathrm{i}\frac{\partial}{\partial \tau} + \frac{1}{2}\frac{\partial^2}{\partial s^2} - \frac{1}{2}\lambda^2 s^2 \right] b(s,\tau) = \frac{1}{f(\rho)}\left[-\frac{1}{2}\nabla_\perp^2 + \frac{1}{2}\rho^2 \right] f(\rho). \tag{14.47}$$

As the left hand side is a function of variables s and τ, whereas the right hand side is only a function of ρ, both sides of the equation are equal to a constant ν_ρ. Thus, the function f satisfies

$$-\frac{1}{2}\nabla_\perp^2 f + \frac{1}{2}\rho^2 f = \nu_\rho f, \tag{14.48}$$

which corresponds to a well-known eigenvalue problem, the two-dimensional Schrödinger equation for a particle in an isotropic harmonic potential. Its ground state, attained for $\nu_\rho = 1$, is $f_0(\rho) = \mathrm{e}^{-\rho^2/2}$ which we will choose as an approximate solution for f.

Multiplying Equation (14.46) by $\rho f_0(\rho)$ and integrating on ρ to eliminate the ρ dependence, we find

$$\mathrm{i}\frac{\partial b}{\partial \tau} = -\frac{1}{2}\frac{\partial^2 b}{\partial s^2} + \frac{1}{2}\lambda^2 s^2 b - Q\,|b|^2\,b + \nu_\rho b, \tag{14.49}$$

where the factor 2 in the nonlinear term has disappeared because

$$\int_0^\infty f_0^4 \rho \mathrm{d}\rho \Big/ \int_0^\infty f_0^2 \rho \mathrm{d}\rho = \frac{1}{2}. \tag{14.50}$$

Finally making the change of unknown function $b(s,\tau) = k(s,\tau)\mathrm{e}^{-\mathrm{i}\nu_\rho \tau}$, we obtain

$$\mathrm{i}\frac{\partial k(s,\tau)}{\partial \tau} + \frac{1}{2}\frac{\partial^2 k(s,\tau)}{\partial s^2} + Q\,|k(s,\tau)|^2\,k(s,\tau) = \frac{1}{2}\lambda^2 s^2 k(s,\tau). \tag{14.51}$$

This is a one-dimensional NLS equation with an additional term on the right hand side. It is important to emphasise that this regime is obtained as the one-dimensional limit of the three-dimensional mean-field theory, generated by averaging in the transverse plane, rather than the one-dimensional mean-field theory, which would be appropriate when the transverse dimension is of the order of the atomic size. The factor Q of the third term in Equation (14.51) instead of $2Q$ in the three-dimensional version (14.46) attests the difference. Let us emphasise that, although the methodology of this section is not entirely self-consistent (Q being set temporarily to zero), the derivation can be rigorously proven [116].

14.5.2 Repulsive interaction: dark and grey solitons

Solitons were first observed [35, 53] in sodium BECs for which the atom–atom interaction is repulsive ($a > 0$). In these experiments, the parameter λ of the trapping potential was very small and for example, the first experiment [35] was performed with $\omega = 2\pi \times 425$ Hz and $\omega_z = 2\pi \times 14$ Hz, which leads to $\lambda \simeq 0.033$. It is therefore legitimate to neglect the effect of the external potential in the s-direction and Equation (14.51) can be rewritten as

$$\mathrm{i}\frac{\partial \chi(s,\tau)}{\partial \tau} + \frac{1}{2}\frac{\partial^2 \chi(s,\tau)}{\partial s^2} - 2\,|\chi(s,\tau)|^2\,\chi(s,\tau) = 0, \tag{14.52}$$

by defining $\chi(s,\tau) = k(s,\tau)\sqrt{\pi a N/\ell}$.

This is a nonlinear Schrödinger equation analogous to Equation (3.25) with the important difference that here, PQ, the product of the dispersion and nonlinearity pre-factors is negative. As we did not consider this case in Chapter 3, we have to look here for a new and important class of solutions.

The method is very similar to the one used in Section 3.2.1 with, however, a significant difference: we shall be looking here for solutions with a constant but nonvanishing amplitude A_0 at large distances since, in Chapter 3, we showed that localised solutions with vanishing amplitude at the boundary do not exist when $PQ < 0$.

Let us look again for solutions in which both the carrier wave φ and the envelope A are permanent profile solutions, but with, a priori, different propagation velocities, u_p for φ and u_e for A, i.e. we are looking for a solution such that

$$\chi = A(s - u_e\tau)\exp[\mathrm{i}(\varphi(s - u_p\tau) + \beta\tau)]. \tag{14.53}$$

Such an ansatz is of course strongly reminiscent of Ansatz (3.32) in the focusing case ($PQ > 0$) with, however, a phase factor $\beta\tau$ which was absent previously, and comes from the present boundary conditions

$$\lim_{|\xi|\to\infty} A(\xi) = A_0 \quad \text{and} \quad \lim_{|\xi|\to\infty} \varphi(\xi) = 0. \tag{14.54}$$

Using them, at the boundaries the ansatz (14.53) reduces to the solution $\chi = A_0\exp(\mathrm{i}\beta\tau)$ which can be a solution of Equation (14.52) if and only if

$$\beta = -2A_0^2. \tag{14.55}$$

One immediately realises that, if the amplitude $A(\xi)$ tends to $A_0 = 0$ when $|\xi|$ tends to infinity, this phase factor disappears, justifying the ansatz considered in Section 3.2.1.

Introducing (14.53) into Equation (14.52) and considering separately the real and imaginary parts, we obtain

$$Au_p\varphi_s + 2A_0^2A + \frac{1}{2}A_{ss} - \frac{1}{2}A\varphi_s^2 - 2A^3 = 0 \tag{14.56}$$

$$-u_eA_s + A_s\varphi_s + \frac{1}{2}A\varphi_{ss} = 0. \tag{14.57}$$

Multiplying Equation (14.57) by $2A$ and integrating, we get

$$-u_eA^2 + A^2\varphi_s = C \;\Rightarrow\; \varphi_s = u_e + \frac{C}{A^2}, \tag{14.58}$$

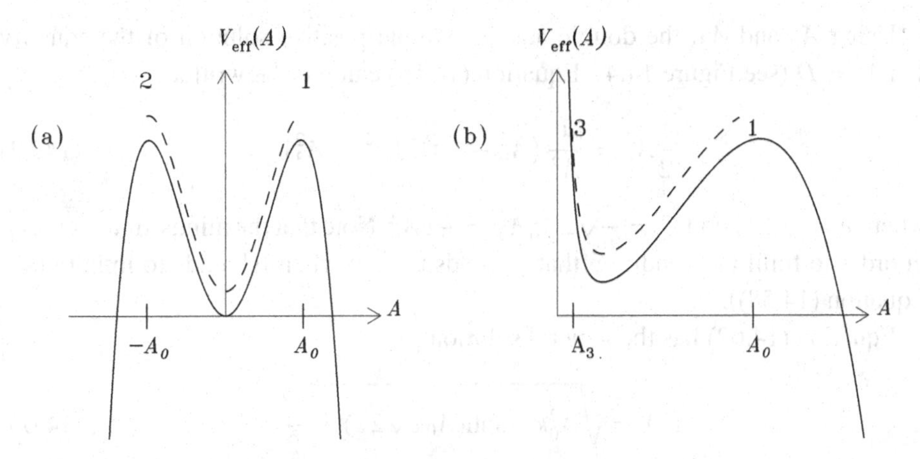

Figure 14.4. Pseudo-potential $V_{eff}(A)$ for (a) $C = 1/2$ and (b) $C = 0$ when the velocity is set to $v = 5.5$.

where C is a constant. Introducing the result (14.58) into Equation (14.56), multiplying the resulting equation by $2A_s$ and finally integrating, one gets the usual equation for a fictitious particle with constant energy D

$$\frac{1}{2}A_s^2 + V_{eff}(A) = D, \tag{14.59}$$

if we define the pseudo-potential energy

$$V_{eff}(A) = -A^4 + \left(u_e u_p - \frac{u_e^2}{2} + 2A_0^2\right) A^2 + C(u_e - u_p)\ln A + \frac{C^2}{2A^2}. \tag{14.60}$$

Several parameters are involved, but it is possible here to choose the simplest solution where the carrier and envelope are propagating at the same velocity $v = u_e = u_p$. This case was forbidden in the focusing case $PQ > 0$. Expression (14.60) can thus be further simplified as

$$V_{eff}(A) = -A^4 + \left(\frac{v^2}{2} + 2A_0^2\right) A^2 + \frac{C^2}{2A^2}, \tag{14.61}$$

which is plotted in Figure 14.4.

As we are looking for solutions with the asymptotic condition for the envelope $\lim_{s\to+\infty} A = A_0$, the motion of the fictitious particles has to start from the unstable equilibrium point $A = A_0$ (labelled 1) to reach either the state $A = -A_0$ (labelled 2 in Figure 14.4(a)) if $C = 0$, or a state (labelled 3) with a small but nonvanishing amplitude $A = A_3$ if $C \neq 0$ (Figure 14.4(b)).

Using A_0 and A_3, the double and the simple positive solution of the equality $V_{\text{eff}}(A) = D$ (see Figure 14.4), Equation (14.59) can thus be written

$$\frac{1}{2}A_s^2 = \frac{1}{A^2}\left(A^2 - A_0^2\right)^2\left(A^2 - A_3^2\right) \tag{14.62}$$

where $v = \sqrt{2}A_3$ and $C = -\sqrt{2}A_0^2 A_3 = -vA_0^2$. Note that the minus sign is chosen in order to fulfil the condition that φ_s tends to zero when $|s|$ tends to infinity (see Equation (14.58)).

Equation (14.62) has the general solution

$$A = \sqrt{A_0^2\kappa^2 \tanh(A_0\kappa\sqrt{2}\,s) + \frac{v^2}{2}}, \tag{14.63}$$

where $\kappa = \sqrt{1 - A_3^2/A_0^2}$ belongs to the interval [0, 1]. Introducing this result into Equation (14.58), we obtain

$$\varphi = \int_0^{s-v\tau} \mathrm{d}s\left(v - v\frac{A_0^2}{A^2}\right) = -\left[\arctan\left(\frac{v}{\sqrt{2}A_0\kappa\tanh(A_0\kappa\sqrt{2}\,s)}\right)\right]_0^{s-v\tau}. \tag{14.64}$$

Collecting all contributions, the solution of Equation (14.52), the one-dimensional Gross–Pitaevskii equation in the absence of a longitudinal potential, is finally

$$\chi(s,\tau) = \sqrt{A_0^2\kappa^2\tanh(A_0\kappa\sqrt{2}\,s) + \frac{v^2}{2}}$$
$$\exp\left[-\mathrm{i}\arctan\left(\frac{v/\sqrt{2}}{A_0\kappa\tanh(A_0\kappa\sqrt{2}(s - v\tau))}\right)\right]\exp\left[-\mathrm{i}2A_0^2\tau\right]. \tag{14.65}$$

Recognising that $\sqrt{a^2+b^2}\exp[-\mathrm{i}\arctan(b/a)] = a - ib$, Equation (14.65) can be rewritten in the more compact form

$$\chi(s,\tau) = \left[A_0\kappa\tanh(A_0\kappa\sqrt{2}(s - v\tau)) - i\frac{v}{\sqrt{2}}\right]e^{-i2A_0^2\tau}. \tag{14.66}$$

In the particularly simple case $\kappa = 1$, the velocity v vanishes so that Solution (14.66) simplifies as

$$\chi(s,\tau) = A_0\tanh(A_0\sqrt{2}\,s)\,\exp\left[-2iA_0^2\tau\right]. \tag{14.67}$$

It is represented with a solid line in Figure 14.5(a) and is called a *dark soliton*. The soliton has zero velocity and zero density at its centre.

For a nonunitary value of κ (or nonvanishing values of C), as emphasised in Figure 14.4(b) by the dashed line, the amplitude A of the envelope decreases but

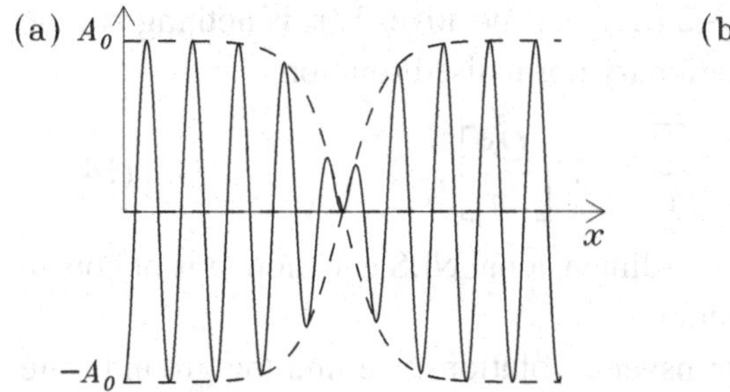

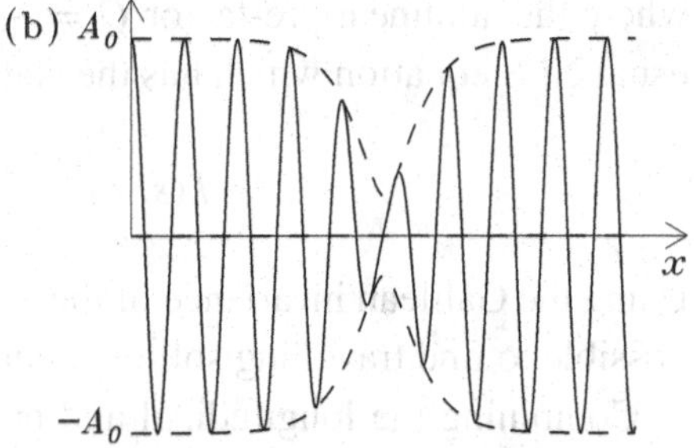

Figure 14.5. In (a), the solid line is the real part of the *dark soliton* solution (14.66) for χ in the case $C = 0$, whereas the dashed ones correspond to the envelopes $\pm A$. In (b), the *grey soliton* solution, the envelopes are represented for $\kappa = 0.98$.

never vanishes. The complete solution (14.66) is thus characterised by a notch in the density profile as shown in Figure 14.5(b). Such a solution is called a *grey soliton*.

A characteristic of these solutions is that their speeds are less than a critical value $v_c = A_0\sqrt{2}$ since one has

$$v = v_c\sqrt{1 - \kappa^2}. \tag{14.68}$$

As κ decreases, the soliton becomes shallower, has a more gradual phase variation and its speed increases approaching the speed of sound since in original units the critical velocity is the Bogoliubov speed of sound c_s.

To observe the soliton propagation [35, 53], the BEC density distribution is measured with absorption imaging after imprinting a phase step. The initial BEC with a uniform phase is modified by exposing it to pulsed off-resonant laser light with a spatially dependent intensity pattern. The resulting imprinted phase distribution of the condensate wave function is measured with a Mach–Zender matter–wave interferometer that makes use of optically induced Bragg diffraction. The soliton propagation is observed by measuring BEC density distributions with absorption imaging.

The dark and the grey soliton solutions that we have derived here are not only interesting for a BEC. As mentioned in Section 3.6.5, they are general solutions of the NLS equations for $PQ < 0$.

14.5.3 Attractive interaction: bright solitons

Let us consider now the solution of Equation (14.51) in the case of a negative scattering length ($a < 0$) for a condensate in a cylindrically symmetric parabolic trap. In the case $\lambda = 0$, Equation (14.51) reduces to

$$\mathrm{i}\frac{\partial k(s,\tau)}{\partial \tau} + \frac{1}{2}\frac{\partial^2 k(s,\tau)}{\partial s^2} + Q\,|k(s,\tau)|^2\,k(s,\tau) = 0, \tag{14.69}$$

where the nonlinear pre-factor $Q = -2\pi a N/\ell$ is positive. This is nothing but the usual NLS equation which has the stationary normalised solution

$$k(s,\tau) = \frac{\sqrt{Q}}{2\pi}\,\mathrm{sech}\left[\frac{Qs}{2\pi}\right]. \tag{14.70}$$

Using the Galilean invariance of the one-dimensional NLS equation, it is of course possible to find travelling soliton solutions.

Combining the longitudinal and transverse solutions, we find the ground state solution

$$\Phi(\rho, z, t) = \frac{N}{\ell^2}\sqrt{\frac{-a}{2\pi}}\,\mathrm{sech}\left[\frac{aN\,s}{\ell}\right]\mathrm{e}^{-\rho^2/2}\mathrm{e}^{-\mathrm{i}\nu_\rho\tau}. \tag{14.71}$$

A more formal derivation [147] using the multiple expansion method leads to the same result, explaining why this solution, which takes into account only the ground state f_0, is close to this approximate profile even in the nonperturbative region.

It is remarkable to note that the width of the cloud $\ell/(aN)$, appearing in the argument of the sech function of Equation (14.71), is inversely proportional to the number of atoms N. Condensates with a small number of particles would be very long, while condensates with more particles would be shorter. If the number of particles is large enough, the condensate is unstable and collapses. Indeed, as shown in Sections 14.4.2 and 14.4.3, an important result related to negative scattering length condensates is that stable solutions of the GP equation exist only under certain conditions for the number of particles and the size of the trap. When these conditions are not satisfied [40], the condensate is unstable and destroyed by the collapse phenomenon. This is a drawback for two reasons: having larger condensates is important to get better experimental observations of BECs; moreover, atom clouds as large as possible are needed for future practical applications (atom interferometers, atom clocks etc.).

To check the existence of the soliton solution, it is possible to numerically compute the ground-state solution of Equation (14.46) for different values of λ. Starting from the solution in the noninteracting case ($Q = 0$),

$$u(\rho, s) = f(\rho)h(s) = \lambda^{1/4}\pi^{-3/4}\exp\left(-\frac{\rho^2}{2} - \frac{\lambda s^2}{2}\right), \tag{14.72}$$

one can use the steepest descent method to minimise the energy

$$E = \int \mathbf{dr}\left[|\nabla u|^2 + (\rho^2 + \lambda^2 s^2)\,|u|^2 - 2Q\,|u|^4\right]. \tag{14.73}$$

Decreasing λ and increasing Q, we should obtain the soliton solution (14.71). Figure 14.6 presents sections of the fundamental state $u(\rho, s)$ for different values of the parameter λ. As it decreases, the solution gets wider, but does not widen

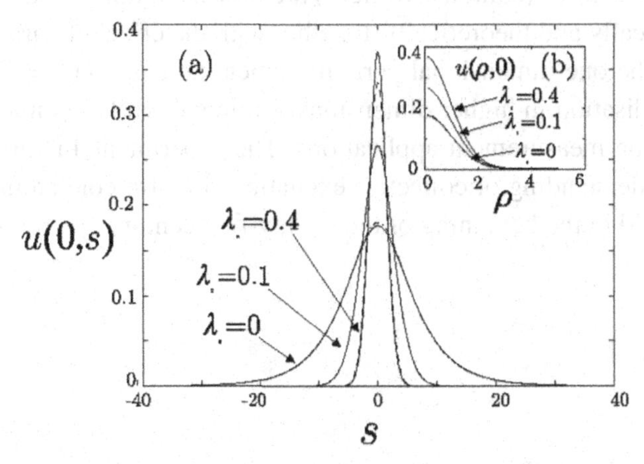

Figure 14.6. The solid curves represent sections of the ground state solution $u(\rho, s)$ for different values of λ when $Q = 5/4$. The dash-dotted line corresponds to the theoretical prediction for $\lambda = 0$ given by the bright soliton while the dashed line corresponds to the Gaussian solution (14.72) for $\lambda = 0.4$. (a) presents s-sections for $\rho = 0$ while (b) shows ρ-sections for $s = 0$ (from reference [147]).

indefinitely when $\lambda = 0$ as it would for positive scattering length a (i.e. negative Q in Equation (14.73)). The agreement with the soliton solution is extremely good, except in the region around $s = 0$. This picture shows in addition that, when the trapping potential is strong ($\lambda = 0.4$), the numerical solution is very close to the Gaussian ground-state solution (14.72), except in the vicinity of the centre of the trap where one notices an enhanced compression of the solution due to nonlinear effects.

These theoretical predictions have been very recently confirmed by pioneering experiments [98, 159]. The essential experimental method used to transform a stable attractive BEC of ^{7}Li into a matter–wave bright soliton was as follows. A BEC with positive scattering length was condensed into a harmonic trap. The trap was adiabatically deformed into a cigar-shaped geometry. The scattering length was then tuned to be small and negative. Note that, contrary to the experiment carried out at Rice University [159] and to the framework of the above analytical description, the resulting condensate of the experiment carried out in Paris [98] was projected onto an *expulsive* harmonic potential ($\omega_z^2 < 0$) in the longitudinal direction z. In addition, the first experimental set up led to the creation of trains of solitons generated through modulational instability, while, in the second one, single solitons were created. Recent work has shown how to analytically treat this case [39].

The key to these experiments, and the beauty of cold atom physics, is the ability to tune the atom–atom interactions smoothly from positive to negative. Avenues for research nowadays include important studies of soliton stabilities and interactions between solitons, as well as other nonlinear dynamics of condensates. Higher dimensional structures (solitons, vortices) are now beginning to be studied analytically, numerically and theoretically, together with the effect of various potentials, in particular the one-dimensional periodic optical lattice, $V(x) = V_0 \sin^2(\pi x/d)$, and its generalisation in higher dimensions. Atom soliton lasers may be also useful for precision measurement applications. The experimental realisation and the theoretical understanding of collective excitations in ultra-cold atomic clouds is a very active field in the beginning of the twenty-first century.

Part IV

Nonlinear excitations in biological molecules

Introduction

Considering the complexity of biological systems at the molecular level, it may seem hopeless to attempt to describe them in physical terms. However there are fundamental phenomena, based on well defined physical and chemical processes that may be amenable to a theoretical description. Establishing fairly simple models which agree with experimental observations can help us determine the basic processes behind the complexity of life. While the biologist is trying to refine his/her knowledge of all the intricate processes which contribute to the properties of biomolecules, the physicist proceeds with a belief that some general principles can be found. To make a comparison with another domain, let us imagine that we open the bonnet of a modern car: the engine with all its parts, tubes, wires etc. looks incredibly complex, but it operates according to a four-stroke process that thermodynamics understands and can help to optimise. Shall we be able to reach this level of understanding for DNA or a protein, which are far more complex than the most elaborate car engine?

Since the beginning of 1990, the interest in physical modelling of biological problems has been growing due to the conjunction of two effects:

(i) Molecular biology has made considerable progress and is now able to describe in detail the structure and sometimes the dynamics of biological molecules, allowing the development of models on precise grounds.
(ii) The development of the analysis of nonlinear systems, which introduced new concepts such as self-organisation, or solitons has provided starting elements for the models, although caution should be used when applying these concepts to biological systems.

Three physical processes play a large role in the function of biological molecules, such as proteins or DNA:

(i) Energy storage and transfer.
(ii) Large conformational changes of the molecules, which are essential for many functions.
(iii) Charge transport (electrons and protons).

The energy which is necessary for protein activity is often provided by the hydrolysis of the adenosine tri-phosphate molecule (ATP), which binds to a receptor site in the protein and reacts with water to release about 0.42 eV. This amount of energy is very large at the molecular scale (about 20 $k_B T$) but it can only be useful if it is not quickly distributed amongst all the degrees of freedom of the protein in a fast energy equipartition process, but instead stays localised. It could be stored in an electronic excited state, but its localisation, and perhaps its transfer, as a localised nonlinear excitation is another possibility. This hypothesis is still controversial for proteins, but recent experiments on molecular crystals, which contain chemical bonds which are very similar to the peptide bonds of proteins, have shown that vibrational energy could be trapped by nonlinearity in these model systems [57]. Moreover the idea of nonlinear trapping of vibrational energy is worth examining because models showing this phenomenon include concepts which could be of general validity. We shall discuss this question in Chapter 15.

Large conformational changes play a role in the function of proteins, and particularly for enzymes, but they are also crucial for DNA because it would be impossible to read the genetic code without a local unwinding of the double helix. Such a local opening implies the breaking of many hydrogen bonds (typically 50) so that it requires a significant focusing of energy in a small region of the molecule. Before the start of the transcription, i.e. the reading of a gene, there is no chemical reaction that can provide energy, but nonlinear energy localisation phenomena might play some role. Transcription itself is probably too complex to be modelled as a physical process with our current knowledge, but there is a purely physical process which shows some similarity to the initiation of transcription, the formation of 'thermal denaturation bubbles' of DNA which are local openings induced by heating. Studying the physical basis of this phenomenon is a first step toward a physical model of transcription, and it can also be used to validate some physical models of DNA. Chapter 16 presents some recent investigations in this domain.

A third problem which is still open is proton transport across or along membranes. A purely diffusive mechanism could be considered but it raises some questions. How could it be compatible with the coherence of some biological processes? How could it be sufficiently efficient when large proton flows are needed? How could we explain that proton transport is almost independent of the pH on both sides of a membrane. This would not be compatible with diffusion in a concentration gradient. Transport could occur along the chains of hydrogen bonds that span membranes, either inside the proton channels of proteins such as gramicidin or within the membrane proteins themselves. Models involving coherent proton transport along such hydrogen bonded chains have been proposed. Their validity in a biological context is still questioned, and this is why they won't be presented in this book, but an introduction to these ideas can be found in reference [149].

In all these applications, the soliton concept is certainly not strictly correct but it nevertheless provides a good basis for the development of models. There is no doubt about the large role of nonlinearity in these biological processes. The energy released locally by ATP hydrolysis is very large with respect to the thermal energy. Similarly the 180° rotations of bases in DNA unwinding are obviously far beyond a harmonic approximation. Moreover the large size of biological molecules allows collective motions of some groups of atoms. Thus the two basic ingredients of the soliton, nonlinearity and cooperativity are indeed present.

15

Energy localisation and transfer in proteins

15.1 The mechanism proposed by Davydov

Proteins are chain-like molecules made of a sequence of amino acids, which have the basic chemical structure shown in Figure 15.1, where the symbol R designates a hydrocarbon chain, which depends on the specific amino acid. There are 20 different amino acids. They are bound together in proteins by a peptide bond which is obtained by the elimination of a water molecule between the acidic group and the –NH group. This leads to long chains containing a repeated motif, as shown in Figure 15.2. These chains fold into the secondary structure of proteins, which is stabilised by the formation of hydrogen bonds between these motifs. The two main elements found in protein secondary structures are α-helices, shown in Figure 15.3, and flat structures called β-sheets. The mechanism proposed by Davydov for energy localisation and transfer in proteins concerns α-helices.

The hydrolysis of ATP

$$\mathrm{ATP}^{4-} + \mathrm{H_2O} \rightarrow \mathrm{ATP}^{3-} + \mathrm{HPO}_4^{2-} + \mathrm{H}^+, \tag{15.1}$$

releases energy which is almost equal to two quanta $\hbar\omega$ of the vibrational mode of the C=O bond. This vibration is therefore a natural candidate for storage of the ATP hydrolysis energy. Moreover, as the oxygen atom of the C=O group is involved in a hydrogen bond which stabilises the geometry of the helix, a coupling between the excitation of the C=O bond and deformation modes of the helix (phonon modes) is likely. This is the essence of the Davydov model, which assumes that the energy transmitted to the C=O bond induces a local distortion of the helix. This distortion slightly shifts the frequency of the excited C=O bond, which no longer resonates with the other C=O bonds in its vicinity. The frequency misfit causes a drop in the rate of the transfer of the energy of the excited C=O to its neighbours, which tends to trap the ATP hydrolysis energy.

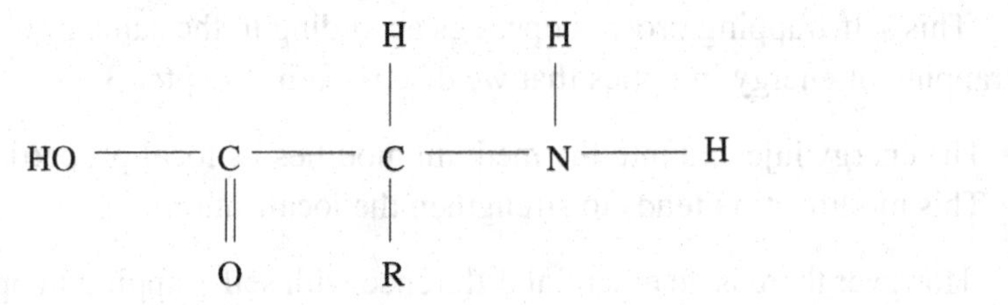

Figure 15.1. Chemical formula of an amino acid.

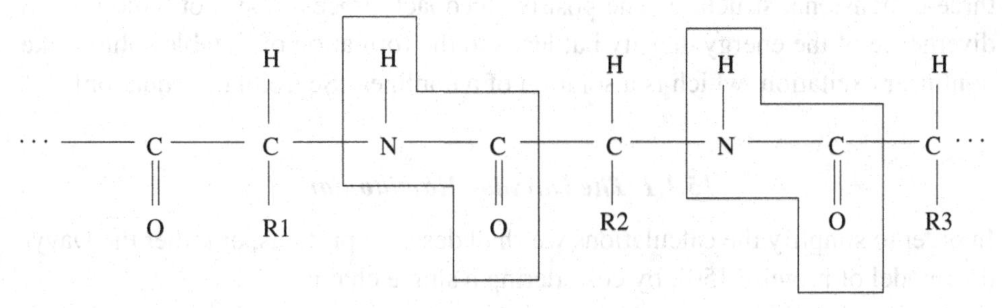

Figure 15.2. A chain of amino acids linked by a peptide bond, part of a protein.

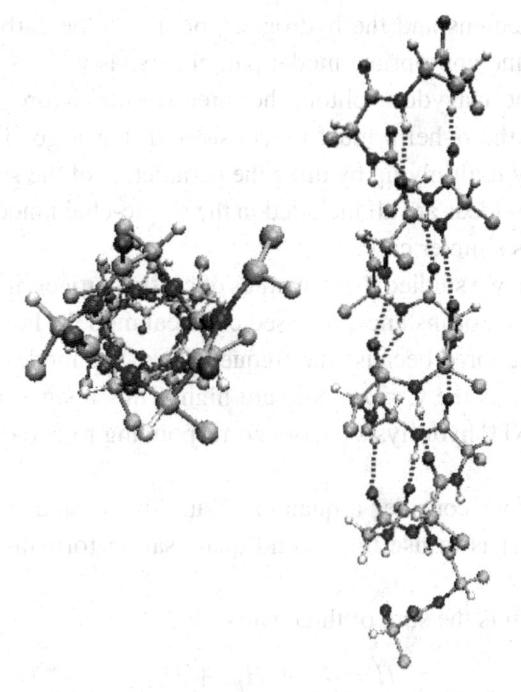

Figure 15.3. Top and side views of a natural α-helix (part of the longest helix of the phosphoglycerate kinase protein). Dotted lines show the hydrogen bonds [184].

This self-trapping process operates according to the same principles as the self-trapping of energy in optics that we discussed in Chapter 3:

- The energy injected into the medium modifies its local properties.
- This modification tends to strengthen the localisation.

However there is an essential difference with self-trapping in optics which takes place in a three-dimensional medium because the hydrogen-bonded chains of proteins are essentially one-dimensional, even if these chain-like molecules fold in a three-dimensional structure. The positive feedback process does not build up to a divergence of the energy density but leads to the formation of a stable soliton-like nonlinear excitation, which is a solution of a nonlinear Schrödinger equation!

15.1.1 The Davydov Hamiltonian

In order to simplify the calculations we shall describe proteins, or rather the Davydov model of proteins [50], by considering a single chain

$$\mathrm{H-N-C=O\ldots}\underbrace{\mathrm{H-N-C=O\ldots}}_{\text{cell } n \text{ of the model}}\mathrm{H-N-C=O\ldots} \tag{15.2}$$

where the R side chains and the hydrogens, bound to the carbon atoms, are not shown. To determine appropriate model parameters, as well as the conditions for the existence of the Davydov soliton, the three parallel hydrogen-bonded chains that extend along the α-helix must be considered, but a good approximation is simply obtained by multiplying by three the parameters of the single-chain model, and, as the physical ideas are all included in the single-chain model, we restrict our presentation to this simpler case.

Up to now, when we studied the dynamics of crystal lattices, in the FPU model or ferroelectric crystals for instance, we used classical models. For proteins quantum effects cannot be ignored because the frequencies of the modes which play a role in the theory, such as the C=O mode, are high. Thus a small number of quanta are involved, the ATP hydrolysis energy corresponding to two quanta of the C=O vibration only.

We must therefore consider a quantum Hamiltonian and, as for the polymer chains, the simplest is to use the second quantisation formalism to describe the C=O bonds.

The Hamiltonian is the sum of three parts:

$$H = H_e + H_{ph} + H_{int}. \tag{15.3}$$

- The first term, H_e, is called the exciton Hamiltonian. It describes the C=O bonds. We denote by B_n^+ the operator which creates a quantum of the C=O vibration in cell n. The operator giving the number of quanta in the nth C=O bond is therefore $B_n^+ B_n$. The energy of a quantum of the C=O vibration is denoted by $E_0 = \hbar\omega_{C=O}$, where the frequency $\omega_{C=O}$ corresponds to a wavenumber of about 1600 cm^{-1}. The exciton Hamiltonian is written as

$$H_e = \sum_n E_0 B_n^+ B_n - J\left(B_{n+1}^+ B_n + B_n^+ B_{n+1}\right). \tag{15.4}$$

Its first term describes the vibrational energy in cell n, while the second term describes the possible transfer of a part of the energy of cell n to the neighbouring cells, which lowers the energy of the system because the coupling coefficient J is assumed to be positive. This corresponds to the natural spreading of the energy in a chain of identical oscillators: the excitation of one of them tends to be transferred to the others. The term $B_{n+1}^+ B_n$ destroys a quantum at site n and creates one at site $n+1$, so that it is indeed the term that describes the transfer from site n to $n+1$ and similarly $B_n^+ B_{n+1}$ describes the transfer from $n+1$ to n.

- The phonon Hamiltonian H_{ph} describes the deformations of the peptide chain. Let us denote by U_n the operator associated with the translation of the nth peptide group. Only its motion along the chain is considered in this one-dimensional model. We denote by P_n the operator corresponding to the momentum conjugate of U_n. Both operators obey the usual commutation rule $[U_n, P_{n'}] = i\hbar\delta_{nn'}$, where δ is the standard Kronecker symbol. It must be stressed that operators corresponding to different sites are acting on different subspaces, so that they commute. This is also true for the operators of H_e.

  The phonon Hamiltonian is written in a harmonic approximation, so that it is given by

$$H_{ph} = \sum_n \frac{P_n^2}{2m} + \frac{K}{2}\left(U_{n+1} - U_n\right)^2, \tag{15.5}$$

where K is the coupling constant between neighbouring peptide groups. Such a Hamiltonian can be diagonalised by going to Fourier space, i.e. by introducing normal modes. To be specific, let us assume that the chain has N cells and choose periodic boundary conditions. Denoting by a_q^+ the operator that creates a phonon with wavevector q, and generalising the usual expressions used for a single quantum harmonic oscillator, we get

$$U_n = \frac{1}{\sqrt{N}} \sum_q \sqrt{\frac{\hbar}{2m\omega_q}}\, e^{i2\pi qn/N} \left(a_{-q}^+ + a_q\right), \tag{15.6}$$

$$P_n = \frac{i}{\sqrt{N}} \sum_q \sqrt{\frac{1}{2} m\hbar\omega_q}\, e^{i2\pi qn/N} \left(a_{-q}^+ - a_q\right), \tag{15.7}$$

where q is an integer such that $-N/2 < q \le N/2$ and $\omega_q = 2\sqrt{K/m}\,|\sin(\pi q/N)|$.

The discrete Fourier transform that we have introduced verifies the usual relations

$$\sum_{n=0}^{N-1} \mathrm{e}^{\mathrm{i}2\pi(\mathrm{q}-\mathrm{q}')\mathrm{n}/\mathrm{N}} = N\delta_{qq'}, \tag{15.8}$$

$$\sum_{q=-N/2+1}^{N/2} \mathrm{e}^{\mathrm{i}2\pi \mathrm{q}(\mathrm{n}-\mathrm{n}')/\mathrm{N}} = N\delta_{nn'}. \tag{15.9}$$

The phonon Hamiltonian is thus written in terms of normal modes as

$$H_{ph} = \sum_q \hbar\omega_q \left(a_q^+ a_q + \frac{1}{2}\right). \tag{15.10}$$

- The interaction Hamiltonian is the crucial part of the Davydov Hamiltonian because it describes the variation of the vibrational energy of the C=O bond when the distance between the peptide groups changes. It occurs because the electronic distribution on the oxygen atom changes when the neighbouring hydrogen atom moves closer or further away. The interaction term is chosen to be

$$H_{int} = \sum_n \chi\,(U_{n+1} - U_n)\, B_n^+ B_n. \tag{15.11}$$

The term $\chi\,(U_{n+1} - U_n)$ appears as a correction to the energy E_0 of the exciton Hamiltonian H_e. The parameter χ is positive. The coupling involves the sites n and $n+1$, and is not symmetrised by introducing $n-1$ too because the nth C=O bond is indeed between the n and $n+1$ peptide groups, characterised by the operators U_n and U_{n+1}.

It should be noticed that the Hamiltonians H_e and H_{ph} are purely harmonic, and the nonlinearity of the model only comes from the coupling between the exciton and phonon degrees of freedom, described by H_{int}.

ALEXANDER S. DAVYDOV (1912–93), was born in Eupatoria in Ukraine, and came to Moscow in 1931 where he started to work as a blue-collar, polishing cars in a car factory, in the perfect Soviet tradition at that time. He followed evening classes at the 'University of the Workers' and succeeded in passing the admission examination of the prestigious State University of Moscow in 1933. He first studied the statistical theory of light diffusion in condensed matter under the supervision of the Russian physicist I. E. Tamm (1895–1971) who got the Nobel prize in 1958 for the understanding of the Cherenkov effect.

During the second world war, Davydov directed a laboratory using X-rays in a plane factory, and then moved to Kiev to study molecular crystals, especially their light absorption properties. Back in Moscow in 1953, he developed a theory of the collective excitations in atomic nuclei, and then returned finally to Kiev in 1964 where he produced intense and internationally recognised scientific activity. He is known, for instance, for the Davydov degeneracy lift, the theory of nonaxial nuclei and, of course, the concept of the molecular soliton in biological molecules (Picture: Larissa Brizhik).

15.1.2 The variational method and the D_2 ansatz

Describing energy transport in a protein requires the solution of an initial value problem: we want to start from a given distribution of the vibrational energy in the C=O bonds, assuming for instance that a bond at the end of the protein has been excited by ATP hydrolysis, and then determine its time evolution.

Therefore we have to solve a time-dependent Schrödinger equation. As the Hamiltonian (15.3) is complex, in particular due to the coupling between the exciton and phonon degrees of freedom, an exact solution cannot be obtained. Several approximate solutions have been proposed. Rather than giving the solution proposed by Davydov himself, we shall describe here the simplest [97], known as the 'D_2 ansatz', which is equivalent to a variational method. The calculation solves the Schrödinger equation in a subspace of the space of the physical states, which is determined by the choice of an approximate solution. This solution, that we shall call the 'trial ket' depends on a set of parameters which must be determined as a function of time.

The first two terms of the Hamiltonian, H_e and H_{ph}, are acting on different subspaces because H_e acts on the variables which characterise the C=O vibration while H_{ph} deals with the motions of the peptide groups. If there were no interaction term, the Hamiltonian would be the sum of two commuting operators, and therefore its eigenstates could be written as a tensor product

$$|\psi(t)\rangle = |\phi_e\rangle \bigotimes |\phi_{ph}\rangle. \tag{15.12}$$

In the D_2 ansatz approximation, this decoupling is kept in the presence of the interaction term. Therefore this ansatz neglects the mixing between excitons and phonons because a general state ket belonging to a product of states $\mathcal{E}_e \bigotimes \mathcal{E}_{ph}$ is not a simple product of a ket of one space by a ket of the other space. The solution using the D_2 ansatz starts from the product of an approximate solution in each of the spaces $\mathcal{E}_e$ and $\mathcal{E}_{ph}$.

For the *exciton part*, we choose the ket

$$|\phi_e\rangle = \sum_n a_n(t) B_n^+ |0\rangle_e, \tag{15.13}$$

where $|0\rangle_e$ is a vacuum state[1] for the C=O excitations. It means that, whatever the site n that we choose, we have $B_n|0\rangle_e = 0$. The operator B_n^+ creates a quantum of vibration at site n. The probability that the nth site is excited by one quantum is

$$\left|\langle 0|_1 \ldots \langle 0|_{n-1} \langle 1|_n \langle 0|_{n+1} \ldots \langle 0|_N \big|\phi_e\rangle\right|^2 = |a_n|^2, \tag{15.14}$$

which gives meaning to the coefficient a_n in the expression of $|\phi_e\rangle$.

Moreover we assume that the chain of C=O bonds is excited *by a single quantum* of the C=O vibration, which imposes the condition $\sum_n |a_n|^2 = 1$. As the ATP hydrolysis releases about two quanta, more complete ansatz have been proposed, but the calculations are significantly more complex while the physical ideas are the same, so that they won't be discussed here.

For the *phonon part*, we choose a phonon coherent state, which corresponds to a quasi-classical state of the harmonic oscillator [77]. Appendix C is a reminder of the properties of coherent states. In the Davydov model, the phonon Hamiltonian is the sum of independent harmonic oscillators of frequencies ω_q, so that the coherent state is chosen as

$$|\phi_{ph}\rangle = \mathrm{e}^{\sum_q (\alpha_q a_q^+ - \alpha_q^\star a_q)} \, |0\rangle_{ph}, \tag{15.15}$$

where $|0\rangle_{ph}$ is the vacuum state in the space $\mathcal{E}_{ph}$, such that $\forall q, a_q|0\rangle = 0$.

It is useful to explicitly show the real and imaginary parts of the complex number α_q for each oscillator by defining

$$\alpha_q = \sqrt{\frac{m\omega_q}{2\hbar}}\, \beta_q + \mathrm{i}\sqrt{\frac{1}{2m\omega_q\hbar}}\, \Pi_q, \tag{15.16}$$

where β_q and Π_q are two real numbers. A direct calculation from the relations connecting U_n and P_n to a_q and a_{-q}^+ leads to

$$-\frac{\mathrm{i}}{\hbar} \sum_n (\beta_n P_n - \Pi_n U_n) = \sum_q \alpha_q a_q^+ - \alpha_q^\star a_q. \tag{15.17}$$

[1] The state denoted by $|0\rangle$ is actually the state $(|0\rangle_1 |0\rangle_2 \ldots |0\rangle_n \ldots)$ if we explicitly write the states of all the sites.

The quantities β_n, Π_n, are related to β_q, Π_q, by a discrete Fourier transform, according to

$$\beta_q = \frac{1}{\sqrt{N}} \sum_n \mathrm{e}^{-\mathrm{i}2\pi qn/N} \beta_n, \tag{15.18}$$

$$\Pi_q = \frac{1}{\sqrt{N}} \sum_n \mathrm{e}^{-\mathrm{i}2\pi qn/N} \Pi_n. \tag{15.19}$$

Therefore we can write

$$|\phi_{ph}\rangle = \mathrm{e}^{-\frac{\mathrm{i}}{\hbar}\sum_n(\beta_n P_n - \Pi_n U_n)} |0\rangle_{ph}. \tag{15.20}$$

The calculation shows that

$$\langle\phi_{ph}|U_n|\phi_{ph}\rangle = \beta_n, \tag{15.21}$$

$$\langle\phi_{ph}|P_n|\phi_{ph}\rangle = \Pi_n. \tag{15.22}$$

These results generalise the usual properties of a coherent state of a single oscillator, but the expressions are more complicated because we are dealing with coupled oscillators.

At this level we have defined the trial ket

$$|\psi\rangle = |\phi_e\rangle \bigotimes |\phi_{ph}\rangle, \tag{15.23}$$

which depends on the parameters a_n, giving the degree of excitation of the C=O bonds, and the quantities β_n, Π_n, which measure the average position and the average momentum of a peptide group. We shall henceforth omit the sign of the tensor product to simplify the notation. The time evolution of this ket $|\psi\rangle$, deduced from the laws of quantum mechanics, determines the time evolution of the parameters, $\beta_n(t)$, $\Pi_n(t)$ and $a_n(t)$.

15.1.3 The dynamic equations for $\beta_n(t)$

As $|\psi\rangle = |\phi_e\rangle|\phi_{ph}\rangle$, and U_n is only acting on $|\phi_{ph}\rangle$, we have

$$\langle U_n\rangle = \langle\phi_e|\langle\phi_{ph}|U_n|\phi_e\rangle|\phi_{ph}\rangle \tag{15.24}$$

$$= \langle\phi_e|\phi_e\rangle\langle\phi_{ph}|U_n|\phi_{ph}\rangle \tag{15.25}$$

$$= 1 \cdot \beta_n(t) \tag{15.26}$$

and similarly $\langle P_n \rangle = \Pi_n$. The equations verified by β_n and Π_n are the usual quantum mechanical equations for the time evolution of the mean value of an observable,

$$\frac{d\beta_n(t)}{dt} = \frac{d\langle U_n \rangle}{dt} = \frac{1}{i\hbar} \langle \psi(t) | [U_n, H] | \psi(t) \rangle, \tag{15.27}$$

$$\frac{d\Pi_n(t)}{dt} = \frac{d\langle P_n \rangle}{dt} = \frac{1}{i\hbar} \langle \psi(t) | [P_n, H] | \psi(t) \rangle. \tag{15.28}$$

The calculation of the commutators is simplified because the phonon and exciton operators act on different subspaces and therefore commute. This is also true for operators corresponding to different sites.

From the value of the commutator

$$[U_n, P_n] = i\hbar, \tag{15.29}$$

we get

$$\frac{d\langle U_n \rangle}{dt} = \frac{1}{m} \langle P_n \rangle = \frac{1}{m} \Pi_n(t), \tag{15.30}$$

which gives the relation between $\Pi_n(t)$ and $\beta_n(t)$, $\Pi_n(t) = m\, d\beta_n(t)/dt$.

Taking into account the operators which commute with P_n, the commutator $[P_n, H]$ reduces to

$$\begin{aligned} [P_n, H] &= \left[P_n, \sum_n \frac{K}{2} (U_{n+1} - U_n)^2 \right] + \left[P_n, \sum_n \chi \,(U_{n+1} - U_n)\, B_n^+ B_n \right] \\ &= C_1 + C_2, \end{aligned} \tag{15.31}$$

where the two terms C_1 and C_2 will be computed separately.

Using $[P_n, U_n^2] = -2i\hbar U_n$ and $[P_n, U_n] = -i\hbar$, we get

$$C_1 = \frac{K}{2} \left[P_n, \left(U_{n+1}^2 - 2U_n U_{n+1} + U_n^2 \right) + \left(U_n^2 - 2U_{n-1} U_n + U_{n-1}^2 \right) \right] \tag{15.32}$$

$$= \frac{K}{2} (2i\hbar U_{n+1} - 2i\hbar U_n - 2i\hbar U_n + 2i\hbar U_{n-1}) \tag{15.33}$$

$$= i\hbar K\, (U_{n+1} + U_{n-1} - 2U_n) \tag{15.34}$$

and

$$C_2 = \left[P_n, \chi\, (U_{n+1} - U_n)\, B_n^+ B_n + \chi\, (U_n - U_{n-1})\, B_{n-1}^+ B_{n-1} \right] \tag{15.35}$$

$$= i\hbar \chi (B_n^+ B_n - B_{n-1}^+ B_{n-1}). \tag{15.36}$$

We have

$$\langle B_n^+ B_n \rangle = \langle \phi_e | \langle \phi_{ph} | B_n^+ B_n | \phi_e \rangle | \phi_{ph} \rangle \tag{15.37}$$

$$= \langle \phi_e | B_n^+ B_n | \phi_e \rangle \langle \phi_{ph} | \phi_{ph} \rangle = \langle \phi_e | B_n^+ B_n | \phi_e \rangle, \tag{15.38}$$

and, from Definition (15.13), we have

$$B_n^+ B_n | \phi_e \rangle = B_n^+ B_n \sum_m a_m(t) B_m^+ |0\rangle_e \tag{15.39}$$

$$= \sum_m a_m(t)\, B_n^+ B_n \underbrace{B_m^+ \, |0\rangle_e}_{=|0\rangle|0\rangle\ldots|1\rangle_m|0\rangle\ldots} \tag{15.40}$$

$$= \sum_m a_m(t)\, B_n^+ B_n \delta_{nm} |0\rangle|0\rangle \ldots |1\rangle_m |0\rangle \ldots \tag{15.41}$$

$$= a_n(t)\, |0\rangle|0\rangle \ldots |1\rangle_n |0\rangle \ldots, \tag{15.42}$$

so that

$$\langle \phi_e | B_n^+ B_n | \phi_e \rangle = \sum_m a_m^\star \, {}_e\langle 0 | \, a_n(|0\rangle \ldots |0\rangle_{n-1} |1\rangle_n |0\rangle_{n+1} \ldots) = |a_n|^2. \tag{15.43}$$

Therefore we get

$$\frac{\mathrm{d}\Pi_n}{\mathrm{d}t} = \frac{\mathrm{d}\langle P_n \rangle}{\mathrm{d}t} = K(\beta_{n+1}(t) + \beta_{n-1}(t) - 2\beta_n(t)) + \chi(|a_n|^2 - |a_{n-1}|^2). \tag{15.44}$$

Using the equality (15.30), we finally get a first set of equations for the coefficients of the trial ket,

$$m\frac{\mathrm{d}^2 \beta_n}{\mathrm{d}t^2} = K\,(\beta_{n+1} - \beta_{n-1} - 2\beta_n) + \chi(|a_n|^2 - |a_{n-1}|^2). \tag{15.45}$$

15.1.4 The dynamic equations for $a_n(t)$

A second set of equations connecting $a_n(t)$ and $\beta_n(t)$ can be deduced from the Schrödinger equation for our trial ket,

$$\mathrm{i}\hbar \frac{\mathrm{d}|\psi\rangle}{\mathrm{d}t} = H|\psi\rangle. \tag{15.46}$$

The left hand side can be expanded into

$$i\hbar \frac{\mathrm{d}|\psi\rangle}{\mathrm{d}t} = i\hbar \frac{\mathrm{d}}{\mathrm{d}t}(|\phi_e\rangle|\phi_{ph}\rangle) \tag{15.47}$$

$$= i\hbar \frac{\mathrm{d}}{\mathrm{d}t}\left\{\left(\sum_n a_n(t)B_n^+|0\rangle_e\right)\left(\mathrm{e}^{-i/\hbar\sum_n(\beta_n(t)P_n-\Pi_n(t)U_n)}|0\rangle_{ph}\right)\right\}$$

$$= \left(\sum_n i\hbar \frac{\mathrm{d}a_n}{\mathrm{d}t} B_n^+|0\rangle_e\right)|\phi_{ph}\rangle$$

$$+ |\phi_e\rangle\left(\frac{\mathrm{d}}{\mathrm{d}t}\mathrm{e}^{-i/\hbar\sum_n(\beta_n(t)P_n-\Pi_n(t)U_n)}|0\rangle_{ph}\right). \tag{15.48}$$

In order to simplify the second term, we can take advantage of the commutation of two operators attached to different sites. Let us introduce the operator

$$T_n = \mathrm{e}^{-i/\hbar(\beta_n(t)P_n-\Pi_n(t)U_n)}. \tag{15.49}$$

It can be used to write the second parenthesis of Equation (15.48) as

$$\frac{\mathrm{d}}{\mathrm{d}t}\prod_n T_n = \sum_n\left(\frac{\mathrm{d}T_n}{\mathrm{d}t}\prod_{n'\neq n} T_{n'}\right). \tag{15.50}$$

In order to go further and calculate $\mathrm{d}T_n/\mathrm{d}t$, we must be careful when we compute the derivative of the exponential because the derivative of the operator that it contains does not commute with the operator itself. However, when two operators C and D commute with their commutator $[C, D]$, we can use the relation

$$\mathrm{e}^{C+D} = \mathrm{e}^{-[C,D]/2}\,\mathrm{e}^C\,\mathrm{e}^D, \tag{15.51}$$

which is reduced to the standard expression $\mathrm{e}^{C+D} = \mathrm{e}^C\,\mathrm{e}^D$, valid for scalars, if the two operators commute.

Here, from $C = -i\beta_n(t)P_n/\hbar$ and $D = i\Pi_n(t)U_n/\hbar$, we obtain the expression of the commutator $[C, D] = -i\beta_n(t)\Pi_n(t)/\hbar$, which indeed commutes with C and D since it is a simple multiplicative scalar.

Therefore we have

$$\frac{dT_n}{dt} = \frac{d}{dt}\left(e^{\frac{i\beta_n(t)\Pi_n(t)}{2\hbar}}\, e^{-i/\hbar\beta_n(t)P_n}\, e^{-i/\hbar\Pi_n(t)U_n}\right) \tag{15.52}$$

$$\begin{aligned} &= \frac{i}{2\hbar}\left(\dot\beta_n\Pi_n + \beta_n\dot\Pi_n\right) e^{\frac{i\beta_n(t)\Pi_n(t)}{2\hbar}}\, e^{-i/\hbar\beta_n(t)P_n}\, e^{-i/\hbar\Pi_n(t)U_n} \\ &\quad + e^{\frac{i\beta_n(t)\Pi_n(t)}{2\hbar}}\left(-\frac{i}{\hbar}\dot\beta_n(t)P_n\right) e^{-i/\hbar\beta_n(t)P_n}\, e^{-i/\hbar\Pi_n(t)U_n} \\ &\quad + e^{\frac{i\beta_n(t)\Pi_n(t)}{2\hbar}}\, e^{-i/\hbar\beta_n(t)P_n}\left(\frac{i}{\hbar}\dot\Pi_n(t)U_n\right) e^{-i/\hbar\Pi_n(t)U_n}. \end{aligned} \tag{15.53}$$

In order to simplify this expression, we can rewrite the last term with the equality

$$e^{\alpha P}\cdot U = U\cdot e^{\alpha P} - [U, e^{\alpha P}] = U\cdot e^{\alpha P} - i\hbar\,\alpha\, e^{\alpha P} = (U - i\hbar\alpha)\, e^{\alpha P}. \tag{15.54}$$

It leads to

$$e^{-i/\hbar\beta_n(t)P_n}\left(\frac{i}{\hbar}\dot\Pi_n(t)U_n\right) = \frac{i}{\hbar}\dot\Pi_n(t)\,(U_n - \beta_n(t))\, e^{-i/\hbar\beta_n(t)P_n}. \tag{15.55}$$

Now using Equation (15.55) in (15.53), we obtain

$$\frac{dT_n}{dt} = \left[\frac{i}{2\hbar}\left(\dot\beta_n\Pi_n + \beta_n\dot\Pi_n\right) - \frac{i}{\hbar}\dot\beta_n(t)P_n + \frac{i}{\hbar}\dot\Pi_n(t)\,(U_n - \beta_n(t))\right] T_n \tag{15.56}$$

$$= \left[\frac{i}{2\hbar}\left(\dot\beta_n\Pi_n - \beta_n\dot\Pi_n\right) - \frac{i}{\hbar}\left(\dot\beta_n(t)P_n - \dot\Pi_n(t)U_n\right)\right] T_n. \tag{15.57}$$

This expression can be introduced in (15.50), which yields

$$\frac{d}{dt}\prod_n T_n = \sum_n\left(\frac{dT_n}{dt}\prod_{n'\neq n} T_{n'}\right) \tag{15.58}$$

$$\begin{aligned} &= \sum_n\left(\left[\frac{i}{2\hbar}\left(\dot\beta_n\Pi_n - \beta_n\dot\Pi_n\right) - \frac{i}{\hbar}\left(\dot\beta_n(t)P_n - \dot\Pi_n(t)U_n\right)\right] T_n\prod_{n'\neq n} T_{n'}\right) \\ &= \left[-\frac{i}{2\hbar}\sum_n\left(\beta_n\dot\Pi_n - \dot\beta_n\Pi_n\right) - \frac{i}{\hbar}\sum_n\left(\dot\beta_n P_n - \dot\Pi_n U_n\right)\right] \\ &\qquad e^{-i/\hbar\sum_n \beta_n(t)P_n - \Pi_n(t)U_n} \end{aligned} \tag{15.59}$$

which provides an explicit expression for the second parenthesis of Equation (15.48).

The first term is an additional term which appears because the operators involved in the derivative of the exponential do not commute, while the second term is the one that would have been obtained if we had derived the exponential of a scalar.

From Equation (15.48), we obtain

$$\mathrm{i}\hbar\frac{\mathrm{d}|\psi\rangle}{\mathrm{d}t} = \left(\sum_n \mathrm{i}\hbar\frac{\mathrm{d}a_n}{\mathrm{d}t}B_n^+|0\rangle_e\right)|\phi_{ph}\rangle$$
$$+|\phi_e\rangle\left\{\sum_n\left(\dot{\beta}_n P_n - \dot{\Pi}_n U_n\right) + \frac{1}{2}\left(\beta_n\dot{\Pi}_n - \dot{\beta}_n\Pi_n\right)\right\}|\phi_{ph}\rangle. \quad (15.60)$$

Fortunately the calculation of the right hand side of Equation (15.46) is simpler! We have

$$H|\psi\rangle = \left(\sum_n E_0 B_n^+ B_n - J\left(B_{n+1}^+ B_n + B_n^+ B_{n+1}\right)\right)|\phi_e\rangle|\phi_{ph}\rangle$$
$$+\left(\sum_q \hbar\omega_q\left(a_q^+ a_q + \frac{1}{2}\right) + \sum_n \chi(U_{n+1} - U_n)B_n^+ B_n\right)|\phi_e\rangle|\phi_{ph}\rangle. \quad (15.61)$$

Using the equality

$$\sum_n B_n^+ B_{n+1}|\phi_e\rangle = \sum_n B_n^+ B_{n+1}\left(\sum_k a_k(t)B_k^+|0\rangle_e\right) \quad (15.62)$$
$$= \sum_n\sum_k a_k(t)B_n^+ \underbrace{B_{n+1}B_k^+}_{\delta_{n+1,k}}|0\rangle_e \quad (15.63)$$
$$= \sum_n a_{n+1}(t)B_n^+|0\rangle_e, \quad (15.64)$$

we get

$$H|\psi\rangle = \sum_n E_0 a_n(t)B_n^+|0\rangle_e|\phi_{ph}\rangle - J\sum_n\left(a_{n-1} + a_{n+1}\right)B_n^+|0\rangle_e|\phi_{ph}\rangle$$
$$+H_{ph}|\phi_e\rangle|\phi_{ph}\rangle + \chi\sum_n a_n B_n^+|0\rangle_e\left(U_{n+1} - U_n\right)|\phi_{ph}\rangle, \quad (15.65)$$

in which we have also used $B_n\sum_k a_k(t)B_k^+|0\rangle_e = a_n(t)|0\rangle_e$.

Let us now write the equality between the left hand side (15.60) and the right hand side (15.65) of the Schrödinger equation, and make the scalar product by $\langle\phi_{ph}|\left(\langle 0|_1\langle 0|_2\ldots\langle 1|_n\ldots\langle 0|_N\right)$. Its second factor selects the coefficients of $B_n^+|0\rangle_e$

on each side. Taking into account Equations (15.21) and (15.22), we get

$$i\hbar\frac{\mathrm{d}a_n}{\mathrm{d}t} = -\underbrace{\frac{1}{2}\left(\sum_k \dot{\beta}_k \Pi_k - \dot{\Pi}_k \beta_k\right)}_{=S} a_n + E_0 a_n - J\left(a_{n-1} + a_{n+1}\right)$$

$$+ \langle \phi_{ph} | H_{ph} | \phi_{ph} \rangle\, a_n + \chi a_n \left(\beta_{n+1} - \beta_n\right). \qquad (15.66)$$

The first term, which we shall denote by S, comes from a partial cancellation of the last two terms of $\mathrm{d}|\psi\rangle/\mathrm{d}t$.

The last step is to calculate the term not yet expanded in Equation (15.66):

$$W = \langle \phi_{ph} | H_{ph} | \phi_{ph} \rangle = \left\langle \phi_{ph} \left| \sum_q \hbar\omega_q \left(a_q^+ a_q + \frac{1}{2}\right) \right| \phi_{ph} \right\rangle. \qquad (15.67)$$

As the ket $|\phi_{ph}\rangle$ is a product of coherent states, for each index q we can use the relation (see Appendix C)

$$|\phi_{ph}\rangle = \mathrm{e}^{\sum_q (\alpha_q a_q^+ - \alpha_q^\star a_q)} |0\rangle = \prod_q |\alpha_q\rangle. \qquad (15.68)$$

We also have

$$\langle \alpha_q | a_q^+ a_q | \alpha_q \rangle = |\alpha_q|^2 \quad \text{with} \quad \alpha_q = \sqrt{\frac{m\omega_q}{2\hbar}}\,\beta_q + \mathrm{i}\frac{1}{\sqrt{2m\omega_q\hbar}}\,\Pi_q. \qquad (15.69)$$

Therefore we obtain

$$W = \sum_q \frac{1}{2m}\left|\Pi_q\right|^2 + \frac{1}{2} m\omega_q^2 \left|\beta_q\right|^2 + \frac{1}{2}\sum_q \hbar\omega_q. \qquad (15.70)$$

This sum in Fourier space can also be written in real space as

$$W = \sum_n \frac{1}{2m}\Pi_n^2 + \frac{1}{2}K\left(\beta_{n+1} - \beta_n\right)^2 + \underbrace{\frac{1}{2}\sum_q \hbar\omega_q}_{W_0}. \qquad (15.71)$$

The first two terms in this formula are very reminiscent of the classical expression of the lattice vibrational energy, while the last term has a purely quantum origin because it is the zero point energy.

To write the expression of the quantity S defined in Equation (15.66) we can use the dynamic equations for β and Π obtained earlier, (15.30) and (15.44), which lead to

$$S = \sum_k \frac{\Pi_k^2}{m} - \beta_k \left[\frac{K}{m}\left(\beta_{k+1} + \beta_{k-1} - 2\beta_k\right) + \frac{\chi}{m}\left(|a_k|^2 - |a_{k-1}|^2\right)\right]. \qquad (15.72)$$

Using the equality

$$\sum_k (\beta_{k+1} + \beta_{k-1} - 2\beta_k)\, \beta_k = -\sum_k (\beta_{k+1} - \beta_k)^2 \,, \tag{15.73}$$

we finally obtain

$$S = \sum_k \frac{\Pi_k^2}{m} + \sum_k \frac{K}{m} (\beta_{k+1} - \beta_k)^2 - \sum_k \frac{\chi}{m} \beta_k \left(|a_k|^2 - |a_{k-1}|^2 \right) . \tag{15.74}$$

When this expression of S is put into Equation (15.66), the first two terms of S cancel with the first two terms of W. Therefore only the zero point energy W_0 subsists. It is a constant for a given lattice so that we finally get

$$\mathrm{i}\hbar \frac{\mathrm{d}a_n}{\mathrm{d}t} = \{\gamma(t) + \chi\, (\beta_{n+1} - \beta_n)\}\, a_n - J\, (a_{n-1} + a_{n+1}) \,, \tag{15.75}$$

where the term $\gamma(t) = E_0 + W_0 + \sum_k (\chi/m)\beta_k(|a_k|^2 - |a_{k-1}|^2)$ does not depend on index n.

Since it is $|a_n|^2$ which has a physical meaning rather than a_n, we can introduce a phase change of a_n in order to simplify the equation without changing the physical results. Let us define

$$a_n(t) = a_n'(t)\, \mathrm{e}^{-\mathrm{i}/\hbar \int_0^t \gamma(t')\mathrm{d}t'} = a_n'\, \mathrm{e}^{\mathrm{i}\theta(t)} . \tag{15.76}$$

If we put this expression into Equation (15.75) and simplify by the overall factor $\mathrm{e}^{\mathrm{i}\theta}$, we get

$$\mathrm{i}\hbar \frac{\mathrm{d}a_n'}{\mathrm{d}t} = \chi\, (\beta_{n+1}(t) - \beta_n(t))\, a_n'(t) - J\left(a_{n-1}'(t) + a_{n+1}'(t) \right) . \tag{15.77}$$

We notice that the finite difference in the last term is analogous to a discretisation of a Laplacian, i.e. a second derivative, but a term is missing. This suggests a new phase change in order to add the missing term $-2a_n'$. Defining

$$a_n'(t) = a_n''(t)\, \mathrm{e}^{\mathrm{i}2Jt/\hbar} = a_n''(t)\, \mathrm{e}^{\mathrm{i}\varphi(t)}, \tag{15.78}$$

and dividing the equation by $\mathrm{e}^{\mathrm{i}\varphi}$, we get

$$\mathrm{i}\hbar \frac{\mathrm{d}a_n''}{\mathrm{d}t} = \chi\, (\beta_{n+1}(t) - \beta_n(t))\, a_n''(t) - J\left(a_{n-1}''(t) + a_{n+1}''(t) - 2a_n''(t) \right) . \tag{15.79}$$

At this point it is possible to simplify the notations and omit the $''$ because the variables $a_n''(t)$ differ from the $a_n(t)$ only by a phase change. Only the square of their modulus, which gives the probability of excitation of a C=O bond, has a physical meaning, so that the difference between a and a'' is irrelevant.

Equations (15.45) and (15.79) are the two equations obtained by Davydov with the D_2 ansatz, but he derived them from a classical approximation. He computed the quantity $\langle\psi(t)|H|\psi(t)\rangle$, using the D_2 ansatz for $|\psi(t)\rangle$, and then used this quantum average of the energy as a classical 'Hamiltonian' for the variables $a_n(t)$ and $\beta_n(t)$. The Hamilton–Jacobi equations resulting from this Hamiltonian also lead to Equations (15.45) and (15.79). The method used by Davydov can be viewed as a variational method applied to a dynamic problem because the classical equations of motion derive from an extremum principle for the Hamiltonian.

15.2 The Davydov equations

The dynamics of the evolution of the energy transferred to a protein by ATP hydrolysis has been reduced to a set of equations for the complex variables $a_n(t)$, which give the state of excitation of the different C=O bonds, and the real variables $\beta_n(t)$ giving the quantum average of the displacements of the peptide groups,

$$m\frac{\mathrm{d}^2\beta_n(t)}{\mathrm{d}t^2} = K\left(\beta_{n+1} + \beta_{n-1} - 2\beta_n\right) + \chi\left(|a_n|^2 - |a_{n-1}|^2\right), \tag{15.80}$$

$$\mathrm{i}\hbar\frac{\mathrm{d}a_n}{\mathrm{d}t} = -J\left(a_{n+1} + a_{n-1} - 2a_n\right) + \chi\left(\beta_{n+1} - \beta_n\right)a_n. \tag{15.81}$$

The analytical solution of this set of coupled nonlinear equations is not known, as it is often the case when we study discrete nonlinear systems. These equations are easy to study numerically, but it is also possible to understand the mechanism of the Davydov soliton by looking for an approximate solution using an adiabatic approximation. The idea is to neglect the delay which is necessary for the lattice to adjust when a C=O bond is excited. It is important to stress that this *does not mean* that we assume that the lattice vibrations are following the vibration of the C=O bond, which has a very high frequency. We simply assume that the lattice distortion appears without delay when the vibration is transferred from one site to the next.

Neglecting the inertial term in the phonon equation (15.80), i.e. neglecting the second-order time derivative, we get

$$K\left(\beta_{n+1} - \beta_n\right) - K\left(\beta_n - \beta_{n-1}\right) = -\chi(|a_n|^2 - |a_{n-1}|^2), \tag{15.82}$$

which has the solution

$$(\beta_{n+1} - \beta_n) = -\frac{\chi}{K}\,|a_n|^2. \tag{15.83}$$

Putting this expression in Equation (15.81), we get

$$\mathrm{i}\hbar\frac{\mathrm{d}a_n}{\mathrm{d}t} = -J\left(a_{n+1} + a_{n-1} - 2a_n\right) - \frac{\chi^2}{K}|a_n|^2 a_n. \tag{15.84}$$

Therefore we find the nonlinear Schrödinger equation once more, with a nonlinearity controlled by the coupling between the C=O vibration and phonons. It is interesting to notice that the sign of the coupling coefficient χ is irrelevant.

However here we get a *discrete* form of the nonlinear Schrödinger equation that we studied earlier. This equation is not exactly solvable, contrary to the very similar equation

$$\mathrm{i}\hbar \frac{\mathrm{d}\psi_n}{\mathrm{d}t} = -J\left(\psi_{n+1} + \psi_{n-1} - 2\psi_n\right) - \frac{\chi^2}{2K}|\psi_n|^2\left(\psi_{n+1} + \psi_{n-1}\right), \quad (15.85)$$

known as the Ablowitz–Ladik equation. In a first-order continuum limit approximation, both give the same NLS equation. The Ablowitz–Ladik equation, which is completely integrable, can be used as a starting point to study Equation (15.84) with a perturbative calculation using the inverse scattering method.

Let us examine Equation (15.84) in the continuum limit approximation. Using notations which are now usual, we get

$$\mathrm{i}\frac{\partial a}{\partial t} + P\frac{\partial^2 a}{\partial x^2} + Q|a|^2 a = 0, \quad (15.86)$$

which has the static solution

$$a(x,t) = a_0 \operatorname{sech}\left(\sqrt{\frac{Q}{2P}}\, a_0 x\right) \mathrm{e}^{\mathrm{i}Qa_0^2 t/2} \quad \text{with} \quad \sqrt{\frac{Q}{2P}} = \sqrt{\frac{\chi^2}{2KJ}}. \quad (15.87)$$

Contrary to the general case of the nonlinear Schrödinger equation for which the amplitude of the soliton is a free parameter, here we must take into account the condition $\sum_n |a_n|^2 = 1$, which says that one quantum of the C=O vibration has been excited. In the continuum limit approximation, it becomes

$$\int |a(x,t)|^2 \mathrm{d}x = \int a_0^2 \operatorname{sech}^2\left(a_0\sqrt{\frac{\chi^2}{2KJ}}x\right) \mathrm{d}x = 1 \quad (15.88)$$

leading to $a_0 = \sqrt{\chi^2/2KJ}/2$.

The solution of Equation (15.86) becomes

$$|a(x,t)|^2 = \frac{\chi^2}{8KJ} \operatorname{sech}^2\left(\frac{\chi^2}{4KJ}\, x\right). \quad (15.89)$$

As discussed earlier, in such an expression, the factor of x is the inverse of the width L of the soliton, i.e. $L = 4KJ/\chi^2$. It determines whether the excitation is free to move as a soliton in the system or whether it is trapped by discreteness as we saw for the kinks of the Frenkel–Kontorova or ϕ^4 models. The analysis of the trapping of an NLS soliton due to discreteness is, however, much more complex than for a kink due to the internal dynamics of the soliton. The exact trapping condition is not known, but numerical simulations confirm our intuition: the soliton is trapped when its width is of the order of one unit cell.

To apply the theoretical results to the problem of the energy localisation and transfer in proteins, we must evaluate the model parameters:

- The value of the coupling constant K of the phonon lattice can be deduced from the experiments that measure the phonon dispersion curves, using for instance neutron inelastic scattering. This value is however not very accurate and the role of the coupling between the three hydrogen-bonded chains in an α-helix has to be clarified. The consensus is that K lies in the range $[39, 58.5]\ \mathrm{N\,m^{-1}}$ when the three parallel chains are taken into account.
- The value of J has been deduced from a calculation of the dipolar interactions between neighbouring C=O bonds. It is evaluated to $J = 1.55 \cdot 10^{-22}$ J.
- The exciton–phonon coupling constant χ is the hardest parameter to determine because it is generally obtained from the temperature dependence of the vibrational C=O modes, observed in Raman or infrared spectroscopy experiments. Some ab initio calculations have been attempted but they are very difficult because only a short piece of the helix can be studied, and the bases used for the eigenfunctions have to be truncated, which is particularly detrimental when hydrogen bonds are involved. The values estimated for χ are in the range $[35, 62] \cdot 10^{-12}$ N.

With these estimates, considered the best presently [168], and the mass $m = 5.7 \cdot 10^{-25}$ kg which takes into account the three peptide groups on the three chains of a helix, the value of L varies between 6 and 30 unit cells. Thus, if the parameters are correct, the Davydov soliton should be able to move along the α-helix with small discreteness effects. This picture suggests that the energy of ATP hydrolysis might be able to move in a protein as an 'energy packet'.

15.3 Does the Davydov soliton exist?

The calculations that we have presented are not sufficient to conclude the existence of the Davydov soliton. Several questions are still open:

(i) What is the validity of the D_2 ansatz used for $|\psi\rangle$?
Many attempts have been made to go beyond this ansatz, and start from a more complete basis for the state ket. For instance, instead of choosing a tensor product for $|\psi\rangle$, a mixing between the phonon and exciton states can be introduced. Calculations with more than one quantum of the vibration of the C=O bond have been made too.

These calculations modify the quantitative results, but they do not invalidate the conclusions of the simple model that we presented. The most serious criticisms of the Davydov model do not concern this point, although the results of D. Brown *et al.* [185] have pointed out that the solution of Davydov is very crude.

(ii) What is the role of thermal fluctuations?
This is a more serious criticism and this problem is not solved. The energy localisation of the Davydov soliton needs a local distortion of the lattice, which could be destroyed by the averaging effect of thermal fluctuations. To determine the stability of the soliton, the value of the parameter χ and the details of the ansatz used to describe the quantum state are crucial, which explains why different studies sometimes lead to contradictory results. The proper description of thermal fluctuations in a quantum system described by a Hamiltonian as complex as the Davydov Hamiltonian is still an open problem.

(iii) Is it possible to create a soliton from an excitation due to ATP hydrolysis on one side of the protein?
Besides the soliton solution, the discrete nonlinear Schrödinger equation (15.84) also has *nonlocalised solutions* in the linear limit. When the number of cells is large enough in the protein, it is possible to satisfy the condition $\sum_n |a_n|^2 = 1$ while keeping a_n small. Thus one may ask whether the energy may actually self-localise to form a soliton. For the continuous NLS equation (3.25) studies show that there is a threshold for the norm $\int \mathrm{d}x |\psi|^2$ beyond which the system self-localises energy to make solitons. This is a difficult question for a discrete system because even a small discretisation can qualitatively change the results.

However, for biological applications, a spontaneous localisation of energy is not required because the energy of ATP hydrolysis is released locally. Numerical simulations show that, in the parameter range where solitons exist, an initial condition localised on a single cell can relax toward a soliton and a small-amplitude radiation, so that we can expect that a large part of the ATP hydrolysis energy may indeed stay localised.

Present calculations cannot say with certainty whether the Davydov soliton exists or not. Only experiments could decide but, up to now, the observation of localised energy transfer in a protein has not been possible. One approach might be to attach chromophores at the two ends of a helix. They are molecular groups which perform a conversion between vibrational energy and light, thanks to an energy exchange between vibrational energy levels and electronic levels. One of the chromophores, excited by a laser, could release vibrational energy at one end of a helix. If this energy propagates as a localised excitation such as the Davydov soliton, a delayed light pulse should be emitted by the second chromophore when it collects the energy.

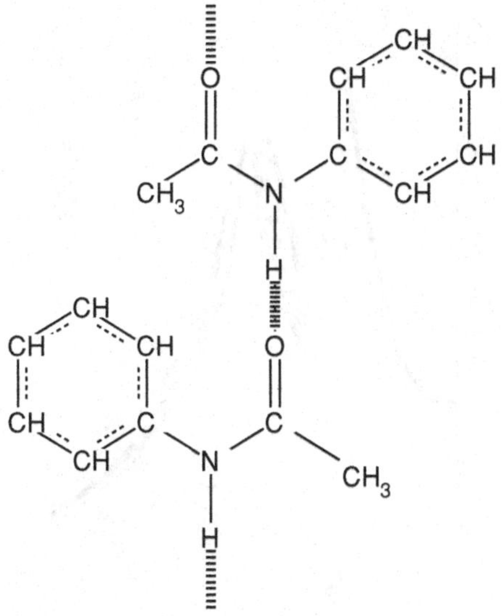

Figure 15.4. Schematic picture of the unit cell of acetanilide. The dotted lines between the H and O atoms show the hydrogen bonds.

This experiment is, however, extremely delicate, because, at the molecular level, it is very hard to excite one end of a protein without touching the other. In principle this can be solved by choosing chromophores which react to different wavelengths. Building this molecular complex is a hard task, but the hardest one would be to achieve a sufficient transfer of energy to the protein to excite its nonlinearities.

While the *transport* of energy by a soliton is still an open question, its *localisation* by a mechanism very close to the process that gives rise to the Davydov soliton has been demonstrated experimentally in a model system, the acetanilide crystal.

15.4 A model physical system: the acetanilide crystal

The experimental study of proteins is difficult, first because they have a complex structure and second because the α-helices are rather short and make up only one part of a protein. They do not have the perfect periodic structure that we have assumed in the calculation, and besides possible theoretical difficulties, this makes the spectroscopic or neutron scattering measurements difficult to perform and analyse.

A suggestion by Giorgio Careri has been to look for crystals which have hydrogen-bonded chains and a bonding structure close to that of proteins. This is the case for acetanilide, which has the formula $(CH_3 - CO - NH - C_6H_5)_n$, shown in Figure 15.4. Its crystal structure shows chains of hydrogen bonds with a geometry similar to those in proteins.

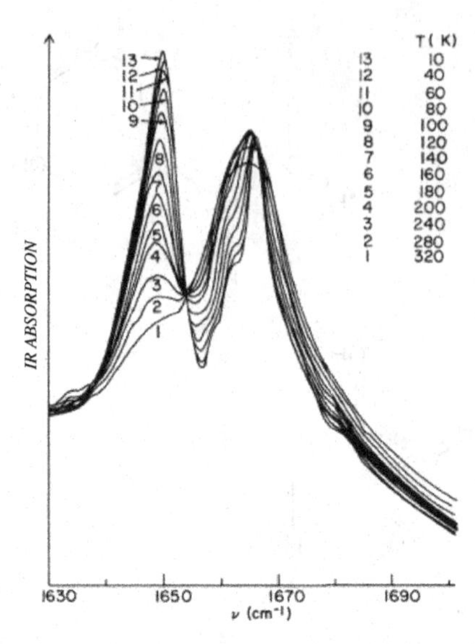

Figure 15.5. Infrared absorption spectrum of acetanilide at different temperatures (from reference [38]).

Although the interpretation of the first experiments stayed controversial for a very long time, recent 'pump–probe' spectroscopy experiments [57], designed to show nonlinear effects only, demonstrate that self-localisation of energy exists in acetanilide crystals, but that its mechanism is more complex than in the model of Davydov because it involves not only the C=O bond but also the N–H bond.

The first experiment which attracted attention to the peculiarities of the dynamics of acetanilide was the observation of the temperature dependence of its infrared spectrum around 1600 cm^{-1}, made by G. Careri and plotted in Figure 15.5. This range includes a mode at 1665 cm^{-1} assigned to the stretching of the C=O bonds, which depends weakly on temperature. But the spectra also show another mode, at a slightly lower frequency, around 1650 cm^{-1}, which is remarkable. Its frequency does not change with temperature, but its amplitude decreases drastically when the temperature increases. It is very strong at 10 K, but almost vanishes at 300 K. This is not consistent with any 'conventional' explanation. The decay of the amplitude of the modes when temperature increases is a common observation in spectroscopy, which is attributed to anharmonicity. But it should be accompanied by a shift of the frequency downward, which is not observed for this mode of acetanilide.

Recent experiments by P. Hamm and J. Edler [57] have shown that this 'anomalous' mode is indeed a localised mode, its self-localisation coming from a coupling between the vibration of the C=O bond and a phonon mode. The observations are made by a pump–probe method. A first laser pulse, the pump, excites the system, and a second pulse, the probe, observes its properties after some delay. This method can completely eliminate the response of the harmonic modes. The response is proportional to the square of the matrix element of the dipole moment μ of the system between the initial and the final state of the transition. The pump takes the harmonic oscillator in its ground state $|0\rangle$ and brings it to state $|1\rangle$. The response of the system to this signal is proportional to $\langle 0|\mu|1\rangle^2$. The probe pulse, which follows with a small delay, acts on the system in state $|1\rangle$ and can lead either to a stimulated decay to the ground state with a transition probability $\langle 0|\mu|1\rangle^2$ or to an excitation to state $|2\rangle$ with a transition probability $\langle 1|\mu|2\rangle^2$.

The experiment measures the difference between the signal with and without the pump pulse, which includes three contributions:

(i) An absorption due to the probe (which is a negative contribution to the signal) proportional to $\langle 1|\mu|2\rangle^2$, concerning the molecules which have been excited to state $|1\rangle$ by the pump.

(ii) A positive contribution due to the stimulated emission of the same oscillators which have been excited by the pump, proportional to $\langle 0|\mu|1\rangle^2$.

(iii) Another positive contribution proportional to $\langle 0|\mu|1\rangle^2$ which appears because the population of state $|0\rangle$ has been lowered by the pump. Therefore there is less absorption of the probe (which would be a negative contribution) in the state with the pump than in the state without it. This actually gives a positive contribution when the difference between the two signals is taken.

But the dipole moment of an oscillator is proportional to its position coordinate. Using the properties of a quantum harmonic oscillator, it is easy to check that $\langle 1|\mu|2\rangle^2 = 2\langle 0|\mu|1\rangle^2$, so that the two positive contributions are exactly balanced by the negative contribution.

It may seem strange to design such a sophisticated technique that gives... no output! In fact it is perfect to study nonlinear phenomena because this cancellation only holds for a harmonic oscillator, not a nonlinear system. Moreover the analysis of the nonlinear contribution can be used to compute a 'participation ratio' which indicates the degree of localisation of a mode. The idea is that, given a signal intensity, if it originates from a delocalised state involving N molecules the amplitude of the motion of each one is $\sqrt{N}$ smaller than if a single site had contributed. Thus a delocalised mode is closer to the harmonic limit. The measurements clearly show that the 1665 cm^{-1} mode is perfectly delocalised, like an ordinary phonon mode, while the 1650 cm^{-1} mode is localised.

It is possible to analyse these results by a theory which is analogous to the study of the Davydov soliton in proteins. The idea can be easily understood in terms of the general properties of the NLS equation for a lattice, with the help of Figure 15.6 which shows that its dispersion relation includes nonlocalised modes

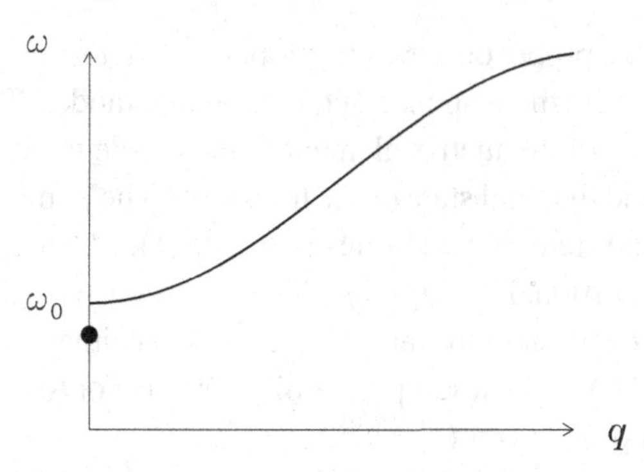

Figure 15.6. Plot of the dispersion relation of the NLS equation on a lattice. The continuous curve corresponds to the optical modes of the C=O bonds, while the symbol • shows the frequency of a localised nonlinear mode.

which belong to the exciton band (according to the usual terminology for proteins) but also localised modes in the gap. The 1650 cm^{-1} mode is such, localised due to nonlinear effects.

The amplitude of this peak decays when the temperature increases because its localisation requires a coherent motion of the lattice, which is destroyed when thermal fluctuations become too large. But its frequency does not change because it is fixed by the condition $\sum_n |a_n|^2 = 1$.

A quantitative explanation can be provided by a Hamiltonian analogous to the Davydov Hamiltonian. The exciton contribution is

$$H_e = \sum_n E_0 B_n^+ B_n - J\left(B_{n+1}^+ B_n + B_n^+ B_{n+1}\right), \tag{15.90}$$

as in the Davydov model. The phonon contribution is written as

$$H_{ph} = \sum_n \frac{1}{2m} P_n^2 + \frac{K}{2} U_n^2. \tag{15.91}$$

It corresponds to a lattice of isolated oscillators, as in the Einstein model in solid state physics. This Hamiltonian has been chosen for acetanilide because, in agreement with experiments, it corresponds to an *optical mode* which has a nonvanishing frequency for a null wavevector. Its dispersion is neglected, so that the phonon Hamiltonian is a sum of independent oscillators. At this level the model of acetanilide differs from the protein model, but this makes sense. In a crystal such as acetanilide, a hydrogen-bonded chain is not free, but linked to neighbours. This is why its vibrational modes do not include the mode at vanishing frequency and vanishing wavevector which is a translation mode, hence the 'optical' character of the phonons involved in the acetanilide Hamiltonian. For this optical mode the small dispersion can be neglected. This greatly simplifies the analysis since it leads to uncoupled oscillators. In the case of the acoustic modes of the Davydov model, this approximation is not possible because a dispersionless phonon branch would only include zero frequency modes.

Finally, the interaction Hamiltonian is written as

$$H_i = \sum_n \chi_0 U_n B_n^+ B_n, \tag{15.92}$$

which is consistent with the uncoupled character of the phonon oscillators.

It is then possible to carry out a calculation analogous to the calculation that we presented for the Davydov model. For $|\phi_e\rangle$ we must introduce the coherent states corresponding to the local oscillators instead of the amplitudes a_q and $a_q^\star$ in Fourier space. The calculation again leads to a discrete nonlinear Schrödinger equation,

$$\mathrm{i}\hbar \frac{\mathrm{d}a_n'}{\mathrm{d}t} + J\left(a_{n+1}' + a_{n-1}'\right) + \frac{\chi_0^2}{K}|a_n'|^2 a_n' = 0. \tag{15.93}$$

In this case the model parameters do not justify a continuum limit approximation, so that a second phase change in order to exhibit the discrete form of a spatial derivative is not necessary. The equation describes the time evolution of $a_n' = a_n\, \mathrm{e}^{\mathrm{i}\Omega_0 t}$, where Ω_0 is the frequency of the nonlocalised exciton mode (the mode at 1665 cm^{-1} in Figure 15.5). We did not drop this phase factor because it is useful here to analyse the infrared experiments since it determines the frequency of the observed peaks.

It is possible to derive a nonpropagative solution of Equation (15.93) by looking for it as

$$a_n'(t) = \mathrm{e}^{-\mathrm{i}\delta t/\hbar}\, \phi_n, \tag{15.94}$$

where ϕ_n does not depend on time. Writing the set of ϕ_n as a column matrix, we get the following eigenvalue equation

$$\delta \begin{pmatrix} \cdot \\ \phi_{n-1} \\ \phi_n \\ \phi_{n+1} \\ \cdot \end{pmatrix} = \begin{pmatrix} \cdot & \cdot & & & & \\ J & Q|\phi_{n-1}|^2 & J & & & \\ & J & Q|\phi_n|^2 & J & & \\ & & J & Q|\phi_{n+1}|^2 & J & \\ & & & \cdot & \cdot & \end{pmatrix} \begin{pmatrix} \cdot \\ \phi_{n-1} \\ \phi_n \\ \phi_{n+1} \\ \cdot \end{pmatrix} \tag{15.95}$$

where we introduce the parameter $Q = \chi_0^2/K$.

This is a nonlinear matrix equation because the tridiagonal matrix includes the terms $|\phi_i|^2$ which depend on the eigenvector that we want to determine. It can be solved iteratively. We shall choose an initial set of ϕ_n which verifies the condition $\sum_n |\phi_n|^2 = 1$. This defines the initial tridiagonal matrix, and we look for a normalised eigenvector of this matrix. This provides a new set of ϕ_n used to refine the matrix for the next iteration, and a first estimate of the value of δ. The process is repeated until a good convergence is achieved. This gives the value of the frequency shift δ between the exciton mode at 1665 cm^{-1} and the localised mode at 1650 cm^{-1} as well as the distribution of the amplitude ϕ_n.

In the case of acetanilide, but also in some proteins, it is possible to extend this method to study the coupling between one exciton and several phonon modes. This leads to discrete nonlinear Schrödinger equations, as before, but they include more complicated coupling terms. Their matrix form includes a matrix which is no longer in the tridiagonal form of Equation (15.95). These generalised discrete NLS equations are known as 'self-trapping equations' and are a useful generalisation of the NLS equation [59]. For acetanilide, the experiments show that the vibration of the C=O bond is not the only motion which is concerned in the localisation process. As shown in Figure 15.4, the N–H bond is also involved in the hydrogen-bond chain. In the experiments, it also exhibits nonlinear satellite peaks which are associated with localised modes. Their structure is much richer than for the C=O mode because up to nine satellites have been observed, corresponding to nine states of excitation of the self-localised mode [58]. This experiment has also shown that the de-excitation of these nonlinear modes when the laser pulse is switched off is 20 times faster for acetanilide in solution than for its crystalline form. This points out the importance of the periodic structure of the crystal, which enters through its phonon modes to increase the lifetime of the nonlinear modes.

16

Nonlinear dynamics and statistical physics of DNA

The famous discovery of the double-helix structure of DNA attracted attention to the relation between structure and function in biological molecules. However, DNA should not be viewed as a frozen structure. Actually the molecule is a highly dynamic object, permanently fluctuating with large-amplitude motions, which can be detected experimentally and play a crucial role in its function. Modelling this nonlinear dynamics in order to understand it better is interesting and offers the opportunity to use models to answer questions of biological interest.

In this chapter we introduce a simple model for DNA, which describes its fundamental properties. But of course, once such a model is proposed, it is necessary to check its validity by comparing its predictions with experimental results. For DNA, most of the experiments are biological studies which involve complex interactions between DNA and other molecules, in particular with enzymes. Presently physics is not yet able to give a satisfactory picture of these biological processes. But there is a physical phenomenon which can be used to test the validity of the DNA model, and to determine its parameters: the thermal denaturation of DNA, i.e. the separation of its two strands by heating. We shall study how a model introduced to study the dynamics of DNA can also be used to investigate this 'melting' of the double helix. This will give us the opportunity to introduce some methods of statistical physics, and to show how the concept of nonlinear excitation can be used to study the thermodynamic properties of a system.

16.1 A simple DNA model

16.1.1 The static structure of DNA

The discovery of DNA structure, was published in April 1953 in a paper [186] by J. D. Watson (1928–) and F. H. C. Crick (1916–2004). It is interesting to note that it was deduced from a *model* and not from a quantitative analysis of an X-ray diffraction pattern. Of course Watson and Crick had been inspired by the

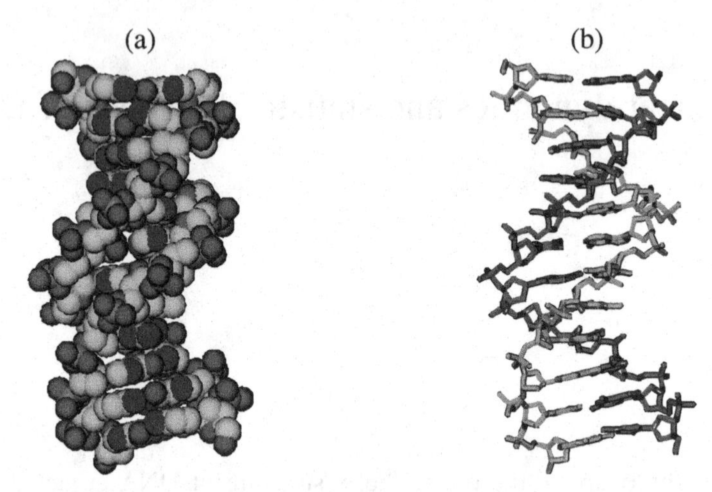

Figure 16.1. Plot [184] of the structure of a DNA sequence (111D) taken from the 'Protein Data Bank' (www.rcsb.org.pdb). (a) Short segment of the molecule showing all the atoms with approximately their actual size. (b) Schematic plot showing the internal details of the structure. Notice the 'bases' which connect the strands forming the helix of the structure. They contain the genetic code.

experimental data of M. H. F. Wilkins (1916–2004) and R. E. Franklin (1920–58), which convinced them that the structure was helicoidal, but it is by building an actual model of the chemical groups which make DNA that they had the idea of the correct structure. This model was in particular essential to suggest the specific pairing of the bases that form the steps of the 'ladder', which is now called the Watson–Crick pairing.

Deoxyribonucleic acid (DNA) is a very large polymer. The human genome would be more than one metre long if it were fully unwound with all its chromosomes put end to end. Each of our 46 chromosomes is made of a single DNA molecule, which has a length of a few centimetres. DNA molecules are so long that they can be individually manipulated in some experiments.

Figure 16.1 shows that the molecule is made of two entangled helices. The strands are made of nucleotides which include a phosphate group $-PO_4$ and an organic ring which is a sugar (deoxyribose). Each sugar is attached to a basic group. There are only four possible bases: adenine, cytosine, guanine and thymine, denoted by A, C, G and T. The bases are located between the backbones of the DNA strands, lying perpendicular to the axis of the molecule.

The series of single chemical bonds between neighbouring nucleotides gives some flexibility to the backbones because rotation around these bonds is possible. But this flexibility is restricted by the three-dimensional structure of DNA. The two polypeptide chains wind around each other in a double helix, and they are connected by hydrogen bonds between the bases, which form pairs between guanine

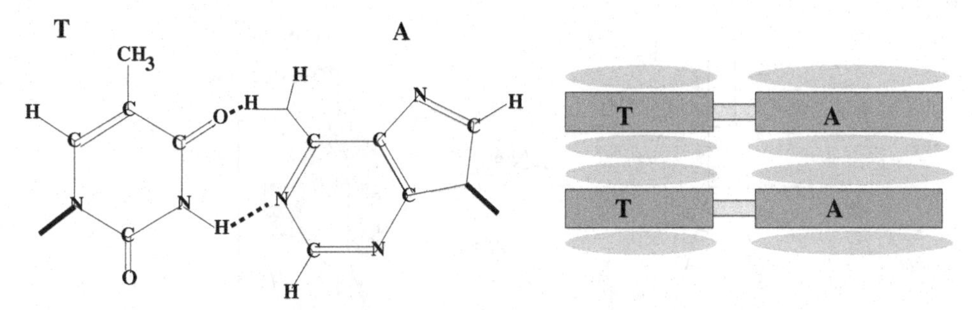

Figure 16.2. Chemical formula of a DNA base pair (T – A) and schematic picture of the geometry of the stacking of two base pairs (dark grey). The interaction of their π electrons (light grey) determines the stacking energy. The heavy lines in the formula show the bonds that connect the bases to the backbone of the double helix.

and cytosine on the one hand and between adenine and thymine on the other. This specific pairing is such that the allowed pairs A – T and G – C have very similar sizes. This gives the same length to all the 'rungs of the ladder' in the double helix. This condition explains the regular structure of the double helix although all the bases are not the same. The diameter of the helix is about 20 Å and two neighbouring bases are separated by 3.3 Å along the axis. The helix has about 11 base pairs per turn.

The Watson–Crick pairing means that the two strands of the double helix have complementary structures. Each of the two entangled strands contains the same genetic information, encoded in the sequence of the bases. It is this peculiar structure which is used in the transmission of the genetic information when cells divide or in the reproduction of an organism.

The stacking of the bases, which are flat chemical groups, is important to stabilise the DNA structure. There are two types of interactions between the bases:

(i) The two bases of a pair are linked by hydrogen bonds.
(ii) Along the strands, the bases are linked by stacking interactions which come from the overlap of the π electrons of the organic rings that make the bases, as schematised in Figure 16.2.

These two types of interaction can be studied experimentally and they have been investigated in detail because they determine how the genetic code is read. The structure and stability of the DNA molecule is also controlled by many *electrostatic interactions*. For instance, the phosphate groups of the backbone are negatively charged. Their repulsion would be strong enough to lead to a separation of the two strands without the screening effect of the positive ions which are added to DNA solutions or which exist in the intracellular medium.

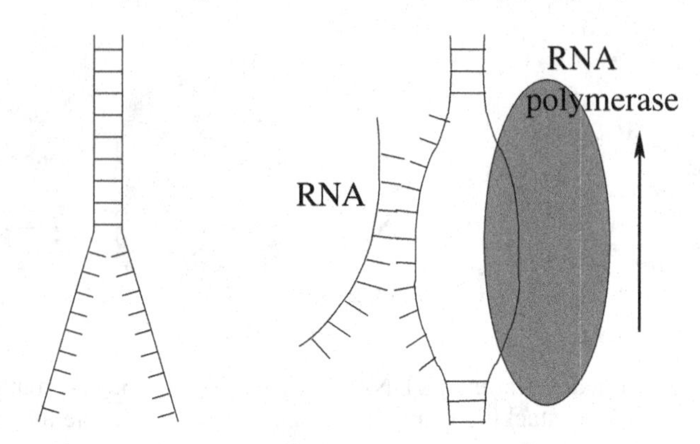

Figure 16.3. Schematic plot of DNA replication (left) and transcription (right)

16.1.2 The dynamic processes

The structure of the molecule (Figure 16.1(a)) shows that the genetic code carried by the bases is buried inside the double helix, where it is well protected. This is useful to avoid its alteration, but this is also a serious limitation when it has to be 'read'. Reading the code actually means performing chemical reactions involving the bases. To access the information encoded in DNA, the double helix has to be locally opened. This is possible because the structure of DNA is not frozen. It undergoes several types of motion. Some of them occur during the biological processes of transcription or replication, while others are large-amplitude fluctuations, known by the biologists as 'DNA breathing', which can occur at any time. Thermal denaturation is another possible structural change of the double helix which leads to a complete destruction of the helicoidal structure by heating.

Replication and transcription

Replication and transcription both lead to an opening of the double helix to allow access to the genetic code, but they are very different. During replication the entire genetic information of a DNA molecule is copied. This is achieved by opening the double helix like a 'zipper' as shown in Figure 16.3.

Transcription is the reading of a single gene, i.e. a few hundred or a few thousand base pairs. It is initiated by an enzyme which binds to DNA in order to synthesise messenger RNA, later used as a template to synthesise proteins. After a sequence of intermediate steps, the initiation leads to a local opening of the double helix, which extends along 10 to 20 bases. This is the rate-limiting step of transcription. Then the opening moves along the gene, together with RNA polymerase, successively exposing all the bases of the gene to chemical reaction (Figure 16.3).

DNA 'breathing'

As the structure of biological molecules is stabilised by weak interactions, large thermal fluctuations can be expected. The existence of open states in natural or synthetic DNA has been demonstrated with proton-exchange experiments. The base pairs open reversibly, exposing the protons of the hydrogen bonds to possible exchange with the protons of the solvent, and then they close again. There are many of these temporary open states in the double helix at physiological temperature.

To study the motion of the bases, it is possible to measure the rate of a reaction between an external reactant and some component of the bases. The rate may be different if the bases are open or closed. For instance, using formaldehyde as a reactant, the reactivity of the nitrogen in the organic ring of the base and the nitrogen of the imino bond have been tested. Other experiments have studied the nitrogens of uridine, a base of RNA, with respect to mercury. The fraction of the time during which the sites of the bases are exposed to the reactant is obtained by comparing the observed rates with the rates of isolated nucleotides, which are not part of a double helix.

The exchange of the protons of the hydrogen bonds linking the bases is a direct and nonperturbing method to study the motion of the bases. It can be observed using tritium labelling, which can be detected by radioactivity measurements, or deuterium labelling which can be measured by its influence on the optical spectrum of the molecule. These methods have been used to study proton exchange in DNA and RNA, using synthetic homopolymers (i.e. artificial sequences in which all the bases are the same) or natural DNA. Nuclear magnetic resonance, which is able to identify the protons which have been exchanged, is now extensively used in these studies.

At the beginning of 1980, NMR [30, 89, 132] and Raman [142, 180] experiments showed long-lived, low-frequency motions in DNA. These motions were compared to 'breathing' (this word was used in connection with DNA for the first time by P. H. Von Hippel) and they were then studied in detail by Prohofsky's group [154, 155], which showed that these 'breathers' could be either highly localised or extended along the molecule. Inspired by the theory of solitons, and particularly of breathers, this group suggested that breathers could be at the origin of the energy localisation in DNA, and lead to local denaturation. Our aim in this chapter is to introduce a theoretical description of these 'DNA bubbles' and to study their dynamics.

Thermal denaturation

Contrary to the motions described above, which are natural motions of DNA in its biological environment, denaturation is due to an external perturbation. This

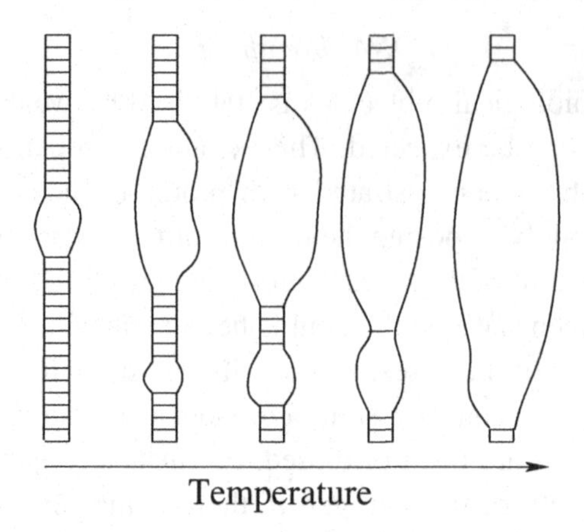

Figure 16.4. Schematic plot of DNA thermal denaturation.

opening of the double helix can be experimentally observed by changing the pH of the DNA solution, which ionises the bases, or by heating the solution. Thermal fluctuations break the hydrogen bonds within the base pairs. Figure 16.4 schematises the process: local openings, called 'denaturation bubbles' appear first, then they grow and merge into bigger bubbles until the complete separation of the two strands is achieved [163].

Denaturation does not break the covalent bonds along the strands which connect the bases but the double helix structure is gradually destabilised. Fluctuations around the average values of the helix parameters lead to a breaking of the hydrogen bonds and to a weakening of the stacking interactions, and finally to a complete change of conformation from the native to the denatured state. The stability of the double helix depends on the sequence. The G – C rich regions are harder to denature than the A – T rich regions because guanine and cytosine are bound by three hydrogen bonds while adenine and thymine are only bound by two, so that the thermal denaturation temperature of DNA can be empirically evaluated as a function of the G – C content of the sequence.

The easiest method to study denaturation is to record the variation of the UV absorption of a dilute solution of DNA while it is slowly heated (Figure 16.5). Light with a wavelength of 260 nm is weakly absorbed by double-stranded DNA, while single-stranded DNA absorbs it strongly. The breaking of the hydrogen bonds leads to a redistribution of the electrons on the bases which is associated with this sharp rise in UV absorption. Measuring the absorption and normalising appropriately, it is possible to calculate the percentage of broken base pairs versus temperature. Experiments show that the denaturation occurs within a narrow temperature range, which extends only about 1 °C for an homopolymer of DNA which has a few thousand base pairs. A detailed study of the experiments performed with long natural

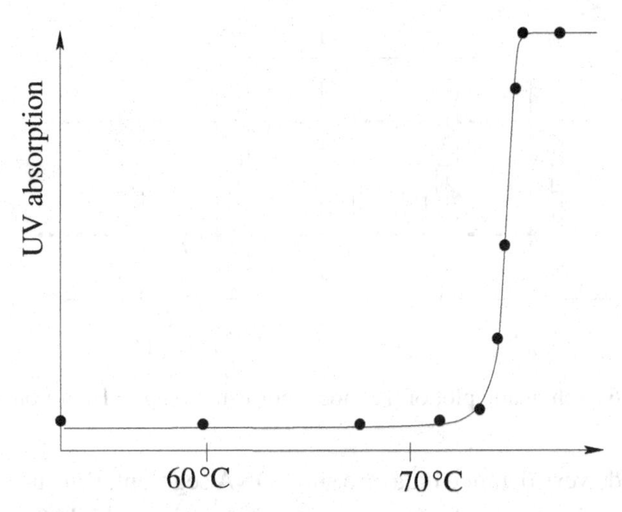

Figure 16.5. Optical absorption of UV light at 260 nm wavelength by DNA. Figure adapted from data obtained on synthetic poly-GC DNA, which has a single type of base pair [92].

DNAs, which have an inhomogeneous base sequence, shows that they denature in steps of a few hundred base pairs that open together in a given environment and temperature, leading to a complex denaturation curve for the whole molecule, characteristic of its base sequence.

16.1.3 A DNA model

To establish a DNA model, the first step is to select the appropriate scale. It depends on the questions that we want to answer with the model. For instance, one may want to analyse a common situation in biology, the diffusion of DNA in a gel, which is used to determine the size of the molecules, the shorter ones diffusing faster than the longer ones. In this case a model in which the double helix is simply viewed as a flexible string is enough. If we are interested in denaturation only, a two-state model for each base pair, which is considered as closed (state 0) or open (state 1) may be sufficient. Such 'Ising models' have been used to study the denaturation transition. One drawback is that their parameters are difficult to evaluate because they determine for instance, the probability that a given base pair opens, according to the open/closed state of its neighbours, which is difficult to connect to the physical parameters of the molecule. Another limitation of such models is that they can only represent the two extreme states of a base pair, but cannot follow the dynamics of its opening because the intermediate states are not described. Another possibility would be to make a very detailed model of the molecule, describing the positions of all the atoms. This approach is used in molecular dynamics studies which are

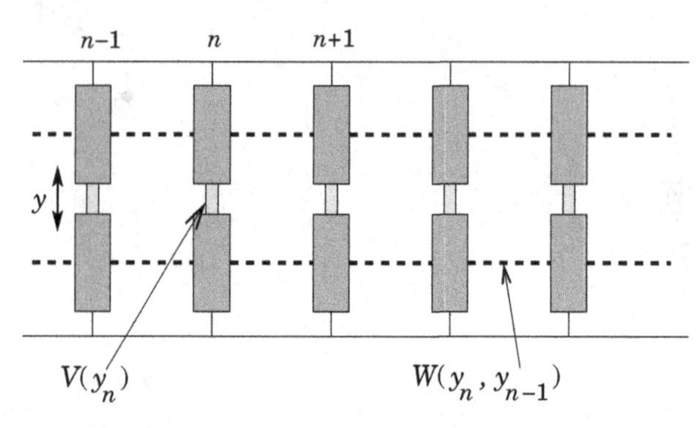

Figure 16.6. Schematic plot of the model corresponding to Hamiltonian (16.1).

useful to provide very detailed data on a short DNA segment. But current computing facilities, although they are far more powerful than the MANIAC used by Fermi, Pasta and Ulam, are however limiting such studies to short sequences and short timescales, which can be a severe restriction because the dynamics of DNA occurs over a broad range of timescales. Light atoms such as hydrogen are moving with characteristic times of 10^{-14} to 10^{-15} s, while large conformational changes in which many atoms are involved, for instance for the opening of a base pair, may be much slower (10^{-11} to 10^{-12} s). In a numerical study, the fastest processes impose a time step which has to be of the order of 10^{-15} s, so that 10^6 steps are necessary to follow the dynamics over 1 ns, which is still very short, since, for instance, a base pair stays closed for about 10 ms between two 'breathing' events.

This is why the studies of biological macromolecules stimulated the development of models which give a more complete description than the Ising models while, however, providing an extremely simplified view of the molecule. They are adapted to the description of the large-amplitude slow motions of these molecules. Here we introduce one of these models, which is intended to describe the nonlinear dynamics of base-pair opening, without including all the atoms. This simplified picture can be viewed as a first step toward the understanding of complex phenomena in DNA. The model that we discuss here can be improved in successive steps that bring it closer to the actual structure of DNA, but even the simplest model turns out to be able to give useful information on DNA, which is beginning to be applied to solving selected biological questions.

The model shown in Figure 16.6 only includes *one* degree of freedom for the nth base pair, the stretching of the hydrogen bonds connecting the two bases of the pair, denoted by y_n. It appears as a natural extension of the Ising models in which the discrete variable $\{0, 1\}$ is replaced by a real number, which can in principle vary in the range $-\infty$ to $+\infty$. Actually large negative values are forbidden by steric hindrance because two bases cannot overlap. In the model this constraint is enforced by the interaction potential between the bases. Large positive values of y_n are allowed, however, because they occur if the two strands separate from each other, i.e. in DNA denaturation. To use a Hamiltonian formalism, we also introduce for each base pair the momentum $p_n = m(\mathrm{d}y_n/\mathrm{d}t)$ conjugate of y_n, where m is the reduced mass of the bases in their relative motion with respect to one another.

The Hamiltonian of the model [150]

$$H = \sum_n \left[\frac{p_n^2}{2m} + W(y_n, y_{n-1}) + V(y_n) \right] \tag{16.1}$$

describes the different contributions to the energy of the molecule.

The first term is simply the kinetic energy of the relative motion of the bases.

The second term is the energy associated with base stacking. Neighbouring base pairs are not independent from each other. If one pair is broken, its bases move out of the stack and tend to pull the neighbouring bases, in particular through the interaction that comes from the overlap of the π electrons, which are on both sides of the organic rings made by the bases (see Figure 16.2). If we assume that the *relative* motion of neighbouring bases is not too large, the stacking energy can be approached by its harmonic expansion

$$W(y_n, y_{n-1}) = \frac{K}{2}(y_n - y_{n-1})^2. \tag{16.2}$$

This is of course only an approximation of the actual potential, and we shall show later that the nonlinearity of the stacking interaction should not be ignored in order to get results that agree with the experiments. The harmonic approximation is, however, a useful first step because it allows analytical calculations.

The third term in Hamiltonian (16.1) describes the interaction between the two bases within a Watson–Crick pair. We shall choose the Morse potential

$$V(y_n) = D\,(\mathrm{e}^{-ay_n} - 1)^2, \tag{16.3}$$

plotted in Figure 16.7, because it has the appropriate qualitative shape. The parameter a is positive so that the potential has a strong repulsive part for $y < 0$, preventing the overlap of the bases. For large $y > 0$, the bases no longer interact and the potential reaches a plateau.

The model describes a DNA homopolymer, i.e. all the bases are assumed to be identical because we want to study the role of nonlinearity in the dynamics and statistical physics of DNA, without the complexity introduced by the sequence. To study an actual DNA sequence, the model could include base-dependent parameters.

The choice of the parameters is delicate. Since the potentials which enter into the Hamiltonian describe the interactions between groups of atoms, one could think that it is sufficient to sum up the 'well-known' potentials of molecular dynamics to get

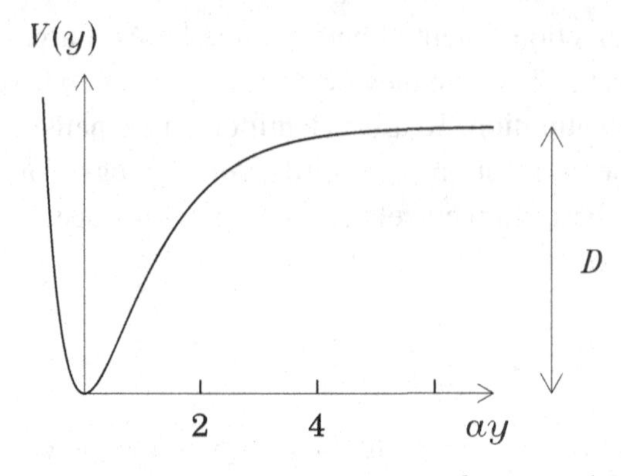

Figure 16.7. The Morse potential $V(y)$ plotted as a function of the dimensionless variable ay.

good effective potentials for our Hamiltonian. But actually it is not so simple. The potentials of molecular dynamics have been fitted to describe the *small-amplitude motions* of the atoms around their equilibrium positions. They have been adjusted on the vibrational modes observed in infrared or Raman spectroscopy. But the knowledge of a function near its minimum does not tell one much about its values far from the minimum. Moreover, the effective potentials of the model do not describe DNA isolated in space. Our goal is to take into account the solvent, and also the ions that it contains, which are screening some electrostatic interactions. For instance, if the Morse potential is established by only summing the potentials of the hydrogen bonds that link the bases, the value of D is one order of magnitude too high. As discussed earlier, the interaction between the bases is not only determined by their attraction due to the hydrogen bonds, but also by the strong repulsion between the negatively charged PO_4 groups of the strands, which is only partly screened by the ions of the DNA solution.

In practice the parameters are obtained a posteriori, by comparing the properties of the model with experimental results. This is why well-controlled physical experiments such as DNA thermal denaturation are important. However, for such a simple model, an uncertainty in the parameter subsists because the model cannot describe all the properties of DNA to high accuracy, which would allow a very accurate determination of the parameters. Such a model should rather be used to obtain a qualitative description of some of the properties of DNA and to reach some understanding of their origin.

We shall choose a dissociation energy $D = 0.03$ eV, a coupling constant $K = 0.06\,\mathrm{eV\AA^{-2}}$, the inverse of the characteristic distance over which the Morse potential varies $a = 4.5\ \mathrm{\AA^{-1}}$. The reduced mass of the bases is very close to 300 atomic mass units, whatever the base pair. The ratio between the coupling energy and the on-site energy is measured by $S = K/(Da^2)$, which can be used, as in Chapter 10 on

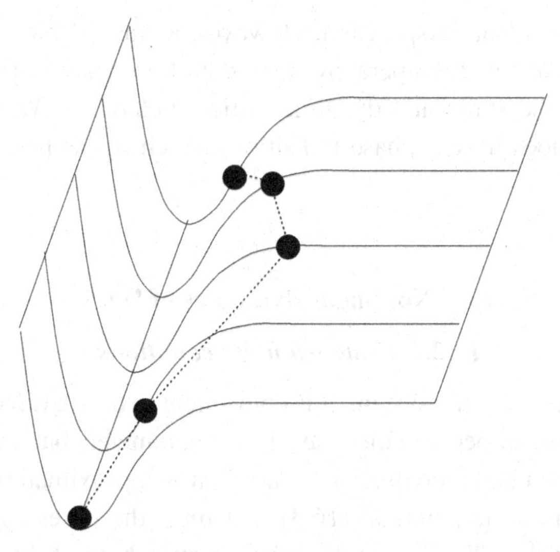

Figure 16.8. Schematic plot of the dynamics of the model (16.1). The base pairs are represented by massive particles moving over the energy surface of their Morse interaction potential. The ground state is the bottom of the Morse potential but, at nonzero temperature, the chain experiences large fluctuations, creating 'denaturation bubbles' in which some of the bases move on the plateau.

ferroelectrics, to distinguish an 'order–disorder' behaviour for $S \ll 1$ from a strong coupling regime when $S \gg 1$. Our parameters give $S = 0.099$, which means that discreteness effects are important for DNA. This agrees with experimental results which observe proton exchanges in some bases and not in the neighbours, indicating that a single base may have a large-amplitude opening while the neighbouring ones do not.

Figure 16.8 illustrates the dynamic behaviour of the model. We shall show that it can be used to study several properties of DNA including:

- Its nonlinear dynamics; we shall show that the model has some solutions which describe the 'breathing' observed in the experiments.
- The phase transition between the closed state (double helix) and the open state (thermally denatured state); in this framework, we shall show that, besides the standard methods of statistical physics, the study of nonlinear localised excitations provides a complementary technique which can be used to proceed analytically beyond the possibilities of the usual methods.

This model shows some similarities with the model of Krumhansl and Schrieffer [106] who studied the role of nonlinear soliton-like excitations in phase transitions, as discussed in Chapter 10. They could show the signature of the domain walls on the thermodynamic properties of the ϕ^4 one-dimensional model and on the central peak of its dynamic spectrum. However, as the ϕ^4 model does not have a phase transition at finite temperature, they could not analyse the role of coherent structures on the static and dynamic critical behaviour. We shall show that the DNA model does have a phase transition, which opens new possibilities for study.

16.2 Nonlinear dynamics of DNA

16.2.1 Dimensionless equations

To study the dynamics of the system, it is convenient to derive dimensionless equations which are simpler because they have fewer parameters but, more importantly, because they allow a better control of the necessary approximations.

The expression of the potential (16.3) that links the bases suggests a natural dimensionless variable, $Y = ay$, to describe the stretching of the hydrogen bonds between the bases. We get a dimensionless Hamiltonian by dividing Hamiltonian (16.1) by D

$$H' = \frac{H}{D} = \frac{1}{2}\frac{m}{Da^2}\left(\frac{\mathrm{d}Y_n}{\mathrm{d}t}\right)^2 + \frac{1}{2}\frac{K}{Da^2}(Y_n - Y_{n-1})^2 + (\mathrm{e}^{-Y_n} - 1)^2. \qquad (16.4)$$

We notice that the dimensionless parameter $S = K/(Da^2)$, which measures the ratio between the coupling term and the on-site potential term, appears. Introducing the dimensionless time $t' = \sqrt{Da^2/m}\, t$, the energy of the model can be written in the dimensionless form

$$H' = \frac{1}{2}\left(\frac{\mathrm{d}Y_n}{\mathrm{d}t'}\right)^2 + \frac{1}{2}S\,(Y_n - Y_{n-1})^2 + (\mathrm{e}^{-Y_n} - 1)^2, \qquad (16.5)$$

which shows that the model actually depends on a single parameter, S. The values that we have chosen for DNA give $S = 0.099$ indicating a weak coupling between the dynamics of successive bases, as previously noticed.

To obtain a correct Hamiltonian form, we can introduce the conjugates of the variables Y_n, the momenta $P_n = \mathrm{d}Y_n/\mathrm{d}t'$.

We henceforth drop the $'$ on the Hamiltonian and time to simplify the notation. Every time we use the dimensionless variables Y_n, it is understood that all quantities in the equations are dimensionless.

The equations of motion which derive from Hamiltonian (16.5) are

$$\frac{\mathrm{d}^2 Y_n}{\mathrm{d}t^2} = S(Y_{n+1} + Y_{n-1} - 2Y_n) + 2\,\mathrm{e}^{-Y_n}(\mathrm{e}^{-Y_n} - 1). \qquad (16.6)$$

16.2.2 Nonlinear solutions of the dynamic equations

We have noticed many times that such a set of coupled nonlinear differential equations seldom has an exact solution. This is again true here and we must make some approximations in order to solve the system (16.6). There are two approximations that we can rule out a priori:

(i) A linearisation of the system, which would neglect one essential feature, the nonlinearity of the Morse potential, which allows large-amplitude motions of the bases.
(ii) The continuum limit approximation which is hardly acceptable for $S \approx 0.1$.

A dynamic solution keeping the full expression of the Morse potential is not known, but it is possible to perform a medium-amplitude expansion of the nonlinear term. To measure the order of magnitude of the different terms, we shall introduce the variable ϕ_n defined by

$$Y_n = \varepsilon\phi_n \quad \text{with} \quad \varepsilon \ll 1. \tag{16.7}$$

Keeping the first two nonlinear terms in the expansion, the system of equations (16.6) becomes

$$\frac{d^2\phi_n}{dt^2} = S(\phi_{n+1} + \phi_{n-1} - 2\phi_n) - 2\left(\phi_n - \frac{3}{2}\varepsilon\phi_n^2 + \frac{7}{6}\varepsilon^2\phi_n^3\right) + \mathcal{O}(\varepsilon^3). \tag{16.8}$$

This set of equations is similar to the equations that we studied to describe nonlinear waves in the pendulum chain in Chapter 3. Due to the low value of S, the continuum limit approximation that we used for the pendulum chain cannot be used here, but the results that we got for this problem can inspire us to look for a solution of the form

$$\phi_n(t) = \left(F_n e^{i\theta_n} + F_n^* e^{-i\theta_n}\right) + \varepsilon\left(G_n + H_n e^{2i\theta_n} + H_n^* e^{-2i\theta_n}\right) + \mathcal{O}(\varepsilon^2), \tag{16.9}$$

with $\theta_n = qn - \omega t$. The idea is that, as for the pendulum chain, we look for a solution which is a 'carrier wave' with a variable amplitude, and, due to the nonlinear terms in the equation, we must also introduce the first overtones of this wave. Due to the sum and differences of frequencies which appear in the overtones, we expect to get some terms proportional to $\exp(\pm 2i\theta_n)$ and others without the exponential factors.

To proceed further we can assume that the amplitude factors F, G and H, vary slowly in space and time, so that we make a continuum limit approximation *for the amplitude factors only*. This is formalised in a multiple scale expansion. This method is called the semi-discrete approximation [158] because it preserves a carrier wave which obeys the discrete dispersion relation, as we shall check, and only the variation of the amplitude is treated in a continuum limit.

In this approximation, F, G and H are assumed to be independent of the 'fast' variables t and n, and their space and time variations are only determined by the 'slow' variables $X_i = \varepsilon^i x$ and $T_i = \varepsilon^i t$, $i \geq 1$. A continuum limit approximation is then made. For instance, $F_n(t)$ is replaced by $F(X_1, X_2, \ldots, T_1, T_2, \ldots)$. $F_{n\pm1}$ is computed at order ε^2 by a Taylor expansion

$$F_{n\pm1} = F \pm \varepsilon \frac{\partial F}{\partial X_1} \pm \varepsilon^2 \frac{\partial F}{\partial X_2} + \frac{\varepsilon^2}{2} \frac{\partial^2 F}{\partial X_1^2} + \mathcal{O}(\varepsilon^3) \tag{16.10}$$

and the time derivative of F_n is given by

$$\frac{\partial F_n}{\partial t} = \varepsilon \frac{\partial F}{\partial T_1} + \varepsilon^2 \frac{\partial F}{\partial T_2} + \mathcal{O}(\varepsilon^3). \tag{16.11}$$

Similar equations are written for H and G and they are used by putting Expression (16.9) into the set of equations (16.8). The calculation is tedious but easy. Then the different orders in ε are collected and we write that the factors of $\exp(\pm \mathrm{i}\theta_n)$, $\exp(\pm 2\mathrm{i}\theta_n)$ and the terms without an exponential term must vanish at each order.

At order ε^0, the cancellation of the factor of $\exp(\pm \mathrm{i}\theta_n)$ gives a linear equation for F_n, which is satisfied if ω and q are related by the dispersion relation

$$\omega^2 = 2 + 4S \sin^2 \frac{q}{2}. \tag{16.12}$$

This is the dispersion relation of the linear waves in the system described by the set of equations (16.8). We observe that this is indeed the *dispersion relation of the discrete model.* This is the appeal of the semi-discrete approximation as opposed to the multiple scale expansion of the continuous equation, as we did in Chapter 3.

At order ε^1, the cancellation of terms proportional to $\exp(\mathrm{i}\theta_n)$ gives

$$\frac{\partial F}{\partial T_1} + v_g \frac{\partial F}{\partial X_1} = 0 \quad \text{with} \quad v_g = \frac{S \sin q}{\omega}, \tag{16.13}$$

which is the group velocity $\mathrm{d}\omega/\mathrm{d}q$ of the waves which obey the dispersion relation (16.12).

At this order, the terms without an exponential dependence and the terms proportional to $\exp(2\mathrm{i}\theta_n)$ give

$$G = 3F F^* \quad \text{and} \quad H = -\frac{1}{2} \frac{F^2}{1 + (8S/3)\sin^4(q/2)}. \tag{16.14}$$

The most interesting equation is provided by the terms proportional to $\exp(\mathrm{i}\theta_n)$ at order ε^2. In the frame moving at speed v_g, it reduces to the nonlinear Schrödinger equation

$$\mathrm{i}\frac{\partial F}{\partial \tau_2} + P \frac{\partial^2 F}{\partial \xi_1^2} + Q|F|^2 F = 0. \tag{16.15}$$

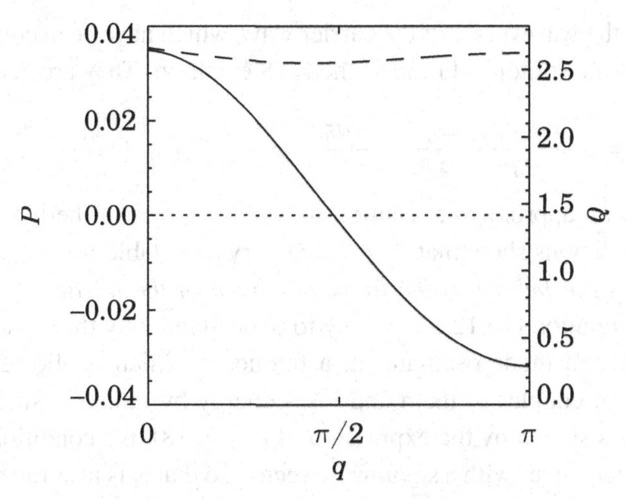

Figure 16.9. Variation of the parameters P (full line) and Q (dashed line) of the NLS equation (16.15) which describes the nonlinear dynamics of DNA, versus the wavevector q of the carrier wave.

The coefficients

$$P = \frac{S\omega^2 \cos q - S^2 \sin^2 q}{2\omega^3} \quad \text{and} \quad Q = \frac{1}{2\omega}\left(11 - \frac{9}{3 + 8S\sin^4(q/2)}\right) \tag{16.16}$$

depend on the wavevector q of the carrier wave, as shown in Figure 16.9. We notice that the product PQ is positive for all the wavevectors smaller than $\pi/2$, so that we can expect spatially localised solutions for all the carrier waves with a small wavevector.

Using the solution of the NLS equation to obtain F, and then calculating the expressions of G and H given by Equation (16.14), the expression of $Y_n = \varepsilon\phi_n$ is obtained as

$$\begin{aligned} Y_n(t) = {} & 2a_0 \operatorname{sech}\left[a_0\sqrt{\frac{Q}{2P}}(n - v_g t - u_e t)\right]\cos(q'n - \omega' t) \\ & + a_0^2 \operatorname{sech}^2\left[a_0\sqrt{\frac{Q}{2P}}(n - v_g t - u_e t)\right]\left(3 - \frac{1}{2}\frac{\cos[2(q'n - \omega' t)]}{1 + (8S/3)\sin^4(q/2)}\right), \end{aligned} \tag{16.17}$$

where a_0 and u_e are the two parameters which characterise the solution of the NLS equation. The factor ε, which had been introduced to track the different orders of the expansion, but does not have a physical meaning, has been included in the amplitude a_0 of the soliton. The quantity u_e is the velocity of the envelope of the soliton solution of the NLS equation. It also includes the factor ε coming from the variable ξ_1 of the NLS equation (16.15). The parameters ω' and q' are the frequency and the wavevector of the carrier wave which have been corrected by the phase factor of the soliton solution of the NLS equation. They are given by

$$\omega' = \omega + \frac{v_g u_e}{2P} + \frac{u_e^2}{4P} - \frac{Q a_0^2}{2} \quad \text{and} \quad q' = q + \frac{u_e}{2P}. \tag{16.18}$$

This is only an approximate solution for the dynamics of the DNA model, but numerical simulations show that it is satisfactory and stable *provided that the frequency ω' does not belong to the frequency band of the linear modes,* given by the dispersion relation (16.12). It is easy to understand why this condition should hold: if a localised mode oscillates at a frequency which is allowed for nonlocalised modes, it couples to them and loses energy by radiating small-amplitude linear waves. As shown by the expression of ω' (16.18), the condition is easier to satisfy for carrier waves with a small wavevector, so that ω is near the bottom of the allowed frequency band $\omega = \sqrt{2}$. Indeed nonlinearity tends to decrease the vibrational frequency, thanks to the $-Qa_0^2/2$ contribution in the expression of ω'. If ω is small, this drives ω' in the gap $\omega' < \sqrt{2}$. Moreover Expression (16.18) indicates that the stability condition is easier to achieve for the solitons which have a small envelope velocity u_e, which is confirmed by numerical simulations.

The existence of a localised solution, which is qualitatively similar to the fluctuational opening of DNA observed in the experiments, is interesting but it is not enough to guarantee the relevance of this solution for the physics of DNA. It may play a role in the properties of the molecule only if it is naturally created. We have already pointed out one feature of solitons which makes them so interesting, their property of being the 'attractors' for a large variety of excitations. The nonlinear solution that we derived for the DNA model is not an exact soliton, but it nevertheless exhibits this property of being easily created in the molecule. It can even spontaneously emerge from thermal fluctuations, as shown by the numerical simulations of the model in contact with a thermal bath.

16.2.3 Dynamics of the model in contact with a thermal bath

A possible method to study the dynamics of the DNA model is to perform a numerical simulation of the dynamic equations (16.6). It is easy to simulate these nonlinear equations without having to approximate them, but this is not enough to study the physics of the molecule because the thermal fluctuations cannot be ignored. We must add a thermostat to the set of equations (16.6) in order to maintain the molecule at the temperature T of interest.

To introduce the method, let us consider the simple example of an atom of mass m, mobile in the x direction, belonging to a molecule which is in a solvent at temperature T. The atom is in a potential $V(x)$ which describes its interaction with other parts of the molecule and it is also subjected to a random force $F(t)$ applied by the molecules of the solvent, which are continuously moving. Its equation of motion is thus

$$m\frac{\mathrm{d}^2x}{\mathrm{d}t^2} = -\frac{\partial V}{\partial x} + F(t). \tag{16.19}$$

The random force can be divided in two contributions. It includes a part, which we shall denote by $-\gamma \mathrm{d}x/\mathrm{d}t$, which is systematically opposed to the velocity of the atomic motion. We can understand its origin by realising that the moving atom is more likely to meet solvent molecules on its 'front side', because it moves toward them, than on its 'rear side'. It is easy to convince ourselves that this is true by thinking about what happens when we ride a bicycle in the rain: we are wet on the chest, and almost dry on the back! At the molecular scale, the phenomenon is the same and it leads to a force that tends to damp the motion. When this systematic contribution is subtracted from $F(t)$, only a completely random force, which we shall denote $f(t)$, remains. Its statistical average is zero. The equation of motion of the atom is therefore

$$m\frac{\mathrm{d}^2x}{\mathrm{d}t^2} = -\frac{\partial V}{\partial x} - \gamma\frac{\mathrm{d}x}{\mathrm{d}t} + f(t). \tag{16.20}$$

To study a system in contact with a thermal bath, one can generally assume that the timescale of the fluctuations of the thermal bath is negligible with respect to the characteristic timescale of the system of interest. This amounts to assuming that the time correlation of the random force $f(t)$ vanishes. The correlation function of $f(t)$ is thus of the form $\langle f(t)f(t')\rangle = q\delta(t-t')$, where q is a constant.

The statistical physics of the thermalised system shows that the two parts of the force $F(t)$ are not independent of each other. They are related by the fluctuation–dissipation theorem [157] which imposes $Q = 2k_BT\gamma/m$, where k_B is the Boltzmann constant. To simulate a thermalised system, a possible method is therefore to solve an equation analogous to Equation (16.20), known as the Langevin equation. The principle of the calculation is simple because, once we have chosen the parameter γ which measures the strength of the coupling between the system of interest and the thermostat, the equations can be simulated with any standard algorithm for the numerical treatment of differential equations [15, 153]. The force $f(t)$ is generated by a Gaussian random number generator. The numerical method uses a discrete time step Δt instead of a continuous time variable. In this case, in order to correctly verify the relation imposed by the fluctuation–dissipation theorem, the variance of the distribution of the values of $f(t)$ must be chosen as $\sqrt{2k_BT\gamma/(m\Delta t)}$ [75].

Using a Langevin equation to simulate a thermalised system is the simplest approach, but it is not the most efficient. A method to exactly obtain the equilibrium properties of a system in contact with a thermal bath has been proposed by Nosé. The original method of Nosé has since been improved in order to ensure a better exploration of the phase space of the system. It is now a very efficient approach to simulate a thermalised system [125].

The idea of the Nosé method is to define an extended Hamiltonian system, which includes the physical system and a few additional variables corresponding to the thermostat. It can be schematised as follows for a physical system described by the variables x_i and their conjugate momenta p_i:

$$\boxed{\begin{array}{l}\text{Physical system}\\ x_i,\, p_i = m_i \dfrac{\mathrm{d}x_i}{\mathrm{d}t}\end{array}} \Leftrightarrow \boxed{\begin{array}{l}\text{Thermostat 1}\\ \eta_1,\, p_{\eta 1},\, Q_1\end{array}} \Leftrightarrow \boxed{\begin{array}{l}\text{Thermostat 2}\\ \eta_2,\, p_{\eta 2},\, Q_2\end{array}} \cdots \boxed{\begin{array}{l}\text{Thermostat M}\\ \eta_M,\, p_{\eta M},\, Q_M\end{array}}\,.$$

The extended Hamiltonian is

$$H_1 = H + \sum_{j=1}^{M} \frac{p_{\eta j}^2}{2Q_j} + N k_B T \eta_1 + \sum_{j=2}^{M} k_B T \eta_j \tag{16.21}$$

where η_j are the thermostat variables and $p_{\eta j}$ their conjugate momenta. The equations of motion which derive from this Hamiltonian are

$$\frac{\mathrm{d}x_i}{\mathrm{d}t} = \frac{p_i}{m_i}, \qquad \frac{\mathrm{d}p_i}{\mathrm{d}t} = -\frac{\partial V}{\partial x_i} - p_i \frac{p_{\eta 1}}{Q_1}, \tag{16.22}$$

$$\frac{\mathrm{d}\eta_1}{\mathrm{d}t} = \frac{p_{\eta 1}}{Q_1}, \qquad \frac{\mathrm{d}p_{\eta 1}}{\mathrm{d}t} = \left[\sum_{i=1}^{N} \frac{p_i^2}{m_i} - N k_B T\right] - p_{\eta 1} \frac{p_{\eta 2}}{Q_2}, \tag{16.23}$$

$$\frac{\mathrm{d}\eta_j}{\mathrm{d}t} = \frac{p_{\eta j}}{Q_j}, \qquad \frac{\mathrm{d}p_{\eta j}}{\mathrm{d}t} = \left[\frac{p_{\eta j-1}^2}{Q_{j-1}} - k_B T\right] - p_{\eta j} \frac{p_{\eta j+1}}{Q_{j+1}}, \tag{16.24}$$

$$\frac{\mathrm{d}\eta_M}{\mathrm{d}t} = \frac{p_{\eta M}}{Q_M}, \qquad \frac{\mathrm{d}p_{\eta M}}{\mathrm{d}t} = \left[\frac{p_{\eta M-1}^2}{Q_{M-1}} - k_B T\right]. \tag{16.25}$$

The parameter Q_j, which plays the role of 'mass' for each of the thermostats, determines how fast the thermostats respond to the temperature variations in the system. The goal is to control temperature without completely killing the fluctuations which exist in any finite physical system. The optimal values are $Q_1 = N k_B T/\omega_M^2$, $Q_j = k_B T/\omega_M^2$, where ω_M is the highest frequency in the dynamics of the physical system of interest (for instance for the DNA model, it is the frequency of the fastest nonlocalised mode, at the top of the band). This method gives very good results, even with a very small number of thermostats ($M = 5$ for instance).

Equation (16.22) shows that, as with the Langevin method, the Nosé method introduces a term proportional to the velocity (or to the momentum) in the equation

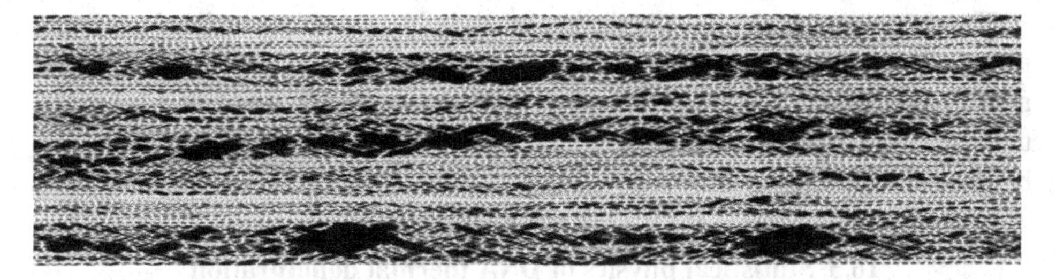

Figure 16.10. Dynamics of the DNA model with Hamiltonian (16.1) in contact with a thermal bath at temperature $T = 340$ K. The amplitude of the motion of each base pair is shown by a grey scale going from white (fully closed) to black (fully open). The vertical axis extends along the molecule and corresponds to the index n of the base pairs ($1 \leq n \leq 256$) and the horizontal axis corresponds to time.

of motion, but, instead of being multiplied by a fixed damping coefficient γ, it has an effective damping coefficient $p_{\eta 1}/Q_1$ which is itself a dynamic variable of the system. Its equation of evolution is such that the method exactly gives the equilibrium values of a canonical system.

When the equations of motions of the DNA model (16.6) are simulated using the Nosé method, with Equations (16.22)–(16.25), the spontaneous formation of localised excitations in the molecule due to the effect of thermal fluctuations is observed. Figure 16.10 shows an example at $T = 340$ K.

In this case where DNA is slightly below its denaturation temperature, the figure shows black regions associated with a simultaneous opening of a set of neighbouring base pairs. They correspond to the 'thermal denaturation bubbles' of Figure 16.4. Moreover, one can also observe dotted lines, which indicate a large amplitude oscillation of a few base pairs which open and close periodically. These dotted lines are the breathing modes of DNA that we described previously. Their analytical expression is approximately given by Equation (16.17) but the coupling with the thermal bath slightly perturbs the regular oscillation of the theoretical solution. It should be noticed that these breathing modes survive for a large number of oscillation periods. The energy density around these modes is higher than in the lighter regions of the figure. Thus there is a temporary breaking of the equipartition of energy along the DNA molecule. Study of the system for a much longer time shows that energy equipartition does indeed exist, because a given localised mode may finally die under the perturbations of the thermal bath and another one may show up elsewhere. This indicates that the validity of thermodynamics is not questioned by nonlinear physics! However, nonlinear energy localisation may have significant consequences because the dynamic properties of the system are very different from those of a linear system.

The simulation also points out the important influence of the localised modes as precursors to the local openings of the molecule. They play a role in the thermodynamics of DNA, but there is also another nonlinear excitation which is essential to understanding the thermodynamics of the DNA model, the domain wall discussed below.

16.3 Statistical physics of DNA thermal denaturation

The statistical physics of the DNA model can be investigated by standard techniques. For this one-dimensional system, the optimal method is again the transfer integral method introduced in Chapter 10 for the ϕ^4 model. We showed that all the information on the thermodynamics was contained in the eigenfunctions ϕ_n and the eigenvalues $\exp(-\beta\varepsilon_n)$ of the transfer operator

$$\int_{-\infty}^{+\infty} \mathrm{d}y \, \mathrm{e}^{-\beta\left[V(x)+V(y)+K(x-y)^2\right]/2} \, \phi_n(y) = \mathrm{e}^{-\beta\varepsilon_n} \, \phi_n(x), \tag{16.26}$$

(with $\beta = 1/(k_B T)$) written here in terms of the dimensional variables.

As shown in Chapter 10, in the thermodynamic limit the free energy per site is given by ε_0 which corresponds to the largest eigenvalue (smallest value of ε) of Equation (16.26), to which the nonsingular term coming from the integration over the momenta should be added, giving

$$f = \varepsilon_0 - \frac{1}{2\beta} \ln\left(\frac{2\pi m}{\beta}\right). \tag{16.27}$$

The normalised eigenfunction ϕ_0 associated with the eigenvalue $\exp(-\beta\varepsilon_0)$ gives the weighting factor to compute the mean value of the stretching of the base pairs $\langle y \rangle$, i.e. the mean separation of the two DNA strands which is the order parameter of the thermal denaturation transition, given by

$$\sigma = \langle y \rangle = \int_{-\infty}^{+\infty} \mathrm{d}y \; y \; |\phi_0|^2. \tag{16.28}$$

Using the continuum limit approximation, valid in the temperature range $D \ll k_B T \ll K/a^2$ [81], we showed in Chapter 10 that the integral equation (16.26) can be reduced to the pseudo-Schrödinger equation

$$\left[-\frac{1}{2\beta^2 K}\frac{\mathrm{d}^2}{\mathrm{d}y^2} + D(\mathrm{e}^{-ay} - 1)^2\right]\phi_n(y) = e_n \, \phi_n(y), \tag{16.29}$$

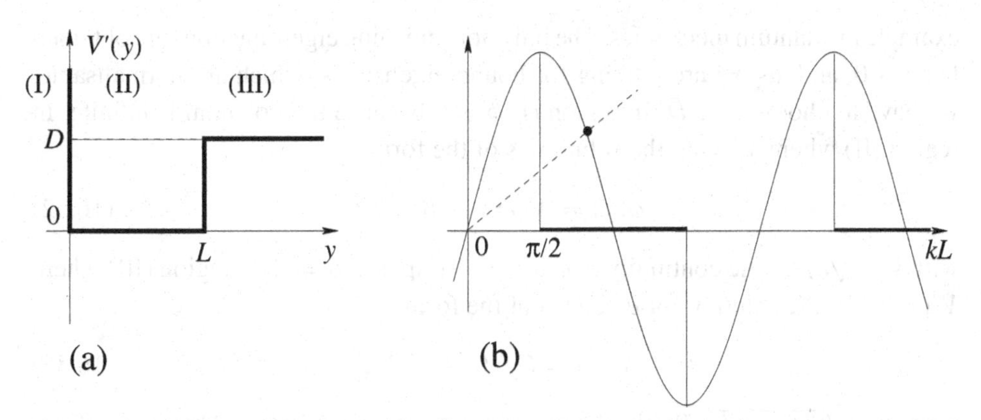

Figure 16.11. (a) Potential $V'(y)$ (heavy line) used for a qualitative study of the statistical physics of DNA. (b) Graphical search of the kL values which give a bound state in the potential $V'(y)$. The values allowed by the condition $\cos(kL) < 0$ are indicated by the thick lines on the horizontal axis. The solution is the intersect of the straight line of slope $\sqrt{\alpha/D}/L$ with the curve $\sin(kL)$.

where we introduce $e_n = \varepsilon_n + \ln(2\pi/\beta K)/2\beta$. The study of the statistical physics of DNA by the transfer integral method [150] is thus reduced to a quantum mechanical problem for a particle in one dimension.

16.3.1 Qualitative study of the phase transition

An exact solution of Equation (16.29) with the Morse potential is possible, but, in order to prevent technical points from hiding the physical ideas, it is useful to start with a potential which has the same qualitative shape as the Morse potential, but allows very simple calculations. Let us consider the square potential well $V'(y)$ of finite depth D and width L, plotted in Figure 16.11,

$$V'(y) = \begin{cases} D & \text{for} \quad y \geq L \\ 0 & \text{for} \quad 0 < y < L \\ +\infty & \text{for} \quad y \leq 0, \end{cases} \tag{16.30}$$

which is qualitatively similar to the Morse potential plotted in Figure 16.7, if we assume that the width L plays the role of $1/a$ in the Morse potential. The Schrödinger equation

$$-\alpha \frac{\mathrm{d}^2\phi}{\mathrm{d}y^2} + V'(y)\,\phi(y) = e\,\phi(y), \tag{16.31}$$

where we defined $\alpha = 1/(2\beta^2 K)$, has to be solved to get the eigenfunctions and eigenvalues of the transfer operator of this simplified model. This is a textbook example in quantum mechanics. The only nonvanishing eigenfunctions are obtained for $e > 0$, and, as we are looking for bound eigenstates, which are normalisable, we have to choose $e < D$. In region (I) $\phi = 0$ because the potential is infinite. In region (II) where $V' = 0$, the solution is of the form

$$\phi(y) = A \sin y + B \cos y, \tag{16.32}$$

with $k = \sqrt{e/\alpha}$. The continuity of ϕ in $y = 0$ imposes $B = 0$. In region (III) where $V'(y) = D$, the solution for $e < D$ is of the form

$$\phi(y) = C\mathrm{e}^{-\rho y} + E\mathrm{e}^{\rho y}, \tag{16.33}$$

with $\rho = \sqrt{(D-e)/\alpha}$. By the definition of a bound eigenstate, we have $\phi \to 0$ for $y \to \infty$ so that $E = 0$.

The continuity of ϕ and its derivative in $y = L$ gives

$$A \sin(kL) = C\mathrm{e}^{-\rho L} \tag{16.34}$$

$$Ak \cos(kL) = -\rho\, C\mathrm{e}^{-\rho L}. \tag{16.35}$$

Thus a bound state exists only if k verifies the condition $\cos(kL)/\sin(kL) = -\rho/k$ with $\cos(kL) < 0$. This relation can be written as

$$1 + \frac{\cos^2(kL)}{\sin^2(kL)} = \frac{1}{\sin^2(kL)} = 1 + \frac{\rho^2}{k^2} = \frac{D}{e} \quad \text{with} \quad \cos(kL) < 0. \tag{16.36}$$

This defines an implicit equation for kL, and therefore e,

$$\sin(kL) = \sqrt{\frac{e}{D}} = \pm\frac{1}{L}\sqrt{\frac{\alpha}{D}}\quad kL. \tag{16.37}$$

Its solution is the intersect between the straight line of slope $(1/L)\sqrt{\alpha/D}$ and the curve $\sin(kL)$, with the additional condition $\cos(kL) < 0$. Figure 16.11 shows that the solution only exists if

$$\frac{1}{L}\sqrt{\frac{\alpha}{D}} < \frac{2}{\pi} \quad \text{i.e.} \quad T < \frac{2L\sqrt{2KD}}{\pi k_B} = T_m, \tag{16.38}$$

taking into account the expression of α.

Thus the calculation shows that the transfer operator has a bound ground state only if the temperature is lower than a critical value T_m given by Equation (16.38). Below this temperature the eigenstate ϕ_0 is normalisable, and the calculation of the order parameter σ gives a finite result. On the other hand, for $T > T_m$ the ground state ϕ_0 is no longer localised, and the mean value of the distances between the bases diverges. Therefore temperature T_m appears as the temperature at which the two strands of DNA separate from each other. If we take into account the approximate equivalence $1/L \sim a$ between the potential $V'(y)$ and the Morse potential we obtain for the DNA model

$$T_m = \frac{2\sqrt{2KD}}{\pi a k_B}. \tag{16.39}$$

This calculation, which is approximate because it has not been made with the Morse potential, indicates that the simplified DNA model that we introduced exhibits a phase transition between the double helix for $T < T_m$ and the denaturated state for $T > T_m$. This is an interesting result because it is actually observed in DNA, but it may seem surprising since the model is one-dimensional and only has nearest neighbour interactions, i.e. short-range interactions, because a 'theorem' often quoted in statistical physics forbids one-dimensional phase transitions in systems which do not have long-range interactions. Actually this theorem does not apply here. The theorem of Gursey [80], proved by van Hove [181] and then generalised by Ruelle [161] is only valid for systems in which the particles are bound by pair interactions. This is not the case for our DNA model due to the on-site potential $V(y_n)$ which depends on a single variable. Another argument against one-dimensional phase transitions was given by Landau [109]. It is based on the possibility that the system splits into regions, separated by interfaces, which rule out any order, even at low temperature. It is valid for systems with a substrate potential such as the ϕ^4 model, but it is, however, not valid for the DNA model because the energy of the domain wall that separates a closed and an open region is infinite, as shown below.

However, to conclude the existence of a transition in the DNA model, it is necessary to study the full system, i.e. to solve the Schrödinger equation (16.29) with the Morse potential. Before showing how it can be done, let us notice that the transition temperature T_m cannot be given by a dimensional analysis. The principle of such an analysis is to build an energy $k_B T_m$ as a function of the model parameters D, K, a and m, which have the dimensions

$$[D] = E \tag{16.40}$$

$$[K] = E \cdot L^{-2} \tag{16.41}$$

$$[a] = L^{-1} \tag{16.42}$$

$$[m] = E \cdot T^2 \cdot L^{-2}, \tag{16.43}$$

where E is an energy, L a length and T a time. According to Buckingham's π theorem, we can look for the transition temperature as a power law of the parameters

$$k_B T_m = c\, D^{\xi} K^{\beta} a^{\gamma} m^{\delta}, \tag{16.44}$$

where c is a numerical factor of order one. Identifying the powers of T, E and L we get

$$2\delta = 0, \qquad \xi + \beta + \delta = 1 \quad \text{and} \quad -2\beta - \gamma - 2\delta = 0, \tag{16.45}$$

so that, according to this dimensional argument, the transition temperature should be such that

$$k_B T_m = c D^{\xi} \left(\frac{K}{a^2} \right)^{1-\xi}, \tag{16.46}$$

which is more general than Expression (16.38). The transition temperature that we obtained is a particular case of the result given by the dimensional analysis for $\xi = 1/2$, but the dimensional analysis itself is not sufficient to determine it.

16.3.2 The Morse oscillator problem

Let us now come back to the original DNA model, with the Morse potential. We have to solve Equation (16.29). This was done by Morse himself. Let us introduce the new variable

$$z = 2\delta \exp(-ay) \quad \text{with} \quad \delta = \frac{\beta}{a}\sqrt{2DK}, \tag{16.47}$$

and the transformation $\phi_n(y) = \mathrm{e}^{-z/2}\, z^s\, w_n(z)$ where $s = \delta\sqrt{1 - e_n/D}$. The equation becomes

$$z \frac{\mathrm{d}^2 w_n}{\mathrm{d}z^2} + (2s + 1 - z) \frac{\mathrm{d}w_n}{\mathrm{d}z} + n w_n = 0, \tag{16.48}$$

where we defined $n = \delta - s - 1/2$. If n is a positive integer, the solution of Equation (16.48) is a Laguerre polynomial [135].

Noting that the eigenfunctions $\phi_n(y)$ stay finite in the whole domain $[0, +\infty]$ only if $s > 0$ determines the spectrum of the bound states

$$\frac{e_n}{D} = 1 - \left(1 - \frac{n + 1/2}{\delta} \right)^2 \quad \text{with} \quad n = 0, 1, \ldots, E(\delta - 1/2), \tag{16.49}$$

where $E(.)$ denotes the function 'integer part'. Therefore, as long as δ is larger than the critical value $\delta_c = 1/2$, the ground state is a bound state. In the quantum mechanical problem this criterion corresponds to the mass threshold below which the particle is kicked out of the potential well by quantum fluctuations. It should be noticed that this is a peculiarity of asymmetrical potential wells. Symmetrical wells, with an even potential function, always have a ground state which is a bound state [34, 134].

Using Equation (16.46), the threshold δ_c defines a critical temperature

$$T_c = \frac{2\sqrt{2KD}}{ak_B}. \tag{16.50}$$

When temperature increases and approaches T_c, the last bound state gets less and less localised and the order parameter $\langle y \rangle$ increases rapidly. This corresponds to the phase transition.

The calculation with the Morse potential confirms the result obtained with the square potential. Up to a factor π, the critical temperature given by the Morse potential is the same as the temperature T_m obtained with the simplified potential $V'(y)$, which was given by Equation (16.38).

Using Equations (16.27) and (16.49) we get the free energy per site $f = e_0 + f_0$, with

$$f_0 = -\frac{1}{2\beta} \ln \left[\left(\frac{2\pi}{\beta} \right)^2 \frac{m}{K} \right] \tag{16.51}$$

and

$$\begin{aligned} e_0 &= D[1 - |t|^2] \qquad \text{if } T < T_c \\ &= D \qquad\qquad\quad \text{otherwise,} \end{aligned} \tag{16.52}$$

where $t = T/T_c - 1$ is a reduced temperature. Therefore the entropy per site contains a nonsingular part

$$S_{\text{nonsing}} = S(T_c) + k_B \log\left(\frac{T}{T_c}\right) \tag{16.53}$$

and a singular part

$$\begin{aligned} S_{\text{sing}} &= \frac{2D}{T_c} t \quad \text{if} \quad T < T_c \\ &= 0 \qquad\ \ \text{if} \quad T > T_c. \end{aligned} \tag{16.54}$$

The result (16.54) indicates that there is a jump in the specific heat at the critical temperature T_c while the critical exponent α, which gives the evolution of the specific heat c in the vicinity of T_c ($T < T_c$) by $c \propto t^{-\alpha}$, is $\alpha = 0$. The transition is second order.

16.3.3 The order parameter for DNA

The quantity which is directly observed in the experimental investigations of DNA thermal denaturation is the fraction of closed base pairs, which determines the UV

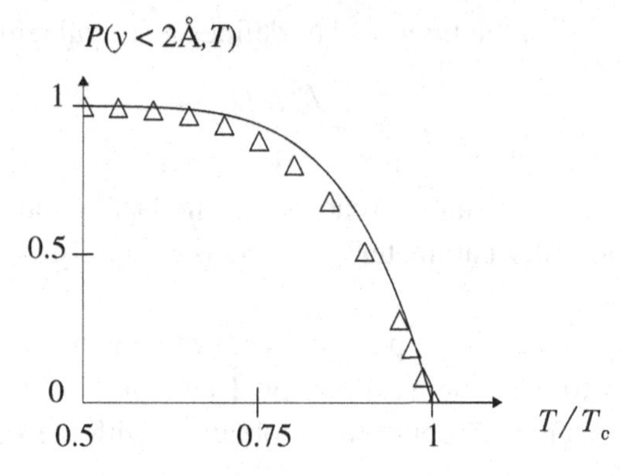

Figure 16.12. Fraction of closed base pairs $P(y < 2\text{Å}, T)$ versus T/T_c in the DNA model. The triangles are the points obtained by a numerical calculation of the eigenstates of the transfer integral operator, while the curve is the theoretical prediction (16.57) derived in the continuum limit approximation.

absorption by the sample. This is a quantity that we can calculate if we define the closed base pairs as the base pairs which have a mean stretching y below a given threshold b. The probability $P(y < b, T)$ that a base pair has an equilibrium stretching smaller than b is given by

$$P(y < b, T) = \int_{-\infty}^{b} \mathrm{d}y \; |\phi_0(y)|^2. \tag{16.55}$$

Using the solution of Equation (16.48), its analytical expression is found to be

$$P(y < b, T) = 1 - \frac{\gamma(2\delta - 1, 2\delta \mathrm{e}^{-ab})}{\Gamma(2\delta - 1)} \tag{16.56}$$

$$\approx (2\delta - 1)\, Ei(2\delta \mathrm{e}^{-ab}) \tag{16.57}$$

which is expressed in terms of the gamma function Γ, the incomplete gamma function γ and the exponential integral function Ei [13]. The approximation of the second line is only valid in the immediate vicinity of T_c. The expression is complex, but the important result is that, with the model defined by Hamiltonian (16.1), the fraction of closed base pairs drops linearly to zero when T approaches T_c (Figure 16.12). This conclusion does not depend on the choice made for b, which does not enter into the slope of the curve.

Comparing Figure 16.12 with an experimental denaturation curve (see Figure 16.5) shows that the theoretical prediction is not in quantitative agreement with the observations because the model predicts a smooth denaturation of the double helix. According to the calculations, a significant part of the molecule should be denaturated for $T = 0.75\ T_c$, while experiments show a sharp denaturation, which looks like a first-order transition. This disagreement appears because we confined ourselves to a highly simplified model of DNA. Without any major change, it is possible to get a denaturation curve in good agreement with experiments if the stacking potential $W(y_n, y_{n-1}) = \frac{1}{2}K(y_n - y_{n-1})^2$ of Hamiltonian (16.1) is improved by going beyond a harmonic expansion. Actually, when a base moves out of the double helix, its energy of interaction with its neighbours decreases. This can be described by a nonlinear stacking interaction.With such an improvement a very sharp denaturation is found [47], but the treatment of the transfer operator has to be done numerically.

16.4 Stability of a domain wall: another approach to denaturation

Up to now we have studied the thermodynamics of the DNA model with standard methods in statistical physics. We shall show that the soliton concept can be used to propose an elegant interpretation of the denaturation by determining with a fairly simple calculation why and when the transition occurs. There is an exact nonlinear solution which connects a nondenatured helicoidal part of DNA to a denaturated part. This is a domain wall between the two phases. In the thermodynamic limit, i.e. in a system where the number of base pairs tends to infinity, its energy is infinite because each base pair of the denatured phase brings a finite contribution to the energy. It is this peculiarity which allows a phase transition in this one-dimensional system. Studying the properties of the domain wall will give us a new method to determine the transition temperature.

16.4.1 The domain wall

To get an analytical solution, we shall use the continuum limit approximation although we know that it is not very good for DNA. We shall show later how to go beyond this approximation. In the continuum limit, the set of dimensionless equations (16.6) give

$$\frac{\partial^2 Y}{\partial t^2} - S\frac{\partial^2 Y}{\partial x^2} + \frac{\partial V(Y)}{\partial Y} = 0 \quad \text{with} \quad V(Y) = (\mathrm{e}^{-Y} - 1)^2, \tag{16.58}$$

using the lattice spacing, i.e. the distance between two bases along the axis, as the unit length in the x direction.

Besides the breathing-mode solutions that we studied in Section 16.2, this nonlinear equation also has exact domain-wall solutions. If we look for a *static* solution, we get

$$-S\frac{\mathrm{d}^2 Y}{\mathrm{d}x^2} + \frac{\partial V(Y)}{\partial Y} = 0 \quad \text{with} \quad V(Y) = (\mathrm{e}^{-Y} - 1)^2. \tag{16.59}$$

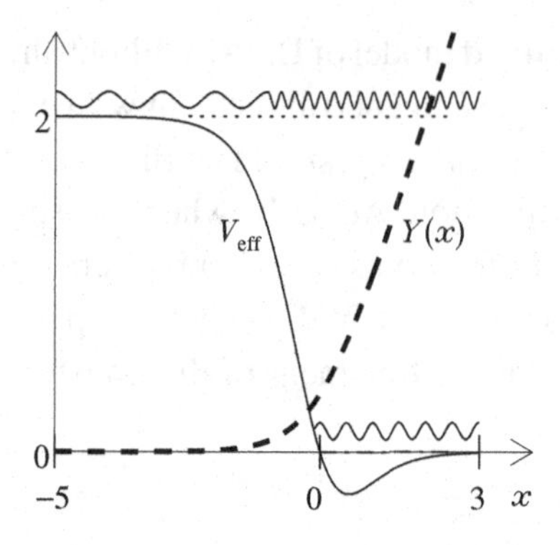

Figure 16.13. The heavy dashed line shows the domain-wall solution (16.63) (solution with the + sign in the exponential). The solid line labelled V_{eff} shows the effective potential of the Schrödinger equation (16.69) and the oscillations are a schematic plot of the vibrational modes on both sides of the domain wall.

It is formally identical to the equation of motion of a particle in the potential $V(Y)$, with the variable x instead of time. It can be integrated by a quadrature. We multiply it by $\mathrm{d}Y/\mathrm{d}x$ and integrate with respect to x to get

$$-\frac{1}{2}S\left(\frac{\mathrm{d}Y}{\mathrm{d}x}\right)^2 + V(Y) = C, \tag{16.60}$$

where C is an integration constant, which is determined by the boundary conditions. If we look for a solution which approaches the equilibrium state $Y = 0$ for $x \to -\infty$, we get $C = 0$. The solution is then computed by separating the variables. Equation (16.60) gives

$$\pm\sqrt{\frac{S}{2}}\frac{\mathrm{d}Y}{\mathrm{d}x} = (\mathrm{e}^{-Y} - 1) \quad \text{or} \quad \frac{\mathrm{d}Y}{\mathrm{e}^Y - 1}\frac{\mathrm{e}^Y}{} = \pm\sqrt{\frac{2}{S}}\,\mathrm{d}x, \tag{16.61}$$

which is readily integrated into

$$\ln(\mathrm{e}^Y - 1) = \pm\sqrt{\frac{2}{S}}(x - x_0), \tag{16.62}$$

which can also be written

$$Y(x) = \ln[1 + \mathrm{e}^{\pm\sqrt{2/S}\,(x - x_0)}], \tag{16.63}$$

where x_0 is an integration constant which determines the position of the domain wall. This solution is plotted in Figure 16.13. It corresponds to a configuration that connects one part of the chain ($x < x_0$) where the base pairs are closed, and another part ($x \gg x_0$) where the separation between the strands grows linearly.

Let us calculate the energy of this solution for a finite DNA chain with N base pairs. The sites which have an index lower than x_0 are such that $Y \simeq 0$. In this region the Morse potential and the stacking interaction between neighbouring sites vanish. Thus this region does not contribute to the energy. For the sites with an index greater than x_0, $Y \gg 1$ so that the Morse potential takes the value 1, while $\mathrm{d}Y/\mathrm{d}x \simeq \sqrt{2/S}$ which corresponds to a stacking energy $\frac{1}{2}S(\sqrt{2/S})^2 = 1$. Therefore any site with an index larger than x_0 brings a contribution $e = 2$ to the energy. The energy of the domain wall is thus

$$E_P^+ = 2(N - x_0) + \mathcal{O}(N^0). \tag{16.64}$$

The term $\mathcal{O}(N^0)$ is the contribution of the core of the domain wall, where the solution gradually changes from the bottom of the Morse potential to the plateau. The exponent + indicates that we have calculated the energy of the solution with the + sign in the exponential. As previously mentioned, the domain-wall energy becomes infinite if N tends to ∞.

At zero temperature, the solution (16.63) is not stable because the wall can lower its energy if x_0 grows, which increases the size of the domain in which the base pairs are closed. This is expected because for $T = 0$, the stable state of DNA is the double helix, in which all bases are paired. Contrary to other field theories of the Klein–Gordon type which have a finite energy for the soliton or kink, the domain wall of DNA does not have a zero-frequency Goldstone mode. Some energy is required to move the domain wall in order to increase the number of open sites. The contrast with the ϕ^4 model that we studied in Chapter 10 lies in the difference in the energy per site for the two states which are connected by the domain wall.

But, if thermal fluctuations exist, i.e. at $T \neq 0$, entropic effects must be added to the energy to determine the stability of the domain wall. To calculate them we need to study the fluctuations around the domain wall.

16.4.2 Fluctuations around the domain wall

Let us look for a solution to Equation (16.59) as

$$Y(x,t) = Y_P(x) + f(x,t), \tag{16.65}$$

where $Y_P(x)$ designates the solution (16.63) and $f(x,t)$ is assumed to be small enough to allow us to linearise the equation with respect to f. Introducing this solution into the equation of motion (16.58), we get

$$\frac{\partial^2 f}{\partial t^2} - S\frac{\partial^2 f}{\partial x^2} - S\frac{\partial^2 Y_P}{\partial x^2} + \mathcal{F}(Y_P) + f(x,t)\left(\frac{\partial \mathcal{F}}{\partial Y}\right)_{Y=Y_P} = 0 \tag{16.66}$$

$$\text{with} \quad \mathcal{F}(Y) = \frac{\partial V(Y)}{\partial Y}.$$

The third and fourth terms of (16.66) cancel each other because Y_P is a static solution of Equation (16.59). The equation for $f(x,t)$ is therefore reduced to

$$\frac{\partial^2 f}{\partial t^2} - S\frac{\partial^2 f}{\partial x^2} + V_{\text{eff}}(x)\ f(x,t) = 0. \tag{16.67}$$

The effective potential V_{eff}, plotted in Figure 16.13, is derived from the expression of $\mathcal{F}(Y)$ and from $\exp(Y_p) = 1 + \exp(z)$ with $z = \sqrt{2/S}\ (x - x_0)$, which results from the solution (16.63) for Y_P. It is given by

$$V_{\text{eff}} = -2\frac{\mathrm{e}^z - 1}{(1+\mathrm{e}^z)^2}. \tag{16.68}$$

Looking for a solution of Equation (16.67) of the form $f(x,t) = \exp(-\mathrm{i}\omega t)\ g(x)$, we get

$$-S\ \frac{\partial^2 g}{\partial x^2} + V_{\text{eff}}(x)\, g(x) = \omega^2\, g(x), \tag{16.69}$$

which is again an equation formally identical to the Schrödinger equation. The shape of the potential V_{eff} is such that it does not have any bound state, but we can predict the existence of two families of extended states:

(i) For $\omega^2 < 2$, the solution $g(x)$ is confined in the region $x > x_0$.
(ii) For $\omega^2 > 2$, the solution is extended over the whole space, but, since $V_{\text{eff}}(x)$ has different values for $x < x_0$ and $x > x_0$, the dispersion relation of waves which are solutions of the equation are different in the two regions.

These extended states are schematised in Figure 16.13.

The expression of the potential $V_{\text{eff}}(x)$ allows an analytical solution of Equation (16.69) but it is tedious to derive. It is easy to get an approximate solution which contains the physics of the phenomena by assuming that the potential has a sharp change from $x < x_0$ to $x > x_0$ as plotted in Figure 16.14. With this simplified view, the bases of index $n \leq x_0$ are assumed to be exactly closed and the effective potential in this region is $V_{\text{eff}}(x) = 2$. The solutions of Equation (16.69) are $g(x) = \exp(\mathrm{i}qx)$ where q and ω are related by the dispersion relation $\omega^2 = 2 + Sq^2$. This is the low q limit (corresponding to the continuum limit approximation) of the dispersion relation (16.12) which was obtained by looking for linearised dynamic solutions of the model.

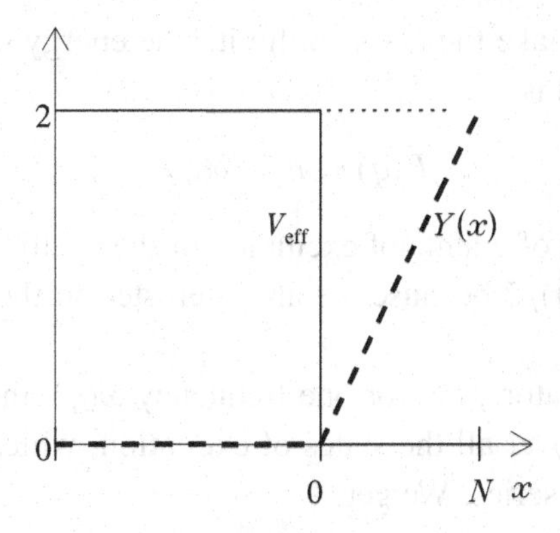

Figure 16.14. Simplified plot of the domain wall (heavy dashed line) and of the effective potential (solid line) that it creates in a chain of N base pairs. This figure is analogous to Figure 16.13.

The base pairs with an index $n > x_0$ are assumed to be fully open so that Y has a value which is on the plateau of the Morse potential. In this region we assume that $V_{\text{eff}}(x) = 0$. The solution for $g(x)$, is still $g(x) = \exp(iqx)$, but now ω and q are related by $\omega^2 = Sq^2$.

16.4.3 Free energy of the domain wall

We now have all the elements to calculate the free energy F of the domain wall, replaced by the simplified solution of Figure 16.14, located at the position x_0. It includes two contributions:

$$F = F_P + F_{\text{fluct}}. \tag{16.70}$$

- The contribution F_P is the free energy of the solution Y_P *for a given* x_0, which is simply the energy of the domain wall since there is no entropic term because x_0 is chosen and Y_P is a given static solution. The contribution F_P is therefore identical to Expression (16.64)

$$F_P = 2(N - x_0), \tag{16.71}$$

  in which we do not include the small correction due to the core of the domain wall since we have considered the approximate solution of Figure 16.14.
- The contribution F_{fluct} is the free energy of the fluctuations around the domain wall, which were calculated in the previous section.

These modes have been obtained in a harmonic approximation so that their energies are the energies of harmonic oscillators with frequencies $\omega(q)$. To get the corresponding free energy, it is convenient to use the properties of quantum oscillators, and then to take the classical limit. The energy of an oscillatory mode around the domain wall is

$$E(q) = n_q\, \hbar\omega(q), \tag{16.72}$$

where n_q is the number of quanta of excitation of the oscillator. We do not add the zero-point energy $\hbar\omega(q)/2$ because we are interested in the classical limit of the result.

For one of the oscillators, i.e. for one frequency $\omega(q)$, the partition function is obtained by summing over all the states of excitation, which is easy because it is the sum of a geometric series. We get

$$Z_q = \sum_{n_q=0}^{\infty} \exp(-\beta\hbar\omega(q)n_q) = \frac{1}{1-\exp(-\beta\hbar\omega(q))}. \tag{16.73}$$

We recover the usual Bose–Einstein formula.

In the classical limit where $\hbar\omega(q) \ll k_B T$, we can replace the exponential by its first-order expansion $\exp(-\beta\hbar\omega(q)) \simeq 1 - \beta\hbar\omega(q)$, so that we simply get

$$Z_q = \frac{1}{\beta\hbar\omega(q)}. \tag{16.74}$$

It should be noticed that the classical limit is perfectly legitimate for the dynamics of the bases of DNA because they are large chemical groups, which move at frequencies of the order of $5 \cdot 10^{11}$ Hz, which corresponds to $\hbar\omega \approx 2 \cdot 10^{-3}$ eV while $k_B T$ is of the order of $25 \cdot 10^{-3}$ eV at room temperature.

The partition function of a set of independent oscillators is the product of the partition function of each oscillator because the partition function Z measures the number of states which are accessible to the system. If we consider a system S_1 with n_1 states and a system S_2 with n_2 states which are independent of S_1, for a system made up of S_1 *and* S_2, each state of S_1 corresponds to the n_2 states of S_2, giving a total number of states $n_1 \times n_2$. Thus the partition function for all the harmonic oscillators corresponding to the fluctuations around the domain wall is

$$Z = \prod_q \frac{1}{\beta\hbar\omega(q)} \tag{16.75}$$

which gives the free energy

$$F_{\text{fluct}} = -k_B T \ln Z = k_B T \sum_q \ln(\beta\hbar\omega(q)). \tag{16.76}$$

This discrete sum is not easy to calculate, but it can be replaced by an integral due to the very large number of modes $\omega(q)$. For a lattice of p particles, the possible wavevectors are $q_\ell = \ell\pi/p$ ($\ell = 0, \ldots, p-1$), so that, from one mode to the next

q varies by $\Delta q = \pi/p$. This gives the density of states $1/\Delta q = p/\pi$ from which we get

$$\sum_q f[\omega(q)] \simeq \frac{1}{\Delta q} \int dq f[\omega(q)], \tag{16.77}$$

where $f[\omega(q)]$ denotes any function of $\omega(q)$.

For a DNA molecule having N base pairs and a domain wall at position x_0, we have x_0 particles having the dispersion relation $\omega(q) = \sqrt{2 + Sq^2}$ and $N - x_0$ particles with the dispersion relation $\omega(q) = \sqrt{S}\, q$. Thus we have

$$F_{\text{fluct}} = k_B T \frac{x_0}{\pi} \int_0^\infty dq \ \ln(\beta\hbar\sqrt{2 + Sq^2}) + k_B T \, \frac{N - x_0}{\pi} \int_0^\infty dq \ln(\beta\hbar\sqrt{S}\, q), \tag{16.78}$$

where the integrals over q have been extended to infinity, which is consistent with the continuum limit approximation that we use in this calculation. Separating terms which depend on the position x_0 of the wall, we get

$$F_{\text{fluct}} = \frac{k_B T}{\pi} x_0 \int_0^\infty dq \ln\left(\frac{\sqrt{2 + Sq^2}}{\sqrt{S}\, q}\right) + F_1, \tag{16.79}$$

where F_1 is a quantity which does not depend on x_0. The integral of Equation (16.79), that we shall denote by I, can be calculated by

$$I = \int_0^\infty dq \ln\sqrt{\frac{2}{Sq^2} + 1} = \frac{1}{2} \int_0^\infty dq \ln\left(1 + \frac{2}{Sq^2}\right) \tag{16.80}$$

if we define $u = q\sqrt{2/S}$ and use

$$\int_0^\infty du \frac{\ln(1 + u^2)}{u^2} = \pi. \tag{16.81}$$

We get

$$F_{\text{fluct}} = k_B T \frac{x_0}{2} \sqrt{\frac{2}{S}} + F_1. \tag{16.82}$$

Collecting the results (16.71) and (16.82), we get the following expression for the free energy of the domain wall

$$F = x_0 \left[\frac{k_B T}{2} \sqrt{\frac{2}{S}} - 2\right] + F_1', \tag{16.83}$$

where F_1' is a contribution *which does not depend on* x_0.

- For low T (or $T = 0$), the expression within the bracket is negative. The molecule can decrease its free energy by increasing x_0. Due to the shape of the domain wall, this increases the size of the region where the base pairs are closed. This result is in agreement with the stability of the closed helicoidal state of DNA at low temperature.
- For large T the expression in the bracket is positive. The free energy of the molecule can be lowered by decreasing x_0, i.e. by opening base pairs that were closed.
- For $(k_B T/2)\sqrt{2/S} = 2$, i.e.

$$T = T_c = \frac{2}{k_B}\sqrt{2S}, \tag{16.84}$$

  the energy of the molecule is *independent of the position* x_0 of the domain wall which separates the closed and open regions. Therefore this particular temperature appears as the *transition temperature* at which the molecule changes from the helicoidal state to the denatured state.

It is important to notice that, if we come back to dimensional variables by replacing S by its value $S = K/(Da^2)$, and taking into account that the dimensional form of the energy $k_B T$ is restored by multiplying by D the dimensionless expression (see Section 16.2.1), we get $T_c = 2\sqrt{2KD}/(ak_B)$ which is *exactly the transition temperature given by the transfer integral calculation* (Equation (16.50)).

16.4.4 Discussion

Therefore we notice that the study of the free energy of a nonlinear localised excitation, the domain wall, can recover the thermodynamic properties of the DNA model. This result points out that an approach based on solitons (or on nonlinear localised excitations because the DNA domain wall is not a soliton) provides an interesting alternative method to study the properties of a physical system.

Actually it is even more powerful than the transfer integral method because it can easily take into account discreteness effects, which could only be done in a painful perturbative calculation for the transfer integral.

We already mentioned that DNA is a weakly coupled system since we got $S = 0.099$ when we introduced realistic parameters. Discreteness effects enter in particular into the dispersion relations of small-amplitude oscillations around the domain wall, which become

$$\omega_q^2 = \begin{cases} 2 + 4S\sin^2(q/2) = 2 + 2S(1 - \cos q) & \text{for } x > x_0 \\ 4S\sin^2(q/2) = 2S(1 - \cos q) & \text{for } x < x_0, \end{cases}$$

as shown for instance by Equation (16.12).

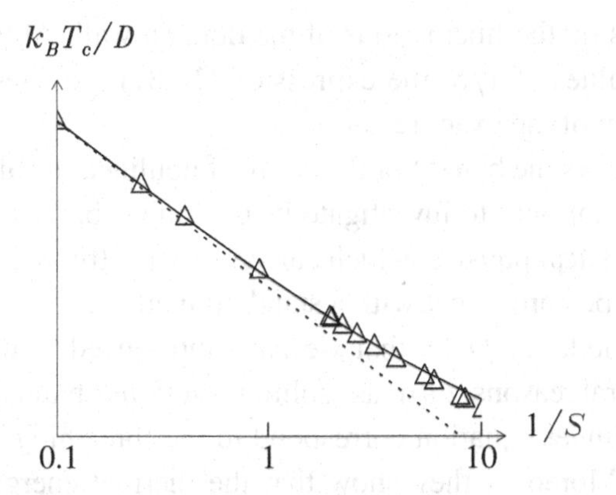

Figure 16.15. Variation of the critical temperature as a function of the discreteness parameter $1/S$. The dotted line is the continuum limit result (16.50). The solid line is the discrete solution (16.87). The triangles show the exact result obtained by a numerical treatment of the transfer integral.

If we assume that the shape of the domain wall itself is not modified by the effect of discreteness, except perhaps in the core, which is confirmed by a numerical search of the domain-wall solution in the discrete case so that discreteness effects only change the fluctuations, the calculation of the previous section can be expressed in terms of the new dispersion relations. The integral over q must be calculated up to $q = \pi$ instead of being extended to infinity. The only change appears in the integral (16.79) which becomes

$$F_{\text{fluct}} = \frac{k_B T}{\pi} x_0 \int_0^{\pi} dq \quad \ln\left(\frac{\sqrt{2 + 2S(1 - \cos q)}}{\sqrt{2S(1 - \cos q)}}\right) + F_1, \tag{16.85}$$

which can be calculated using

$$\int_0^{\pi} du \ln\left[1 - \cos u + \frac{1}{S}\right] = 2\pi \ \ln\left[\sqrt{1/(2S)} + \sqrt{1 + 1/(2S)}\right]. \tag{16.86}$$

We finally get the new value of the critical temperature in dimensional variables as

$$T_c = \frac{2D}{k_B \ \ln\left[\sqrt{1/(2S)} + \sqrt{1 + 1/(2S)}\right]}. \tag{16.87}$$

It tends to the value obtained previously (Equation (16.50)) in the limit of large S, which is the strong coupling limit for which the continuum limit approximation is valid.

Figure 16.15 compares the exact result for the transition temperature of the DNA model, which can be obtained by a numerical treatment of the transfer integral, with the analytical results obtained in the continuum limit (obtained either by the transfer integral method or with the free energy of a domain wall) and by taking into account discreteness effects on the fluctuations of the domain wall. Although it is still not perfect for large values of $1/S$, the expression (16.87) provides, however, a very good approximation of the exact result.

This example shows the benefit of the study of nonlinear localised excitations in a physical system, not only to investigate its dynamics, but also to provide a new approach to its statistical physics, which can be very fruitful because it can provide a result that would be hard to get with a standard method.

The nonlinear model of DNA that we have introduced in this chapter is thus interesting for several reasons. Its quasi-soliton solutions obtained by a reduction to a nonlinear Schrödinger equation correspond to the 'breathing of DNA' observed by the biologists. Moreover they show that the thermal energy can temporarily self-localise due to nonlinear phenomena. This may turn out to be relevant in some biological processes, but it can also play a role as a precursor of thermal denaturation. This model is also interesting for statistical physics because it is probably the simplest model which has a phase transition and which can be treated exactly analytically, in a strong coupling limit [48].

In this chapter we discussed some of the simplest properties of the model. Other aspects can be studied, such as the static structure factor, which exhibits a strong component at very low wavevectors near T_c. It is probably due to nonlinear excitations, as with the central peak of ferroelectrics studied in Chapter 10 with the ϕ^4 model. The study is however more complex for the DNA model because, contrary to the kinks, which have a single degree of freedom, the fluctuational openings of DNA have an internal oscillation. Moreover, near T_c, they play the role of nucleation centres for open regions, which can be approximately viewed as made of two domain walls, corresponding to the two interfaces between a closed and an open region which make the boundaries of a 'denaturation bubble'. From the viewpoints of statistical physics and nonlinear science, in spite of its simplicity, this DNA model still leaves many questions open.

Conclusion

Physical solitons: do they exist?

In this book we travelled through physics, from the macroscopic to the microscopic scale, and we met solitons at all scales. It is so because the conditions for their existence are very general: the coexistence of dispersion and nonlinearity in a spatially extended system. The soliton concept is not confined to a particular domain in physics, and this is why it is so interesting.

Nonlinearity is everywhere in physics, while it is linearity which is an exception so that it is a pity that we have to resort to a negation, 'non'linearity to name this ubiquitous property. Likewise, dispersion is very common because components at different wavelengths in a complex signal seldom travel at the same speed in a medium.

Therefore it is not surprising that these common properties, which generally have opposite influences, because nonlinearity tends to localise signals while dispersion spreads them, may, in some regimes, balance each other and lead to solitons in a large variety of systems.

More than the results that we presented, what has to be remembered is the method that can be used to approach nonlinear dispersive systems, so often met in physics.

The first step is to determine the relevant phenomena and degrees of freedom. This is of course the usual approach in physics, whether we are interested in solitons or not, but it becomes more crucial if we study the nonlinear properties of a system because we expect a priori to get equations that may be hard to solve. Thus it is important to look for all possible simplifications from the first step of the modelling process.

The second important point is to make sure that space is not lost in the analysis. Contrary to the tendency inherited from the experience of linearised approximations, it is not wise to switch to reciprocal space by a spatial Fourier transform. We saw how this type of reasoning, stemming from normal mode perturbative expansions, blocked the solution of the Fermi–Pasta–Ulam problem for ten years.

For most of the systems, after the preliminary physical analysis, we get nonlinear dispersive equations that we are not able to solve. It is at this level that it is useful to have in mind the main classes of soliton equations, which are limiting models to which many systems can be reduced in some approximate description, along the lines that we presented, using for instance a continuum limit approximation or a multiple scale expansion.

- ▷ As these soliton equations are only approximate for physical systems, we can say that *physical solitons do not exist.*
- ▷ However, we have seen that *solitons are important for physics because they provide clues to understand many experimental observations* from the exceptional stability of some waves in the oceans to the structure factor of ferroelectrics or magnetic materials, measured by neutron scattering.

The solution of this paradox comes from the exceptional properties of solitons. When we move from the idealised description provided by a soliton equation to a realistic model which adds many extra terms to the equation, we lose the exact soliton equation but most of the properties of the nonlinear excitations are generally preserved. This has some consequences for the most appropriate method to study a problem. One should not be afraid of using approximations that look severe, to reduce the physical equations to one of the main class of soliton equations. What is deduced from this equation will still be useful for the real system. Then it will be possible to study the perturbed equation by treating solitons as entities which have individuality. We can define specific parameters for them, such as a position, a velocity. This is the substance of the collective coordinate methods which are very helpful... provided a good preliminary study has been made to select the appropriate coordinates.

Of course, while it is wrong to say that *physical solitons do not exist* because a physical system is never exactly described by an equation having true soliton solutions, it is equally wrong to say that *solitons are everywhere*. The limitations of the models should not be forgotten, and this is particularly important at the microscopic scale. We have met the example of dislocations where discreteness effects are so large that free propagation is impossible. Another topic which has been the subject of a lot of controversy is the role of solitons in biological molecules. The section 'News and Views' in the journal *Nature* attested to these debates, with articles ranging from enthusiastic approval [120, 121] to very severe criticisms [71] justified by the overstatements of some theoreticians. Nowadays opinions on this subject have become less passionate and, as experiments have made a lot of progress, nonlinear localisation has been found in systems which are very close to biological systems [57]. In biological molecules themselves, the question is still open. Rather than solitons carrying energy along, it is likely that nonlinear energy localisation may be the dominant nonlinear effect.

This book has of course left many questions on the side. One of them is the role of dissipation, often invoked as a possible cause of the destruction of solitons. At a macroscopic scale, dissipation can of course cause a decay of nontopological solitons, or stop the propagation of topological solitons, but we have seen that nature sometimes has a solution to this problem without injecting any energy in the system: for instance, the conicity of the arteries is one remedy to the decay of the localised blood pulse due to blood viscosity. In many other systems, dissipation, even if it exists, is not sufficient to drastically modify the properties of the solitons. This is often the case in hydrodynamics and the solitons which travel hundreds of kilometres in the Andaman sea provide a remarkable example of that. At the microscopic level the situation is more complex although the dynamics are not dissipative. A coupling with other degrees of freedom, or the thermal bath due to a solvent for instance, may cause some energy loss for the solitons. On the other hand, thermal energy can provide the energy to create highly nonlinear excitations, as we saw for DNA.

Another point which has been left out is the case of nonlinear excitations in multidimensional systems. This is a vast research area, which is less advanced than soliton studies although many investigations have been devoted to it. Some of the methods that we have introduced in this book are not restricted to one dimension. For instance, the multiple scale expansion is the same as the amplitude equation approach used in two or three dimensions. The collective coordinate approach that we introduced for solitons is also valid for vortices in magnetic films, for instance.

This book does not claim to be exhaustive and we hope that its readers will be tempted to go ahead by themselves into the world of nonlinear coherent structures, where there are still many unknown territories to explore.

Part V

Appendices

A

Derivation of the KdV equation for surface hydrodynamic waves

Surface-water waves are very interesting for a physicist because they are readily accessible to observation, from sea waves and the wake of a boat to tsunamis, and are associated with a large variety of wave equations depending on the boundary conditions, from linear nondispersive to strongly nonlinear and highly dispersive waves.

In this appendix we derive the Korteweg–de Vries equation for shallow-water waves from the equations of hydrodynamics. The calculation is presented in detail because it was at the root of understanding the first observation of a soliton, and also because it deserves some attention since a proper justification of the approximations is not simple.

A.1 Basic equations and boundary conditions

Let us consider a perfect fluid, incompressible and inviscid. Its dynamics are described by the Euler equation

$$\rho \frac{\mathrm{d}\vec{v}}{\mathrm{d}t} = \rho \vec{g} - \vec{\nabla} P \tag{A.1}$$

where ρ is the density of the fluid, $\vec{v}$ the velocity field, $\vec{g}$ the gravity field and P the pressure in the fluid. In this equation, the time derivative is a total derivative. When it is expanded, we get

$$\rho \frac{\partial \vec{v}}{\partial t} + \rho(\vec{v} \cdot \vec{\nabla})\vec{v} = \rho \vec{g} - \vec{\nabla} P \tag{A.2}$$

which must be completed by the mass conservation relation, which reduces to

$$\vec{\nabla} \cdot \vec{v} = 0 \tag{A.3}$$

for an incompressible fluid. To derive a dynamic equation for surface-water waves we must supplement these equations by the relations imposed by the boundary conditions.

A.1.1 Kinematic boundary condition

We shall consider the surface defined by equation

$$F(\vec{r}, t) = 0$$

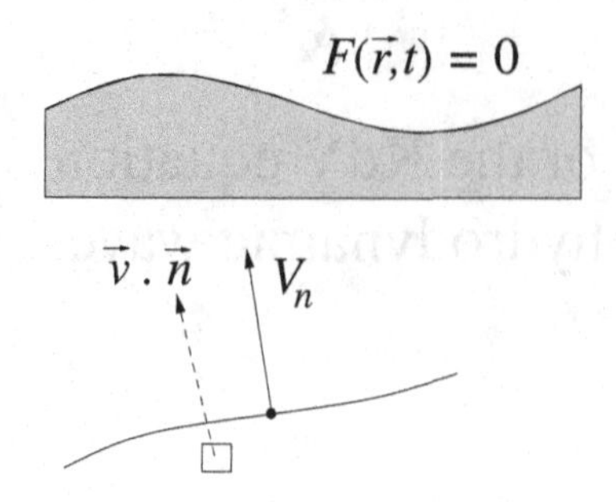

which separates a liquid from a gas which imposes a uniform pressure above it. Physically the surface is defined by the fact that the molecules do not cross it. Therefore the component orthogonal to the surface, $V_n(\vec{r_0}, t)$, of the velocity of a point belonging to the geometric surface must be equal to the component orthogonal to the surface of the velocity of a small fluid element infinitely close to the surface, which can be deduced from the relation

$$\lim_{\vec{r} \to \vec{r_0}} \vec{v}(\vec{r}, t) \cdot \vec{n} = V_n(\vec{r_0}, t). \tag{A.4}$$

The unit vector orthogonal to the surface is deduced from the surface equation by $\vec{n} = \vec{\nabla} F(\vec{r_0}, t)/|\vec{\nabla} F(\vec{r_0}, t)|$, and the velocity $\mathrm{d}\vec{r_0}/\mathrm{d}t$ of a point belonging to the surface is given by

$$\frac{\mathrm{d}F}{\mathrm{d}t} = 0 = \frac{\partial F}{\partial t} + \frac{\mathrm{d}\vec{r_0}}{\mathrm{d}t} \cdot \vec{\nabla} F(\vec{r_0}, t). \tag{A.5}$$

Therefore we get

$$V_n(\vec{r_0}, t) = \vec{n} \cdot \frac{\mathrm{d}\vec{r_0}}{\mathrm{d}t} = \frac{1}{|\vec{\nabla} F(\vec{r_0}, t)|} \vec{\nabla} F(\vec{r_0}, t) \cdot \frac{\mathrm{d}\vec{r_0}}{\mathrm{d}t} \tag{A.6}$$

$$= -\frac{1}{|\vec{\nabla} F(\vec{r_0}, t)|} \frac{\partial F}{\partial t}. \tag{A.7}$$

In a point $\vec{r_0}$ on the surface, the kinematic condition (A.4) is therefore

$$\frac{\partial F(\vec{r_0}, t)}{\partial t} + \vec{v}(\vec{r_0}, t) \cdot \vec{\nabla} F(\vec{r_0}, t) = 0. \tag{A.8}$$

It can be put in a different form, useful for the following calculations. Let u, v and w henceforth denote the components of $\vec{v}$ in any point inside the fluid. If the equation of the surface is given as $z = \eta(x, y, t)$, which is equivalent to $F = \eta(x, y, t) - z$, the calculation of $\vec{\nabla} F$ for a point belonging to the surface expresses the kinematic condition as

$$w = \frac{\partial \eta}{\partial t} + u\frac{\partial \eta}{\partial x} + v\frac{\partial \eta}{\partial y}. \tag{A.9}$$

A.1.2 Physical boundary condition

The condition (A.9) is purely geometrical, but it does not contain the physical condition which makes the surface special, the fact that it has a *surface tension*. As the molecules at the surface are not surrounded by others, contrary to molecules in the bulk, they have an excess energy. Thus the surface behaves like an elastic membrane, which tends to shrink to reduce the surface energy. With such a surface tension, a curvature of the surface is associated with a difference of pressure between its two sides. It is this phenomenon which explains why a fluid moves higher in a narrow tube than in a large vessel, by capillarity. However, for hydrodynamic solitons, the radius of curvature of the surface is usually large enough to make this effect negligible. Thus the presure in the fluid in the immediate vicinity of the surface can be assumed to be equal to the pressure P_A of the gas above.

A.2 Mathematical formulation of the problem

The most general study of surface waves is very difficult. We shall select a particular problem, mathematically well posed, and corresponding to a real physical situation without being too complicated.

To solve it, it will be convenient to introduce dimensionless variables. The dimensional physical variables are identified by a $'$.

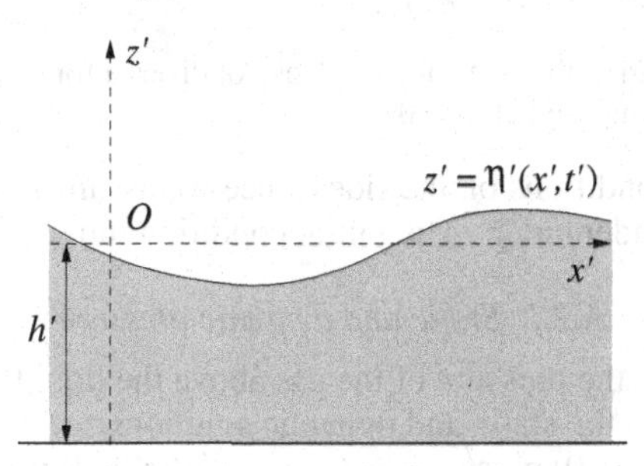

Let us consider a layer of an inviscid and incompressible fluid, which has the *mean depth* h', lying above a horizontal plane situated at altitude $z' = 0$. We assume that the *motion of the fluid is two-dimensional*, i.e. all the properties of the system are independent of the coordinate y' and the component of the velocity along y' is zero.

We denote by $u'(x', z', t')$, 0 and $w'(x', z', t')$, the components of the velocity field inside the fluid, and by $z' = h' + \eta'(x', t')$ the equation of its surface.

Above the fluid there is a gas at pressure P'_A, which is constant and uniform in space. The fluid experiences the force $\overrightarrow{f} = \rho \overrightarrow{g}$ per unit volume due to gravity.

A.2.1 Equations defining the problem

They include:

(i) The Euler equation inside the volume of the fluid. If we divide it by ρ, and express its components, it gives

$$\frac{\partial u'}{\partial t'} + u'\frac{\partial u'}{\partial x'} + w'\frac{\partial u'}{\partial z'} = -\frac{1}{\rho}\frac{\partial P'}{\partial x'} \tag{A.10}$$

$$\frac{\partial w'}{\partial t'} + u'\frac{\partial w'}{\partial x'} + w'\frac{\partial w'}{\partial z'} = -\frac{1}{\rho}\frac{\partial P'}{\partial z'} - g, \tag{A.11}$$

where P' is the pressure in the fluid. These equations must be completed by the mass conservation requirement

$$\frac{\partial u'}{\partial x'} + \frac{\partial w'}{\partial z'} = 0. \tag{A.12}$$

(ii) The boundary condition at the bottom, which says that the velocity of the fluid when $z' = 0$ does not have any vertical component,

$$w'(z' = 0) = 0. \tag{A.13}$$

(iii) The kinematic boundary condition at the surface,

$$w' = \frac{\partial \eta'}{\partial t'} + u' \frac{\partial \eta'}{\partial x'} \quad \text{for} \quad z' = h' + \eta'(x', t'). \tag{A.14}$$

(iv) The physical condition at the surface

$$P_A - P' = 0 \quad \text{for} \quad z' = h' + \eta'(x', t'). \tag{A.15}$$

(v) The condition which specifies that the average depth of the fluid is h', i.e. the surface corresponding to $\eta' = 0$ is, on average, at altitude $z' = h'$,

$$\lim_{K \to \infty} \frac{1}{2K} \int_{-K}^{+K} \eta'(x', t')\, \mathrm{d}x' = 0, \tag{A.16}$$

completed by the condition that the solution does not diverge for $|x'| \to \infty$.

(vi) The initial conditions, imposed at $t' = 0$.

There are no boundary conditions on the sides since we assumed that the problem is two-dimensional, independent of y'. The initial condition must also verify this condition.

A.2.2 Static and dynamic pressure

It is convenient to remove the pressure of the gas above the liquid from the equations. This can be done by comparing the static and dynamic solutions.

The static solution ($\overrightarrow{v} = 0$) gives $-g - (1/\rho)\partial P'/\partial z' = 0$ from Equation (A.11). This determines the *static pressure* $P_0' = -g\rho(z' - h') + P_A$ after integration and application of the boundary condition (A.15) for $z' = h'$, which is the surface at rest. We recover the fundamental equation of hydrostatics as expected.

We define the *dynamic pressure as* $p'(x', z', t') = P'(x', z', t') - P_0'(z')$. Putting this expression into the Euler equations, Equation (A.10) is not changed while $-g$ is removed from Equation (A.11). Moreover Equation (A.15) which indicates that the pressure at the surface of the fluid is equal to the external pressure becomes

$$p' = \rho g \eta'(x', t') \quad \text{for} \quad z' = h' + \eta'(x', t'). \tag{A.17}$$

A.2.3 Dimensionless equations

Writing dimensionless equations is important to reduce the number of parameters in the problem because it also ensures that the parameters are independent of each other. Moreover it is essential to allow correct control of the approximations, giving a real meaning to the notion of a 'small' term. To get dimensionless quantities, we must select a characteristic length and a characteristic time, which will be used as references. For quantities homogeneous to a length, it is actually useful to use different length scales to describe the *position* inside the fluid (coordinates x' and z') and the *displacement* of the fluid because these variables play a very different role in the problem.

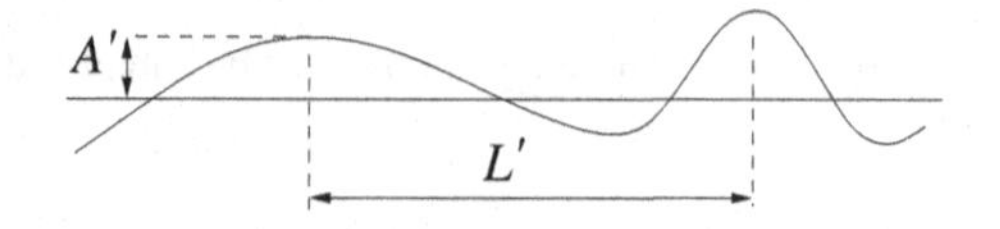

We shall select a characteristic amplitude A' for the deformations of the surface. It will be used to measure the displacement of the fluid along z', and to define the components of the velocity of the fluid. Therefore we should have $\eta'/A' \approx 1$. If the solution happened not to verify this condition, another scale for the displacement along z' would have to be chosen, which could change the terms that we can neglect. We shall select a second characteristic length L' along the direction x'. The study of the linearised equation introduces a speed c_0', which is the speed of the long-wavelength waves, i.e. the waves which have a wavevector which tends to zero. We use L' and c_0' to define a characteristic time $t_0' = L'/c_0'$ which will be used to measure time.

Thus we select the dimensionless variables

$$t = \frac{t'}{t_0'}, \quad x = \frac{x'}{L'}, \quad z = \frac{z'}{L'}, \quad \eta = \frac{\eta'}{A'}, \tag{A.18}$$

and we define

$$u = \frac{u'}{A'/t_0'}, \quad w = \frac{w'}{A'/t_0'}, \quad p = \frac{p'}{\rho A' L'/t_0'^2}, \tag{A.19}$$

which are dimensionless quantities. It is important to notice that we measure the *position* (including the altitude) with the characteristic length L', but the *displacements* of the fluid with the length A'. Finally we introduce the dimensionless numbers

$$F = \frac{g t_0'^2}{L'} \text{ (Froude number)}, \quad \varepsilon = \frac{A'}{L'} \quad \text{and} \quad \delta = \frac{h'}{L'}. \tag{A.20}$$

Putting these quantities into the equations defining the problem, we get the set of equations

$$\frac{\partial u}{\partial t} + \varepsilon\left(u\frac{\partial u}{\partial x} + w\frac{\partial u}{\partial z}\right) = -\frac{\partial p}{\partial x}, \tag{A.21}$$

$$\frac{\partial w}{\partial t} + \varepsilon\left(u\frac{\partial w}{\partial x} + w\frac{\partial w}{\partial z}\right) = -\frac{\partial p}{\partial z}, \tag{A.22}$$

$$\frac{\partial u}{\partial x} + \frac{\partial w}{\partial z} = 0, \tag{A.23}$$

$$w = 0 \qquad \text{for} \quad z = 0, \tag{A.24}$$

$$w = \frac{\partial \eta}{\partial t} + \varepsilon u \frac{\partial \eta}{\partial x} \qquad \text{for} \quad z = \delta + \varepsilon\eta, \tag{A.25}$$

$$F\eta = p \qquad \text{for} \quad z = \delta + \varepsilon\eta. \tag{A.26}$$

Moreover we *assume an irrotational flow* which has two consequences:

(i) The curl of the velocity ($\vec{\nabla} \wedge \vec{v}$) vanishes so that

$$\frac{\partial u}{\partial z} - \frac{\partial w}{\partial x} = 0. \tag{A.27}$$

(ii) We can define a velocity potential $\phi(x, z, t)$ such that $u = \partial\phi/\partial x$ and $w = \partial\phi/\partial z$.

Putting the expression of p from Equation (A.26) into Equation (A.21), and then replacing $\partial u/\partial z$ by $\partial w/\partial x$ from Equation (A.27), we get a new expression for the surface boundary condition, which will be useful later

$$\frac{\partial u}{\partial t} + \varepsilon\left(u\frac{\partial u}{\partial x} + w\frac{\partial w}{\partial x}\right) + F\frac{\partial \eta}{\partial x} = 0. \tag{A.28}$$

A.2.4 *Scaling hypothesis*

The limit $\varepsilon \to 0$ corresponds to the linear limit. We want to go beyond, however we shall stay in the case of weak nonlinearities by assuming $\varepsilon \ll 1$ but not zero.

We shall select the case of *shallow-water waves* by imposing $\delta = h'/L' \ll 1$ which means that the depth is small with respect to the spatial extent of the wave.

The parameters ε and δ are two *independent* small parameters: δ is determined by the characteristics of the system (such as the depth of the water layer) while ε is fixed by the boundary conditions.

However, in order to make expansions in which the orders of magnitude are properly taken into account, we must *select the relative scale of these small parameters.* This is often a delicate step in the selection of approximations. We shall assume that

$$\varepsilon \sim \delta^2, \tag{A.29}$$

which means that ε and δ^2 have similar orders of magnitude, without being equal since they are, a priori, independent from each other. *The calculations will be carried out up to the order $\varepsilon\delta$* or δ^3.

It is convenient to measure the displacement η of the surface in units of δ by defining $\varphi = \eta/\delta$. Thus the altitude of the surface is equal to $z = \delta(1 + \varepsilon\varphi)$ while Equation (A.28) becomes

$$\frac{\partial u}{\partial t} + \varepsilon\left(u\frac{\partial u}{\partial x} + w\frac{\partial w}{\partial x}\right) + F\delta\frac{\partial \varphi}{\partial x} = 0. \tag{A.30}$$

We assume that φ is of order 1, so that $\eta'/h' = \varepsilon\varphi$ is of order ε, which is consistent with our choice of a weak nonlinearity.

A.2.5 *The velocity potential*

Assuming an irrotational flow allows us to work with the velocity potential, which makes the calculation possible even in the nonlinear case. As a function of the velocity potential, the incompressibility condition (A.23) gives

$$\frac{\partial^2 \phi}{\partial x^2} + \frac{\partial^2 \phi}{\partial y^2} = 0. \tag{A.31}$$

The 'shallow-water' hypothesis $z \sim \delta$ suggests that we can look for a solution of the Laplace equation (A.31) as an expansion in the powers of z

$$\phi(x, z, t) = \sum_{n=0}^{\infty} z^n \phi_n(x, t). \tag{A.32}$$

Putting this expansion into the Laplace equation, we get

$$\sum_{n=0}^{\infty} z^n \left[\frac{\partial^2 \phi_n}{\partial x^2} + (n+2)(n+1)\phi_{n+2}\right] = 0. \tag{A.33}$$

As this equation has to be verified for any z inside the fluid, the bracket must vanish, which gives a recurrence relation between the ϕ_ns,

$$\phi_{n+2} = -\frac{1}{(n+2)(n+1)} \frac{\partial^2 \phi_n}{\partial x^2}. \tag{A.34}$$

The boundary condition in the bottom $w(z=0)=0$ gives $\partial\phi/\partial z = 0$ for $z=0$, therefore $\phi_1(x,t)=0$, which implies that all ϕ_{2p+1} vanish. The expression of $\phi(x,z,t)$ is thus reduced to

$$\phi(x,z,t) = \phi_0(x,t) - \frac{1}{2} z^2 \frac{\partial^2 \phi_0}{\partial x^2} + \frac{1}{24} z^4 \frac{\partial^4 \phi_0}{\partial x^4} + \cdots, \tag{A.35}$$

which determines u and w

$$u = \frac{\partial \phi}{\partial x} = \frac{\partial \phi_0}{\partial x} - \frac{1}{2} z^2 \frac{\partial^3 \phi_0}{\partial x^3} + \frac{1}{24} z^4 \frac{\partial^5 \phi_0}{\partial x^5} + \cdots, \tag{A.36}$$

$$w = \frac{\partial \phi}{\partial z} = -z \frac{\partial^2 \phi_0}{\partial x^2} + \frac{1}{6} z^3 \frac{\partial^4 \phi_0}{\partial x^4} + \cdots \tag{A.37}$$

We shall henceforth use the notation $\partial\phi_0/\partial x = f(x,t)$. Moreover, as $z \sim \delta$, and since the calculation is only carried up to order δ^3, we have

$$u(x,z,t) = f(x,t) - \frac{1}{2} z^2 f_{xx}, \tag{A.38}$$

$$w(x,z,t) = -z f_x + \frac{1}{6} z^3 f_{xxx}. \tag{A.39}$$

A.3 The linear limit

It is useful to examine the linear limit to select the characteristic velocity c_0' which was used in the definition of the dimensionless variables.

In this limit $\varepsilon = 0$ so that $z_{\text{surface}} = \delta + \varepsilon\eta \simeq \delta$. Therefore, on the surface $w \simeq -\delta f_x$ and $u \simeq f$ (neglecting terms of order δ^2). The kinematic condition at the surface is thus reduced to

$$w = \frac{\partial \eta}{\partial t} = \delta \frac{\partial \varphi}{\partial t}. \tag{A.40}$$

The boundary condition (A.28) simplifies to

$$\frac{\partial u}{\partial t} + F\delta \frac{\partial \varphi}{\partial x} = 0. \tag{A.41}$$

In terms of function f, these equations give $-\delta f_x = \delta \varphi_t$, so that $f_{xx} = -\varphi_{xt}$ and $f_t = -F\delta\varphi_x$, so that $f_{tt} = -F\delta\,\varphi_{xt} = F\delta f_{xx}$. In the linear limit, the equation verified by f at the surface is therefore

$$\frac{1}{F\delta} f_{tt} - f_{xx} = 0, \tag{A.42}$$

which has *nondispersive wave* solutions that propagate at speed $c = \sqrt{F\delta}$. This dimensionless speed, characteristic of small-amplitude waves, can be used to select the speed c_0' that we used to define the characteristic time t_0'. The dimensionless speed c is the ratio between a variation Δx of a coordinate and a time interval Δt. It is connected to the dimensional velocity of the linear waves by

$$c = \frac{\Delta x}{\Delta t} = \frac{\Delta x'/L'}{\Delta t'/t'_0} = \frac{\Delta x'/L'}{c'_0\,\Delta t'/L'} = \frac{c'_{\rm lin}}{c'_0}. \tag{A.43}$$

Taking into account the definitions of F and δ, and the relation $c'_0\, t'_0 = L'$, we get $c'^{\,2}_{\rm lin} = c'^{\,2}_0 c^2 = F\delta\, c'^{\,2}_0 = g\, t'^{\,2}_0\, h'/L'^{\,2} c'^{\,2}_0 = gh'$

We henceforth select $c'_0 = c'_{\rm lin} = \sqrt{gh'}$. This particular choice of scales is equivalent to setting $F\delta = 1$, which simplifies the expression of the boundary condition (A.30).

A.4 The nonlinear equation in shallow water

In the nonlinear case, the procedure is similar: first we express u and w at the surface as a function of f, and then we put them into the kinematic and physical boundary conditions at the surface (the boundary condition at the bottom was used when we expressed u and w as a function of the velocity potential). We keep terms up to order $\varepsilon\delta$ or δ^3.

With $z_{\rm surface} = \delta(1+\varepsilon\varphi)$, from Equations (A.38) and (A.39) we get

$$u_{\rm surface} = f - \frac{1}{2}\,\delta^2\, f_{xx} \tag{A.44}$$

$$w_{\rm surface} = -\delta\,(1+\varepsilon\varphi)\, f_x + \frac{1}{6}\,\delta^3\, f_{xxx}. \tag{A.45}$$

With $\eta = \delta\varphi$, the kinematic boundary condition (A.25) gives

$$w_{\rm surface} = \delta\,\frac{\partial\varphi}{\partial t} + \varepsilon\delta\; u_{\rm surface}\,\frac{\partial\varphi}{\partial x}. \tag{A.46}$$

We put $u_{\rm surface}$ from (A.44) into the above equation. As it is multiplied by $\varepsilon\delta$, it is sufficient to use $u_{\rm surface} = f$ to have all the contributions up to orders $\varepsilon\delta$ or δ^3. We get

$$w_{\rm surface} = \delta\varphi_t + \varepsilon\delta f\,\varphi_x. \tag{A.47}$$

The equality between the expressions (A.45) and (A.47) of $w_{\rm surface}$ gives a relation between f and φ,

$$\varphi_t + f_x + \varepsilon\varphi f_x + \varepsilon f\,\varphi_x - \frac{1}{6}\,\delta^2\, f_{xxx} = 0. \tag{A.48}$$

The next step is to use the physical boundary condition (A.30) (with $F\delta = 1$). This expression can be simplified because $w \simeq \delta$, so that $\varepsilon w(\partial w/\partial x) \simeq \varepsilon\delta^2$ can be neglected because its order is higher than the orders that we decided to keep. In the term $\varepsilon u(\partial u/\partial x)$, we only use $u = f$ to stay within our order of approximation. Thus we get

$$f_t - \frac{1}{2}\,\delta^2 f_{xxt} + \varepsilon f f_x + \varphi_x = 0. \tag{A.49}$$

Equations (A.44), (A.45), (A.48) and (A.49) make the system of nonlinear equations which describe shallow-water waves.

Let us first solve the system of equations (A.48) and (A.49) with respect to $f(x,t)$ and $\varphi(x,t)$ by a perturbative expansion:

(i) At order 0 in ε and δ, the system (A.48) and (A.49) is reduced to

$$\varphi_t + f_x = 0 \quad \text{and} \quad f_t + \varphi_x = 0 \tag{A.50}$$

which has the solution $f = \varphi$, provided that we impose the condition $f_t + f_x = 0$.

(ii) At order ε, δ^2, we look for a solution of the form

$$f = \varphi + \varepsilon\alpha + \delta^2\beta \tag{A.51}$$

where $\alpha(x,t)$ and $\beta(x,t)$ are functions yet to be determined, and which must be such that the condition on f obtained at order 0 is satisfied, i.e.

$$\alpha_t + \alpha_x = 0 + \mathcal{O}(\varepsilon, \delta^2), \quad \beta_t + \beta_x = 0 + \mathcal{O}(\varepsilon, \delta^2) \quad \text{and} \quad \varphi_t + \varphi_x = 0 + \mathcal{O}(\varepsilon, \delta^2) \tag{A.52}$$

where $\mathcal{O}(\varepsilon, \delta^2)$ denotes quantities of the order of ε or δ^2.

Putting the expression of f into (A.48) and keeping only terms up to order ε or δ^2, we get

$$\varphi_t + \varphi_x + \varepsilon\alpha_x + \delta^2\beta_x + 2\varepsilon\varphi\varphi_x - \frac{1}{6}\delta^2\varphi_{xxx} = 0. \tag{A.53}$$

Putting the expression of f into (A.49), we get

$$\varphi_t + \varepsilon\alpha_t + \delta^2\beta_t + \varphi_x + \varepsilon\varphi\varphi_x - \frac{1}{2}\delta^2\varphi_{xxt} = 0. \tag{A.54}$$

Subtracting (A.54) from (A.53), we obtain

$$\varepsilon(\alpha_x - \alpha_t) + \delta^2(\beta_x - \beta_t) + \varepsilon\varphi\varphi_x - \frac{1}{6}\delta^2\,\varphi_{xxx} + \frac{1}{2}\delta^2\,\varphi_{xxt} = 0. \tag{A.55}$$

Using Relations (A.52), we get

$$\varepsilon(2\alpha_x + \varphi\varphi_x) + \delta^2\left(2\beta_x - \frac{2}{3}\varphi_{xxx}\right) = 0. \tag{A.56}$$

But, although they have the same order of magnitude, ε and δ^2 are *independent of each other* because ε is determined by the boundary condition, while δ is determined by the properties of the system, and particularly the depth of the water. Therefore, Equation (A.56) can only be satisfied for any set (ε, δ^2) if we have

$$2\alpha_x + \varphi\varphi_x = 0 \quad \text{and} \quad 2\beta_x - \frac{2}{3}\varphi_{xxx} = 0. \tag{A.57}$$

A spatial integration gives

$$\alpha = -\frac{1}{4}\varphi^2 + C_1(t) \quad \text{and} \quad \beta = \frac{1}{3}\varphi_{xx} + C_2(t), \tag{A.58}$$

where $C_1(t)$ and $C_2(t)$ depend on time only. Thus we get an expression of function f which allows us to calculate the velocity potential as a function of φ which determines the equation of the surface,

$$f = \varphi - \frac{\varepsilon}{4}\varphi^2 + \frac{\delta^2}{3}\varphi_{xx} + \varepsilon\, C_1(t) + \delta^2 C_2(t). \tag{A.59}$$

Consequently this equation links the dynamics of the fluid to the boundary condition φ that we wish to determine. It is this aspect which explains the difficulty of this problem: *we are trying to solve the dynamic equation for a fluid in a domain which has one of its boundaries as an unknown in the problem. And this unknown itself is determined by the dynamics of the fluid.* It means that the problem has to be solved in a self-consistent way. But it is now possible to get an equation for φ by putting the expression of f into Equation (A.48). At the order of our calculation, it gives

$$\varphi_t + \varphi_x + \frac{3}{2}\varepsilon\varphi\varphi_x + \frac{1}{6}\delta^2\varphi_{xxx} = 0. \tag{A.60}$$

Thus we have got a nonlinear equation which describes the motion of the surface for shallow-water waves. It becomes simpler if we write it in a frame moving at velocity 1, which, in dimensional variables, corresponds to a frame moving at velocity $c_0' = \sqrt{gh'}$.

In this frame the variables are $\xi = x - t$ and $\tau = t$ so that $\partial/\partial x = \partial/\partial\xi$ and $\partial/\partial t = \partial/\partial\tau - \partial/\partial\xi$. Thus $\varphi_t + \varphi_x$ is reduced to φ_τ. The equation becomes

$$\boxed{\varphi_\tau + \frac{3}{2}\,\varepsilon\,\varphi\varphi_\xi + \frac{1}{6}\delta^2\varphi_{\xi\xi\xi} = 0} \tag{A.61}$$

which is the equation derived by Korteweg and de Vries in 1895.

The path that leads to it looks very long because we detailed all the steps. It is possible to find 'elegant' derivations to establish the KdV equation quickly, but the price to pay is that the approximations are not controlled. To reach this equation, we had to make approximations which look numerous, but all of them rely on the same hypothesis $\varepsilon \simeq \delta^2 \ll 1$. This hypothesis is generally well verified experimentally because the solitary waves which are observed have a spatial extent which is much larger than their amplitude (this is the condition $\varepsilon = A'/L' \ll 1$) and the experiments are indeed carried out in shallow water ($\delta^2 = (h'/L')^2 \ll 1$). This is why the KdV equation has turned out to be so fruitful in the analysis of shallow-water hydrodynamic waves, such as the solitons of the Andaman sea.

B

Mechanics of a continuous medium

Most of the systems which have soliton-like excitations are continuous media. The Lagrangian and Hamiltonian formalism of the mechanics of discrete systems [78] must be extended to treat such media. This appendix is a reminder of the main results of the Lagrangian and Hamiltonian mechanics of continuous media, restricted to one space dimension to simplify the notations, without reducing the generality because the extension to a multidimensional system is straightforward.

B.1 Lagrangian formalism

Let us consider a continuous medium, characterised by a field $\eta(x, t)$, which could, for instance, be the displacement field of a vibrating string. We shall introduce a quantity $\mathcal{L}$, called the Lagrangian density of the system. Usually $\mathcal{L}$ does not depend only on the field η and its time derivative η_t, as with the Lagrangian of a particle which depends on its position and velocity, but it is also a function of η_x, and may also depend on the independent variables x and t,

$$\mathcal{L} = \mathcal{L}\left(\eta, \frac{\mathrm{d}\eta}{\mathrm{d}x}, \frac{\mathrm{d}\eta}{\mathrm{d}t}, x, t\right). \tag{B.1}$$

The Lagrangian of the system is the integral of $\mathcal{L}$ over the domain $[x_1, x_2]$ on which the spatial variable x can vary

$$L = \int_{x_1}^{x_2} \mathrm{d}x\, \mathcal{L}, \tag{B.2}$$

while the action between times t_1 and t_2 is

$$S = \int_{t_1}^{t_2} \mathrm{d}t\, L = \int_{t_1}^{t_2} \mathrm{d}t \int_{x_1}^{x_2} \mathrm{d}x\, \mathcal{L}. \tag{B.3}$$

The equations of motion are obtained from Hamilton's principle which says that the dynamics of the system between t_1 and t_2 are such that the variation of the action is zero along the correct path of the motion. It means that the action keeps the same value, except for a first-order infinitely small quantity, for all the 'paths', i.e. the functions $\eta(x,t)$ which differ by an infinitesimal quantity from the path actually followed by the system. This is expressed by

$$\delta S = 0 = \int_{t_1}^{t_2} \mathrm{d}t \int_{x_1}^{x_2} \mathrm{d}x \left[\frac{\partial \mathcal{L}}{\partial \eta}\,\delta\eta + \frac{\partial \mathcal{L}}{\partial \eta_t}\,\delta\eta_t + \frac{\partial \mathcal{L}}{\partial \eta_x}\,\delta\eta_x \right]. \tag{B.4}$$

Two integration by parts of the right hand side lead to

$$\begin{aligned}\delta S = &\int_{t_1}^{t_2} \mathrm{d}t \int_{x_1}^{x_2} \mathrm{d}x \left[\frac{\partial \mathcal{L}}{\partial \eta}\,\delta\eta \right] + \int_{x_1}^{x_2} \mathrm{d}x \left(\left[\frac{\partial \mathcal{L}}{\partial \eta_t}\,\delta\eta \right]_{t_1}^{t_2} - \int_{t_1}^{t_2} \mathrm{d}t \frac{\mathrm{d}}{\mathrm{d}t}\left(\frac{\partial \mathcal{L}}{\partial \eta_t} \right) \delta\eta \right) \\ &+ \int_{t_1}^{t_2} \mathrm{d}t \left(\left[\frac{\partial \mathcal{L}}{\partial \eta_x}\,\delta\eta \right]_{x_1}^{x_2} - \int_{x_1}^{x_2} \mathrm{d}x \frac{\mathrm{d}}{\mathrm{d}x}\left(\frac{\partial \mathcal{L}}{\partial \eta_x} \right) \delta\eta \right).\end{aligned} \tag{B.5}$$

This condition of stationarity is written with fixed conditions at the ends, which means that $\delta\eta$ is zero at ends 1 and 2. Thus the expression of δS is reduced to

$$\delta S = 0 = \int_{t_1}^{t_2} \mathrm{d}t \int_{x_1}^{x_2} \mathrm{d}x \left[\frac{\partial \mathcal{L}}{\partial \eta} - \frac{\mathrm{d}}{\mathrm{d}t}\left(\frac{\partial \mathcal{L}}{\partial \eta_t} \right) - \frac{\mathrm{d}}{\mathrm{d}x}\left(\frac{\partial \mathcal{L}}{\partial \eta_x} \right) \right] \delta\eta. \tag{B.6}$$

As the stationarity condition must be verified for *all possible variations* $\delta\eta$, Equation (B.6) can only be satisfied if the bracket vanishes, which leads to the Lagrange equation

$$\boxed{\frac{\mathrm{d}}{\mathrm{d}t}\left(\frac{\partial \mathcal{L}}{\partial \eta_t} \right) + \frac{\mathrm{d}}{\mathrm{d}x}\left(\frac{\partial \mathcal{L}}{\partial \eta_x} \right) = \frac{\partial \mathcal{L}}{\partial \eta}}\,. \tag{B.7}$$

While a system with N degrees of freedom would have led to a system of N Lagrange equations, for the continuous system, which has an infinite number of degrees of freedom, we have a single equation. But this is not a paradox because the Lagrange equation (B.7) is a partial differential equation for the *two* variables x and t while the N Lagrange equations for a discrete system are differential equations of the *single* time variable t.

This formalism can easily be generalised to more than one spatial dimension. It is interesting to notice that the time variable t and the space variables x, y and z play the same role in both Hamilton's principle and the Lagrange equations.

If the Lagrangian density depends on higher derivatives of η, the Lagrange equation (B.7) must be completed. This is, for instance, the case for the Lagrangian (1.24) of the KdV equation. Performing a double integration by parts shows, along the same lines as before, that the Lagrangian density $\mathcal{L}(\eta, \eta_x, \eta_t, \eta_{xx}, x, t)$ leads to the Lagrange equation.

$$\frac{\mathrm{d}}{\mathrm{d}t}\left(\frac{\partial \mathcal{L}}{\partial \eta_t} \right) + \frac{\mathrm{d}}{\mathrm{d}x}\left(\frac{\partial \mathcal{L}}{\partial \eta_x} \right) - \frac{\mathrm{d}^2}{\mathrm{d}x^2}\left(\frac{\partial \mathcal{L}}{\partial \eta_{xx}} \right) = \frac{\partial \mathcal{L}}{\partial \eta}. \tag{B.8}$$

Notice the minus sign in front of the term $\mathrm{d}^2/\mathrm{d}x^2$. It is easy to remember it by thinking that each integration by parts introduces a sign change.

B.2 Hamiltonian formalism

Once we have a Lagrangian, we can define a Hamiltonian H, or a Hamiltonian density $\mathcal{H}$. We must first define a canonical momentum density π, which is conjugate from the field η by

$$\pi = \frac{\partial \mathcal{L}}{\partial \eta_t}. \tag{B.9}$$

The Hamiltonian density is deduced from the canonical transform

$$\boxed{\mathcal{H}(\eta, \pi, x, t) = \pi \frac{\mathrm{d}\eta}{\mathrm{d}t} - \mathcal{L}}\,. \tag{B.10}$$

Here, contrary to the Lagrangian formalism, the time variable plays a special role and the symmetry between the time and space variables in the equations is lost.

The Hamiltonian is then obtained by a spatial integration

$$H = \int_{x_1}^{x_2} \mathrm{d}x\, \mathcal{H} = \int_{x_1}^{x_2} \mathrm{d}x \left(\pi \frac{\mathrm{d}\eta}{\mathrm{d}t} - \mathcal{L} \right), \tag{B.11}$$

and it can be used to derive the equations of motion, called Hamilton's equations.

From the definition of the Hamiltonian density (B.10), we immediately get the first equation

$$\boxed{\frac{\partial \mathcal{H}}{\partial \pi} = \frac{\mathrm{d}\eta}{\mathrm{d}t}}\,. \tag{B.12}$$

The second equation requires more work. Similarly to the derivation of the equation of motion of a single particle, we must look separately for the position and the velocity, which are independent variables in the variational method. For the continuous medium, the variables η and π are two quantities which must be treated as independent of each other, which gives

$$\frac{\partial \mathcal{H}}{\partial \eta} = 0 + 0 - \frac{\partial \mathcal{L}}{\partial \eta} \tag{B.13}$$

$$= -\frac{\mathrm{d}}{\mathrm{d}t} \underbrace{\left(\frac{\partial \mathcal{L}}{\partial \eta_t} \right)}_{=\pi} - \frac{\mathrm{d}}{\mathrm{d}x} \left(\frac{\partial \mathcal{L}}{\partial \eta_x} \right), \tag{B.14}$$

if we use the Lagrange equation (B.7). Moreover we have

$$\frac{\partial \mathcal{H}}{\partial \eta_x} = -\frac{\partial \mathcal{L}}{\partial \eta_x}. \tag{B.15}$$

Combining Equations (B.14) and (B.15), we finally get

$$\boxed{\frac{\partial \mathcal{H}}{\partial \eta} - \frac{\mathrm{d}}{\mathrm{d}x}\left(\frac{\partial \mathcal{H}}{\partial \eta_x}\right) = -\frac{\mathrm{d}\pi}{\mathrm{d}t}} \tag{B.16}$$

which is the second Hamilton equation.

This standard derivation uses the essential hypothesis that the variables η and its conjugate π are only related by a derivative operator and must be treated as independent variables in the variational approach. The phase space (η, π) is thus correctly defined because its different 'components' are independent of each other. If this is not true (see Section 3.3.2 for an example), the first terms of Equation (B.13) do not necessarily vanish. The Hamilton equations must be derived by taking this constraint into account.

C

Coherent states of a harmonic oscillator

As a function of the annihilation a and creation $a^\dagger$ operators, the Hamiltonian of a harmonic oscillator is

$$H = a^\dagger a + \frac{1}{2}. \tag{C.1}$$

Its eigenstates $|n\rangle$ are associated with the eigenvalues $E_n = n + 1/2$.

The coherent states of this oscillator are the eigenstates of the annihilation operator a. Thus they are defined by

$$a\,|\alpha\rangle = \alpha\,|\alpha\rangle, \tag{C.2}$$

where α is a complex number because the operator a is not Hermitian. They can be derived by looking for their expansion on the basis of states $|n\rangle$, but it is also possible to derive their expression in a compact form [77], which we use in Chapter 15.

Let us assume that there exits a unitary operator $D(\beta)$, which depends on a complex parameter β, and which is such that

$$D^\dagger(\beta)\,a\,D(\beta) = a + \beta. \tag{C.3}$$

Property

Let us prove the property

$$a\,D^{-1}(\beta)\,|\alpha\rangle = (\alpha - \beta)\,D^{-1}(\beta)\,|\alpha\rangle \tag{C.4}$$

which means that the operator $D^{-1}(\beta)$ transforms the coherent state $|\alpha\rangle$ into another coherent state, with a shifted eigenvalue.

Proof

As we assume that D is unitary, we can use the equality $D^{-1} = D^\dagger$. Applying D to Equation (C.3), we get

$$D(\beta)\,D^\dagger(\beta)\,a\,D(\beta) = a\,D(\beta) = D(\beta)\,a + D(\beta)\beta \tag{C.5}$$

which can also be written as

$$D(\beta)\, a = a\, D(\beta) - \beta D(\beta). \tag{C.6}$$

If we apply this operator to $D^{-1}(\beta)\,|\alpha\rangle$, we get

$$\begin{aligned} D(\beta)\, a\, D^{-1}(\beta)\,|\alpha\rangle &= a\, D(\beta)\, D^{-1}(\beta)\,|\alpha\rangle - \beta D(\beta)\, D^{-1}(\beta)\,|\alpha\rangle & \text{(C.7)} \\ &= a\,|\alpha\rangle - \beta|\alpha\rangle & \text{(C.8)} \\ &= (\alpha - \beta)\,|\alpha\rangle & \text{(C.9)} \end{aligned}$$

if we use Equation (C.2). Finally, if we apply the operator $D^{-1}(\beta)$ to this expression, we obtain

$$D^{-1}(\beta)\, D(\beta)\, a\, D^{-1}(\beta)\,|\alpha\rangle = D^{-1}(\beta)(\alpha - \beta)|\alpha\rangle \tag{C.10}$$

which proves the property (C.4).

Consequence

In the particular case $\alpha = \beta$, we get $a D^{-1}(\alpha)\,|\alpha\rangle = |0\rangle$. Since the eigenstate $|0\rangle$ of a harmonic oscillator is not degenerate, it implies that $D^{-1}(\alpha)\,|\alpha\rangle = C|0\rangle$ where C is a constant to be determined. Thus we get $|\alpha\rangle = C\, D(\alpha)\,|0\rangle$. As the states $|\alpha\rangle$ and $|0\rangle$ are normalised and D is unitary, it implies $C = 1$. Thus we have

$$|\alpha\rangle = D(\alpha)|0\rangle. \tag{C.11}$$

This expression is useful because it gives us a way to generate $|\alpha\rangle$ in a systematic way if we know the operator $D(\alpha)$.

In order to determine this operator D, which is defined by Equation (C.3), let us examine the particular case of an operator D for an infinitesimal shift $\mathrm{d}\alpha$, and carry out the calculation to first order. We can check that

$$D(\mathrm{d}\alpha) = 1 + a^\dagger\, \mathrm{d}\alpha - a\, \mathrm{d}\alpha^*, \tag{C.12}$$

because this relation leads to

$$a\, D(\mathrm{d}\alpha) = a + a\, a^\dagger\, \mathrm{d}\alpha - a^2\, \mathrm{d}\alpha^*. \tag{C.13}$$

Taking into account the relation $D^\dagger = 1 + a\, \mathrm{d}\alpha^* - a^\dagger\, \mathrm{d}\alpha$ which derives from (C.12), we have

$$\begin{aligned} D^\dagger(\mathrm{d}\alpha)\, a\, D(\mathrm{d}\alpha) &= (1 + a\, \mathrm{d}\alpha^* - a^\dagger\, \mathrm{d}\alpha)(a + a\, a^\dagger\, \mathrm{d}\alpha - a^2\, \mathrm{d}\alpha^*) & \text{(C.14)} \\ &= a + a\, a^\dagger\, \mathrm{d}\alpha - a^2\, \mathrm{d}\alpha^* + a^2\, \mathrm{d}\alpha^* - a^\dagger\, a\, \mathrm{d}\alpha + \mathcal{O}\,(\mathrm{d}\alpha^2) & \text{(C.15)} \\ &= a + [a, a^\dagger]\, \mathrm{d}\alpha + \mathcal{O}\,(\mathrm{d}\alpha^2) & \text{(C.16)} \\ &= a + \mathrm{d}\alpha, & \text{(C.17)} \end{aligned}$$

because $[a, a^\dagger] = 1$. This relation confirms that Expression (C.12) is indeed the operator D defined by Equation (C.3) which corresponds to the infinitesimal shift $\mathrm{d}\alpha$.

This result can be used to derive a differential equation for the operator $D(\alpha)$ by considering a variation of α which results from a product by the real number λ,

$$D((\lambda + \mathrm{d}\lambda)\alpha) = D(\alpha \mathrm{d}\lambda)\, D(\alpha\lambda) \tag{C.18}$$

$$= (1 + a^{\dagger}\mathrm{d}\lambda\alpha - a\mathrm{d}\lambda\alpha^{*})\, D(\alpha\lambda), \tag{C.19}$$

which leads to

$$\frac{D((\lambda + \mathrm{d}\lambda)\alpha) - D(\alpha\lambda)}{\mathrm{d}\lambda} = \alpha\, a^{\dagger} - \alpha^{*}\, a. \tag{C.20}$$

If we integrate this expression with respect to λ, and evaluate the solution for $\lambda = 1$, we get the expression for D,

$$D(\alpha) = \exp(\alpha\, a^{\dagger} - \alpha^{*}\, a), \tag{C.21}$$

that we are looking for. Using Equation (C.11), we get the coherent state as

$$|\alpha\rangle = \mathrm{e}^{(\alpha\, a^{\dagger} - \alpha^{*}\, a)}\, |0\rangle. \tag{C.22}$$

Moreover the expressions of the position X and momentum P operators as functions of a and $a^{\dagger}$, immediately give the two equalities

$$\langle\alpha|X|\alpha\rangle = \sqrt{\frac{2\hbar}{m\omega}}\, \mathrm{Re}\,(\alpha) \tag{C.23}$$

$$\langle\alpha|P_x|\alpha\rangle = \sqrt{2m\omega\hbar}\, \mathrm{Im}\,(\alpha), \tag{C.24}$$

for a coherent state. They are used in Chapter 15.

References

We have separated the books specifically devoted to solitons from other books and research papers. Those are listed in the second part, in alphabetical order.

Books on solitons

[1] Ablowitz, M. J., Segur, H., *Solitons and the Inverse Scattering Transform*, SIAM Studies in Applied Mathematics, **4**, Society for Industrial and Applied Mathematics (1981).
[2] Braun, O. M., Kivshar, Y. S., *The Frenkel–Kontorova Model. Concepts, Methods and Applications*, Springer (2004).
[3] Dodd, R. K., Eilbeck, J. C., Gibbon, J. D., Morris, H. C., *Solitons and Nonlinear Wave Equations*, Academic Press (1982).
[4] Drazin, P. G., Johnson, R. S., *Solitons: an Introduction*, Cambridge University Press (1993).
[5] Eilenberger, G., *Solitons: Mathematical Methods for Physicists*, Springer (1981).
[6] Infeld, E., Rowlands, G., *Nonlinear Waves, Solitons and Chaos*, Cambridge University Press (1990).
[7] Lamb, G. L., *Elements of Soliton Theory*, John Wiley (1980).
[8] Newell, A. C., *Solitons in Mathematics and Physics*, SIAM (1985).
[9] Remoissenet, M., *Waves Called Solitons*, Springer (1996).
[10] Scott, A. C., *Nonlinear Science*, Oxford University Press (1999).
[11] Toda, M., *Theory of Nonlinear Lattices*, Springer (1978).

Books and research papers

[12] Ablowitz, M. J., Kaup, D. J., Newell, A. C., Segur, H., The inverse scattering transform – Fourier analysis for nonlinear problems, *Studies in Applied Mathematics*, **53** (1974), 249–315.
[13] Abramowitz, M., Stegun, I. A., *Handbook of Mathematical Functions*, Dover (1965).
[14] Agrawal, G. P., *Nonlinear Fiber Optics*, Academic Press (2001).
[15] Allen, M. P., Tildesley, D. J., *Computer Simulations of Liquids*, Clarendon (1987).
[16] Anderson, M. H., Ensher, J. R., Matthews, M. R., Wieman, C. E., Cornell, E. A., Observation of Bose–Einstein condensation in a dilute atomic vapor, *Science*, **269** (1995), 198–201.

[17] Asano, T., Nojiri, H., Inagaki, Y., Boucher, J. P., Sakon, T., Ajiro, Y., Motokawa, M., ESR investigation on the breather mode and the spinon-breather dynamical crossover in Cu benzoate, *Physical Review Letters*, **84** (2000), 5880–3.
[18] Ashcroft, N. W., Mermin, N. D., *Solid State Physics*, Holt-Saunders International Editions (1976).
[19] Aubry, S., A unified approach to the interpretation of displacive and order-disorder systems: I. Thermodynamical aspect, *Journal of Chemical Physics*, **62** (1975), 3217–29.
[20] A unified approach to the interpretation of displacive and order-disorder systems: II. Displacive systems, *Journal of Chemical Physics*, **62** (1976), 3392–402.
[21] The new concept of transition by breaking of analyticity in a crystallographic mode, in *Solitons and Condensed Matter*, A. R. Bishop, T. Schneider (Eds.), Springer (1978), 264–77.
[22] Structure incommensurables et brisure de la symétrie de translation, in *Structures et Instabilités*, C. Godrèche (Ed.), EDP Sciences (1986).
[23] Barone, A., Paterno, G., *Physics and Applications of the Josephson Effect*, Wiley (1982).
[24] Bazin, H., Recherches expérimentales relatives aux remous et à la propagation des ondes, in *Recherches Hydrauliques*, H. Darcy, H. Bazin (Eds.), Imprimerie Impériale (1865).
[25] Benjamin, T. B., Bona, J. L., Mahony, J. J., Model equations for long waves in nonlinear dispersive systems, *Philosophical Transactions of the Royal Society of London*, **272** (1972), 47–8.
[26] Benjamin, T. B., Feir, J. F., The disintegration of wave trains on deep water, *Journal of Fluid Mechanics*, **27** (1967), 417–30.
[27] Blackburn, J. A., Smith, H. J. T., Experimental study of an inverted pendulum, *American Journal of Physics*, **60** (1992), 909–11.
[28] Blackburn, J. A., Smith, H. J. T., Gronbech-Jensen, N., Stability and Hopf bifurcations in an inverted pendulum, *American Journal of Physics*, **60** (1992), 903–8.
[29] Boesch, R., Stancioff, P., Willis, C. R., Hamiltonian equations for multiple-collective-variable theories of nonlinear Klein–Gordon equations: a projector operator approach, *Physical Review B*, **38** (1988), 6713–35.
[30] Bolton, P. H., James, T. L., Fast and slow conformational fluctuations of RNA and DNA – sub-nanosecond internal motion correlation times determined by P-31-NMR, *Journal of the American Chemical Society*, **102** (1980), 25–31.
[31] Bose, S. N., Plancks Gesetz und Lichtquantenhypothese, *Zeitschrift für Physik*, **26** (1924), 178–81.
[32] Boucher, J. P., Nonlinear soliton excitations in antiferromagnetic chains, *Hyperfine Interactions*, **49** (1989), 423–38.
[33] Boussinesq, J., Théorie des ondes et des remous qui se propagent le long d'un canal rectangulaire horizontal, en communiquant au liquide contenu dans ce canal des vitesses sensiblement pareilles de la surface au fond, *Journal de Mathématiques Pures et Appliquées*, **17** (1872), 55–108.
[34] Buell, W. F., Shadwick, B. A., Potentials and bound states, *American Journal of Physics*, **63** (1995), 256–8.
[35] Burger, S., Bongs, K., Dettmer, S., Ertmer, W., Sengstock, K., Sanpera, A., Shlyapnikov, G. V., Lewenstein, M., Dark Solitons in Bose–Einstein condensates, *Physical Review Letters*, **83** (1999), 5198–201.

[36] Campbell, D. K., Peyrard, M., Sodano, P., Kink-antikink interactions in the double sine-Gordon equation, *Physica D*, **19** (1986), 165–205.
[37] Campbell, D. K., Schonfeld, J. F., Wingate, C. A., Resonance structure in kink-antikink interactions in ϕ^4 theory, *Physica D*, **9** (1983), 1–32.
[38] Careri, G., Buontempo, U., Galluzi, F., Scott, A. C., Gratton, E., Shyamsunder, E., Spectroscopic evidence for Davydov-like solitons in acetanilide, *Physical Review B*, **30** (1984), 4689.
[39] Carr, L. D., Castin, Y., Dynamics of a matter-wave bright soliton in an expulsive potential, *Physical Review A*, **66** (2002), 063602.
[40] Castin, Y., Bose–Einstein condensates in atomic gases: simple theoretical results, in *Coherent atomic matter waves*, R. Kaiser, C. Westbrook, F. David (Eds.), EDP Sciences and Springer Verlag (2001), 1–136.
[41] Cohen-Tannoudji, C., Diu, B., Laloë, F., *Mécanique Quantique*, Hermann, Tome 1, Chap. 4, complément G (1977); translated in *Quantum Mechanics*, Wiley (1977).
[42] Condat, C. A., Guyer, R. A., Miller, M. D., Double sine-Gordon chain, *Physical Review B*, **27** (1983), 474.
[43] Courant, R., Hilbert, D., *Methods of Mathematical Physics*, Wiley (1962).
[44] Dafolvo, F., Giorgini, S., Pitaevskii, L. P., Stringari, S., Theory of Bose–Einstein condensation in trapped gases, *Review of Modern Physics*, **71** (1999), 463–512.
[45] Dalibard, J., Coherence and superfluidity of gaseous Bose–Einstein condensates, in *Dynamics and Thermodynamics of Systems with Long-range Interactions*, T. Dauxois, S. Ruffo, E. Arimondo, M. Wilkens (Eds.), Springer (2002), 293–311.
[46] Dashen, R. F., Hasslacher, B., Neveu, A., Non-perturbative methods and extended-hadron models in field theory: II. Two dimensional models and extended hadrons, *Physical Review D*, **10** (1974), 4130–8.
[47] Dauxois, T., Peyrard, M., Bishop, A. R., Entropy driven DNA denaturation, *Physical Review E*, **47** (1993), R44-7 (Rapid Communication).
[48] Dauxois, T., Theodorakopoulos, N., Peyrard, M., Thermodynamic instabilities in one dimension: correlations, scaling and solitons, *Journal of Statistical Physics*, **107** (2002), 869–91.
[49] Davis, K. B., Mewes, M. O., Andrews, M. R., van Druten, N. J., Durfee, D. S., Kurn, D. M., Ketterle, W., Bose–Einstein condensation in a gas of sodium atoms, *Physical Review Letters*, **75** (1995), 3969–73.
[50] Davydov, A. S., The theory of contraction of proteins under their excitation, *Journal of Theoretical Biology*, **38** (1973), 559–69.
[51] de Bouard, A., Ghidaglia, J.-M., Saut, J. C., Histoires d'eau – histoires d'ondes, *Images des Mathématiques*, **95** (1995), 23–30.
[52] de Vries, G., *Bidrage tot de Kebbis der lange golven*, Doctoral Thesis, University of Amsterdam (1894).
[53] Denschlag, J., Simasarian, J. E., Feder, D. L., Clark, C. W., Collins, L. A., Cubizolles, J., Deng, L., Hagley, E. W., Helmerson, K., Reinhardt, W. P., Rolston, S. L., Schneider, B. I., Philipps, W. D., Generating solitons by phase engineering of a Bose–Einstein condensate, *Science*, **287** (2000), 97–101.
[54] Dias, F., Quand les vagues deviennent dévastatrices, *La Recherche*, **345** (2001), 50–1.
[55] Dirac, P. A. M., *The Principles of Quantum Mechanics*, Oxford University Press (1981).
[56] Donley, E. A., Claussen, N. R., Cornish, S. L., Roberts, J. L., Cornell, E. A. Wieman, C. E., Dynamics of collapsing and exploding Bose–Einstein condensate, *Nature*, **412** (2001), 295–9.

[57] Edler, J., Hamm, P., Self-trapping of the amide I band in a peptide model crystal, *Journal of Chemical Physics*, **117** (2002), 2415–24.
[58] Edler, J., Hamm, P., Scott, A. C., Femtosecond study of self-trapped vibrational excitons in crystalline acetanilide, *Physical Review Letters*, **88** (2002), 067403.
[59] Eilbeck, J. C., Lombdahl, P. S., Scott, A. C., Soliton structure in crystalline acetanilide, *Physical Review B*, **30** (1984), 4703–12.
[60] Einstein, A., Quantentheorie des einatomigen idealen gases, *Sitzungsberischte der Preussische Akademie der Wissenschaften* (1924), 261–7; Quantentheorie des einatomigen idealen gases, ibid (1925), 3–14.
[61] Elkin, L. O., Rosalind Franklin and the double helix, *Physics Today*, **56** (2003), 42–8.
[62] Emmerson, G. S., *John Scott Russell, A Great Victorian Engineer and Naval Architect*, John Murray (1977).
[63] Falcon, E., Fauve, S., Laroche, C., Observation of depression solitary surface waves on a thin fluid layer, *Physical Review Letters*, **89** (2002), 204501.
[64] Fei, Z., Kivshar, Y., Vasquez, L., Resonant kink–impurity interactions in the sine-Gordon model, *Physical Review A*, **45** (1992), 6019–30.
[65] Fei, Z., Konotop, V., Peyrard, M., Vasquez, L., Kinks in the periodically modulated ϕ^4 model, *Physical Review E*, **48** (1993), 548–54.
[66] Fermi, E., Pasta, J., Ulam, S., Studies of nonlinear problems. I., *Los Alamos report* LA-1940 (1955), published later in *Collected Papers of Enrico Fermi*, E. Segré (Ed.), University of Chicago Press (1965); also in *Nonlinear Wave Motion*, A. C. Newell (Ed.), Lecture in Applied Mathematics **15** AMS, (1974); also in *The Many-Body Problem*, C. C. Mattis (Ed.), World Scientific (1993).
[67] Feynman, R. P., *The Feynman Lectures on Physics, Quantum Mechanics*, Addison-Wesley (1965).
[68] Fogel, M. B., Trullinger, S. E., Bishop, A. R., Krumhansl, J. A., Dynamics of sine-Gordon solitons in the presence of perturbations, *Physical Review B*, **15** (1977), 1578–92.
[69] Ford, J., Equipartition of energy for nonlinear systems, *Journal of Mathematical Physics*, **2** (1961), 387–93.
[70] Ford, J., Waters, J., Computer studies of energy sharing and ergodicity for nonlinear oscillator systems, *Journal of Mathematical Physics*, **4** (1963), 1293–306.
[71] Frank-Kamennetskii, M., Physicists retreat again, *Nature*, **328** (1987), 108.
[72] Frenkel, Y., Kontorova, T., On the theory of plastic deformation and twinning, *Journal of Physics*, **1** (1939), 137–49.
[73] Frenkel, V. Y., *Yakov Ilich Frenkel: His Work, Life and Letters*, Birkhäuser Verlag (1996).
[74] Gardner, C. S., Greene, J. M., Kruskal, M. D., Miura, R. M., Korteweg–deVries equations and generalizations: methods for exact solutions, *Communications in Pure and Applied Mathematics*, **XXVII** (1974), 97–133.
[75] Gillespie, D. T., Fluctuation and dissipation in Brownian motion, *American Journal of Physics*, **61** (1993), 1078.
[76] Ginzburg, V. L., Pitaevskii, L. P., On the theory of superfluidity, *Soviet Physics Journal of Experimental and Theoretical Physics*, **7** (1958), 858–61.
[77] Glauber, R. J., Coherent and incoherent states of the radiation field, *Physical Review*, **131** (1963), 2766–88.
[78] Goldstein, H., *Classical Mechanics*, Addison-Wesley (1980).
[79] Gross, E. P., Structure of a quantized vortex in boson systems, *Nuovo Cimento*, **20** (1961), 454–77.

[80] Gursey, F., Classical statistical mechanics of a rectilinear assembly, *Proceedings of the Cambridge Philosophical Society*, **46** (1950), 182–94.
[81] Guyer, R. A., Miller, M. D., The sine-Gordon chain: equilibrium statistical mechanics, *Physical Review A*, **17** (1978), 1205–17.
[82] Hammack, J., McCallister, D., Scheffner, N., Segur, H., Two-dimensional periodic waves in shallow water. Part 2. Asymmetric waves, *Journal of Fluid Mechanics*, **285** (1995), 95–122.
[83] Hammack, J., Scheffner, N., Segur, H., Two-dimensional periodic waves in shallow water, *Journal of Fluid Mechanics*, **209** (1989), 567–89.
[84] Hasegawa, A., Kodama, Y., Signal transmission by optical solitons in monomode fiber, *Proceedings IEEE*, **69** (1981), 1145–50.
[85] Hasegawa, A., Tappert, F., Transmission of stationary nonlinear optical pulses in dispersive dielectric fiber: II. Normal dispersion, *Applied Physics Letters*, **23** (1973), 171–2.
[86] Heeger, A. J., Kivelson, S., Schrieffer, J. R., Su, W. P., Solitons in conducting polymers, *Review of Modern Physics*, **60** (1988), 781–850.
[87] Heinrich, P., Roche, R., Mangeney, A., Boudon, G., Modéliser un raz de marée créé par un volcan, *La Recherche*, **318** (1999), 66–71.
[88] Hirooka, H., Saito, N., Computer studies on the approach to thermal equilibrium in coupled anharmonic oscillators: I. Two dimensional case, *Journal of the Physical Society of Japan*, **26** (1969), Supplement, 624–30.
[89] Hogan, M. E., Jardetzky, O., Internal motions in deoxyribonucleic-Acid II, *Biochemistry*, **19** (1980), 3460–8.
[90] Huang, K., *Statistical Mechanics*, Wiley (1987).
[91] Ikezi, H., Experiments on ion-acoustic solitary waves, *Physics of Fluids*, **16** (1973), 1668–75.
[92] Inman, R. B., Baldwin, R. L., Helix-random coil transitions in DNA homopolymer pairs, *Journal of Molecular Biology*, **8** (1964), 452–69.
[93] Inouye, S., Andrews, M. R., Stenger, J., Miesner, H.-J., Stamper-Kurn, D. M., Ketterle, W., Observation of Feshbach resonances in a Bose–Einstein condensate, *Nature*, **392** (1998), 151–4.
[94] Jackson, J. D., *Classical Electrodynamics*, John Wiley & Sons, Chap. 10 (1975).
[95] Josephson, B. D., Possible new effect in superconductive tunnelling, *Physics Letters*, **1** (1962), 251–3.
[96] Kadomtsev, B. B., Petviashvili, V. I., On the stability of solitary waves in weakly dispersing media, *Soviet Physics Doklady*, **15** (1970), 539–41.
[97] Kerr, W., Lombdahl, P., Quantum mechanical derivation of the equations of motion for Davydov solitons, *Physical Review B*, **35** (1987), 3629–32.
[98] Khaykovich, L., Schreck, F., Ferrari, G., Bourdel, T., Cubizolles, J., Carr, L., Castin, Y., Salomon, C., Formation of a matter-wave bright soliton, *Science*, **296** (2002), 1290–3.
[99] Kittel, C., *Introduction to Solid State Physics*, Wiley Chap. 20 (1995).
[100] *Quantum Theory of Solids*, Wiley (1963).
[101] Kivshar, Y. S., Dark solitons in nonlinear optics, *IEEE Journal of Quantum Electronics*, **28** (1993), 250–65.
[102] Kivshar, Y. S., Malomed, B., Dynamics of solitons in nearly integrable systems, *Review of Modern Physics*, **61** (1989), 763–915.
[103] Kodama, Y., Hasegawa, A., Nonlinear pulse propagation in a monomode dielectric guide, *IEEE Journal of Quantum Electronics*, **23** (1987), 510–24.

[104] Korteweg, D. J., de Vries, G., On the change of form of long waves advancing in a rectangular canal and on a new type of long stationary waves, *Philosophical Magazine 5th Series*, **36** (1895), 422–43.
[105] Kroll, D. M., Lipowski, R., Universality classes for the critical wetting transition in two dimensions, *Physical Review B*, **28** (1983), 5273–80.
[106] Krumhansl, J. A., Schrieffer, J. R., Dynamics and statistical mechanics of a one-dimensional model Hamiltonian for structural phase transition, *Physical Review B*, **11** (1975), 3535–45.
[107] Lamb, H., *Hydrodynamics*, Cambridge University Press (1932).
[108] Landau, L. D., Lifshitz, E. M., *Quantum Mechanics*, Pergamon Press (1977).
[109] *Statistical Physics*, Pergamon Press (1969).
[110] Lax, P. D., Integrals of nonlinear equations of evolution and solitary waves, *Communications on Pure and Applied Mathematics*, **21** (1968), 467–90.
[111] Legett, A. J., Bose–Einstein condensation in the alkali gases: some fundamental concepts, *Review of Modern Physics*, **73** (2001), 307–56.
[112] Leroy, B., Nonlinear evolution equations without magic: I. The Korteweg–de Vries equation, *European Journal of Physics*, **10** (1990), 82–6.
[113] Nonlinear evolution equations without magic: II. The cubic nonlinear Schrödinger equation, *European Journal of Physics*, **10** (1990), 87–92.
[114] Levy, L. P., *Magnétisme et Supraconductivité*, InterÉditions & CNRS Éditions (1997); translated in *Magnetism and Supraconductivity*, Springer (2000).
[115] Li, Y., Raichlen, F., Non-breaking and breaking solitary wave run-up, *Journal of Fluid Mechanics*, **456** (2002), 295–318.
[116] Lieb, E. H., Seiringer, R., Yngvason, J., One-dimensional bosons in three-dimensional traps, *Physical Review Letters*, **91** (2003), 150401–4.
[117] Lipowski, R., Critical effects at complete wetting, *Physical Review B*, **32** (1985), 1731–50.
[118] Lombdahl, P. S., Soerensen, O. H., Christiansen, P. L., Soliton excitations in Josephson tunnel junctions, *Physical Review B*, **25** (1982), 5737–48.
[119] London, F., The λ-phenomenon of liquid helium and the Bose–Einstein degeneracy, *Nature*, **141** (1938), 643–4.
[120] Maddox, J., Physicists about to hi-jack DNA? *Nature*, **324** (1986), 11.
[121] Towards the calculation of DNA, *Nature*, **339** (1989), 577.
[122] Magyari, E., Thomas, H., Kink instability in planar ferromagnets, *Physical Review B*, **25** (1982), 531–3.
[123] Makhankov, V. G., Dynamics of classical solitons (in non integrable systems), *Physics Reports*, **35** (1978), 1.
[124] Manneville, P., *Structures, Dissipatives, Chaos et Turbulence*, Aléa Saclay (1991).
[125] Martyna, G. J., Klein, M. L., Tuckerman, M., Nosé–Hoover chains: The canonical ensemble via continuous dynamics, *Journal of Chemical Physics*, **97** (1992), 2635–43.
[126] Maurye, M. F., *The Physical Geography of the Sea and its Meteorology*, Harper (1861).
[127] Messiah, A., *Mécanique Quantique*, Dunod (1995); First edition (1959) translated in *Quantum Mechanics*, North-Holland Publishing Co. (1961).
[128] Metropolis, N., The Age of Computing: a personal memoir, *Daedalus*, **121** (1992), 119–30.
[129] Mikeska, H. J., Nonlinear dynamics of classical one-dimensional anti-ferromagnets, *Journal of Physics C*, **13** (1980), 2913–23.

[130] Mikeska, H. J., Steiner, M., Solitary excitations in one-dimensionnal magnets, *Advances in Physics*, **40** (1991), 196–356.
[131] Miles, J. W., The Korteweg–de Vries equation: a historical essay, *Journal of Fluid Mechanics*, **106** (1980), 131–47.
[132] Millar, D. P., Robbins, R. J., Zewail, A. H., Direct observation of the torsional dynamics of DNA and RNA by picosecond spectroscopy, *Proceedings of the National Academy of Sciences (USA)*, **77** (1980), 5593–7.
[133] Mollenauer, L. F., Stolen, R. H., Islam, M. N., Experimental observation of a picosecond pulse narrowing and solitons in optical fibers, *Physical Review Letters*, **45** (1980), 1095–8.
[134] Morse, P. M., Feshbach, H., *Methods of Theoretical Physics*, McGraw-Hill (1953).
[135] Morse, P. M., Stueckelberg, E. C. G., Diatomic Molecules According to the Wave Mechanics I: Electronic Levels of the Hydrogen Molecular Ion, *Physical Review*, **33** (1929), 932–47.
[136] Nayfeth, A. H., *Perturbation Methods*, John Wiley (1973).
[137] Newell, A. C., Nonlinear tunnelling, *Journal of Mathematical Physics*, **19** (1978), 1126–33.
[138] Newell, A. C., Moloney, J., *Nonlinear Optics*, Addison Wesley (1992).
[139] Novoselov, K. S., Geim, A. K., Dubonos, S. V., Hill, E. W., Grigorieva, I. V., Subatomic movements of a domain wall in the Peierls Potential, *Nature*, **426** (2003), 812–16.
[140] Orlando, T. P., Delin, K. A., *Foundations of Applied Superconductivity*, Addison Wesley (1991).
[141] Osborne, A. R., Burch, T. L., Internal solitons in the Andaman sea, *Science*, **208** (1980), 451–60.
[142] Painter, P. C., Mosher, L., Rhoads, C., Low frequency modes in the Raman spectrum of DNA, *Biopolymers*, **20** (1981), 243–7.
[143] Paquerot, J-F., *Dynamique non linéaire des ondes de pression sanguine dans les grosses artères*, Doctoral Thesis, University of Bourgogne (1995).
[144] Paquerot, J-F., Lambrakos, S. G., Monovariable representation of blood-flow in a large elastic artery, *Physical Review E*, **49** (1995), 3432–9.
[145] Peierls, R. E., *Quantum Theory of Solids*, Oxford University Press (1956).
[146] Pérez-Garía, V., Michinel, H., Cirrac, J. I., Lewenstein, M., Zoller, P., Low energy excitations of a Bose–Einstein condensate: a time-dependent variational analysis, *Physical Review Letters*, **77** (1996), 5320–3.
[147] Pérez-Garía, V., Michinel, H., Herrero, H., Bose–Einstein solitons in highly asymmetric traps, *Physical Review A*, **57** (1998), 3837–42.
[148] Perry, P. R., Schimke, G. R., Large amplitude internal waves observed off the northwest coast Sumatra, *Journal of Geophysical Research*, **70** (1965), 2319–24.
[149] Peyrard, M., *Nonlinear Excitations in Biomolecules*, Les Éditions de Physique, Les Ulis (1995).
[150] Peyrard, M., Bishop, A. R., Statistical mechanics of a nonlinear model for DNA denaturation, *Physical Review Letters*, **62** (1989), 2755–8.
[151] Peyrard, M., Kruskal, D., Kink dynamics in the highly discrete sine-Gordon system, *Physica D*, **14** (1984), 88–102.
[152] Pitaevskii, L. P., Vortex lines in an imperfect Bose gas, *Soviet Physics Journal of Experimental and Theoretical Physics*, **13** (1961), 451–4.
[153] Press, W. H., Teukolsky, S. A., Vetterling, W. T., Flannery, B. P., *Numerical Recipes. The Art of Scientific Computing*, Cambridge University Press (1992).

[154] Prohofsky, E. W., Lu, K. C., van Zandt, L. L., Putnam, B. F., Breathing modes and induced resonant melting of the double helix, *Physics Letters A*, **70** (1979), 492-4.
[155] Putnam, B. F., van Zandt, L. L., Prohofsky, E. W., Lu, K. C., Mei, W. N., Resonant and localized breathing modes in terminal regions of the DNA double helix, *Biophysical Journal*, **35** (1981), 271–87.
[156] Rasmussen, J. J., Rypdal, K., Blow up in nonlinear Schrödinger equations, a general review, *Physica Scripta*, **33** (1986), 481–97.
[157] Reif, F., *Statistical and Thermal Physics*, McGraw-Hill (1965).
[158] Remoissenet, M., Low-amplitude breather and envelope solitons in quasi-one-dimensional physical models, *Physical Review B*, **33** (1986), 2386–92.
[159] Strecker, K. E., Partide, G. B., Truscott, A. G., Hulet, R. G., Formation and propagation of matter-wave solitons trains, *Nature*, **417** (2002), 150–3.
[160] Roche, S., Peyrard, M., Impurity effects on soliton dynamics in planar ferromagnets, *Physics Letters A*, **172** (1993), 236-42.
[161] Ruelle, D., Statistical mechanics of a one-dimensional lattice gas, *Communications in Mathematical Physics*, **9** (1968), 267–78.
[162] Russell, J. S., Report on Waves, *Report of the Fourteenth Meeting of the British Association for the Advancement of Science*, John Murray (1844), 311–90.
[163] Saenger, W., *Principles of Nucleic Acid Structure*, Springer (1984).
[164] Saito, N., Hirooka, H., Long-time behavior of the vibration in one-dimensional harmonic lattice, *Journal of the Physical Society of Japan*, **23** (1967), 157–66.
[165] Computer studies of ergodicity in coupled oscillators with anharmonic interaction, *Journal of the Physical Society of Japan*, **23** (1967), 167–71.
[166] Satsuma, J., Yajima, S., Initial value problem of one dimensional self-modulation of nonlinear waves in dispersive media, *Supplement of the Progress of Theoretical Physics*, **55** (1974), 284–306.
[167] Schaffer, H. E., Friend, R. H., Heeger, A. J., Localized phonons associated with solitons in polyacetylene: coupling to the nonuniform mode, *Physical Review B*, **36** (1987), 7537–41.
[168] Scott, A. C., Davydov's soliton, *Physics Reports*, **217** (1992), 1–67.
[169] Stolen, R. H., Mollenauer, L. F., Tomlinson, W. J., Observation of pulse restoration at the soliton period in optical fibers, *Optics Letters*, **8** (1983), 186–8.
[170] Stolen, R. H., Mollenauer, L. F., Gordon, J. P., Tomlinson, W. J., Extreme picosecond pulse narrowing by means of soliton effect in single-mode optical fibers, *Optics Letters*, **8** (1983), 289–91.
[171] Su, W. P., Schrieffer, J. R., Heeger, A. J., Soliton excitations in polyacetylene, *Physical Review B*, **22** (1980), 2099–111.
[172] Sundermeyer, K., *Constrained Dynamics*, Springer (1982).
[173] Synolakis, C. E., The runup of solitary waves, *Journal of Fluid Mechanics*, **185** (1987), 523–45.
[174] Theodorakopoulos, N., Dauxois, T., Peyrard, M., Order of the phase transition in models of DNA thermal denaturation, *Physical Review Letters*, **85** (2000), 6–9.
[175] Tinkham, M., *Introduction to Superconductivity*, McGraw Hill (1996).
[176] Toda, M., Vibration of a chain with nonlinear interaction, *Journal of the Physical Society of Japan*, **22** (1967), 431–6.
[177] Mechanics and statistical mechanics of a nonlinear chain, *Journal of the Physical Society of Japan*, **26** (1969), Supplement, 235–7.
[178] Tran, M. Q., Ion acoustic solitons in a plasma: a review of their experimental properties and related theories, *Physica Scripta*, **20** (1979), 317–27.
[179] Ulam, S. M., *Adventures of a Mathematician*, Charles Scribner's Sons (1976).

[180] Urabe, H., Tominaga, Y., Low-lying collective modes of DNA double helix by Raman-spectroscopy, *Biopolymers*, **21** (1982), 2477–81.
[181] van Hove, L., Sur l'intégrale de configuration pour les systèmes de particules à une dimension, *Physica*, **16** (1950), 137–43.
[182] Vardeny, Z., Ehrenfreund, E., Brafman, O., Horowitz, B., Fujimoto, H., Tanaka, J., Tanaka, M., Detection of soliton shape modes in polyacetylene, *Physical Review Letters*, **57** (1986), 2995–8.
[183] Villain, J., Physics on surfaces and with interfaces, in *Structures et Instabilités*, C. Godrèche (Ed.), EDP Sciences (1986).
[184] Virtual molecular dynamics (VMD): logicial developed by the Theoretical and Computational Biophysics Group from Beckman Institute for Advanced Science and Technology, University of Illinois at Urbana-Champaign.
[185] Wang, X., Brown, D. W., Lindenberg, K., Quantum Monte Carlo simulation of the Davydov model, *Physical Review Letters*, **62** (1989), 1796.
[186] Watson, J. D., Crick, F. H. C., A structure for deoxyribose nucleic acid, *Nature*, **171** (1953), 737–8.
[187] Wysin, G. M., *Classical kink dynamics and quantum thermodynamics in easy-plane magnetic chains with an applied magnetic field*, Doctoral Thesis, Cornell University (1985).
[188] Wysin, G. M., Bishop, A. R., Kumar, P., Soliton dynamics on a ferromagnetic chain, *Journal of Physics C*, **15**, (1982) L337–43; Soliton dynamics on an easy-plane ferromagnetic chain, *Journal of Physics C*, **17** (1984), 5975–91.
[189] Wysin, G. M., Bishop, A. R., Oitmaa, J., Single kink dynamics in an easy-plane classical antiferromagnetic chain, *Journal of Physics C*, **19** (1986), 221–33.
[190] Zabusky, N. J., Kruskal, M. D., Interaction of solitons in a collisionless plasma and the recurrence of initial states, *Physical Review Letters*, **15** (1965), 240–3.
[191] Zabusky, N. J., Nonlinear lattice dynamics and energy sharing, *Journal of the Physical Society of Japan*, **26** (1969), Supplement, 196–202.
[192] Computational Synergetics and Mathematical Innovation, *Journal of Computational Physics*, **43** (1981), 195–249.
[193] Zakharov, V. E., Shabat, A. B., Exact theory of two-dimensional self-focusing and one-dimensional self-modulation of waves in nonlinear media, *Soviet Physics Journal of Experimental and Theoretical Physics*, **34** (1972), 62–9.
[194] Zakharov, V. E., Synakh, V. S., The nature of the self focusing singularity, *Soviet Physics Journal of Experimental and Theoretical Physics*, **41** (1976), 465–8.

Index

Printed in the United States
By Bookmasters

Printed in the United States
By Bookmasters